STEM Education:

Concepts, Methodologies, Tools, and Applications

Information Resources Management Association
USA

Volume II

Information Science
REFERENCE
An Imprint of IGI Global

Managing Director:	Lindsay Johnston
Managing Editor:	Austin DeMarco
Director of Intellectual Property & Contracts:	Jan Travers
Acquisitions Editor:	Kayla Wolfe
Production Editor:	Christina Henning
Development Editor:	Chris Shearer
Multi-Volume Book Production Specialist:	Deanna Jo Zombro
Cover Design:	Jason Mull

Published in the United States of America by
Information Science Reference (an imprint of IGI Global)
701 E. Chocolate Avenue
Hershey PA, USA 17033
Tel: 717-533-8845
Fax: 717-533-8661
E-mail: cust@igi-global.com
Web site: http://www.igi-global.com

Library of Congress Cataloging-in-Publication Data

STEM education : concepts, methodologies, tools, and applications / Information Resources Management Association, editor.
volumes cm
Includes bibliographical references and index.
Summary: "This reference brings together an impressive array of research on the development of Science, Technology, Engineering, and Mathematics curricula at all educational levels"-- Provided by publisher.
ISBN 978-1-4666-7363-2 (hardcover : v. 1) -- ISBN 978-1-4666-7366-3 (print & perpetual access) -- ISBN 978-1-4666-7364-9 (ebook) 1. Science--Study and teaching--United States. 2. Technology--Study and teaching--United States. 3. Engineering--Study and teaching--United States. 4. Mathematics--Study and teaching--United States. 5. Curriculum planning--United States. 6. Instructional systems--United States--Design. I. Information Resources Management Association. II. Title: Science, technology, engineering, and mathematics education.
Q181.S6944 2015
507.1'073--dc23

2014040310

British Cataloguing in Publication Data
A Cataloguing in Publication record for this book is available from the British Library.

List of Contributors

Table of Contents

Section 2
Tools and Technologies

This section presents an extensive coverage of various tools and technologies available in the field of STEM Education that practitioners and academicians alike can utilize to develop different techniques. These chapters enlighten readers about fundamental research on the many tools facilitating the burgeoning field of STEM Education. It is through these rigorously researched chapters that the reader is provided with countless examples of the up-and-coming tools and technologies emerging from the field of STEM Education. With 13 chapters, this section offers a broad treatment of some of the many tools and technologies within the STEM Education field.

Section 3
Frameworks and Methodologies

This section provides in-depth coverage of conceptual architecture frameworks to provide the reader with a comprehensive understanding of the emerging developments within the field of STEM Education. Research fundamentals imperative to the understanding of developmental processes within STEM Education are offered. From broad examinations to specific discussions on methodology, the research found within this section spans the discipline while offering detailed, specific discussions. From basic designs to abstract development, these chapters serve to expand the reaches of development and design technologies within the STEM Education community. This section includes 15 contributions from researchers throughout the world on the topic of STEM Education.

Section 4
Cases and Applications

This section discusses a variety of applications and opportunities available that can be considered by practitioners in developing viable and effective STEM Education programs and processes. This section includes 19 chapters that review topics from case studies in STEM Education. Contributions included in this section provide excellent coverage of today's education and IT communities and how research into STEM Education is impacting the social fabric of our present-day global village.

Section 5
Issues and Challenges

This section contains 10 chapters, giving a wide variety of perspectives on STEM Education and its implications. The section also discusses new ethical considerations within transparency and accountability. Within the chapters, the reader is presented with an in-depth analysis of the most current and relevant issues within this growing field of study. Crucial questions are addressed and alternatives offered.

Section 6
Emerging Trends

This section highlights research potential within the field of STEM Education while exploring uncharted areas of study for the advancement of the discipline. Introducing this section are chapters that set the stage for future research directions and topical suggestions for continued debate, centering on the new venues and forums for discussion. A pair of chapters on the usability and effectiveness research makes up the middle of the section of the final 10 chapters, and the book concludes with a look ahead into the future of the STEM Education field, with "Why Immersive, Interactive Simulation Belongs in the Pedagogical Toolkit of 'Next Generation' Science." In all, this text will serve as a vital resource to practitioners and academics interested in the best practices and applications of the burgeoning field of STEM Education.

Preface

The constantly changing landscape of STEM Education makes it challenging for experts and practitioners to stay informed of the field's most up-to-date research. That is why Information Science Reference is pleased to offer this three-volume reference collection that will empower students, researchers, and academicians with a strong understanding of critical issues within STEM Education by providing both broad and detailed perspectives on cutting-edge theories and developments. This reference is designed to act as a single reference source on conceptual, methodological, technical, and managerial issues, as well as provide insight into emerging trends and future opportunities within the discipline.

STEM Education: Concepts, Methodologies, Tools, and Applications is organized into six distinct sections that provide comprehensive coverage of important topics. The sections are: (1) Fundamental Concepts and Theories, (2) Tools and Technologies, (3) Frameworks and Methodologies, (4) Cases and Applications, (5) Issues and Challenges, and (6) Emerging Trends. The following paragraphs provide a summary of what to expect from this invaluable reference tool.

Section 1, "Fundamental Concepts and Theories," serves as a foundation for this extensive reference tool by addressing crucial theories essential to the understanding of STEM Education. Introducing the book is "STEM in Early Childhood Education," a great foundation laying the groundwork for the basic concepts and theories that will be discussed throughout the rest of the book. Another chapter of note in Section 1 is titled "Creating Open Source Lecture Materials." Where Section 1 leaves off with fundamental concepts, Section 2 discusses tools and technologies in place for STEM Education.

Section 2, "Tools and Technologies," presents extensive coverage of the various tools and technologies used in the implementation of STEM Education. Section 2 begins where Section 1 left off, though this section describes more concrete tools at place in the modeling, planning, and applications of STEM Education. The first chapter, "Technology in Mathematics Education," lays a framework for the types of works that can be found in this section, a perfect resource for practitioners looking for the types of technologies currently in practice in STEM Education. Section 2 is full of excellent chapters like this one, including such titles as "Teaching Mathematics with Tablet PCs," "Using Educational Computer Games for Science Teaching," and "Online Simulator Use in the Preparing Chemical Engineers," to name a few. Where Section 2 described specific tools and technologies at the disposal of practitioners, Section 3 describes frameworks and methodologies within the field.

Section 3, "Frameworks and Methodologies," presents in-depth coverage of the conceptual design and architecture of STEM Education. Opening the section is "Multiple Perspectives for the Study of Teaching." This section is vital for developers and practitioners who want to measure and track the progress of STEM Education through the multiple lens of parametric design. Through case studies, this section lays excellent groundwork for later sections that will get into present and future applications for STEM Education, including, of note: "Learning about the Different Dimensions of Sustainability by Applying the Product Test Method in Science Classes" and "Providing Elementary and Middle School Science

Teachers with Content and Pedagogical Professional Development in an Online Environment." The section concludes with another excellent work, titled "Linking Education to Creating a Knowledge Society."

Section 4, "Cases and Applications," describes how the broad range of STEM Education efforts has been utilized and offers insight on and important lessons for their applications and impact. Section 4 includes the widest range of topics because it describes case studies, research, architectures, theory, analysis, and guides for implementation. The first chapter in the section is titled "3D Multi-User Virtual Environments in Science Education" Section 4 concludes with an excellent view of a case study in a new program, "Video Gaming for STEM Education."

Section 5, "Issues and Challenges," presents coverage of academic and research perspectives on STEM Education tools and applications. The section begins with "Self-Regulated Learning as a Method to Develop Scientific Thinking." The section concludes with "Developing an Online Mathematics Methods Course for Preservice Teachers," a great transitional chapter between Sections 5 and 6 because it examines an important trend going into the future of the field. The last chapter manages to show a theoretical look into future and potential technologies, a topic covered in more detail in Section 6.

Section 6, "Emerging Trends," highlights areas for future research within the field of STEM Education, opening with "Opening Both Eyes: Gaining an Integrated Perspective of Geology and Biology." Section 6 contains chapters that look at what might happen in the coming years that can extend the already staggering amount of applications for STEM Education. Other chapters of note include "Environmental Science Education in the 21st Century" and "New Trends in Service Science and Education for Service Innovation." The final chapter of the book looks at an emerging field within STEM Education, in the excellent contribution, "Why Immersive, Interactive Simulation Belongs in the Pedagogical Toolkit of 'Next Generation' Science."

Although the primary organization of the contents in this multi-volume work is based on its six sections, offering a progression of coverage of the important concepts, methodologies, technologies, applications, social issues, and emerging trends, the reader can also identify specific contents by utilizing the extensive indexing system listed at the end of each volume. Furthermore to ensure that the scholar, researcher, and educator have access to the entire contents of this multi volume set as well as additional coverage that could not be included in the print version of this publication, the publisher will provide unlimited multi-user electronic access to the online aggregated database of this collection for the life of the edition, free of charge when a library purchases a print copy. This aggregated database provides far more contents than what can be included in the print version, in addition to continual updates. This unlimited access, coupled with the continuous updates to the database ensures that the most current research is accessible to knowledge seekers.

As a comprehensive collection of research on the latest findings related to using technology to providing various services, *STEM Education: Concepts, Methodologies, Tools, and Applications*, provides researchers, administrators, and all audiences with a complete understanding of the development of applications and concepts in STEM Education. Given the vast number of issues concerning usage, failure, success, policies, strategies, and applications of STEM Education in countries around the world, *STEM Education: Concepts, Methodologies, Tools, and Applications* addresses the demand for a resource that encompasses the most pertinent research in technologies being employed to globally bolster the knowledge and applications of STEM Education.

Chapter 28
A Tool for Analyzing Science Standards and Curricula for 21st Century Science Education

Danielle E. Dani
Ohio University, USA

Rola Khishfe
American University of Beirut, Lebanon

Sara Salloum
Long Island University, USA

Saouma BouJaoude
American University of Beirut, Lebanon

ABSTRACT

Twentieth century curricula are no longer sufficient to prepare students for life and work in today's diverse, fast-paced, technologically driven, and media saturated world of the 21st century. This chapter presents a new framework for analyzing science standards and curricula to determine the extent of alignment with 21st Century essential understandings and skills. The Tool for Analyzing Science Standards and Curricula (TASSC) was developed using the conceptual frameworks proposed by the Partnership for 21st Century Skills, the Organization for Economic Co-Operation and Development, and the typology of knowledge proposed by Jurgen Habermas. Development of TASSC relied on an iterative process of refinement, testing, and discussions resulting in an instrument with three sections and related rating scales: content, skills, and additional curricular components. TASSC was piloted using middle school science standards and curricula in the context of two US states (Ohio and New York) and two Arab countries (Lebanon and Qatar). The analysis procedure and individual case study results are presented and discussed in the chapter.

INTRODUCTION

For more than twenty years, reforms in science education have been calling for the preparation of scientifically literate students. However, standards and curricula promoting 20th Century conceptualizations of scientific literacy (see American Association for the Advancement of Science, 1993; Bybee, 1997; National Research Council, 1996) may not be sufficient to prepare students for the global economy of the 21st century. Today's students live in a world that is extremely fast-paced, constantly changing, increasingly culturally diverse, technologically driven, and media saturated (Wan & Gut, 2011). The kinds of skills students need to be prepared for of the

DOI: 10.4018/978-1-4666-7363-2.ch028

21st century are different from those needed 20 years ago (P21 Skills, 2009). To prepare for this "*second Renaissance period*" (Treadwell, 2011), we need to bring what we teach and how we teach into the 21st Century. Essential understandings and skills that are necessary for success in today's world must be included.

As Nations and States around the world revise existing curricula or develop new standards, a framework and instrument for analyzing science standards and curricula and determining the extent of alignment with 21st Century essential understandings and skills is necessary. BouJaoude (2010) created such a framework for the analysis of education programs (FAEP, see Table 1). In preparation for a National Association for the Research in Science Teaching symposium in 2011, the authors of this chapter decided to use FAEP to analyze and evaluate the standards and curricula of US states and several countries from around the world. After determining that FAEP was not adequate for analyzing science standards and curricula, the authors decided to develop a new framework. This chapter describes the development of this new framework entitled Tool for Analyzing Science Standards and Curricula (TASSC). The chapter begins with a discussion of the conceptual basis that supports the call for 21st Century essential understandings and skills. The chapter ends with a presentation of case studies using TASSC in multiple contexts.

CONCEPTUAL BASIS

The essential understandings and skills necessary for a college and career ready populace have been posited by several organizations. For example, the Partnership for 21st Century Skills (P21) proposed the Framework for 21st Century Learning. Similarly, the Organization for Economic Co-operation and Development (OECD[1]) proposed a set of competencies needed for a successful life and well-functioning 21st Century society.

Table 1. Framework for analyzing education programs

Item	Evident				Not Evident
1. The program focuses on:					
• helping students understand core academic content at high levels					
• developing information and communication skills					
• developing thinking and problem-solving skills					
• developing interpersonal and self-directional skills					
• developing reflective skills					
• developing autonomy					
2. The program encourages the development of social skills (living peacefully with others, cooperation, empathy...)					
3. Teaching and learning seem to be implemented in a twenty first century context (learning academic content through real-world examples, applications and experiences)					
4. Assessment seems to use appropriate tools and approaches to measure students' performance on twenty-first century content and skills.					
5. Systems seem to consider the fact that students think and process information differently from their predecessors (accounting for students' ways of knowing)					

BouJaoude (2010) additionally used Habermas' types of knowledge (1971; as cited by Cranton, 2002) as he developed FAEP. These conceptual frameworks are described in this section.

Framework for 21st Century Learning

The partnership for 21st century skills (2009) developed a framework for 21st century learning delineating a set of learning outcomes that specify the knowledge, skills, expertise, and literacies needed for success in work and life. In addition to the core subjects of science, the framework identifies a set of 21st century interdisciplinary themes that, when integrated into core subjects, will result in much higher levels of learning. The themes consist of global awareness, financial, economic, business, and entrepreneurial literacy, civic literacy, health literacy, and environmental literacy.

The framework divides the necessary skills into three types,

1. Learning and innovation,
2. Information, media, and technology literacies, and
3. Life and career skills.

Learning and Innovation skills include the four C's: creativity and innovation, critical thinking and problem solving, communication, and collaboration. Life and career skills include flexibility and adaptability, initiative and self-direction, social and cross-cultural skills, productivity and accountability, and leadership and responsibility. The Partnership for 21st Century Skills further defines these outcomes in their P21 Framework Definitions (2009).

OECD's Key Competencies for the 21st Century

OECD (2005) proposed three broad categories of key competencies: Using tools interactively, interacting in heterogeneous groups, and acting autonomously. Today's world is reliant on a variety of tools ranging from physical ones, such as information technology, to socio-cultural ones such as the use of language. For this reason, individuals need to be able to use this wide variety of tools for interacting effectively with the environment and they need to understand them well enough to adapt them for their own purposes. Because today's world is also increasingly interdependent, individuals need to be autonomously able to engage with others of different backgrounds. Finally, individuals in today's world must be able to take responsibility for managing their own lives and situating their lives in the broader social context and act autonomously.

Reflective skills are central to OECD's competency framework. They involve the ability to deal with change effectively, learn from experience, and take a critical stance toward issues. Such a critical stance will only occur in an appropriate learning environment where individuals are not satisfied with applying what they learn blindly, but rather learn to be dynamic and responsive actors in society.

Habermas' Types of Knowledge

Scholars such as Merizow (1997) and Cranton (1997, 2001, 2002) have utilized Habermas' (1971) conceptualizations of human interests and ensuing knowledge to outline knowledge types within the context of adult learning. Habermas (1971; as cited by Cranton, 2001, 2002) proposes three types of knowledge: Instrumental, communicative, and emancipatory. According to Habermas, knowledge types are acquired in response to three basic human interests. A technical interest to explain and manipulate our environment leads to *instrumental knowledge*. Instrumental knowledge is objective and empirically derived allowing for prediction and control of the environment and to an understanding of how the world works (Cranton, 1997). Our practical interest to live within social settings leads us to acquire *communicative knowledge*[2].

Such knowledge is acquired through language and involves knowledge of societal and cultural norms (at the micro, macro, and global levels). Validity of communicative knowledge is determined by "consensus within a group and a sense of rightness or morality" (Cranton, 2001, p. 12); for example, agreed-upon norms and knowledge in one culture may not hold in another culture. Finally, our emancipatory interests lead to *emancipatory knowledge*. Emancipatory knowledge is motivated by an interest in growth, self-awareness, and autonomy. In addition, emancipatory knowledge springs from an interest to free ourselves from the oppression of unquestioned social norms, values, and beliefs (Cranton, 2001). It is developed through critical reflection be it on self or the society.

Habermas' three types of knowledge are deemed important in our global age since:

The essential learning required to prepare a productive and responsible worker for the twenty-first century must empower the individual to think as an autonomous agent in a collaborative context rather than to uncritically act on the received ideas and judgments of others. Workers will have to become autonomous, socially responsible thinkers (Merizow, 1997, p. 8).

A similar argument can be made for teaching and learning at the secondary school level. In order to be compatible with a highly technological and global age, students need to develop instrumental and communicative knowledge. Concurrently, they need to develop emancipatory forms of knowledge that would allow them to become autonomous independent thinkers who critically reflect on their own thinking, values and beliefs and those of the surrounding culture.

TOOL FOR ANALYZING SCIENCE STANDARDS AND CURRICULA (TASSC)

The Tool for Analyzing Science Standards and Curricula (TASSC, see Table 2) was developed using an iterative process of refinement, testing, and discussions over a period of two years. TASSC consists of 13 items distributed over the three sections of content, skills, and additional curricular components. Each section and its individual rating scale will be described below. The process began with the development of a framework to analyze programs in terms of 21st century skills (BouJaoude, 2010). The various steps in the development of TASSC are described in the following sections.

A Framework for Analyzing Education Programs

Saouma BouJaoude (2010) developed and used the Framework for Analyzing Education Programs (FAEP, see Table 1) to analyze the secondary educational systems in England, Germany, Finland, Japan, and the International Baccalaureate in order to derive lessons that might be helpful in improving secondary education in the Arab States. BouJaoude's (2010) framework is based on an alignment and critical examination of the proposals of the Partnership for 21st Century Skills and the OECD using Habermas' types of knowledge (described above) and Prensky's (2001) discussion of student thinking and ways of processing knowledge. According to Prensky, digital "Natives" process knowledge differently from "Immigrants." "Natives" are used to receiving information fast, like to parallel process and multi-task, prefer their graphics before their

Table 2. Tool for analyzing science standards and curricula (TASSC)

The Standards/Curriculum Focuses On:	Rating
Content	
1. Helping students understand core academic content at high levels	
a. Science Content	
b. 21st Century Themes (Environmental literacy, global awareness, economic literacy, civic literacy, and health literacy	
Skills	
2. Developing information skills	
3. Developing communication skills	
4. Developing creativity and innovation skills	
5. Developing thinking and problem-solving skills	
6. Developing reflective skills	
7. Developing autonomy and self-directional skills	
Other	
8. Teaching and learning activities involve a twenty first century context (learning academic content through real-world examples, applications and experiences)	
9. Assessment seems to use appropriate tools and approaches to measure students' performance on twenty-first century content and skills.	
10. Systems consider the fact that students think and process information differently from their predecessors (accounting for students' ways of knowing)	
a. Students are perceived as **active learners** (learning theories)	
b. Curriculum supports a rich technological environment (availability of rich graphics and interaction with technological tools and media)	
c. Curriculum affords utilization of technological tools allowing for parallel processing	

text rather than the opposite, and function best when networked. Conversely, "Immigrants" have learned–and so choose to teach–slowly, step-by-step, one thing at a time, and individually.

BouJaoude's (2010) analysis suggests that advocates of the recommendations of P21 emphasize "Instrumental Knowledge" with some emphasis on "Communicative Knowledge." Prensky's recommendations, on the other hand, seem to focus more on student characteristics and methods of acquiring knowledge and less on the types of knowledge needed by all learners. The competencies advocated by OECD, however, move more closely to what Habermas advocates but with a preference for the communicative and the emancipatory knowledge. BouJaoude concludes that it might be productive to adopt an integrative approach to determine the knowledge, skills, and dispositions needed for students to live and work in the twenty-first century. This integrative approach would not neglect any of the three types of knowledge but rather integrates the competencies advocated by different groups. Moreover, such an approach focuses on preparing knowledgeable, autonomous, and reflective critical thinkers who:

- Are capable of using information to solve academic and everyday life problems,
- Use technology meaningfully,
- Interact well with others in work and other social settings,
- Reflect on their actions and knowledge to improve the quality of their lives, and
- Construct knowledge through research.

This integrative approach rooted the development of FAEP and was subsequently adopted in the making of TASSC.

Development of the TASSC

As mentioned previously, TASSC was developed after it was determined that the FAEP was not adequate for analyzing science standards and curricula. The development of TASSC began with revisiting the conceptual base used to develop FAEP. In a first step, the framework for 21st Century Skills and the OECD key competencies were respectively aligned with Habermas' types of knowledge, and compared to determine areas of similarities and difference. Only areas of similarities that aligned with Habermas' instrumental, communicative, and emancipatory types of knowledge were used in TASSC. For example, a change that was made to reflect Habermas' work was separating the FAEP item "developing information and communication skills" into two distinct items.

Next, FAEP was revisited in light of the results of the analysis performed in step 1. Four main changes were made:

- Distinguishing between knowledge and skills, which are subsumed under Core Academic Content by creating separate sections to address each in the instrument,
- Creating a third section of the instrument to include additional curricular components such as teaching and learning activities, assessment, and overall educational systems,
- Further defining of the Core Academic Content as content knowledge and 21st century themes, and
- Including creativity and innovation skills in the skills section of the instrument.

In the Core Academic Content section, content knowledge consists of the unifying concepts and processes of science, physical science, life science, earth and space science, understandings about scientific inquiry, understandings about science and technology, science in the personal and social perspectives, and the history and nature of science. Twenty-first Century Themes include environmental literacy, global awareness, economic literacy, civic literacy, and health literacy. Conversely, Skills consist of information and communication, creativity and innovation, thinking and problem solving, reflective skills, autonomy, and self-direction.

Then, the rating scales were revisited and further defined. The skills rating scale used in the NRC (2010) report about the intersection of science education and 21st century skills was adopted because it assessed the degree to which a skill is present, complete, explicit, and/or optional. The Skills Rating Scale consists of the following levels:

4-Strong whole skill: The skill is found almost in its entirety in the curriculum in a strong form likely to produce high levels of performance if the curricular goals are met.

3-Weak whole skill: The skill is found almost in its entirety in the curriculum in a weak form, either because it is made optional or described vaguely.

2-Strong component skill: Only one or two components of the larger skill are found in the curriculum, but those elements are met to a high degree.

1-Weak component skill: Only one or two components of the larger skill are found in the curriculum, and even then only a weak form, either because they are made optional or described vaguely or are implicit in the activities of a listed curriculum.

0-None: The skill is completely absent.

The skills rating scale was used as a model for the development of a new content rating scale. The new rating scale was designed to assess the degree to which core academic content and 21st century themes are present, complete, explicit, integrated, and/or optional. The Content Rating Scale consists of the following levels:

4-Comprehensive High Level: The content is found almost in its entirety and explicitly described in curriculum activities in a manner likely to produce high levels of learning if the curricular goals are met.
3-Comprehensive Low Level: The content is found almost in its entirety and is vaguely described or made optional in curriculum activities.
2-Partial High Level: Only one or two content themes are explicitly described and integrated in curriculum activities.
1-Partial Low level: Only one or two content themes are vaguely described or implicit in the activities of the curriculum.
0-None: The content themes are completely absent.

Finally, the rating for the additional curricular components section consisted of "1" to represent "mostly present" or "0" to represent "mostly absent."

In short, the development of TASSC was cooperatively accomplished by the authors through an iterative process involving the coalescing and separation of categories based on conceptual and empirical considerations. The 13 items of the instrument are partly based on how the group conceptualized the categories and the distinct nature of each. For example, to account for Prensky's (2001) notions of digital natives and immigrants, the item targeting overall educational systems was further defined using the following prompts:

- Students are perceived as active learners.
- Curriculum supports a rich technological environment (availability of rich graphics and interaction with technological tools and media), and
- Curriculum affords utilization of technological tools allowing for parallel processing.

The process also involved using versions of the instrument to analyze segments of curricula and noting which categories yielded useful distinctions and which were redundant. For example, an earlier version of the TASSC combined interpersonal and self-directional skills in one item. Conceptually, the authors considered these skills to be distinct from each other, a notion that was supported by preliminary analyses. As a result, these two skills were separated. Furthermore, self-directional skills were found to overlap with autonomy and were therefore compiled together as one item in the final version to avoid redundancy. The result was an instrument that was detailed enough to give an adequate and clear picture of the strengths and limitations of the standards and core curricula, while at the same time succinct so as to be practical to use.

CASE STUDIES USING TASSC

TASSC was used to analyze the standards and curricula of two US states (Ohio and New York) and two Arab countries (Lebanon and Qatar). Middle school rather than high school was chosen as the focus of the analysis for two reasons. First, middle school science is more unified and it is also required for all students. At the high school level, students have a choice of several science electives; for example, fewer students in urban settings take physical science electing to satisfy science graduation requirements with life and earth science courses. Second, middle school years are instrumental in developing student attitudes and dispositions towards science. This section of the chapter describes how the analysis process was accomplished and then presents the individual case studies.

Analysis Procedure

The standards and curricular analysis process used in the individual case studies was conducted in multiple steps and at different levels. First, standard and curricular documents and websites were first read in their entirety, including introductions, prefaces, and appendices, to discern *content, skills*, and *additional curricular component features* as outlined in the framework. Then, the first level of analysis involved color-coding key statements according to the broad categories of Content, Skills and Other. In some instances, cross coding was allowed but it was indicated which category was more prominent. An example of a cross coding instance from the New York City standards is:

M1.1a: identify independent and dependent variables.
M1.1b: identify relationships among variables including: direct, indirect, cyclic, constant; identify non-related material.

The above performance indicators require knowledge and understanding of the concepts of dependent and independent variables and therefore were designated as both process skills and content. In this case, Skills was more prominent.

For the second level of analysis, each item in the framework was assigned a code (see Table 3). Next, *Content, Skills,* and *Other* statements identified in the level one analysis were coded and separated into categories based on the items of the framework. Frequencies were calculated for each code. No rating was given initially for the first two levels of analyses. Table 4 presents an example of level 2 coding for the following State of Ohio grade 6 Model Curriculum statement:

Make a geological map of the local community. Use existing geologic data, historic geologic data, and field explorations to analyze types of formations that are present. Use the finished map to evaluate possible land and resource uses. Present map and recommendations to an authentic audience.

For the third level of analysis (see Table 4), statements within each level two *Content* and *Skill* coding category were re-examined and separated into subcategories based on the degree to which each statement presented

Table 3. Level 2 analysis codes

Content	Code
1. Helping students understand core academic content at high levels a. Science Content	SC
b. **21st Century Themes:** Environmental literacy Global awareness Economic literacy Civic literacy Health literacy	 EnvL GA EcL CL HL
2. Teaching and learning activities involve a twenty first century context (learning academic content through real-world examples, applications and experiences)	21Con
3. Assessment seems to use appropriate tools and approaches to measure students' performance on twenty-first century content and skills.	AsP
4. Systems consider the fact that students think and process information differently from their predecessors (accounting for students' ways of knowing)	IP:--
a. Students are perceived as **active learners** (learning theories)	AL
b. Curriculum supports a rich technological environment (availability of rich graphics and interaction with technological tools and media)	TR
c. Curriculum affords utilization of technological tools allowing for **parallel processing**	PP
Skills	
1. Developing **information skills**	Inf
2. Developing **communication skills**	Com
3. Developing **creativity and innovation skills**	CInn
4. Developing **thinking and problem-solving skills**	ThP
5. Developing **autonomy** and **self-directional skills**	ASD

Table 4. Analysis levels of curriculum statements

Level 1	Level 2		Level 3
Content	SC		
	CL	Use the finished map to evaluate possible land and resource uses	Implicit, integrated
Skills	INF	Use existing geologic data, historic geologic data and field explorations	Explicit, required
	THP	to analyze types of formations that are present	Explicit, required
	COM	Present map and recommendations to an authentic audience	Explicit, required
Other	21CON	Make a geological map of the local community	
	TR	Use existing geologic data, historic geologic data and field explorations AND Make a geological map of the local community	

- Explicit, implicit, integrated, or optional content, or
- Explicit, implicit, optional, and/or vague skill.

The subcategories were derived from the TASSC rating scales described previously. Frequencies were calculated for each subcategory. Finally, a rating for each of the TASSC items was holistically determined using the frequencies calculated in the level 2 and level 3 analyses.

The Case of the Ohio Standards and Model Curriculum

The Ohio Department of Education (ODE) began a standard revision process in 2009 in response to then Governor Strickland's (2009) proclamation that preparing students to be informed decision makers in the global economy of the 21st century is an Ohio state imperative. In addition to revising its standards, ODE has developed a Web-based Model Curriculum, of which 21st century skills are an essential component:

Throughout all grades, PreK-12, [21st century] skills are explicitly addressed through scientific inquiry and applications. The integration of skills and content aligns with the Framework for 21st Century Learning. (ODE, 2010)

Structure of the Ohio Middle Level Standards and Model Curriculum

The revised standards for grades 6-8 are organized around Strands and Topics for each grade level. Strands represent the disciplines of life science, physical science, and earth and space science. Topics consist of the main focus for each strand and provide the basis for further defined Content Statements, which delineate the science content to be learned. Strands are connected by Themes that progressively increase in complexity. The integration of strands and topics within each grade are illustrated by Grade Level Connections. Finally, the standards describe the inquiry skills and applications that must be developed by all students during grades 6-8.

The Model Curriculum is a Web-based resource that illustrates how the material in the Content Statements may be taught by providing content elaborations, expectations for learning, classroom examples, and instructional strategies and resources. A noteworthy characteristic of the Model Curriculum consists of the Cognitive Demands that are used to identify expectations for learning. At the most basic level, students are expected to recall accurate science. At the next level, students are expected to interpret and communicate science concepts. At the third level, students are expected to demonstrate science knowledge. At the highest level of cognitive demand, students are expected to design technological and engineering solutions using science concepts.

Results

Results of the TASSC analysis of the Ohio grades 6-8 revised standards and model curriculum are presented in this section for core academic content, skills, and additional curricular components.

Core Academic Content

Core academic content consists of science content and 21st century themes. Results of the analysis process indicate that the science content component of the core academic content is mostly addressed in the description of the content strands, topics, content statements, themes, grade band connections, and the inquiry skills and applications. Strands, topics, and content statements explicitly delineate earth and space science, life science, and physical science essential understandings for each grade level. Furthermore, unifying concepts and processes are given prominent positions in grade band themes and strand connections. For example, the 7th grade band theme is "order and organization," and the strand connections relate to "systems." Similarly, understandings about and abilities to do scientific inquiry are explicitly included and integrated throughout the middle grade standards as evidenced by, among others, the following statements:

- Identify questions that can be answered through scientific investigations; and
- Recognize and analyze alternative explanations and predications.

Other content components, including science in the Personal and Social Perspectives (PSP) and the History and Nature of Science (HNOS), are addressed in the model curriculum rather than the standards. Statements aligned with these components were found in the sections of content elaborations, instructional strategies, and expectations for learning as illustrated by the following examples:

It is important to provide the background knowledge regarding how scientists know about the structure and composition of the interior of Earth (without being able to "see" it). [HNOS implicit, required – 8th grade content elaboration]

In order to fully grasp plate movement, students must investigate Earth's history and the evidence used in determining plate movement...Interpreting paleomagnetic data for the geologic periods demonstrates how scientists have determined where the plates were over time. [HNOS vague, optional – 8th grade instructional resources]

Analyze and evaluate scientific tradeoffs (environmental, projected research required to move from current knowledge to application) for use of microbes to produce alternative energy or clean-up environmental spills. [PSP implicit, required – 6th grade expectations for learning]

Even though the content statements identified in the content elaboration and expectations for learning sections are required, references to the HNOS and PSP content are vague or implicitly made. Moreover, content statements identified in the instructional resources section of the model curriculum seem optional because the section in question is intended as "additional support and information for educators...not intended to be a prescriptive list of lessons."

The 21st century themes of the core academic content are also addressed in the model curriculum. Results of the analysis process indicate that environmental and civic literacy are the most frequently and explicitly integrated themes in the Ohio model curriculum. Some examples of statements targeting these two themes include:

Research cloning in the food industry. Select one practice and determine whether or not it is an environmentally healthy practice. Justify your position with scientific evidence. [EnvL explicit, integrated – 8th grade expectations for learning]

Develop, test, and evaluate plans outlining a specific method to reduce storm water flow at a specific site in the local community (for example a housing construction project or the school parking lot). Present findings/plan to school administrators or local government. [CL, explicit, integrated – 7th grade expectations for learning]

The themes of economic literacy and global awareness are addressed less often in the Ohio model curriculum. In most instances, references to these themes are vague or made optional. For example:

Ethanol, a plant product, is used in place of fossil fuels. Discuss the pros and cons of using biomass products such as ethanol versus traditional fossil fuels. [EcL vague – 7th grade expectations for learning]

Examining student-based (classroom data) soil sample results can be a good way to compare soil types by regions. The GLOBE program can allow a connection to other classrooms, but can also be used when to analyze data beyond the local area to draw conclusions about specific criteria for soil formation. [GA optional – 6th grade instructional resources]

The theme of health literacy was not addressed in the Ohio standards and model curriculum. In conclusion, the Ohio standards and model curriculum core academic content was present at a partially high level because only two of the 21st century themes were explicitly described and integrated into curriculum activities.

Skills

Results of the analysis process indicate that several 21st century skills are emphasized in the Ohio science standards. Namely, inquiry and its application incorporates the skills of thinking, problem solving, communication, creativity, and innovation. Of these skills, thinking, problem solving, and communication are explicitly included and integrated throughout the middle grade standards and model curriculum as evidenced in Table 4 and, among others, the following statements:

- Design and conduct a scientific investigation;
- Analyze and interpret data;
- Develop descriptions, models, explanations and predictions;
- Think critically and logically to connect evidence and explanations;
- Communicate scientific procedures and explanations.

Creativity and innovation skills are also present in the Ohio standards and model curriculum both explicitly and implicitly in the practice of inquiry. An example of a statement that explicitly supports the development of creativity and innovation skills in the 7th grade Earth and Space Science expectations for learning is:

Design, build, and test a buoy that can sample water temperatures (or other water quality test, such as pH or turbidity levels) of a local lake, pond, pool, or stream. Deploy the buoy and collect/analyze data. Compare and discuss results with the class.

Finally, no statements explicitly or implicitly describing student engagement with reflection, autonomy, and self-direction skills were identified.

Additional Curricular Components

Results of the analysis of the additional curricular components described in TASSC indicate that the Ohio revised standards and model curriculum for grades 6-8 support rich technological environments (see Table 4) and include teaching and learning activities set in real-world examples, applications, and experiences. For example, the following recommended instructional strategy and resource suggest an activity set in a local context that invites investigation of a real-world issue:

Choosing local issues that involve water and conducting field studies and research about the movement of water and/or contamination can lead to deeper understanding of how the cycles work. An example could be researching acid mine drainage problems in southeastern Ohio. The Monday Creek Organization provides research and data for southeastern Ohio and acid mine drainage cleanup efforts.

Furthermore, the standards and model curriculum account for the different ways of student knowing through an attention to learning progressions described in content elaborations. For example, the following excerpt from the 6th grade model curriculum highlights the past and future learning of students with respect to rocks:

Prior concepts related to rocks: PreK-2: Objects have physical properties, properties of objects can change, and Earth's nonliving resources have specific properties. Grades 3-5: rocks and soil have characteristics, soil contains pieces of rocks, rocks form in different ways, and objects are composed of matter and may exhibit electrical conductivity and magnetism.

Future Application of Concepts: Grades 7 and 8: Sedimentary, metamorphic, and igneous environments, and the history of Earth (including the changing environments) from the interpretation of the rock record are studied. High school: The formation of elements, chemical bonding, and crystal structure are found in the Physical sciences. In 11/12th grade Physical Geology, depositional environments, volcanics, characteristics of rocks, and mineralogy are explored at depth.

The model curriculum also accounts for active student learning through references to misconceptions. Resources that describe common misconceptions in core content are provided in the instructional strategies and resources section. For example, one such reference for the 6th grade life science model curriculum states, "The Annenberg Media series Essential Science for Teachers: Life Science Session 1: Children's Ideas provides greater insight to misconceptions children hold about the origin of living things."

Finally, the Ohio model curriculum proposes the use of appropriate tools and approaches to measure students' performance on twenty-first century content and skills. This focus on assessment is evident in the description of the Cognitive Demand expectations for learning described in the section addressing structure of the curriculum. Examples of these expectations for learning are provided in the previous two sections.

The Case of the New York State Standards

New York City (NYC) schools are among the nation's most diverse schools with large percentages of minorities, recent immigrants, and English Language learners (NYC Department of Education-NYC DOE[3]). NYC Public schools follow the New York State (NYS) Core Curriculum[4]. The NYC DOE website gives the following general statement on the goal of science instruction:

[Providing students] with opportunities to interact directly with the natural world and gives them the skills they need to explain the world that surrounds them. The City's approach allows students to

practice problem-solving skills, develop positive science attitudes, learn new science content, and increase their scientific literacy (NYC DOE).

Structure of the NYS Science Core Curriculum

The NYS core curriculum outlines seven Standards for Science, Mathematics, and Technology, of which five are designated as applying specifically to science. Table 5 lists the standards and explanatory statements that outline learning expectations for each.

The science core designates Standards 1, 2, 6, and 7 as the process skills standards and Standard 4 as the science content standard with two main strands: The Living Environment and The Physical Setting. For reasons not explicit in the document, Strand 5, Technology (Students will apply technological knowledge and skills to design, construct, use, and evaluate products and systems to satisfy human and environmental needs) is not designated for science and consequently is not included in the science core curriculum document. Each standard is further expanded by a number of key ideas. Then each key idea is expanded with one or two levels of associated performance indicators. For example Standard 1(Analysis, Inquiry, and Design) subsumes the following key idea and subsequent performance indicators:

Key Idea 2: Beyond the use of reasoning and consensus, scientific inquiry involves the testing of proposed explanations involving the use of conventional techniques and procedures and usually requiring considerable ingenuity.

S2.1: Use conventional techniques and those of their own design to make further observations and refine their explanations, guided by a need for more information.

S2.1a: demonstrate appropriate safety techniques

S2.1b: conduct an experiment designed by others

S2.1c: design and conduct an experiment to test a hypothesis

S2.1d: use appropriate tools and conventional techniques to solve problems about the natural world, including:

- Measuring
- Observing
- Describing
- Classifying
- Sequencing

Depending on its scale and expansiveness, a standard may be divided into broad themes that include several Key Ideas. For example standard 1 (Analysis, Inquiry, and Design) incorporates the themes of Mathematical Analysis, Scientific Inquiry, and Engineering Design.

Table 5. NYC science specific standards

Standard	Expectation
Standard 1: Analysis, Inquiry, and Design	Students will use mathematical analysis, scientific inquiry, and engineering design, as appropriate, to pose questions, seek answers, and develop solutions.
Standard 2: Information Systems	Students will access, generate, process, and transfer information using appropriate technologies.
Standard 4: Science	Students will understand and apply scientific concepts, principles, and theories pertaining to the physical setting and living environment and recognize the historical development of ideas in science.
Standard 6: Interconnectedness: Common Themes	Students will understand the relationships and common themes that connect mathematics, science, and technology and apply the themes to these and other areas of learning.
Standard 7: Interdisciplinary Problem Solving	Students will apply the knowledge and thinking skills of mathematics, science, and technology to address real-life problems and make informed decisions.

Results

The analysis results for Standards 1, 2, 4, and 7 will be presented below to demonstrate the use of TASSC on the curriculum document.

Core Academic Content

Standard 4 is the standard that outlines the science concepts and principles to be covered at the middle school level for New York State students. The major unifying concepts and processes of science are covered at comprehensive high level in standard 4, especially as they pertain to Living Environment and Physical setting.

Standard 1 also addresses science content and was assigned a rating of 'partial high level' coverage for Science Content for understandings about scientific Inquiry and its features. This is somewhat expected, as the standard itself explicitly cites 'inquiry' as a focus in its statement and includes 'scientific inquiry' as one of its themes. Therefore there are several key ideas and performance indicators that incorporate understandings about inquiry, for example, "The central purpose of scientific inquiry is to develop explanations of natural phenomena in a continuing, creative process" (Key Idea). Other performance indicator examples include:

S1.2c: Differentiate among observations, inferences, predictions, and explanations

S3.1: Design charts, tables, graphs, and other representations of observations in conventional and creative ways to help them address their research question or hypothesis.

Of the 21st Century Themes, content within Environmental literacy is covered at a more comprehensive level than Health literacy. For example, two of the seven key ideas in the Living Environment explicitly include aspects of Environmental Literacy, which is covered at comprehensive high level:

KI 6: Plants and animals depend on each other and their physical environment.

KI 7: Human decisions and activities have had a profound impact on the physical and living environment.

Health Literacy is covered at a partial-high level in Key Idea 5 in Living Environment. The component with the strong presence involves the role of nutrients and diet as is apparent in the following performance indicator:

5.2: Describe the importance of major nutrients, vitamins, and minerals in maintaining health and promoting growth, and explain the need for a constant input of energy for living organisms.

No reference is made to global awareness and only minimal references are made to Economic and Civic literacy. For example, the following key idea presents a vague reference to civic literacy issues:

3.2: Describe applications of information technology in mathematics, science, and other technologies that address needs and solve problems in the community.

Skills

Communication, thinking, and problem solving skills are explicitly integrated in the NYS science curriculum, specifically for scientific inquiry. For example, standard 1 presents a strong skill for the items 'thinking and problem solving skills' and 'communication skills' as the latter relates to presenting work associated with scientific inquiry or sharing understandings. Similarly, developing information skills are at a strong skill level in standard 2, clearly requiring that students manage and evaluate information from different sources. The following performance indicators illustrate this finding:

1.4c: Use the collected data to communicate a scientific concept

3.1: Use graphical, statistical, and presentation software to present projects to fellow classmates.

Autonomy, self-direction, creativity, and innovation skills are included but in weak form and mostly as they pertain to carrying out scientific inquiry. Autonomy skills are covered by the expectations that students carry out aspects of scientific inquiries 'independently.' Creativity and innovation skills are implicit to inquiry processes such as formulating hypotheses and designing experiments.

Additional Curricular Components

The strong focus on inquiry in Standard 1 supports students as active learners. Perceptions of students as active learners was evident by the strong emphasis in standard 1 on themes and key ideas under analysis, inquiry, and design, including "students are expected to use inquiry process skills to learn about the world." Parallel processing, on the other hand is not explicitly addressed in any place in the curriculum document.

The NYS core curriculum emphasizes teaching and learning in 21st Century teaching contexts. For example in standard 7, the notion of using content and skill to address everyday problems is explicitly stated in all Key Ideas. Furthermore, two performance indicators of standard 1 mention explicitly solving "real world problems" (M3.1) and the "investigation of situations of general social interest" (T1.1), therefore this item was designated as mostly absent.

Technologically rich environments are also supported in the NYS core curriculum, specifically in standard 2. Standard 2 is concerned with Information Systems and states that "Students will access, generate, process, and transfer information using appropriate technologies." Upon examining all the key ideas within the standard, technologically rich context was strongly emphasized and present.

Finally, assessment appropriate to a 21st Century context is not explicitly addressed in the standards. However, in the appendix of the NYS Core Curriculum it is explicitly mentioned that 20- 25% of the exam grade is on "real world problems," with this being the only reference made to assessment in middle school science. Vagueness about assessment is somewhat disconcerting considering the way by which the 'tested curriculum' (Cuban, 1995) percolates to classroom practices of science teachers.

The Case of Lebanon

Lebanon underwent educational reform in 1998 in order to meet the demands and requirements of the present society, and adequately prepare students for the future. The preparation of the resultant Lebanese curriculum[5] was inspired by the rapid expansion of science and technology in the present century, the new international and global tendencies towards science teaching, and the important role in all aspects of our everyday life. The intermediate (middle level) stage of the Lebanese curriculum consists of grades 7-9. Grade 6 is presently compulsory for all Lebanese students, and all students are provided with a common content until grade 10.

Structure of the Lebanese Science Curriculum

The focus of science teaching in the Lebanese science curriculum is defined by the overall general objectives. Some examples of the overall general objectives include:

- Develop learners' intellectual and practical scientific skills.
- Deepen learners' awareness in the ability of humans to understand, invent, and create.

- Understand the nature of science and technology, their historical development, and their impact on human thought.
- Insure that learners have acquired the facts, concepts, and principles that are essential to understand natural phenomena.
- Acquire knowledge and accordingly behave about issues of health, environment, and safety practices.
- Realize that some natural resources can be depleted and advocate for the role of science in sustaining these resources.
- Encourage learners to use scientific knowledge and skills in novel everyday life situations.
- Emphasize the role of scientists in the advancement of human kind.
- Encourage learners to be open to the ideas of scientists from different cultures and to their contributions to the advancement of science.
- Encourage learners to work independently and cooperatively to solve scientific problems.
- Make learners aware of career possibilities in different science related areas.

The subjects that are addressed in the Lebanese intermediate science curriculum are physics, chemistry, life, and earth sciences. The analysis was performed on all components of the intermediate science curriculum including the introduction for each of the three subjects, objectives of each subject, instructional objectives, and activities of each subject. The objectives of each subject are conceptual (technical and methodological) and advance a relationship between teaching and evaluation. The analysis was also performed on the overall general objectives of science education.

Results

The results of the TASSC analysis of the Lebanese Intermediate science curriculum are presented below as core academic content, skills, and additional curricular components.

Core Academic Content

The core academic content is composed of the science content and the 21st century themes. The TASSC analysis showed that the standards are highly comprehensive and strong for developing the science content for the earth and life sciences, chemistry, and physics.

As for the 21st century themes, the themes of the environmental and health literacy are emphasized in the introduction of the earth and life sciences curriculum with statements such as:

The program deals with a global perspective about health and the environment. This helps the student to maintain his health and the health of others, to develop a responsible behavior towards the environment, and to comprehend scientific messages transmitted by media.

The theme about the environment literacy is further carried along in several objectives of the earth and life sciences, including:

Allow the student to identify the essential biological, physical, and geological components of the environment and to understand their relations that initiate the development of an open and a responsible behavior toward environmental problems: management of media and resources, preservation of species, and risk prevention.

Particularly for Grade 8, there is a chapter in the content that targets the environment, "earth and the environment." That chapter deals with the concrete geodynamic manifestations of the earth activity, leading to the study of changes starting from rocks to landscape that constitutes the life framework for humans. The chapter further addresses the study of geology with a focus on the relationship between man and earth in order to develop among the students a sense of a long-time responsibility towards the environment.

Similarly within the chemistry curriculum, there is a separate chapter "chemistry and the environment" at the grade 9 level. It aims at providing students with a better understanding of the role of chemistry in causing and solving environmental problems. Environmental application and problems are presented as:

- Dilution of toxic wastes and pollution addressed at grade 7,
- Acid rain, antacids, fertilizers addressed at grade 8,
- Pollution problems of air, water, and soil addressed at grade 9.

Likewise, developing a responsible attitude towards the environment is manifested as a theme in the instructional objectives within the physics curriculum. However, it is not clear how that is addressed in the instructional activities and guidelines.

Skills

The TASSC analysis showed that the skills for developing creativity and innovation, as well as the reflective skills are completely absent in the Lebanese curriculum. Furthermore, the curriculum is not very explicit with respect to developing information, communication, and autonomy skills. For example, the introduction of the life and earth sciences curriculum emphasizes the importance of communication skills by indicating that the program needs to focus on teaching scientific processes in a problem situation to allow the learner to acquire communication skills. However, the presence of communication skills is not reflected in the instructional objectives and activities of the curriculum.

The skill of developing autonomy is apparent in the introduction of the life and earth sciences curriculum as noted in the following statement, "The curriculum fosters the development of autonomy and responsibility of personal behavior in learners that are characteristics of a free citizen." Autonomy is further addressed in the instructional objectives of life and earth sciences curriculum as follows:

Develop in students a scientific attitude and motivate them to acquire a better autonomy.

This statement affirms that autonomy is strongly adopted in the curriculum. Yet and again, this is only addressed in the introduction and instructional objectives in only the life and earth sciences curriculum.

Promoting thinking and problem-solving skills is considered a critical component in the Lebanese Intermediate curriculum. It is explicitly present in the general science objectives and also emphasized as one of the instructional objectives in the life and earth sciences curriculum for the intermediate grades:

Permit students to acquire scientific processes, specifically by developing an experimental approach and problem solving activities

Additional Curricular Components

Results of the analysis of the additional curricular components described in TASSC indicate that the Lebanese intermediate curriculum vaguely describes the criteria of assessment and learning the academic content through real-world examples, applications, and experiences. For example, assessment is emphasized in the introduction of the life and earth sciences curriculum as "evaluation must not be limited to recalled information but should also assess the skills required to be achieved." However, it is not clear how that is manifested in terms of the assessment at the level of instructional activities or guidelines.

Furthermore, the Lebanese curriculum was found to account for the different ways of student knowing to various degrees. First, the curriculum makes no mention of experiences that allow for parallel processing. Second, it only references students as active learners. For example, the description in the introduction section of the life and earth sciences curriculum presents the learner as an "active participant" and it is also indicated that the use of different approaches encourages students to "construct" their knowledge themselves.

Third, the curriculum explicitly supports a technologically rich environment. For example, the introduction of the life and earth sciences curriculum states, "The program tends to establish a relationship between technology and society with which the student has strong ties." Technology is also referred to in the instructional objectives of the life and earth sciences curriculum, where it is indicated that there is a need to progressively and continually initiate in students "a scientific knowledge, technical performances and actual research in science and technology." Moreover, it is explicitly stated in the instructional objectives of the chemistry curriculum that students should "perceive the close relationship among chemistry, technology, and industry." This is also referred to in the instructional objectives of the physics curriculum about having students understand the relation between physics, technology, and society.

The Case of Qatar

A new reform initiative, *Education for a New Era*, was launched in Qatar in 2005. This initiative is built on four principles:

- **Autonomy:** Allowing an innovative approach for schools and teachers to meet the needs of students and parents within the curriculum standards framework.
- **Accountability:** Implementing an objective and transparent assessment system to hold school leaders, teachers, and parents responsible for the students' success.
- **Variety:** Encouraging different kinds of schools and instructional programs.
- **Choice:** Allowing parents to select the school that best fits their child's needs and seeking their input on important schools decisions.

Structure of the Educational System

The curriculum of the Qatari preparatory (middle level) cycle emphasizes basic literacy and numeracy skills and is follows the *Early Years Education Foundation Curriculum*. The *Foundation Curriculum* is believed to be future-oriented in attending to the qualities and characteristics needed in young citizens in 2020 in the State of Qatar with the goal of preparing the Qatari youth to compete in the 21st century global economy. The curriculum is also aligned with the vision of *Education for a New Era,* which aims for schools to educate students who will:

- Make a positive contribution to society by being literate and numerate,
- Be effective communicators in Arabic and English,
- Be creative and critical thinkers,
- Have the dispositions and strategies to explore and inquire, and
- Work well in teams and demonstrate pride in being Qatari.

The Qatari curriculum for grades 7-9 includes the subjects of inquiry, life science, materials, earth and space, and physical processes. TASSC analysis was performed on the following components of the Qatari curriculum: overall general goals of science, summary of students' performance for each of the five subjects, and the instructional objectives and activities of each subject at grades 7, 8, and 9. It is important to note that the Qatari standards are explicit in outlining the different instructional activities, questions, and examples for each of the subjects.

Results

The results of the TASSC analysis of the Qatari preparatory curriculum are presented below as core academic content, skills, and additional curricular components.

Core Academic Content

The TASSC analysis showed that the standards are highly comprehensive and strong for developing the science content. As for the 21st century themes, attention to the environment is noted in the overall aims of the Qatari science standards (Supreme Education Council, 2002) as follows:

Students should (a) have a sound and systematic knowledge of important scientific facts, concepts and principles, and possess the skills needed to apply these in new and changing situations in a range of personal, domestic, industrial and environmental contexts, (b) recognize the importance of the application of scientific knowledge in the modern world and be aware of the moral, ethical, social and environmental implications

Further, one of the topics of the life science subject addresses living things and their environment. This topic is addressed in grade 7 and 8 but applications and examples related to the environment seem to be scarce. That is not the case for grade 9, where environmental literacy is a focus along with a relation to the economy of Qatar. For example, the life science curriculum states the expectation that students discuss the economic dimension of genetic engineering and its implications. According to the materials subject, grade 9 students need to "explain the importance of fossil fuels to the economy of Qatar."

Global awareness is manifested in the Qatari standards in grade 9 in the context of the materials subject. For example, students need to explain the causes of global warming and discuss various proposals about the removal of carbon dioxide from the atmosphere as proposed under the Kyoto protocol. Another instance is represented by having students "make a study of the consequences of acid rainfall in some other parts of the world (e.g., northern Europe)."

Health literacy is represented at various instances in the life science curriculum for grades 7 and 8 only. For example, students in grade 7 need to know the importance of good nutrition and hygiene to the health of babies. Students in grade 8 study the effects of smoking on health and the problems of diabetes and obesity, as follows:

Know that smoking damages the lungs and reduces the efficiency of gas exchange.

Know the symptoms, causes, and problems of diabetes and obesity. Chart the statistics on the frequency of diabetes and obesity in Qatar and compare with other countries.

Skills

The TASSC analysis showed that developing information and communication skills is explicitly addressed and described in the Qatari standards for grades 7-9. Actually, one of the topics in the subject of scientific inquiry is processing and communicating information, which is addressed in grades 7 to 9.

The development of creativity and innovation skills is also attended to in the standards. Having students develop the ability to work creatively is noted to be critical in the overall aims of the Qatari standards, and is explicitly reflected in the instructional activities described for grades 8 and 9, as shown in the following examples:

Construct a card game to illustrate what eats what in a range of environments (life science, grade 8)

Design a thermal solar panel and investigate its effectiveness (materials, grade 9)

Build model bridges from simple materials such as mud bricks (high compressive strength but low tensile strength), spaghetti (good tensile strength but easily broken), and cotton (flexible and good tensile strength). Use different kinds of bridge construction (e.g. arch, cantilever, girder, suspension) that make the best use of the properties of the materials. Test the bridges to destruction (physical processes, grade 9)

Developing problem-solving skills among students is considered important as indicated in the overall aims of the Qatari standards. And that is clearly reflected in the instructional activities of the curriculum of grades 7, 8, and 9.

Developing the interpersonal and self-directional skills, the reflective skills, and autonomy do not appear to be manifested in the Qatari standards.

Additional Curricular Components

Based on the TASSC analysis, results show that the instructional activities in the Qatari curriculum harness a 21st century context in the use of many real-world examples, applications and experiences at grades 7, 8, and 9. Following are some examples of the instructional activities at grade 7 targeting the subjects about materials and earth and space:

Collect information on the distillation of seawater to provide drinking water for Qatar (materials)

Show, using a flow-chart, how Qatar gas is used (earth and space).

More examples, which target real-world applications and experiences, are evident from the instructional activities at grade 8 across the subjects of materials and physical processes:

Discuss the occurrence of uncombined metals at the bottom of the reactivity series and methods used to recover alluvial gold. Link this natural low reactivity with the use of these metals in jewelry (materials)

Study the formation of mirages on a hot day, noting how they can be explained by refraction (physical processes)

Still other examples are found from the instructional activities for grade 9 in the subjects of life science, materials, and physical processes:

Collect information on the frequency of common inherited disorders in Qatar (life science).

Make a study of the cooling of industrial processes in Qatar, including a visit to a cooling plant such as the one at the Ras Laffan (materials).

Furthermore, Qatari science assessment seems to use appropriate tools and approaches to measure students' performance on twenty-first century content and skills. Science assessments have three objectives (Supreme Education Council, 2002):

- Knowledge and understanding; application of knowledge and understanding, analysis and evaluation of information.
- Scientific enquiry skills and procedures.
- The weightings of the assessment objectives change at various grade levels.

As students' scientific proficiency and experience develop, greater emphasis is placed on assessing the application of knowledge to solve problems in new situations.

Finally, the Qatari standards were found to vaguely account for students' ways of knowing to various degrees. First, there seems to be no indication that the curriculum affords the utilization of technological tools allowing for parallel processing. Second, the perception that students are active learners is implicit in the strong emphasis on the doing and understanding about scientific inquiry. Third, the standards somewhat support a technologically rich environment. The effective use of information and communications technology in the pursuit and communication of science, as well as the relationship between technology and society is emphasized in the Qatari science standards. However, that is not clearly addressed in the instructional activities of the curriculum.

DISCUSSION

In the previous section, individual case studies present the results of using TASSC to analyze the standards and curricula of Ohio, NYS, Lebanon, and Qatar. In this section, we summarize the findings from the four case studies with respect to core academic content, skills, and additional curricular components. We end this section with a discussion of next steps necessary in the development and use of TASSC.

Core Academic Content

Science content is found almost in its entirety in Ohio and NYS standards and curricula. It is explicitly described in curriculum activities in a manner likely to produce high levels of learning if the curricular goals are met. Specifically, unifying concepts and processes of science, inquiry, earth and space, life, and physical science content were integrated in the targeted US standards.

Twenty-first century themes are present in Ohio and NYS standards and curricula at partially high levels. Environmental Literacy is integrated in the activities of both US standards and curricula. Additionally, civic and health literacy are explicitly integrated in the Ohio and NYS curricula respectively. Other 21st century themes are vaguely referenced (economic literacy) or absent (e.g. global literacy in NYS curriculum and health literacy in Ohio).

Similarly, the science content is found in its entirety in the Lebanese and Qatari curricula. The central theme of the environment permeates all the topics covered in middle level curricula. However, the manifestation of that theme varies within the different subjects. Of the remaining 21st century themes, global and health literacy are explicitly integrated in the Qatari standards while only health literacy is alluded to vaguely in the Lebanese curriculum.

Skills

Twenty-first century skills are present in the Ohio and NYS standards and curricula to varying degrees. Communication, thinking, and problem solving skills are present as strong skills in both states. Creativity and innovation are present in both state documents as weak skills in the NYS standards and strong skills in the Ohio ones. Reflective skills, autonomy, and self-direction are present in the NYS curriculum as weak skills. Reflective skills, autonomy and self-direction seem to be absent from the Ohio standards.

Interestingly, all 21st century skills, with the exception of reflection and autonomy, are explicitly integrated in the Qatari curriculum. Reflection and autonomy are not addressed. Conversely, only thinking and problem solving skills are explicitly integrated in the Lebanese curriculum. Information, Communication, and autonomy skills are vaguely alluded to in the introductory comments only. Creativity, innovation, and reflective skills seem to be absent from the Lebanese curriculum.

Additional Curricular Components

Teaching and learning activities in the Ohio and NYS standards and curricula involve a twenty-first century context and promote the learning of academic content through real-world examples, applications and experiences. These Curricula also support a rich technological environment through the availability of rich graphics and interaction with technological tools and media. Activities within the Ohio and NYS standards and curricula account for students as active learners, but do not explicitly attend to students' ability to parallel process. However, while the assessments advocated in the Ohio standards and model curriculum seem to promote appropriate tools and approaches to measure students' performance on twenty-first century content and skills, the NYS standards only vaguely address assessment.

The majority of additional curricular components of interest in this study are vague or absent from the Lebanese and Qatari curricula. Only vague references in the general statements of both curricula address students as active learners or appropriate tools and approaches for the assessment of 21st century content and skills. Neither attends to students' ability to parallel process. While the Qatari curriculum consistently promotes the learning of academic content through real-world examples, applications and experiences, such recommendations are absent from the Lebanese curriculum. Conversely, the Lebanese curriculum explicitly supports technologically rich environments. References to such environments are vague in the Qatari curriculum.

Next Steps for TASSC Development Process

Even though TASSC was successfully used to analyze the standards and curricula described in the four case studies presented in this report, further refining of the instrument is warranted. First, TASSC needs to be revisited to remove additional areas of overlap that were identified in the development of the case studies. For example, statements depicting thinking and problem solving skills in the standards and curricula were also coded as instances of active learning in context.

Next, attention to the rating scale is needed. For example, additional instructions are necessary to make determinations about the extent themes and skills are explicit. Furthermore, considerations about how many times an instance of a particular theme or skill needs to be identified in the text to warrant a particular rating.

Finally, the content section of the instrument can benefit from the further definition of the science content. Such a definition will allow the instrument to provide more discriminating data by providing a rating for each content category. Similarly, skills can be further subdivided or operationalized for greater accuracy. For example, communication skills could be divided into five component skills as described by Schunn in the 2010 National Research Council report (NRC, 2010):

- Select key pieces of a complex idea to express in words, sounds, and images, in order to build shared understanding,
- Social perceptiveness,
- Persuasion and negotiation,
- Instructing, and
- Service orientation.

Such operationalization allows us to determine if a skill is present in its entirety warranting a rating of "whole," or if parts of the skill are present warranting a rating of "component." This operationalization may also allow us to distinguish between the general education components of each skill and more subject-specific ones. For example, persuasion and negotiation are science specific components of the communication skills that are represented in the science education community as argumentation skills.

CONCLUSION AND IMPLICATIONS

The use of TASSC to analyze the standards and curricula of Ohio, NYC, Lebanon, and Qatar provided useful information for the evaluation of these curricula with respect to their incorporation of 21st century essential understandings and skills. It is evident from the case studies that these states are at various points in their journey to meet the imperative of preparing students for the 21st century. The use of TASSC allows reformers and curriculum designers to identify areas for further development and continuous improvement.

The case studies also illustrate that TASSC can be used in multiple contexts. The intent of TASSC development is not the creation of an instrument for rank-ordering states with respect to the degree to which they incorporate 21st century essential understandings and skills. The use of TASSC is intended to be formative; results can enhance and inform the standard and curriculum development process. Finally, TASSC is most useful for the evaluation of standards, curricula and long-term planning efforts at the local levels.

REFERENCES

American Association for the Advancement of Science. (1993). *Benchmarks for science literacy: Project 2061*. Oxford, UK: Oxford University Press.

BouJaoude. S. (2010). *Competencies and educational structures needed to prepare secondary students for the 21st century*. Paper presented at a Symposium on Secondary Education Organized by the Ministry of Education in the United Arab Emirates in Cooperation with the Arab Bureau of Education for the Gulf States. Dubai, UAE.

Bybee, R. (1997). *Achieving scientific literacy: From purposes to practices*. Portsmouth, NH: Heinemann Educational Books.

Cranton, P. (1997). Higher education: A global community. *New Directions for Teaching and Learning, Winter, 72*, 5-9.

Cranton, P. (2001). Interpretive and critical evaluation. *New Directions for Teaching and Learning, 88*, 11–18. doi:10.1002/tl.33

Cranton, P. (2002). Teaching for transformation. In O. M. Ross-Gordon (Ed.), *Contemporary Viewpoints on Teaching Adults Effectively: New Directions for Adult and Continuing Education*. San Francisco, CA: Jossey-Bass.

Habermas, J. (1971). *Knowledge and human interests*. Boston, MA: Beacon.

Mezirow, J. (1997). Transformative learning: Theory to practice. *New Directions for Adult and Continuing Learning, 74*, 5–12. doi:10.1002/ace.7401

National Research Council. (1996). *The national science education standards*. Washington, DC: National Academy Press.

National Research Council. (2010). *Exploring the intersection of science education and 21st century skills: A workshop summary*. Washington, DC: National Academies Press.

Ohio Department of Education. (2010). *Science education standards revision and model curriculum development*. Retrieved from http://www.ode.state.oh.us/GD/Templates/Pages/ODE/ODEDetail.aspx?page=3&TopicRelationID=1705&ContentID=76585

OECD. (2005). *Definition and selection of key competencies*. Retrieved from www.oecd.org/dataoecd/47/61/35070367.pdf

Partnership for 21st Century Skills. (2009). *Framework for 21st century learning*. Retrieved from http://www.p21.org/index.php?option=com_content&task=view&id=254&Itemid=119

Prensky, M. (2001). *Digital natives, digital immigrants*. Retrieved from http://www.marcprensky.com/writing/Prensky%20- %20Digital%20Natives,%20Digital%20Immigrants%20-%20Part1.pdf

Strickland, T. (2009). *Reforming Ohio's education system for the 21st century: Governor Ted Strickland's education reform and funding plan*. Retrieved from http://www.conversationoneducation.org/information/

Supreme Education Council. (2002). *Science standards*. Retrieved February 16, 2012, from http://www.english.education.gov.qa/section/sec/education_institute/cso/science_standards

Treadwell, M. (2011). Whatever happened? In G. Wan & D. Gut (Eds.), *Bringing Schools into the 21st Century* (pp. 7–40). New York, NY: Springer. doi:10.1007/978-94-007-0268-4_2

Wan, G., & Gut, D. (Eds.). (2011). *Bringing schools into the 21st century*. New York, NY: Springer. doi:10.1007/978-94-007-0268-4

ENDNOTES

1 http://www.oecd.org

2 Habermas coined this knowledge as practical knowledge and Merizow (1997) called it communicative knowledge.

3 http://schools.nyc.gov/Academics/Science/default.htm

4 http://www.p12.nysed.gov/ciai/cores.html#MST

5 http://www.crdp.org/crdp/all%20curriculum/Sciences/Sciences_english/science_english.pdf

This work was previously published in "Approaches and Strategies in Next Generation Science Learning" edited by Myint Swe Khine and Issa M. Saleh, pages 265-289, copyright 2013 by Information Science Reference (an imprint of IGI Global).

Chapter 29
Self-Regulated Learning as the Enabling Environment to Enhance Outcome-Based Education of Undergraduate Engineering Mathematics

Roselainy Abdul Rahman
Universiti Teknologi Malaysia, Malaysia

Yudariah Mohamad Yusof
Universiti Teknologi Malaysia, Malaysia

Sabariah Baharun
Universiti Teknologi Malaysia, Malaysia

Sharifah Alwiah S. Abdur Rahman
Universiti Teknologi Malaysia, Malaysia

ABSTRACT

The paper discusses the results of an action research to improve teaching practice which applied the philosophy of Knowledge – Experiential – Self-regulated (KES) at the Malaysia-Japan International Institute of Technology, Universiti Teknologi Malaysia. The teaching and learning strategies were implemented to achieve the following objectives; (i) to enhance students' mathematical competencies; (ii) to support self-regulated learning; and (iii) to improve the teaching practice of Engineering Mathematics 3 i.e. Advanced Calculus. A modification to a previously developed framework by Roselainy et al. (2012a) was performed to encourage students to adopt self-regulated learning behaviour in an active learning environment. The teaching, learning and assessment activities were aligned constructively based on the theory of Biggs & Tang (2010). Data was collected, analysed and later employed to modify the teaching and learning activities. The findings found that it is imperative for teachers to design an appropriate learning environment and apply suitable strategies in encouraging and supporting students to embrace and take charge of their own learning.

INTRODUCTION

An important outcome in mathematical learning for Malaysian engineering undergraduates is the ability to apply the mathematical knowledge learnt in order to solve complex problems (EAC, 2012). However, previous research and the authors' experience have shown that some students faced difficulties in manipulating concepts, coordinating multiple procedures, manipulating symbols in a

DOI: 10.4018/978-1-4666-7363-2.ch029

flexible way and in answering non-routine questions (Tall & Razali, 1993; Anthony, 2000; Croft & Ward, 2001; Hoch & Dreyfus, 2005, Roselainy, 2009, Roselainy, Yudariah & Sabariah, 2012a). Thus, in an effort to promote holistic students' achievement, Malaysia-Japan International Institute of Technology (MJIIT), has adopted a teaching and learning culture by focusing on Knowledge, Experiential and Self-regulated learning (KES). MJIIT is a government to government initiative that aims to provide Japanese style education in a Malaysian setting. This paper will discuss the strategies that were developed to support Outcome-Based Education in the Engineering Mathematics 3 course by promoting self-regulated learning and focusing on the development of mathematical competencies (SEFI, 2011). To ensure the achievement of the course learning outcomes, the constructive alignment theory (Biggs & Tang, 2010) was employed. This was performed to assess whether the outcomes, teaching and learning activities, as well as the assessment of students' learning were aligned and contributed effectively to learning.

In the authors' earlier work, a framework was developed to support and enhance students' awareness of their own mathematical thinking powers as well as to highlight independent learning, communication and team working skills. The framework is also described in greater detail in the works of the following authors: Roselainy et al. (2007); Baharun et al. (2008); Roselainy (2009) and Roselainy et al. (2012a). Various theoretical inputs were considered (Mason et al., 1982, 2010; Watson & Mason, 1998; Tall, 1991, 1995) and strategies were developed and applied to make the thinking processes, structures of mathematics explicit as well as increased students' awareness on the use of their thinking powers. Appropriate teaching tasks were designed to support, develop and extend students' abilities in working on mathematical problem solving. For this study, a new model was designed to further extend the framework and included strategies to support students' awareness of their Self-Regulated Learning (SRL). There were three main objectives of the research;

1. To enhance students' mathematical competencies;
2. To support self-regulated learning; and
3. To improve the teaching practice of Engineering Mathematics 3 (i.e. Advanced Calculus).

This article discusses the implementation of the research as well as the strategies that were used in the classroom.

THE TEACHING AND LEARNING SITUATION MODEL

The new model proposed in this paper was refined from an earlier model by Roselainy et al. (2012a). The teaching and learning situation was categorized in three components,

1. The learning outcomes of the course,
2. The focus of the teaching, learning and assessment (TLA) activities and tasks, and
3. The environment which supports the SRL processes.

First, in designing the course, the outcomes were identified and these were determined by considering the programme outcomes of the engineering programmes. The outcomes were made up of selected mathematical competencies and undergraduate attributes. Mathematical competencies were defined following Niss (2003, 2011). The eight mathematical competencies which he proposed was later adopted by the European Society for Engineering Education (SEFI, 2011) in their recommendations of mathematics education of engineers. The competencies were, *Thinking Mathematically*, *Reasoning Mathematically*, *Mathematical Solving and Posing Problems*, *Modeling Mathematically*, *Representing Mathe-*

matically, *Communicating Mathematically*, *Using Symbolic and Formalism* and, *Mathematical Aids and Tools*. For the purpose of this study, only four mathematical competencies were selected namely the abilities to use representations and symbols, to think mathematically, to handle problems (solving and posing problems) and to communicate mathematically. As part of a more holistic approach to higher education, UTM has outlined seven graduate attributes and soft skills which must be embedded in the academic programs offered by the respective faculties including MJIIT (UTM, 2011). These attributes include the graduates' ability to communicate effectively, to think critically and solve problems, to become good team workers, to manage information and engage in lifelong learning, to recognise entrepreneurial opportunities, to become good leaders and to be reliable and ethical in life. Each attribute requires a list of identified skills to be developed throughout the programme. In this study, three attributes and the related soft skills were selected; the ability to work in a team, to communicate and to undertake lifelong learning. The focus of the teaching and learning were on the processes of developing the soft skills, knowledge construction, mathematical thinking and problem solving. An appropriate environment to support the achievements of the outcomes was established by supporting the Self-Regulated Learning (SRL) processes through active learning (Keyser, 2000).

CREATING THE SELF-REGULATED LEARNING ENVIRONMENT

Figure 1 illustrates the three main elements to support the SRL processes and these were:

1. Informal Cooperative Learning (ICL),
2. Constructive alignment of the TLA, and
3. Continuous Quality Improvement (CQI).

SRL is demonstrated through the students' ability to independently plan, monitor, and assess their learning. However, many students have to learn and develop these skills to be more effective learners. Research findings showed that through SRL, students adopted better learning habits and study skills (Wolters, 2011). They worked towards better academic performance by applying appropriate learning strategies (Harris, Friedlander, Sadler, Frizzelle, & Graham, 2005); have greater awareness in monitoring their per-

Figure 1. Model of the teaching and learning situation

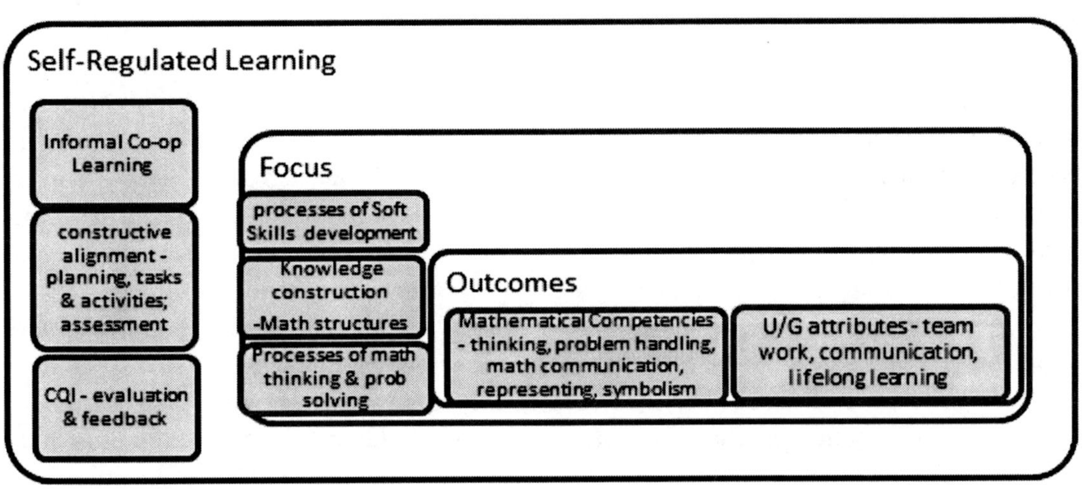

formance (Harris et al., 2005), and were able to evaluate their own academic progress (De Bruin, Thiede & Camp, 2001). The authors' previous work (Roselainy, 2009; Roselainy et al., 2012b) suggested some factors that influenced students' learning behavior which included beliefs about how to study, habits adopted from their pre-university experience, rewards, motivation and a tendency to depend on teachers' guidance. Factors such as motivation and rapport with lecturers have significantly influenced students' awareness and desire to change their learning behaviour. These factors were considered in designing the classroom activities and tasks.

A popular model identified the three phases of SRL; *Forethought and planning*, *Performance monitoring*, and *Reflections on performance* (Pintrich & Zusho, 2002; Zimmerman, 2000). This model is suitable for the students involved in this study who have to use higher order thinking skills to cope with their tertiary studies and engineering mathematics. Unfamiliarity with SRL has prompted the authors to design TLA activities to support the awareness of SRL processes and engagement in SRL. Earlier work by the authors (Roselainy et al., 2012a) introduced the application of active learning approach to encourage students' participation in the classroom. In this study, various informal cooperative learning strategies were applied to promote active learning in order to develop:

- Positive interdependence,
- Individual accountability,
- Face-to-face interaction,
- Interpersonal skills, and
- Self- assessment of group functioning (Felder & Brent, 2008).

These strategies will also help the SRL processes. The CQI is also included as part of the process due to the cyclic manner of action research, to improve the teaching practice and to promote active learning among the students.

Teaching Strategies to Enhance SRL Environment

In promoting teaching and learning to enhance students' awareness on their knowledge construction and their own thinking powers, the following strategies were used to increase students' awareness of the mathematical concepts, ideas and facility with mathematical techniques.

Structure of a topic: Topics were developed to make explicit the structures in the mathematical concepts (definitions, facts, theorems, properties, techniques, examples, etc.) and the mathematical powers used (specializing, generalizing, conjecturing, characterizing, organizing, reasoning, etc.). The tasks were designed to encourage students to be aware of how and when the mathematical thinking powers were used and their connections to the topics were made explicit.

Structured examples: Examples were structured in a manner that led towards a generalization. Students worked on specific examples first, then continued with more general examples. To further strengthen their understanding and knowledge, simple examples were provided and students were asked to create their own examples. The students' examples on what they expected to be taught on the topics were then assessed and this helped the lecturers uncover what they did not understand.

Students' use of their own thinking powers: The tasks also encouraged students' use of their own thinking powers. The tasks allowed students to experience the mathematical thinking activities as well as opportunities of expressing mathematical ideas and objects in words, pictures, and symbols as well as in written and verbal form. Thus, in this way, mathematical communication of concepts and ideas were naturally introduced.

Using mathematical themes, prompts, and questions: Mathematical themes such as 'invariance amidst change' - the basis for many mathematical theorems and technique; 'doing and undoing'- helped students to identify features or structures that should be the focus of attention were used. The 'prompts and questions' were specially constructed to highlight the internal structures of mathematics to focus students' attention to the structures and processes of mathematical thinking, to provoke and invoke students' awareness of their own mathematical thinking powers, and to provide students with a simple vocabulary to generate mathematical discussion.

Using a workbook: The structured examples, prompts and questions, mathematics tasks and activities were created to introduce students to mathematical ideas, thinking activities as well as provided questions to promote discussions. A workbook with five distinctive features was designed to help students to become more familiar with the teaching style namely *Illustrations*, *Structured Examples*, *Making Sense*, *Reflections*, *Review Exercises* and *Further Exercises*.

The SRL Processes

To encourage independent learning, students have to work on the topics by reading the text or notes on their own, discussed the concepts and examples as well as worked on the exercises in their groups. Lecturers worked in pairs to monitor the class progress, acted as facilitators and attended to the students in their groups. However, some examples were explained to students in a lecture setting and if any topic was found difficult by the students, a lecture may be delivered. The teaching sessions were conducted twice a week for two hour duration. The following are the process involved in encouraging students to adopt the SRL.

Forethought and Planning: Students were given the *Course Outline* at the beginning of the semester that contained information about the course outcomes, weekly schedule, Students' Learning Time calculation and the detailed assessment information. In addition, they were also provided *Learning Guides* containing information on topics outcomes, time allocation for each topic and sub-topic and the assignment questions. They were encouraged to manage their own learning by using these guides.

Performance Monitoring: A section referred as *Making Sense* was introduced. Questions were asked to increase students' awareness of how to monitor their knowledge and mathematical skills development. The questions focused on students' understanding of the concepts taught and their awareness on the mathematical procedures and techniques. Answers to the questions were reviewed by the lecturer who later compared the students' performance in the assessment tasks. If a particular question in the assignment was wrongly or poorly answered by a number of students, the misunderstanding or mistakes were then addressed by the lecturer in class. Feedback, both verbal and written was given to students to help them improve learning.

Reflections Monitoring: Another section in the book, named *Reflections* contains extra questions that inquired the students' ability to manage their study, time, emotional responses to the study, what they considered as effective learning strategies, learning difficulties and strategies to overcome those difficulties and perceptions throughout the sessions. These questions provoked the students to evaluate their strengths, weaknesses, preferences about the course, teaching and learning as well as suggestions on aspects of the course they would like to be changed. These self-reflections acted as a catalyst to influence students' planning and goals setting in their learning. The responses were collected to inform the lecturer of the students' concerns and allow them to address accordingly in the next class.

CONSTRUCTIVE ALIGNMENT

The various components in the teaching and learning environment were combined in order to create a strong framework that students can relate to. The objective was to enhance their self-regulation strategies. Biggs and Tang's (1999, 2010) theory on constructive alignment (see Figure 2) was adopted to align the teaching and learning activities and assessment tasks towards the achievement of the course outcomes. The course outcomes addressed four main areas: knowledge and skills, problem solving, teamwork, communication and lifelong learning. These were mapped to the programme outcomes and were compliant with the Malaysian Quality Framework and Engineering Accreditation Council requirements.

The tasks were created to enable the students to utilise their thinking powers and enhance their awareness of utilising the powers. Prompts and questions were used to draw their attention to the thinking processes. Another example was the use of a mini project where students solved non-routine problems in exercising their problem solving skills. The teaching and learning tasks were carried out in small groups to encourage the students to rely on their team members. The constructive alignment was also used to monitor students' performance and achievement of the outcomes.

Figure 2. Constructive alignment

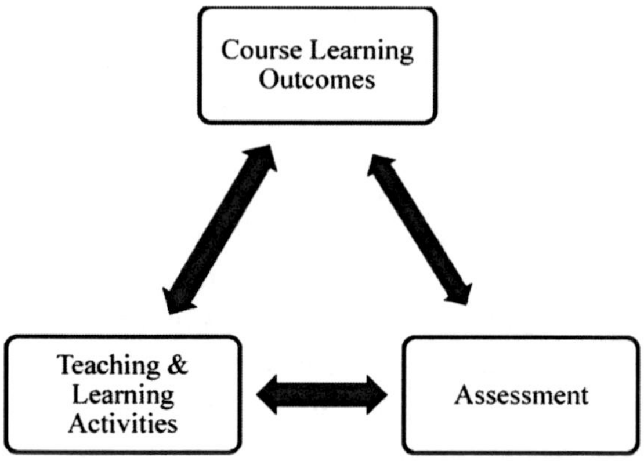

METHODOLOGY

The study was carried out in Semester 1, academic session 2012/2013. It involved two classes of students taking the Engineering Mathematics 3 course. The total number of students was 65 comprising of 33 students from the second year Electrical Engineering (SMJE) and 32 students from the Mechanical Engineering (SMJM) Programmes. The action research methodology was used to study the impact of teaching the classes with the proposed strategies as it has several appropriate features suitable for the context. Firstly, action research was a systematic examination of personal practice with a commitment to educational improvement (Mcniff, Lomax & Whitehead, 1996). Secondly, it was also more user-friendly and thirdly, it is a practical approach to conducting research with the main purpose of improving teaching and learning (Slavin, 2006). The research was carried out within the cycle of action research which involved:

1. The Planning of Teaching, Learning and Assessment (TLA) activities and tasks;
2. The Analysis and Reflection of the implementation results; and
3. Review and modifications of the TLA, if necessary (see Figure 3).

Figure 3. Action research cycle

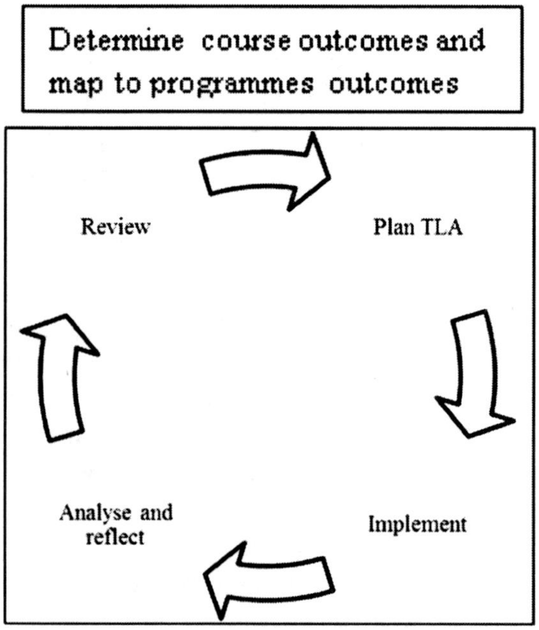

Data was collected through various methods such as:

- Students' observations,
- Students' work,
- Students' reflections on their learning,
- Students' responses on how they made sense of the mathematical knowledge learnt, and
- Students' examination performance.

Engineering Mathematics 3 was the first course where the students were asked to work and read through the chapters, individually and in their groups. A workbook and text book was used in the course. The workbook was written by the authors and incorporated several features to support the development of mathematical thinking and independent learning (Yudariah et al., 2009). A textbook was then chosen along with a reference lists on the subject. The mathematics courses in UTM are taught for 14 weeks in the semester. The course covers 5 main chapters namely Multivariable Functions, Partial Differentiation, Multiple Integrals, Vector Calculus and Line and Surface Integrals. The teaching and learning tasks were based on various examples and questions set out in the workbook. Additional examples for classroom work were used when needed especially at the end of a chapter, after which a Review session was conducted to ensure that students had grasped the mathematical concepts and techniques. The assessment tasks that contributed to their grades were chapter assignments, tests, mini projects and the final examination. Several non-graded tasks were also given in the class when students were asked to share with their peers about the problems faced during the course undertaking.

Data was interpreted as they were collected and helped the researchers to adapt or modify the TLA. Students' responses in *Making Sense* and *Reflections* questionnaires helped the researchers in making comparisons between what was observed during class sessions and performance in the assignments and tests. The whole process was reviewed regularly by considering the various phases of the cycle. The deliberation of the data collected was incorporated within the regular sessions of the teaching and learning after the conclusion of each chapter.

RESULTS AND DISCUSSION

At the end of the semester, the Course Review Report (CRR) which reported the students' overall performance in the assessment tasks and the final examination was prepared and compared against the course and programme outcomes. All the students passed the course with the results shown in Table 1.

The grades were distributed in all the categories. The students' performance was slightly better than their performance in Engineering Mathematics 2. The CRR also reports the students' attainment in the course and programme outcomes. A closer analysis at the achievement of the Key Performance Index (KPI) of the course outcomes gave a better picture of their achievement. The Faculty had set the KPI at 0.65 which corresponded to a B grade. Generally, SMJE students achieved the KPI target for all the course outcomes except for the

Table 1. Students' final results in engineering mathematics 3

Grades	A+	A	A-	B+	B	B-	C+	C	C-	D+
%	6.11	10.75	4.55	12.36	21.55	19.89	12.36	9.33	1.57	1.57
Total	21.41%			53.8%			23.26%			

topic of Multiple Integration (0.63). In contrast, SMJM students had slightly lower achievements in Multiple Integration (0.53) as well as Vector Calculus (0.63). However, the performance of both groups of students for the project were exceptionally good (0.9) but did not match their achievements in knowledge (0.80) and mathematical skills (0.85). This indicated that there was a need to review the problem solving questions and the marking scheme of the problem.

The following are some of the students' responses in terms on the teaching and learning sessions. Some of the responses were very positive in terms of the delivery method, "Very interesting and fun course.," "Challenging and interesting" and "Make me think more and make me always revise before doing any questions." However, there were also students who were not too happy and wanted "More question with answer scheme. More stress with example of question for final exam," and "Make this course less sentence but has more number and no complicated graph and include all the formula in exam paper." More positive responses were recorded from the students than the less enthusiastic responses. The students were very concerned with their performance in examinations. Throughout the course, the researchers have to negotiate with the students between using these independent learning strategies and the traditional 'chalk and talk' method. Responses from the students' reflections indicated their awareness on the SRL processes in terms of Performance and Reflections Monitoring. For example, a student would like to change "My laziness. My time management" while another highlighted attention to "Time management. My attitude." When asked about what they would do differently when working through a topic, some students indicated that that they would 'read more', 'do more exercises', 'group discussion' and 'search the internet'.

A simple criterion was used to categorise the "Making Sense' responses, which is, "ability to display correct mathematics, clear explanations and the correct use of symbols and notations. A rubric in a range of 1 to 4 was used with 4 referring to 'Very Good' and 1 referring to 'Poor'. Most of the students' responses were in the range of 3 and 4. The responses were used to identify students who had difficulties with the mathematical concepts and techniques. These were then addressed in class by the lecturer and any misconceptions or applications of the techniques were also highlighted during lectures.

In answering questions about team working, most of the responses were very favorable:

- "Helps us to cooperate amongst ourselves. Widens our view."
- "Help ourselves understand better. Yes, member is supportive and helping each other."
- "Very good and helpful. Encourage me to work harder. Very supportive and helpful."

Despite unfamiliarity with the new learning style, the students became very adept at managing their own work. The observations indicated that students would get down working in their groups when the class began. In the beginning, reminders were given for them to review the amount of time spent on a topic and to check the Learning Guides. As the semester progressed, they were more able to manage their time and the Learning Guides were employed to pace their work.

The findings indicated that some awareness of the SRL processes have occurred as well as the adoption of cooperative learning habits. The students were quite happy in holding discussions, reaching out for help, more willing to share their concerns and generally more communicative about mathematics problems and their learning needs or difficulties. The classes were active and dynamic, in their own words:

- "You did a great job teaching us."
- "Class not boring. Lecturer understanding the students."

CONCLUSION

The study was conducted to systematically collect the data and develop students' mathematical competencies and awareness of their own thinking powers as well as their SRL skills. Competencies and skills were transferable to other learning situations and thus these will help students become more flexible, independent, and able to take charge of their own learning. Various strategies were used, which harnessed the components from an older framework to strategies that introduced SRL processes for enhancing mathematical learning. The results showed that students demonstrated their efforts to adopt more efficient learning behaviour if they were supported through a more conducive learning environment as well as the constructive alignment of the teaching and learning activities to the assessment tasks.

ACKNOWLEDGMENT

We would like to extend our gratitude to Malaysia-Japan International Institute of Technology, UTM for allowing us to carry out these TLAs and Universiti Teknologi Malaysia for providing grants under the Centre for Teaching and Learning.

REFERENCES

Anthony, G. (2000). Factors influencing first year students' success in mathematics. *Int Jnl of Math Edn in Sc & Tech*, *31*(1), 3–14. doi:10.1080/002073900287336

Baharun, S., Mohd Yusof, Y., & Abdul Rahman, R. (2008). Facilitating thinking and communication in mathematics. In DG 24, presented at ICME 11.

Bahru, J. Malaysia. Abdul Rahman, R. (2009). *Changing my own and my students attitudes to calculus through working on mathematical thinking*. Unpublished PhD thesis, Open University.

Biggs, J., & Tang, C. (1999). *Teaching for quality learning at university*. Open University Press & McGraw-Hill Education.

Biggs, J., & Tang, C. (2010). Teaching for quality learning at university (3rd ed.). Open University Press & McGraw-Hill Education.

Croft, A., & Ward, J. A. (2001). Modern and interactive approach to learning engineering mathematics. *British Journal of Educational Technology*, *32*(2), 195–207. doi:10.1111/1467-8535.00190

de Bruin, A. B., Thiede, K. W., Camp, G., & Redford, J. (2001). Generating keywords improves metacomprehension and self-regulation in elementary and middle school children. *Journal of Experimental Child Psychology*, *109*(3), 294–310. doi:10.1016/j.jecp.2011.02.005 PMID:21397251

Engineering Accreditation Council Malaysia (EAC). (2012). *Engineering programme accreditation manual*.

Felder, R. M., & Brent, R. (2008). *Active and cooperative learning, Workshop notes*. Penn State University.

Harris, K. R., Friedlander, B. D., Saddler, B., Frizzelle, R., & Graham, S. (2005). Self-monitoring of attention versus self-monitoring of academic performance: Effects among students with ADHD in the general education classroom. *The Journal of Special Education, 39*(3), 145–156. doi:10.1177/00224669050390030201

Hoch, M., & Dreyfuss, T. (2005). Students' difficulties with applying a familiar formula in an unfamiliar context. In ChickH. L.VincentJ. L. (Eds.), *Proceedings of the 29th Conference of the International Group for the Psychology of Mathematics Education (PME),* Melbourne, Australia (Vol. 3, pp. 145-152).

Keyser, M. (2000). Active learning and cooperative learning: Understanding the difference and using both styles effectively. *Research Strategies, 17*(1), 35–44. doi:10.1016/S0734-3310(00)00022-7

Mason, J., Burton, L., & Stacey, K. (1982). *Thinking mathematically.* Addison-Wesley Publishing Company, Inc.

Mason, J., Burton, L., & Stacey, K. (2010). *Thinking mathematically* (2nd ed.). Addison-Wesley Publishing Company, Inc.

McNiff, J., Lomax, P., & Whitehead, J. (1996). *You and your action research project.* Routledge & Hyde Publications. doi:10.4324/9780203281291

Meyers, C., & Jones, T. B. (1993). *Promoting active learning: Strategies or the college classroom.* San Francisco, CA: Jossey-Bass Publishers.

Mohd, Y., Yusof, S. B., & Rahman, R. A. (2001). *Mathematics education in UTM: Learning from experience.* Universiti Teknologi Malaysia, Johor: Jurnal Teknologi.

Mohd, Y., Yusof, S. B., & Rahman, R. A. (2011). Multivariable calculus for independent learners (Revised 2nd ed.). Pearson Malaysia Sdn Bhd.

Niss, M. (2003). Mathematical competencies and the learning of mathematics: The Danish KOM project. In *Proceedings of the 3rd Mediterranean Conference on Mathematical Education.* Athens, Greece: Hellenic Mathematical Society.

Niss, M., & Højgaard, T. (2011). *Competencies and Mathematical Learning: Ideas and inspiration for the development of mathematics teaching and learning in Denmark. IMFUFA.* Roskilde University.

Pintrich, P. R., & Zusho, A. (2002). The development of academic self-regulation: The role of cognitive and motivational factors. In A. Wigfield & J. Eccles (Eds.), *Development of achievement motivation* (pp. 249–284). San Diego, CA: Academic Press. doi:10.1016/B978-012750053-9/50012-7

Roselainy, Y., & Sabariah. (2012a). *Realizing desired learning outcomes in undergraduate mathematics.* In Khairiyah et al. (Eds.), Outcome-based science, technology, engineering, and mathematics education. *Innovative practices* (pp. 182-206). Hershey, PA: IGI Global.

Roselainy, Y. (2012b). Factors affecting students' change of learning behaviour. *Procedia: Social and Behavioral Sciences, 56,* 213–222. doi:10.1016/j.sbspro.2012.09.648

Roselainy, Y., & Sabariah. (2007). Enhancing thinking through active learning in engineering mathematics. In *Proceedings of Fourth Regional Conference on Engineering Education*

SEFI. (2011). *A framework for mathematics curricula in engineering education. A Report of the Mathematics Working Group.* European Society for Engineering Education.

Slavin, R. E. (2006). *Educational psychology: Theory and practice* (8th ed.). Boston, MA: Allyn & Bacon.

Tall, D. (1995). *Cognitive growth in elementary and advanced mathematical thinking, Plenary lecture* (Vol. 1, pp. 161–175). Recife, Brazil: Conf. of International Group of PME.

Tall, D., & Razali, M. R. (1993). Diagnosing students' difficulties in learning mathematics. *Int Jnl of Math Edn in Sc & Tech, 24*(2), 209–222. doi:10.1080/0020739930240206

UTM. (2011). *Generic skills and UTM graduate attributes*. UTM Publications.

Watson, A., & Mason, J. (1998). *Questions and prompts for mathematical thinking*. Derby: ATM.

Wolters, C. A. (2011). Regulation of motivation: Contextual and social aspects. *Teachers College Record, 113*(2), 265–283.

Zimmerman, B. J. (2000). Attaining self-regulation: A social cognitive perspective. In M. Boekaerts, P. R. Pintrich, & M. Zeidner (Eds.), *Handbook of self-regulation*. San Diego, CA: Academic Press. doi:10.1016/B978-012109890-2/50031-7

This work was previously published in the International Journal of Quality Assurance in Engineering and Technology Education (IJQAETE), 3(2); edited by Arun Patil, pages 43-53, copyright 2014 by IGI Publishing (an imprint of IGI Global).

Section 3

Frameworks and Methodologies

This section provides in-depth coverage of conceptual architecture frameworks to provide the reader with a comprehensive understanding of the emerging developments within the field of STEM Education. Research fundamentals imperative to the understanding of developmental processes within STEM Education are offered. From broad examinations to specific discussions on methodology, the research found within this section spans the discipline while offering detailed, specific discussions. From basic designs to abstract development, these chapters serve to expand the reaches of development and design technologies within the STEM Education community. This section includes 15 contributions from researchers throughout the world on the topic of STEM Education.

Chapter 30
Multiple Perspectives for the Study of Teaching:
A Conceptual Framework for Characterizing and Accessing Science Teachers' Practical-Moral Knowledge

Sara Salloum
Long Island University – Brooklyn, USA

ABSTRACT

This chapter outlines a framework that characterizes science teachers' practical-moral knowledge utilizing the Aristotelian concept of phronesis/practical wisdom. The meaning of phronesis is further explicated and its relevance to science education are outlined utilizing a virtue-based view of knowledge and practical hermeneutics. First, and to give a background, assumptions about teacher knowledge from a constructivist and sociocultural perspective are outlined. Second, the Aristotelian notion of phronesis (practical wisdom) is explicated, especially in terms of how it differs from other characterizations of practical knowledge in science education and how it relates to practical-moral knowledge. Finally, the authors discuss how the very nature of such practical-moral knowledge makes it ambiguous and hard to articulate, and therefore, a hermeneutic model that explores teachers' practical-moral knowledge indirectly by investigating teachers' commitments, interpretations, actions, and dialectic interactions is outlined. Implications for research and teacher education are outlined. Empirical examples are used to demonstrate certain points. A virtue-based view of knowledge is not meant to replace others, but as a means to enrich the understandings of the complexity of teacher knowledge and to enhance the effectiveness of teacher educators.

INTRODUCTION

Teachers (practitioners) often hold visions of 'good' teaching that differ from those of researchers, teacher educators, and reform documents (e.g., AAAS, 1990; NRC, 1996). Such disparity is a facet of the gap between theory and practice in education, and entails significant difficulties for research and reform efforts aimed at getting teachers to embrace and enact visions of 'good'

DOI: 10.4018/978-1-4666-7363-2.ch030

teaching valued by researchers and teacher educators (Kennedy, 2006; Wildy & Wallace, 1995). Despite concerted efforts, closing this gap has been more elusive than initially imagined (Carr, 1995; Crawford, 2007), where substantial reform efforts in science education have met with limited success (Lynch, 2001; Smith & Southerland, 2007). We argue that a major reason for such limited success stems from educational research and reform that aims at changing teacher *actions* without ample understanding of underlying teacher knowledge.

Teacher actions have been the focus of research efforts since the early 1960s. Initially teacher actions were studied in an attempt to isolate 'effective' teaching techniques (process-product research) (e.g., Medley, 1979; Doyle, 1977; as cited in Issler, 1983; Woolfolk & Galloway, 1985). More recently, teacher actions (including language) have been scrutinized in the broader context of examining teacher knowledge and beliefs; and their influence on student learning, attitudes, skills, and classroom dynamics (e.g., Anderson & Mitchener, 1994; Borko & Putnam, 1996; Haney & McArthur, 2002; Moje, 1995; Mulholland & Wallace, 2008; Tsai, 2002). The latter focus on teacher knowledge, beliefs, and practices has uncovered yet further complexities of the gap between theory and practice. In science teaching, changing teachers' beliefs proved difficult (Smith & Southerland, 2007), and even when teachers held or shifted towards reform consistent-beliefs, dissonance often emerged between actual classroom practices and stated beliefs about teaching, learning, and nature of science (Bell, Lederman, Abd-El-Khalick, 2000; Simmons, et al., 1999; Southerland, Gess-Newsome, & Johnston, 2003). Evidently, translation of teacher beliefs and knowledge (specifically theoretical) into practice is more complex than initially perceived. Due to such complexity, the nature of teacher knowledge remains a rich area for exploration (Mulholland & Wallace, 2008).

PURPOSE

A broad aim of this chapter is to further elucidate a framework for understanding teaching practice as more than an arena for the application of theoretical knowledge and sets of skills (craft), but as a practice where teachers continuously engage a form of non-theoretical practical-moral knowledge. Acknowledging the role of non-theoretical knowledge in teaching has gained momentum in science education and several terms have been used to refer to it: practical knowledge (e.g., Duffee & Aikenhead, 1992; Fenstermacher, 1994; Lotter, Hardwood, & Bonner, 2007; Mulholland & Wallace, 2008; van Driel, Beijaard, & Verloop, 2001); practical-moral knowledge (Salloum & Abd-El-Khalick, 2010); and personal practical theories (Smith & Southerland, 2007). A complicating aspect of studying non-theoretical teacher knowledge though is elucidating and conceptualizing its character: Is it form of knowledge, reasoning, or an aspect of one's 'being' (e.g., Breire & Ralphs, 2009; Feldman, 2002)? Is a set of conceptions, skills, values, and beliefs that teachers develop with experience (e.g., van Driel, et al., 2001)? How can we study such knowledge? These issues have more practical importance than their esoteric nature suggests (Southerland, Sinatra, & Mathews, 2001), specifically since models promoted and utilized in educational research and teacher education are greatly influenced by conceptualizations of teacher knowledge and its nature.

In this chapter, teachers' practical-moral knowledge is characterized utilizing the Aristotelian concept of phronesis or practical wisdom (e.g., Breier & Ralphs, 2010; Clark, 2005; Flyvbjerg, 2001; Korthagen, & Kessels, 1999; Korthagen, Loughran, & Russel, 2006; Salloum & Abd-El-Khalick, 2010; Schwandt, 1996, 2005). However, as Breier and Ralphs (2010) stated, the concept of phronesis and practical wisdom is gaining popularity in education but not always clarity, especially when it comes to its role within specific disciplines such as science teaching and how to study and develop it. We argue in this a chapter that a greater understanding of phronesis will contribute to our understanding of science pedagogy and teaching practices. The aim of this chapter is to further explicate the meaning of phronesis and outline its relevance to science education and educational research. To do so, a virtue-based view of knowledge (Zagzebski, 1996) and practical hermeneutics (Gadamer, 1989) are utilized. First, and to give a background, assumptions about teacher knowledge from a constructivist and sociocultural perspectives are outlined. Second, the Aristotelian notion of phronesis (practical wisdom) is explicated, especially in terms of how it differs from other characterizations of practical knowledge in science education and how it relates to practical-moral knowledge. Finally, we discuss how the very nature of practical-moral knowledge deems it ambiguous and hard to articulate; and therefore a hermeneutic model that explores teachers' practical-moral knowledge *indirectly* by investigating teachers' commitments, interpretations, actions, and dialectic interactions between them is outlined. Implications for research and teacher education are outlined. Even though this is a conceptual chapter, empirical examples from my research will be used to demonstrate certain points (Salloum, 2006; Salloum & Abd-El-Khalick, 2010, Salloum, et al., 2010; Collier & Salloum, 2011).

ASSUMPTIONS ABOUT TEACHER KNOWLEDGE

Constructivist and sociocultural traditions have been invaluable in informing the majority of research studies in science education (Anderson, 2007). Cognitive and social constructivism have been greatly informed by Piaget and Vygotsky's theories. Piaget's and Vygotsky's theories are developmental, where learners progress through stages to ultimately develop the 'adult abilities' related to logical and abstract thinking. Thus, a basic assumption of cognitivists is that individuals are rational problem solvers (Lave, 1988). Naturally, teachers are assumed to be at a stage where they can engage in logical thinking to solve problems, and model such reasoning to their students. Teachers need to help students develop rational thought and use this reasoning to solve problems. Learners are perceived as actively constructing and restructuring knowledge, and so teachers play valuable roles in providing experiences and social settings to aid such re-structuring. Accordingly, 'good' teachers need to know how to create settings and activities for students to restructure ideas and reach higher levels of thinking and knowledge and promote both verbal and practical (through experimentation and inquiry methods) thoughtfulness.

Constructivists utilized Piagetian theory to help teachers restructure their knowledge of teaching and to embrace constructivist pedagogies:

Just as young learners, so, too, do teachers. Teacher education programs based in constructivist view of learning need to do more than offer a constructivist perspective in a course or two. Teachers' beliefs need to be illuminated, discussed, and challenged (Fosnot, 1996, p. 216).

An embedded assumption was that knowledge for teaching science is normative/theoretical and belief-based involving formal-rational reasoning (thus the interests in teacher beliefs). For example, it is assumed that by following logical paths (challenge and confrontation), teachers will change their beliefs about teaching and consequently their actions. A purely logical/rational model has been critiqued for being too narrow to account for teacher knowledge and reasoning (Noel, 1999), and has lead constructivists to take an either/or stance about 'good' science teaching (Wildy & Wallace, 1995). This stance is often criticized for neglecting situational and contextual aspects of teaching:

For science teachers one cannot underestimate the importance of understanding the cultural context of schools, having clear and consistent view of the subject matter, building a learning community of trust, and adapting the curriculum to accommodate the knowledge, needs, and aspirations of the students. Surely, this is what good science teaching is all about (Wildy & Wallace, p. 154).

A sociocultural perspective on science education was critical of constructivism, particularly cognitive constructivism, for neglecting the historical, social, cultural, and physical contexts of learning processes (e.g., Lemke, 2001; O'Loughlin, 1992). Sociocultural perspectives were informed by the work of Lave (1988) and Wertsch (1991). A main aspect adopted from Lave's work was that individuals are embodied selves actively engaged in relational activities with their world. Sociocultural researchers argued that constructivism framed learners in a decontextualized manner with no reference to their environment, which is problematic because "it denies the essentially collaborative and social nature of meaning making; and it privileges only one form of knowledge, namely, the technical rational" (O'Loughlin, 1992, p. 791). Thus, socioculturalists questioned normative theoretical/rational views of knowledge including the assumption of dualism and separation between mind and body (e.g. Brickhouse, 2001; Lave, 1988; Lemke, 2001). For example, Lemke (2001) critiqued the conceptual change model claiming, "there is more at stake than rational choice among competing theories. Changing our mind is not simply a matter of rational decision making. It is a social process with social consequences" (p. 301).

Even though learners' knowledge has been problematized within sociocultural perspectives, both constructivists and socioculturalists still seem to hold normative views of teacher knowledge, assuming a "direct linkage between cognition and action" (Feldman, 2002, p. 1038), thus deeming knowledge for teaching as primarily theoretical-rational (Feldman, 2002; van Manen, 1995). Teachers are perceived as cognizing agents, who make and manage decisions by referring to generalized rules and codes, cohesive sets of beliefs, or a knowledge base (Feldman, 2002; van Manen, 1995). As mentioned above, this lead to extensive research on teacher beliefs to understand teacher knowledge and address difficulties in putting constructivist and sociocultural pedagogies into action (e.g., Anderson & Mitchener, 1994; Borko & Putnam, 1996; Haney & McArthur, 2002; Lederman, 1992; Meyer, 2004; Tsai, 2002; Windschitl, 1999). Difficulty in changing teachers' beliefs and dissonance between stated beliefs and actual practices showed that focus on a belief-based form of knowledge is not adequate to understand teacher knowledge. Sociocultural perspectives did extend research on beliefs by exploring the influence of external and cultural factors on translation of beliefs. Nonetheless, over-emphasis on outside cultural factors to explain adherence to uniformity can be problematic as it fails to take into account the personal and moral strife of individuals who actually transform cultures (Lave, 1988; Vianna & Stensenko, 2006). The aforementioned perspectives may have (unintentionally) undervalued the non-theoretical and non-cognitive dimensions of teachers' knowledge. In the next section, Aristotelian virtue ethics and a virtue-based view of knowledge (Zagzebski, 1996) are utilized to extend views of teacher knowledge. A virtue-based view is not meant to replace other views but to enrich our understandings of teacher knowledge, its study, and development.

TEACHING AND TEACHER KNOWLEDGE: FURTHER CONCEPTUALIZATIONS

Teaching seems to be the sort of occupation in which professional effectiveness is greatly en-

hanced by the possession and exercise of personal qualities and practical dispositions that are not entirely (if at all) reducible to academic knowledge or technical skills (Carr, 2007, p. 369).

Calls for looking outside theoretical and technical models to characterize teacher knowledge are becoming more and more popular in education and science education (e.g. Feldman, 2002; Carr, 2005, 2007; Eisner, 2002; Mulholland & Wallace, 2008; Noel, 1999; Shulman, 2007; van Driel, et al., 2001; van Manen, 1995, 1999). Several features of such knowledge are its *tacit* and *context-dependent* nature (non-propositional); its intimate relation to one's *character* and who one is as a person; and its role in moral *action*, or simply said in doing what is 'right' in a certain situation. In Salloum and Abd-El-Khalick (2010), we referred to such knowledge as practical-moral knowledge: practical since it is associated with everyday actions and ways teachers respond to situations; and moral since actions are reflective of who we are as individuals and our values. A fundamental assumption here, as Carr (2007) stated above, is that theoretical knowledge of content and pedagogy and technical skills are not enough to account for what constitutes 'good' practice. Rather, how and when practitioners make use of theoretical knowledge is influenced by their practical-moral knowledge (Kessels & Korthagen, 1996; Pendlebury, 1995). To better understand practical-moral knowledge, it is characterized by the Aristotelian concept of phronesis. There is a need first, however, to elucidate phronesis/practical wisdom itself.

Phronesis

Aristotle proposed phronesis as one of the intellectual virtues meant to define 'truth disclosing' active conditions (Aristotle, 2002). Aristotle distinguished between two types of intellectual virtues: "one with which we contemplate those things whose first principles are *invariable*" (Aristotle, 1976, p. 204) and ones with which we contemplate things that can be *otherwise*. Phronesis and techne were designated as ones by which we contemplate what can be otherwise. Techne (craft or art) is on things that are *made* and phronesis (practical wisdom) involves *moral actions*. With phronesis, Aristotle was concerned with outlining the "right estimation of the role that reason has to play in moral action" (Gadamer, 1989, p. 310). The other truth-seeking condition of interest is episteme (scientific knowledge), which Aristotle defined as "a kind of judgment that concerns things that are universal" (Aristotle, 2002, p. 107); episteme is context-independent knowledge with invariable first principles. Episteme can be seen as the basis of theoretical and scientific knowledge. Compared to episteme or scientific knowledge, which is universal, conceptual, and decontextualized, phronesis is perceptual, situated, and involves concerns that are variable by nature (Eisner, 2002; Kessles & Korthahgen, 1996). As for differences with techne (craft or art), Aristotle stated that the ends for *making* are different than those for *acting*. In 'making' the ends are in producing something tangible separate from the 'making' process, whereas the end of acting well is acting well.

Aristotle identified practical wisdom as the ability to deliberate correctly or 'beautifully' about what is good and advantageous for oneself and what is conducive to living well in general (Aristotle, 1976, 2002). Practical wisdom is defined as "a truth-disclosing active condition involving reason about human goods that governs action" (Aristotle, 2002, p. 107). According to Aristotle, phronesis is the intellectual virtue necessary to realize *praxis* or morally committed and morally informed action concerned with how humans conduct their life as members of society (Kemmis, 2009, 2010; Schwandt, 2005).

Phronesis or practical wisdom is concerned with:

- What can be deliberated about and can be otherwise,
- Realizing a desired end, and
- Reaching the end is by *voluntary action* based on deliberation.

Practical wisdom is perceptual rather than conceptual (to discern what is to be deliberated on) and entails knowledge of particulars: practical wisdom "is not only about what is universal, but needs to discern particulars as well, since it has to do with action and action is concerned with particulars" (Aristotle, 2002, p. 109). Knowledge of particulars can be more important than knowledge of universals to be effective in action. An example to that effect is a teacher who knows the general principle stating that engagement through connecting science concepts to prior knowledge is essential for deep conceptual learning, but does not know what constitutes relevant prior knowledge for a particular group of students and/or for that particular topic. Such teacher would be less effective in action. Consequently, development of practical wisdom requires experience and thoughtful reflection on experience.

Functions of Phronesis

Aristotle distinguished phronesis from techne (craft and art) and episteme (scientific knowledge) and gave it prominence as the intellectual virtue that governs action and one that cultivates and mediates other virtues (such as temperance, honesty, courage, etc.). The relation between practical wisdom and virtuous character is interdependent: It is not possible to be good in the governing sense without practical wisdom, nor to have practical wisdom without virtue of character (Aristotle, 2002). Virtues are demarcated as acquired excellences that take time and effort to develop, and are distinguished from natural capacities and skills. Zagzebski (1996) utilized Aristotle's virtue ethics to outline a virtue-based theory of knowledge that binds knowledge intimately with moral concepts. Within such a view, virtues can be moral or intellectual: The difference between moral and intellectual virtues is that all intellectual virtues act as motivation to develop new knowledge and skills (Zagzebski, 1996). For example, the virtues of intellectual sobriety, including careful inquiry, and intellectual fairness would motivate a teacher to continuously gain new knowledge about students and ways to represent content in multiple ways to provide access to all of them (Collier & Salloum, 2011).

Phronesis mediates the different intellectual and moral virtues by serving several functions (Zagzebski, 1996). One is to determine the virtuous course required or called for in a particular situation. For example, how empathic should a teacher be in different situations and with different students? The question is not to be empathic or not as two extremes, but what degree of empathy is appropriate considering the situation at hand or students' particular needs. Another function of phronesis is to balance choices among virtues that may lead to conflicting lines of action. For example, in certain situations both empathy for students and responsibility for them learning abstract and complex content are called for; however, engaging in one may conflict with the other. Teachers need to engage phronesis or practical wisdom to resolve such tension. A teacher would need to know how to push students as they learn abstract science concepts and master them, without reaching a point where their frustration can turn them off the content. A third function for phronesis is to coordinate different virtues (and understandings) into a course of action or a line of thought that would lead to action. A course of action resulting from exercising practical wisdom will always involve indeterminacy and actions cannot be entirely accounted for using purely rational arguments (Gadamer, 1989; Zagzebski, 1996).

Practical Wisdom and Teaching

Phronesis and practical wisdom as an intellectual virtue encompasses features and functions that seem to characterize several aspects of knowledge engaged in good teaching. Pendlebury (1995) stated that:

Practical wisdom is the sovereign virtue of a good teacher–a virtue whose realization in teaching requires a subtle interplay between several binary oppositions: reason and imagination, experience and innocence, cleanness of argument and richness of story, respect for principles and attunement to particulars (p. 50).

Situations in the classroom are way too complex to involve one form of knowledge or to entail a clear set of discrete goals and desired ends. Teachers always find themselves having to balance several ones (Kennedy, 2006): inquiry-based teaching for meaningful and conceptual understanding in science *and* test-prep teaching for high achievement on standardized test; serving students with diverse abilities and needs *and* moving the classroom forward within a mandated and tested curriculum; simultaneously the teacher strives to build a respectful and trusting classroom community (Kennedy, 2006; Salloum & Abd-El-Khalick, 2010; Salloum, et al., 2010). The desired ends above do not just involve theoretical or cognitive aspects, but at their heart are moral and intellectual virtues inherent to good teaching (Carr, 2007): *fairness* to students with diverse needs and abilities and to science as a discipline, *responsibility* for students' learning, their attitudes towards science, and their achievement in gatekeeper tests. For example, in interviews, teachers always rationalized their practices and actions drawing on notions such as being responsible towards students and leading them to a "safe haven" when it came to passing high stakes tests (Salloum & Abd-El-Khalick, 2010); or as being *fair* to diverse students by providing access either to difficult content or to opportunities for a better future (e.g. college admission) (Salloum, et al., 2010; Collier & Salloum, 2011).

Moreover, and specifically for science and science teaching, virtues such as the following seem especially relevant and necessary for promoting inquiry in science and deep understandings of nature of science: intellectual sobriety (careful inquiry and accepting what is warranted by evidence); honesty; open mindedness and impartiality; intellectual humility; intellectual courage; perseverance, determination, and thoroughness; and fairness in evaluating arguments of others (Zagzebski, 1996). Probably the most famous intellectual virtues in teaching are the attitudes outlined by Dewey (1933) for reflective practice: wholeheartedness, responsibility, and open-mindedness. Other relevant virtues for teaching are thorough inquiry and reflection on inquiry, the social virtues of being communicative, intellectually honest, and empathic (Carr, 2007; Zagbeski, 1996). What qualifies that above virtues as 'intellectual' is the fact that their enactment entails motivating teachers to develop new understandings, knowledge, and skills.

The enactment of virtues as ones outlined above and knowledge development motivated by them cannot be prescriptive nor is it an easy undertaking. Such enactment will involve a whole range of tensions and dilemmas among virtues themselves, requirements of specific content, and external factors such as mandated curricula and testing. Practical wisdom is essential in resolving tensions and in balancing desired ends: How do we go about managing situations that entail different and sometimes competing desired ends and virtues? How do we go about finding the appropriate mean of a virtue between excess and deficiency? It is such questions that practical wisdom is involved in.

Practical Wisdom and Science Teaching

In science education, non-theoretical practical knowledge has also emerged as an important aspect of science teachers' practice and change (e.g., Duffee & Aikenhead, 1992; Lotter, Hardwood, & Bonner, 2007; Mulholland & Wallace, 2008; van Driel, Beijaarrd, & Verloop, 2001). However, upon examining literature on practical knowledge in science education, it is noted that practical knowledge is characterized by craft knowledge rather than practical wisdom and prudence towards moral action (e.g., Mulholland & Wallace, 2008; van Driel, et al., 2001). Craft is extended to include reflective practice, which entails intellectual virtues such as openmindedness, wholeheartedness, and responsibility. Nonetheless, "reflective" craft knowledge still does not adequately capture the enactment of virtues nor the engagement in strong judgments to serve embedded 'goods' in teaching with respect to students' interests, integrity towards content, and one's own development as a practitioner.

Knowledge as craft is based on techne, one of Aristotle's intellectual virtues. Both phronesis and techne are contextual and pragmatic, but techne is exhibited in production, whereas phronesis is essential for moral practice (Carr, 2007). Gadamer's (1989) elaboration on distinctions between phronesis/practical wisdom and techne/craft is relevant to show limitations of techne/craft in characterizing teacher knowledge. Techne can be encapsulated in sets of skills that can be directly learned to produce something tangible. These are at our disposal to use or not to use (Gadamer, 1989). A craftsperson may choose to acquire them and later they can be forgotten. Whereas phronesis or practical wisdom is involved in situations where we have to act towards desired ends pertaining to teaching science well. Knowledge informing our actions to do 'well' is bound with images of who we are as individuals/science educators and our ideas of right and wrong (e.g., intellectual honesty, open-mindedness, responsibility, empathy, etc.). For example, a teacher can acquire skills in creating efficient lab set-up, clear lab directions and worksheets, and multi-level and diverse assessments. The teacher may choose to use or not use these skills based on the group of students or the topic area. However, a teacher cannot choose not to be intellectually honest without this affecting her character and how she sees herself. Rather her practical wisdom would inform her on what level of intellectual honesty is appropriate in a specific situation. An empirical example will be used to demonstrate this point (Salloum, 2006). Ms. Yasmine is a middle school physical science teacher. She attended a professional development workshop, where they were told that the connotation $HCl_{(aq)}$ is not chemically acceptable, since HCl does not exist in water but rather it is $H^{+}_{(aq)}$ and $Cl^{-}_{(aq)}$ ions that are found in the solution. Ms. Yasmine was not satisfied with the proposed concept and actually checked the grade 8 book and found $HCl_{(aq)}$ used several times. When asked about her personal opinion of what is a more correct connotation, she said it is easier for the students to see $HCl_{(aq)}$ as one entity, especially for solving displacement reactions. I asked her to elaborate on how she would deal with this issue in her teaching:

> *as a chemist I want what is found in the solution (here referring to having the ions and not HCl in an aqueous state), But as a teacher, I will never tell the students something not scientifically accurate, but I start with the correct (and easier) and then give the more correct, I never tell them something that is 'wrong!'*

In the preceding excerpt, Ms. Yasmine's was trying to resolve the 'extent' to which she needs be intellectually honest and at the same time help students understand the material. Ms. Yasmine deliberated on a tension between herself as a 'good' teacher and as a 'good' chemist and her resolution was to give students the *less correct* but *more understandable*, at least as a starting point. She felt she had to maintain being intellectually honest (because this is how she sees herself), but the extent was determined by her assessment of students' needs. In the case above, Ms. Yasmine engaged her practical wisdom to determine the appropriate extent of intellectual honesty needed in a particular situation. As mentioned above on functions of practical wisdom, the question is not whether she needs to be intellectually honest or not as two extremes, but rather the question entailed determining the extent appropriate in light of the situation at hand.

Another distinction between phronesis and techne is that techne can have well defined ends, and means to achieving them are separate from the ends (Aristotle, 2002; Gadamer, 1989). Whereas with phronesis, virtues makes the end 'right,' and practical wisdom makes the means and things related to it 'right' (Aristotle, 1976). With phronesis, the ends are boarder and somewhat uncertain since they pertain to several 'goods' that science teaching is to serve. Moreover, "the considerations of the means is itself a moral consideration and it is this that concretizes the moral rightness of end" (Gadamer, 1989, p. 319); therefore, phronesis is about choosing the 'right' means to get to the 'right' end. An example from science education can be the phenomena of 'teaching to the test.' A teacher can maintain 'students passing the test' as a particular end and can become very skillful in teaching students techniques to tackle test problems, with or without students understanding and retaining the underlying science concepts[1]. Even when students only learn science through rote or the 'plug and chug' method, the desired end of doing well on certain tests is still achieved with techne. With phronesis or practical wisdom the desired end cannot just be about passing the test, rather the ends would be broader and pertain to several 'goods' that science teaching needs to serve: e.g., students passing gatekeeper exams *and* promoting understandings of science concepts and science as discipline. Practical wisdom would be about finding the means to balance the sometimes competing desired ends, that is, teaching to the test and enhancing students' conceptual understandings and reasoning simultaneously. In Salloum and Abd-El-Kahlick (2010), we see Ms. Yasmine doing so by associating deep conceptual understandings of science concepts with better concept retention for doing well in tests.

Finally, teacher knowledge as characterized by phronesis is seen as simultaneously a way of being and thinking/acting without a clear separation between the two (Schwandt, 2008, personal communication) or as Gadamer points out, "... knowledge, not detached from being that is becoming, but determined and is determinative of it" (Gadamer, 1989, p. 310). Accordingly, several assumptions about knowledge emerge: knowledge in this sense is not governed by rules or encapsulated in a theoretical or formal knowledge base or propositional beliefs. Rather, practitioners' knowledge actualizes by being immersed in situations, and by "interpreting formal knowledge[2], beliefs, past experiences, and commitments according to perceived situations" (Salloum & Abd-El-Khalick, 2010).

Practical Wisdom and Practical-Moral Knowledge

In Salloum and Abd-El-Khalick (2010), the authors utilized the term Practical-Moral Knowledge (P-MK) to emphasize its intimate relation with moral concepts within a practice and the cultivation of intellectual and moral virtues such as honesty, fairness, responsibility, open-mindedness etc. The main reason for using this term rather than practical wisdom is that I see teacher's practical-moral knowledge as potentially developing into practical wisdom, but not as equivalent to it. Practical wisdom requires experience and advanced kinds of thoughtful reflection and understanding: the kind of wisdom that a master science teacher would have. Practical wisdom may not characterize knowledge of all teachers (either due to lack of experience or more seriously the lack of passion or an essential virtue). However, it can be suggested that *all* teaching involves engaging practical-moral knowledge even when such knowledge is misguided and does not lead to wise choices. We believe that such practical-moral knowledge needs to be further investigated for better teacher education and that we need to especially explore insights on how to cultivate practical-moral knowledge into practical wisdom.

Different forms of knowledge are all important and essential to teaching (and good teaching). The view suggested here is meant to enrich our understandings of teacher knowledge rather than replace other views. Knowledge for and of teaching will always involve different levels and domains of knowledge: theoretical knowledge of subject matter and pedagogy (episteme), practical-moral knowledge, and craft knowledge (techne). However, what is suggested here is that practical-moral knowledge with its essentially moral nature is how teachers manage different knowledge domains and make value-judgments in action; and those judgments are aligned with virtues they see as defining their practice. In the next section, we discuss general implications for characterizing teacher knowledge as practical-moral knowledge and practical wisdom.

GENERAL IMPLICATIONS

Difficulties in Exploring Practical-Moral Knowledge

Practical-moral knowledge, similar to other forms of practical and craft knowedge, is tacit and cannot be captured by propostions, therefore is illusive and hard to examine and assess; nor can it be directly demonstrated and 'taught' to others (Aristotle, 2002). Practitioners may not be able to articulate what such knowledge is about or how they engage it (Breier & Ralphs, 2009; Lotter, Hardwood, & Bonner, 2007; van Driel, et al., 2001; van Manen, 1999). van Driel et al. explained that both narrative research approaches and research on teacher beliefs have been utilized to elucidate practical forms of knowledge (e.g., Craig, 2006; Lotter, et al., 2007; Xu & Connelly, 2009). They added that narrative research, though valuable, needs to be theoretically interpreted lest teachers' narratives become a set of stories that lack theoretical grounding. On the other hand and as mentioned above, research on beliefs does not always capture teachers' strong judgments in enacting conflicting beliefs.

Another concern in research on practical-moral knowledge is its perceived content-less nature[3]. For instance, notions such as pedagogical tact (van Manen, 1991, 1995) capture the embodied, perceptive, and moral aspects of non-theoretical knowledge, but what does it means to be "tactful" towards physics or chemistry. The role of content knowledge is especially important in science education and consequently interconnections between content and virtues. Therefore, it is especially important in science education to construct a conceptual framework that allows for empirical exploration and analyses in both academic-theoretical and practical-moral domains. In Salloum and Abd-El-Khalick (2010), the authors proposed a hermeneutic framework that empirically explores and illuminates practical-moral knowledge by examining teachers' interpretations, commitments, actions, and interactions among them (see Figure 1). This framework is further developed below to show connections between virtues and teacher interpretations and commitments. But first we explain briefly why interpretations (from a hermeneutic perspective) and commitments are central to understanding practical-moral knowledge

Figure 1. A framework for investigating science teachers' practical-moral knowledge (Salloum & Abd-El-Khalick, 2010)

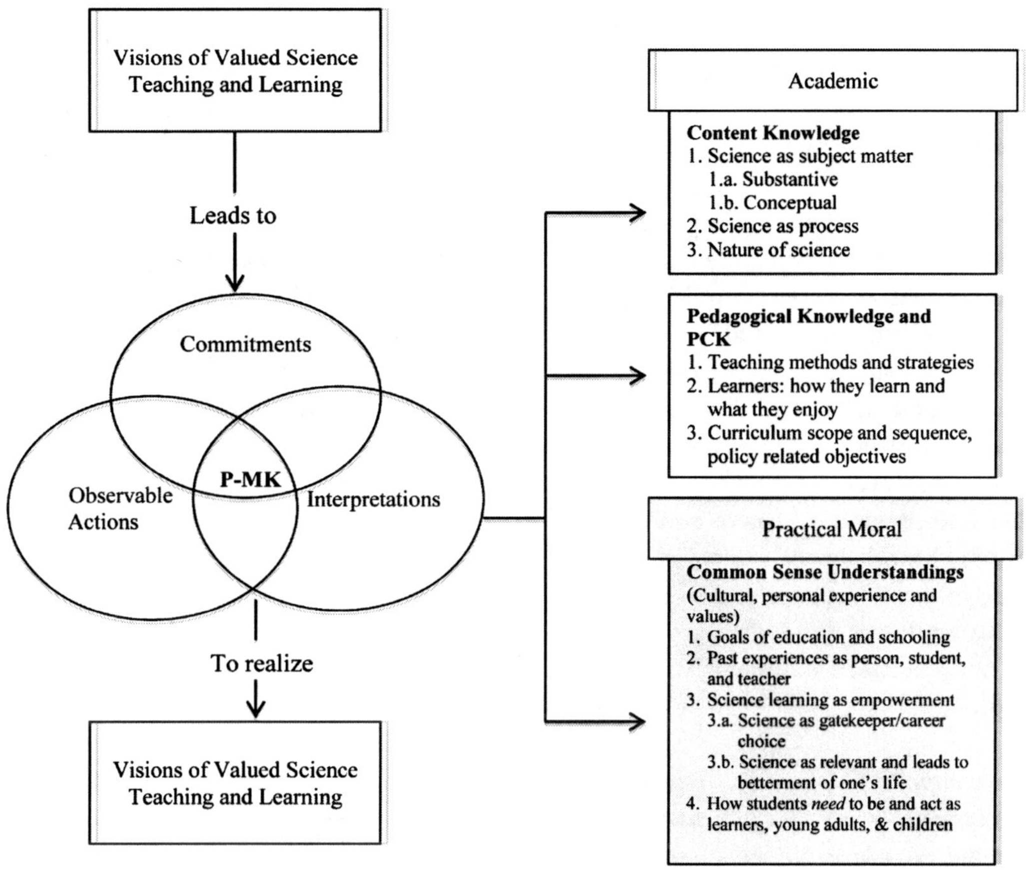

Centrality of Interpretations and Commitments to Practical-Moral Knowledge

From a hermeneutic perspective, interpretation is closely bound to understanding and therefore knowledge development, "Interpretation is not an occasional, post facto supplement to understanding; rather understanding is always interpretation, and hence interpretation is the explicit form of understanding" (Gadamer, 1989, p. 306). Moreover, Schwandt (2000) stated that "Philosophical hermeneutics argues that understanding is not, in the first instance, a procedure- or rule-governed undertaking; rather it is a very condition of being human" (p. 194). Interpretation as a human condition is a process we always find ourselves in: we cannot totally control and it is always constrained by a point of view (Gadamer, 1989). Interpretation as a way of being entails insight, discernment of salient features and prior knowledge, and emotions. It accrues in new understandings and knowledge. From a hermeneutic perspective, interpretation goes beyond the concept of 'beliefs,' which are defined as cognitive imperatives expressed in propositions (Pajares, 1992). Interpretation, on the other hand, is not necessarily propositional but "is carried out primordially not in a theoretical statement but an *action*" (Heidegger, 1962; as cited by Gallagher, 1992, p. 43, emphasis added).

Like practical wisdom, practical-moral knowledge is perceptual and thus involves interpretation of situations in light of prior knowledge and emotions about issues and individuals. From interpretation, understandings will emerge that lead to certain lines of action. Based on how a teacher interprets and frames the situation, the line of action may be constructive or counterproductive. For example, a student or group of students in a classroom may act up; depending on who the students are, the material being presented, and/or context of day and time, the teacher may interpret the situation as one requiring either punishing and reprimanding students or as one requiring reengaging students by slowing and revising instruction. The two lines of actions will have implications on how the lesson proceeds and what goals are achieved.

Teacher interpretations involve coordinating not just formal knowledge of subject matter and students (cognitive aspects), but also managing emotions, commitments, and virtues, we see definitive of our practice; accordingly, Carr (2007) maintained:

In fact, the compelling appeal of virtue ethics is that it shows precisely how the often distinguished and separated cognitive, affective, social, and motivational aspects of moral life may be coherently re-connected (p. 373).

Events of interpretation and understanding are central to the development and cultivation of teachers' practical-moral knowledge. Alternatively, practical-moral knowledge itself influences further events and acts of interpretations (the hermeneutic circle).

Story of a Teacher

An empirical example will be used below to demonstrate how perceptions, interpretation, and handling cognitive, affective, and social aspects interacted in a science classroom and the positive and negative implications that ensued (Salloum, 2006). Ms. Dunia[4] taught middle school physical science at a medium sized village public school in Mount Lebanon. Village settings tend to be more conservative and attached to tradition; people are involved in each other's lives and know details and intricacies of each other's lives and conditions. On one hand, this enhances the sense of community in schools, and on the other, it blurs boundaries between personal and professional spaces. The school served students of lower middle class socioeconomic status, with some families being more economically challenged than others. The author spent five months observing Ms. Dunia's science class and a total of nine months interacting with her and collecting data about her teaching and how she makes sense of her practice (I still maintain a personal relation with Ms. Dunia). Ms. Dunia's practice showed a strong commitment to 'challenging' students. This 'challenge,' however, took two forms: academic and social. The academic form of 'challenging' took a positive form evidenced by Ms. Dunia's use of inductive methods and careful questioning to have students develop science understandings independently. For example, Ms. Dunia was asked to go through one day's field notes and elaborate on what she saw as an important event. During that lesson, Ms. Dunia had used careful questions to have students come up with formula for pressure ($P=F/A$) from their qualitative understandings of the relation between pressure, force and area. Ms. Dunia stressed the value of students deriving the formulae themselves and her building on their qualitative understanding of the concepts to reach the formulaic quantitative expression ($P=F/A$):

The statement (qualitative) was produced by them (students); but they have to change to formula because physics is formulae and they have to work with formulae ...first they understand and then they change it into a formula.

In a different interview, Ms. Dunia elaborated about the importance of challenging students to derive new knowledge themselves and thus promoting their academic independence:

Students can memorize anything, but they need to discover things in science, especially if related to their daily life.

Unfortunately, this academic and positive form of Ms. Dunia's commitment to 'challenging' students did not always translate into higher student achievement and positive attitudes towards science. Even though Ms. Dunia sought conceptual understandings of science concepts, her strife for conceptual understanding did not always actualize and many students did not do as well as she hoped, nor did they have positive attitudes towards science. Of course, many factors contribute to the issue, including low second language proficiency (science in Lebanon in taught in English or French) and lower aspirations and support for students' of lower socioeconomic status. Yet, classroom observations, interviews with Ms. Dunia, and conversations with students and the administration uncovered an additional factor that may have contributed to the situation. Ms. Dunia, who is a dedicated teacher, demonstrated a different form of 'challenging' students. This second form of challenge took a social shape that seemed to lead to negative effects. The excerpt below from grade 8 field notes displays the two forms of 'challenge' Ms. Dunia held on to:

Students reach the question where they need to indicate the reaction type. Ms. D asks them not to look at the books. She says the reactants and has the student at the board find the products and write symbols properly, then balance it. After that, students have to indicate the type of reaction. Ms. D does not accept when they say the type of reaction without proper justification. She uses questions to guide them through the different reactions, and for the most part, she wants the student on the board themselves to reach the acceptable answer: Ziad is at the board Ms. D asks him to underline the ions of the reaction after he wrote it:

$$\text{Na OH} + \text{H Cl} \rightarrow \text{Na Cl} + \text{H}_2\text{ O}$$

A B C D A D C B

Ms. D: *What part does the displacement: H or H_2? (no response)*
Ms. D: *Na replaces what?*

Ziad looks at the products and says Na and Na …

Then he realizes that Na replaces H and says it. Ms. D keeps on asking till he says confidently that Na replaced H.

Ms. D: *what do we call the reaction?*
Ziad: *double displacement.*
Rami: *now the idea is clearer! (I think he refers to underlining the ions on the board)*
Ms. D: *isn't this how we explained it before?! (With exasperation, in a challenging tone)*

Above we see that Ms. Dunia 'challenged' Ziad (student on the board) to independently come up with answers and identify the reaction type, which is in line with academic 'challenge.' Some students had asked earlier for further explanation of reaction types and that is why Rami remarked: *Now the idea is clearer!* Ms. Dunia challenged him: *Isn't this how we explained it?!* Here the challenge took a social form of pointing out to students' lack of attention the first time the concept was explained. Ms. Dunia confronted Rami (who is usually the class joker) with her statement. Ms. Dunia's social 'challenge' of students was evidenced by her harshness in dealing with students, especially ones she perceived as not giving enough effort to do well. Some students had expressed annoyance to me at what they saw as harsh and sometimes demeaning language towards them. I suggest that Ms. Dunia's harshness and rigidity are grounded in a commitment to 'challenge' students in a local social and disciplinarian view. Such (common) social view maintains that when individuals are faced with their 'failure' and 'inadequacy' by an authority figure, they will be motivated to better themselves[5].

From a virtue perspective, Ms. Dunia seemed to uphold the importance of students' *intellectual independence,* which is related to valuing intellectual sobriety and inquiry; at the same time, she upheld to a moral commitment of being a strict disciplinarian responsible for shaping students' behavior. A commitment as a disciplinarian championed teacher authority and devalued students' *social independence,* which is perceived as disrespect. Managing those two virtues (intellectual independence and responsibility as a disciplinarian) proved somewhat difficult for Ms. Dunia and this was demonstrated in a very tense episode that occurred towards the end of the year when grade 9 students took mock national examinations. In Lebanon, Grade 9 students sit for high stakes national examinations to determine if they continue in an academic track or switch to a vocational track. In preparation for the national exams, schools conduct several mock exams towards the end of the year. It is important to note that students' performance in the national exams is used as a source of prestige for schools.

Students did poorly in the mock exams and this resulted in high tension between Ms. Dunia and students, who she was especially harsh to; and between her and administration, who saw her as being too harsh on students. When discussing the episode with the author, Ms. Dunia framed the situation using interpretations of students lacking care and hard work. For example, she recounted that a lower performing student did well on the physics exam because he studied with his cousin Nasma (one of Ms. Dunia's best students). She added that Nasma is not much smarter than the others, but her hard work and her parent's keenness on her getting a good education make her an excellent student. She added by drawing on personal and cultural interpretations to support her arguments:

They (meaning Ms. Dunia's parents) told us this is your future (good education) and your inheritance; this is what we heard growing up...

I asked about her harsh words to students the day before, specifically telling them that she will be happy for the ones who pass and will also be happy for the ones who do not:

Sara: *Don't you think that students may interpret this as you not caring for them?*

Ms. D: *They should not think so, because when we come every day and give them all the lessons and we try that they miss nothing. We come even when there are strikes, days that we do not get paid for, we come to school and teach; why would they think and feel like that?*

During that particular interview, Ms. Dunia used a combination of cultural, personal, subject matter, and pedagogical interpretations to make sense of students' performance and her strictness and harshness. Her frustration materialized in feelings of indignation and disappointment at both students and administration and actual physical discomfort (stomach ache). More importantly, it seemed that her indignation also rested in her being challenged back by students, and to a certain extent by the administration, who she saw as not being appreciative. As a researcher, I cannot interpret Ms. Dunia's practice as 'uncaring,' as her care for the students and the school community was evident throughout the fieldwork, but there is evidence that her management of the situation in light of the conflicting aspects in her practice, that is valuing academic independence but resenting social independence, was not fruitful to her or to others.

Ms. Dunia's practical-moral knowledge was engaged in trying to make sense of the situation. Unfortunately, outcomes did not serve students nor did they serve her, as she was not able to exercise her commitment to 'challenge' students in ways she valued. It is in situations such as the one described above where the engagement and cultivation of practical wisdom becomes essential to manage the complexity of a situation for the benefit of oneself and others (Schwartz & Sharpe, 2010; Zagzebski, 1996). Yet, cultivation of practical wisdom requires modeling and mentorship and open dialogues. Ms. Dunia needed support to question and revise her interpretations and commitments, especially in light of the cognitive, social, and emotional needs of students; and this was not always available, especially as educational setting become more and more test-driven.

Virtues, Interpretation, and Commitments

Teacher interpretations can be within academic domains such as understandings of content, its nature, pedagogical theories, etc. (see Figure 1). Interpretation can also be within moral domains where teachers interpret academic knowledge according to situations at hand and attempt to exercise discretion and 'wise' judgment in daily teaching actions based on commitments seen as inherent to good teaching. As we saw in the example above, those two domains overlap. Interpretation within moral domains are closely connected to teachers' interpretations of the 'goods' science education ought to serve (e.g., promoting scientific inquiry, curiosity, and literacy; science as gatekeeper to career choices, etc.). Accordingly, teachers construct visions and images of valued science education and of themselves as worthy practitioners. Images of what a person ought to be are integral in actions resulting from practical wisdom and practical-moral knowledge: for example images of fairness, responsibility, openness to the ideas of others, empathy, intellectual honesty, and perseverance, etc. (Gadamer, 1989). Teachers maintain standing commitments to such 'goods[6].' Commitments teachers verbalize and act upon reflect virtues they see as important to them as teachers. For example, a teacher who sees her major role as having students achieve good grades in national or state exams to get into certain colleges and majors, will become committed as such and work diligently towards such end as a 'responsible' and 'fair' teacher (as in one of the case studies in Salloum and Abd-El-Khalick, 2010). Also as we saw above, Ms. Dunia valued herself as a disciplinarian as well as a science teacher who challenges students as they 'discover' science concepts; she saw these two aspects of her practice as 'goods' that needed to be served. The dialectical interactions of interpretations and commitments are expressed in actions and classroom practices.

Implications for Educational Research

In order to empirically explore practical-moral knowledge and virtues embodied in it, it is asserted that:

- Teachers need to be engaged in dialogues and inquiries about their interpretations and commitments as practitioners; and
- Investigations of teachers' interpretations and commitments, verbalized as narrative and enacted, along with actions will illuminate practical-moral knowledge and embodied virtues to both practitioners and researchers.

Such inquiry, though, needs to be done *with* teachers rather than on them (Schwandt, 1996). Schwandt proposed that the aim of such an inquiry would not be to produce theoretical/scientific knowledge to be prescribed to practitioners or to replace their practical-moral knowledge, rather the inquiry would be to engage practitioners in critical reflections and revision of their own practical-moral knowledge through scrutinizing and questioning interpretations and commitments.

Teacher-generated data and narratives resulting from such inquiries would illuminate interpretations, commitments, and ultimately virtues. However, teachers' narratives themselves may not lead to pedagogical growth and the cultivation of practical wisdom. What is important are critical and collaborative reflections (by teachers and teacher educators) on interpretations, commitments, and virtues embodied in practices; especially whether these were the most appropriate and conducive to serving students and the content area. Practical-moral knowledge develops into practical wisdom when practitioners become better capable at regulating and mediating multiple interpretations, beliefs, emotions and prior knowledge, and utilizing ones that help them interpret situations in ways conducive to practices that serve oneself and others (Schwartz & Sharpe, 2010; Zagzebski, 1996). The last section discusses how clinically rich teacher education and practitioner inquiry can help promote such critical reflections, growth in different forms of knowledge including practical-moral knowledge, and cultivation of practical-moral knowledge into practical wisdom.

Implications for Teacher Education

An assumption of the conceptual framework suggested here is that even mundane teaching actions involve certain practical-moral concerns. For example, in Salloum and Abd-El-Khalick (2010), teachers' practical arguments always invoked ideas such as being fair and responsible for student achievement (even if only for the deserving group). Therefore, it can be argued that teacher practical-moral knowledge develops in ways aligned with what is 'good' to be as a teacher and what is 'right' to do in a certain situation. The aim is not to mystify teaching, but to emphasize how teaching involves sense making that allows teachers to represent themselves as good teachers/individuals. Teacher educators need to explore ways of making good teaching from research and reform perspectives parallel to teachers' notions of 'good' and to connect reform with virtues teachers see as defining their practice. Teacher education can help teachers develop new interpretations and balance new commitments, and hence cultivate virtues conducive to good and equitable teaching.

Korthagen, Loughran, and Russel (2006) stated, "the development of practice in light of competing demands requires an approach that revolves around the need to create meaningful collaboration in learning and teaching, collaboration of peers and collaboration of teacher educators and student teachers" (p. 1027). Teachers' (in-service and pre-service) interpretations need to be explored and understood by themselves and by teacher educators. As advocated by Korthagen, Loughran, and Russel (2006), and Schwandt (1996) above, teachers and teaching candidates need to be given adequate opportunities for collaborative inquiries and critical reflections on how their practical-moral knowledge is being engaged and developed, and to further understand, value, and question their interpretations and commitments.

A virtue-based view of teacher knowledge supports teacher education models that meaningfully integrate clinically rich teacher apprenticeships/residency (co-teaching and mentoring), collaborative and disciplined practitioner inquiry/action research, along with a strong base in theoretical and skills-based education (Darling-Hammond, 2008; Korthagen, Loughran, & Russel, 2006; Tobin & Roth, 2005, van Driel, et al. 2001). Virtues are acquired excellences that take time and effort; therefore, the expectation in teacher education is for teachers and teaching candidates to understand and cultivate virtues that potentially will enable them and further their knowledge as good teachers. Developing phronesis/practical wisdom and intellectual and moral virtues relevant to teaching involves a social dimension, where good practice needs to be modeled and teacher experiences and narratives shared and reflected upon (Breier & Ralphs, 2009; Kemmis, 2010). It is important for new teachers, from outset of their teacher preparation programs, to observe, experience, and systematically and critically reflect on how master teachers draw on various knowledge, interpretations and commitments. Furthermore, they need to be mentored and supported for sustaining such critical reflections on their own practices into induction years through collaborative practitioner inquiry.

It is important to keep in mind that handling and reflecting on practical experiences is qualitatively different when undertaken from within a phronetic perceptual model than an epistemic rational model (Kessels & Korthagen, 1996). Within a phronetic model, preservice and in-service teachers verbalize and document their interpretations of the intricacies and particularities of their teaching and learning experiences rather than elicit the appeal to abstract rules and theoretical concepts; or as Kessels and Korthagen (1996) stated:

To be able to develop this wider perception-based type of knowledge... what we need is not so much theories, articles, books, and other conceptual matters, but first and for most, concrete situations to be perceived, experiences to be had, persons to be met, plans to be exerted, and their consequences to be reflected upon (p. 21).

This is not to say that theoretical concepts are not important. Considerable time needs to be allotted to addressing theoretical concepts in science content, pedagogy and Pedagogical Content Knowledge (PCK), because it is in trying to apply concepts to specific situations that teachers become better aware of their interpretations, commitments, and how their practical-moral sense of being good teachers is inherent in practice. Teaching candidates need to build solid theoretical knowledge (episteme) and skills (techne) that will allow them to be effective in action as they enact relevant virtues. However, when immersed in extensive field experiences and systematic inquiries from the outset of their teacher preparation, they would be developing the different forms of understandings, knowledge, and skills (phronesis, episteme, and techne) simultaneously and dialectically. Simultaneous and dialectic development stands opposed to a teacher education model that starts with emphasis on theoretical knowledge and concepts that are to be later 'applied.'

Being immersed in experiences may not be enough though for developing their practical-moral knowledge, and has to be augmented with disciplined and collaborative teacher inquiry in order for new teachers, master teachers, and teacher educators to critically reflect on experiences. It is important for new and veteran educators to engage in a systematic and step by step inquiry process, where they plan for and aim some kind of change; act and observe consequences of change; reflect on the process and consequences, and then re-plan; act, observe, reflect, and so on (Kemmis & McTaggart, 2001). Disciplined and collaborative teacher inquiry allows new and experienced practitioners to:

- Create documentation (data) and narratives of teaching,
- Share documentation and narratives with perceptions and interpretations, and
- Develop new understandings, knowledge, and competencies.

Through sharing, teachers can critically reflect on interactions among academic (content and pedagogy) interpretations and moral situated ones. They also become cognizant of the interaction of commitments, interpretations, and actions, thus discerning virtues underlying them. As importantly, they will be able to question the appropriateness of their interpretation, commitments, and actions in serving students and the subject matter. Kemmis (2010) has argued that reflection on individual and collective consequences, such as what occurs in practitioner inquiry, is how we cultivate practical wisdom (phronesis), "action research can help us learn phronesis, the disposition to live wisely and well, by facilitating our reflection on our individual and collective praxis" (p. 422).

Through systematic inquiry, pre-service and novice teachers can cultivate virtues that will guide their practice and act as motivation for new knowledge, skill development, and continual pedagogical growth. Indeed, responsibility, open-mindedness, and wholeheartedness are inherent to reflective practice (Dewey, 1939) and, as mentioned above, these are intellectual virtues that act as motivation for developing new knowledge for teaching. For example, by cultivating virtues such as intellectual honesty, responsibility, fairness, and perseverance, teachers would see the need to learn about multiple representations of abstract and difficult science concepts and how to establish their relevance to diverse groups of students. Similarly, the aforementioned virtues would act as motivation to learning about and utilizing diverse forms of assessment to ensure that students' knowledge is properly represented. Becoming knowledgeable about multiple representations, functions, and assessments of science concepts are important components of Pedagogical Content Knowledge (PCK) and so systematic inquiry not just cultivates virtues but enhances teacher competency and effectiveness in practice. Two examples will be presented below from my work with novice urban teachers in a practitioner inquiry graduate course to demonstrate types of growth through teacher inquiry (Collier & Salloum, 2011; Salloum, et al., 2010).

Multiple Representations and Access

In Collier and Salloum (2011), Traci (co-author) a science teacher in an urban middle school sought to investigate ways to enhance conceptual understandings in science for her diverse students. Schools in urban settings are distinguished by high levels of students' diversity in terms of culture and sociolinguistics, and so it is important to devise instruction that acknowledges this diversity and perceives it as a strength. Traci conducted a teacher inquiry on "how to use discrepant events with higher order and scaffolding questioning to develop students' conceptual understanding?" She also explored "how can students' comprehension and conceptual knowledge be measured through teacher-created materials using academic language of standardized testing?"

What Traci did was to broaden her conception of discrepant events as not just the use intriguing science demonstrations and hand-on activities, but as also utilizing forms of popular culture that students identify with, such as popular songs (e.g. Cha Cha slides to teach the cross-cross method in chemistry) and comic strips (e.g. Dr. Birdley Teaches Science series). She used discrepant events, as an expanded concept, and different levels of questions to enhance students' science understandings and problem-solving competencies. In her final inquiry paper, Traci wrote:

Overall, the clarification of expectations through type of questioning and visual presentation [pop-culture based discrepant events] allowed for the majority of the students to understand each topic's aim ...

The ability for them (students) to see science as academic, transforming, and creative would allow them to see science as less foreign and unattainable.

She later added:

My inquiry provided a step-by-step process where I began to understand how students learn and remember.

Traci came to realize that "Students are no longer passive recipients of information therefore must be engaged in learning," and that "the inquiry's focus on discrepant events provided a springboard for discussion and discovery." As she conducted her inquiry with two different groups, she also noted that certain discrepant events and activities were more effective for one group more than the other. Inherent to Traci's inquiry were the notions of responsibility and fairness towards students by providing access to students and thus best serving them. From a virtue perspective, Traci's inquiry helped her cultivate and mediate virtues such as intellectual fairness and responsibility, in addition to careful inquiry about students' strengths and challenges to better serve them. Concurrently she was building her pedagogical content knowledge on properties of matter.

Teacher Inquiry and Empathy

Gizelle[7] was a high school biology and special education teacher, her teacher inquiry was about developing a sense of "personal and scientific inquiry" in her students, especially students with special needs (Salloum, et al., 2010). Working with her science co-teachers, she aimed to make science learning more inquiry-based and relevant to students' lives, thus encouraging their natural curiosity. The following excerpt is from an interview about her teacher inquiry and how it affected her practice:

Sara: *Do you feel your definition of what inquiry-based biology–or how it would look like in the classroom–do you feel this may have changed when you were actually trying to implement it in your classroom and collecting data about.*

Gizelle: *I think it changed. I knew my students were very involved with art. And they loved the process of incorporating art. But I didn't think it was that much involved. And this was something that would really connect all students, regardless of their grade levels or their reading levels. The one thing they all shared was the concept of art. And they love sharing their opinions. I think they were great debaters. And they also loved the process of the kinetic activity of doing something with their hands. And I didn't think that was going to be such a big deal in my inquiry.... And when we were looking at past activities, we began to see a pattern that the students were more engaged. The less we talked, the more they worked. So the more kinetic activities that there were, they were able to obtain the information and present the information and research information and question their own–instead of us questioning their work, their presentation, their classmates were the ones asking them questions about their work....*

As we see from the excerpt above, Gizelle's inquiry lead to new interpretations and understandings about students and their needs and strengths. Generating data (e.g. observations and narrative, student interviews, analyzing student work, etc.) and reflecting on data expanded her concept of 'scientific inquiry' and how to actualize an ideal of inquiry-based learning for *all* students, including special populations. From a virtue perspective, Gizelle's inquiry was successful as it helped her cultivate and mediate several virtues inherent to good teaching: open mindedness, careful inquiry, and empathy; and enact them in her teaching.

CONCLUDING THOUGHTS AND CHALLENGES AHEAD

Creating meaningful collaboration among practitioners, teacher educators, and new teacher poses many challenges and cannot adopt a 'one size fits all.' For example as collaborative inquiry, practitioner inquiry, and reflective practice are being pushed in teacher education programs, they themselves may become prescriptive and hegemonic, and so lose some of their meaningfulness for new teachers. They need to be carried out in ways that address particular needs of teaching candidates. The author is currently generating data in that respect and even though generally pre-service and novice teachers find the inquiry experience rewarding, certain issues are surfacing (Salloum, 2009; Salloum, et al., 2010). Some issues are ideological such as how teachers see the validity and quality of their inquiries. Others are practical, such as the added burden teachers see in engaging in practitioner inquiry, which is a not a *choice* but rather a requirement for credit.

Collaborative inquiry needs to be dialectical and involve questioning our own perceptions, interpretations, and commitments as practitioners in educational research and teacher education, or as Loughran (2002) suggested of practicing what we preach. Gadamer (1989) has indicated that Aristotle extended phronesis/practical wisdom with 'sympathetic understanding' or 'being understanding.' Sympathetic understanding is meant to be exercised by the one who does need to act, but by one who is trying to understand the actions and practical wisdom of another (or understanding practical-moral knowledge as in the case of novice teachers). A person with 'sympathetic understanding' would judge another by transposing him/herself fully into the concrete situation of the person who has to act, "the person who is understanding does not know and judge as one who stands apart and unaffected but rather he thinks along with the other from the perspective of a specific bond of belonging, as if too he was affected" (Gadamer, 1989, p. 320). Sympathetic understanding can help us as teacher educators better facilitate the teacher inquiry experiences. I work with teaching candidates in urban settings, where candidates are either novice teachers going through an alternative certification route or are pre-service teacher. Each group requires a different approach and framing of the inquiry process itself. For novice teachers struggling with demands of a new practice, it is better framed as an extension to formative assessment and as a way to make assessment more authentic and fair. This way they can build on students' strengths and address their challenges, planning instruction that provides multiple access points for diverse students. For pre-service teachers, the process can be more structured to entail careful inquiry for enhancing candidates' pedagogical content knowledge, empathy for student, and their understandings of students as learners and thinkers.

Continuously developing our sympathetic understanding can help us as researchers and teacher educators better understand the complex nature of teacher knowledge in general, and specifically practical-moral knowledge and practical wisdom (including our own). Along with pre-service, novice, and experienced teachers, we can all develop our practical wisdom by engaging in inquiries that examines how a teaching practice involves the complex interaction of three modalities: the actual, the necessary, and the possible (Heidegger, 1993; as cited in Shulman, 2007). With careful and reflective investigation of how practical-moral knowledge and practical wisdom becomes concretized we will hopefully become better informed on how to bridge the gap between educational theory and practical and how to better manage our interpretations and commitments and those of classroom practitioners.

REFERENCES

AAAS. (1990). *Science for all Americans*. Oxford, UK: Oxford University Press.

Anderson, C. (2007). Perspectives on science learning. In S. Abell & N. Lederman (Eds.), *Handbook of Research on Science Education* (pp. 3–30). Mahwah, NJ: Lawrence Erlbaum.

Anderson, R. D., & Mitchener, C. P. (1994). Research on science teacher education. In D. L. Gabel (Ed.), *Handbook of Research on Science Teaching and Learning* (2nd ed., pp. 3–44). New York, NY: McMillan.

Aristotle,. (2002). *The Nicomachean ethics* (J. Sachs, Trans.). New York, NY: Riverhead Books.

Bell, R. L., Lederman, N. G., & Abd-El-Khalick, F. (2000). Developing and acting upon one's conception of the nature of science: A follow-up study. *Journal of Research in Science Teaching*, *37*, 563–581. doi:10.1002/1098-2736(200008)37:6<563::AID-TEA4>3.0.CO;2-N

Borko, H., & Putnam, R. T. (1996). Learning to teach. In D. C. Berliner & R. C. Calfee (Eds.), *Handbook of Educational Psychology* (pp. 673–708). New York, NY: McMillan.

BouJaoude, S., Salloum, S., & Abd-El Khalick, F. (2004). Relationships between selective cognitive variables and students' ability to solve chemistry problems. *International Journal of Science Education*, *26*, 63–84. doi:10.1080/0950069032000070315

Breier, M., & Ralphs, A. (2009). In search of phronesis: Recognizing practical wisdom in the recognition (assessment) of prior learning. *British Journal of Sociology of Education*, *30*, 479–493. doi:10.1080/01425690902954646

Brickhouse, N. W. (2001). Embodying science: A feminist perspective on learning. *Journal of Research in Science Teaching*, *38*, 282–295. doi:10.1002/1098-2736(200103)38:3<282::AID-TEA1006>3.0.CO;2-0

Carr, D. (2005). Personal and interpersonal relationships in education and teaching: A virtue ethical perspective. *British Journal of Educational Studies*, *53*, 255–271. doi:10.1111/j.1467-8527.2005.00294.x

Carr, D. (2007). Character in teaching. *British Journal of Educational Studies*, *55*(4), 369–389. doi:10.1111/j.1467-8527.2007.00386.x

Carr, W. (1995). *For education*. Berkshire, UK: Open University Press.

Clark, C. (2005). The structure of educational research. *British Educational Research Journal*, *31*, 289–308. doi:10.1080/0141192050082128

Collier, T., & Salloum, S. (2011). *The use of discrepant events and higher order/scaffolding questions for deeper science learning*. Paper presented at the Annual Ethnography in Education Research Forum. Philadelphia, PA.

Crawford, B. A. (2007). Learning to teach science inquiry in the rough and tumble of practice. *Journal of Research in Science Teaching*, *44*, 613–642. doi:10.1002/tea.20157

Darling-Hammond, L. (2008). A future worthy of teaching for America. *Phi Delta Kappan*, *89*, 730–735.

Dewey, J. (1933). *How we think: A restatement of the relation of reflective thinking to the educative process*. Boston, MA: D. C. Heath.

Doyle, W. (1977). Paradigms for research on teacher effectiveness. In L. S. Shulman (Ed.), *Review of Research in Education* (Vol. 5, pp. 163–179). Washington, DC: American Educational Research Association.

Duffee, L., & Aikenhead, G. (1992). Curriculum change, student evaluation, and teacher practical knowledge. *Science Education*, *76*, 493–506. doi:10.1002/sce.3730760504

Eisner, E. W. (2002). From episteme to phronesis to artistry in the study and improvement of teaching. *Teaching and Teacher Education*, *18*, 375–385. doi:10.1016/S0742-051X(02)00004-5

Feldman, A. (2002). Multiple perspectives for the study of teaching: Knowledge, reason, understanding, and being. *Journal of Research in Science Teaching*, *39*, 1032–1055. doi:10.1002/tea.10051

Fenstermacher, G. D. (1994). The knower and the known in teacher knowledge research. In L. Darling-Hammond (Ed.), *Review of Research in Education* (Vol. 20, pp. 3–56). Washington, DC: American Educational Research Association.

Flyvbjerg, B. (2001). *Making social sciences matter*. Cambridge, UK: Cambridge University Press.

Fosnot, C. T. (1996). Teachers construct constructivism: The center for constructivist teaching/teacher preparation project. In C. T. Fosnot (Ed.), *Constructivism: Theory, Perspectives, and Practices* (pp. 205–216). New York, NY: Teachers College Press.

Gadamer, H.-G. (1989). *Truth and method* (2nd ed.). (J. Weinsheimer & D. G. Marshall, Trans.). New York, NY: Crossroad.

Gallagher, S. (1992). *Hermeneutics and education*. New York, NY: SUNY.

Haney, J. J., & McArthur, J. (2002). Four case studies of prospective science teachers' beliefs concerning constructivist teaching practices. *Science Education*, *86*, 783–802. doi:10.1002/sce.10038

Heidegger, M. (1962). *Being and time* (J. Macquarrie & E. Robinson, Trans.). New York, NY: Harper and Row.

Issler, K. (1983). A conception of excellence in teaching. *Education*, *103*, 338–344.

Kemmis, S. (2009). Action research as a practice-based practice. *Educational Action Research*, *17*, 463–474. doi:10.1080/09650790903093284

Kemmis, S. (2010). What is to be done? The place of action research. *Educational Action Research*, *18*, 417–427. doi:10.1080/09650792.2010.524745

Kemmis, S., & McTaggart, R. (2000). Participatory action research. In N. Denzin & Y. Lincoln (Eds.), *Handbook of Qualitative Research* (2nd ed., pp. 567–605). Thousand Oaks, CA: Sage.

Kennedy, M. M. (2006). Knowledge and vision in teaching. *Journal of Teacher Education*, *57*, 205–211. doi:10.1177/0022487105285639

Kessels, J., & Korthagen, F. (1996). The relationship between theory and practice: Back to the classics. *Educational Researcher*, *25*, 17–22.

Kortahgen, F., & Kessels, J. (1999). Linking theory and practice: Changing the pedagogy of teacher education. *Educational Researcher*, *28*, 4–17.

Korthagen, F., Loughran, J., & Russell, T. (2006). Developing fundamental principles for teacher education programs and practices. *Teaching and Teacher Education*, *22*, 1020–1041. doi:10.1016/j.tate.2006.04.022

Lave, J. (1988). *Cognition in practice*. Cambridge, UK: Cambridge University Press. doi:10.1017/CBO9780511609268

Lederman, N. G. (1992). Students' and teachers' conceptions of the nature of science: A review of the research. *Journal of Research in Science Teaching*, *29*, 331–359. doi:10.1002/tea.3660290404

Lemke, J. L. (2001). Articulating communities: Sociocultural perspectives on science education. *Journal of Research in Science Teaching*, *38*, 296–316. doi:10.1002/1098-2736(200103)38:3<296::AID-TEA1007>3.0.CO;2-R

Lotter, C., Harwood, W. S., & Bonner, J. J. (2007). The influence of core teaching conceptions on teachers' use of inquiry teaching practices. *Journal of Research in Science Teaching*, *44*, 1318–1347. doi:10.1002/tea.20191

Loughran, J. (2002). Effective reflective practice: In search of meaning in learning about teaching. *Journal of Teacher Education*, *53*(1), 33–43. doi:10.1177/0022487102053001004

Lynch, S. (2001). Science for all is not equal to one size fits all: Linguistic and cultural diversity and science education reform. *Journal of Research in Science Teaching*, *38*, 622–627. doi:10.1002/tea.1021

Medley, D. M. (1979). Effectiveness of teachers. In P. L. Peterson & H. J. Walberg (Eds.), *Research on Teaching: Concepts, Findings, and Implications*. Berkley, CA: Mc-Cutchan.

Meyer, H. (2004). Novice and expert teachers' conceptions of learners' prior knowledge. *Science Education*, *88*, 970–983. doi:10.1002/sce.20006

Moje, M. (1995). Talking about science: An interpretation of the effects of teacher talk in a high school science classroom. *Journal of Research in Science Teaching*, *32*, 349–371. doi:10.1002/tea.3660320405

Mulholland, J., & Wallace, J. (2008). Computer, craft, complexity, change: Exploration into science teacher knowledge. *Studies in Science Education*, *44*, 41–62. doi:10.1080/03057260701828135

National Research Council. (1996). *National science education standards*. Washington, DC: National Academic Press.

Noel, J. (1999). On the varieties of phronesis. *Educational Philosophy and Theory*, *31*, 273–289. doi:10.1111/j.1469-5812.1999.tb00466.x

O'Loughlin, M. (1992). Rethinking science education: Beyond Piagetian constructivism toward a sociocultural model of teaching and learning. *Journal of Research in Science Teaching*, *29*, 791–820. doi:10.1002/tea.3660290805

Pajares, M. F. (1992). Teachers' beliefs and education research: Cleaning up a messy construct. *Review of Educational Research*, *62*, 307–332.

Pendlebury, S. (1995). Reason and story in wise practice. In H. McEwan & K. Egan (Eds.), *Narrative in Teaching, Learning and Research* (pp. 50–65). New York, NY: Teachers College Press.

Salloum, S. (2006). *Teaching as practice: Blending intellectual and the moral in pursuit of science teachers' practical knowledge*. (Unpublished Doctoral Dissertation). University of Illinois at Urbana-Champaign. Champaign, IL.

Salloum, S. (2009). *Pedagogical growth and practitioner inquiry: ESL urban teachers' perceptions of their growth while conducting inquiries into their practice*. Paper presented at the Annual Meeting of the American Educational Research Association. San Diego, CA.

Salloum, S., & Abd-El-Khalick, F. (2010). Practical knowledge in teaching: Case studies from physical science classrooms. *Journal of Research in Science Teaching*, *47*, 929–951.

Salloum, S., Jennings, M., Arrabito, N., Schmidt, M., McCall, C., & Frederick, T. ... Benn-Scantlebury, A. (2010). *Novice urban teachers engaging in practitioner inquiry: Lessons learned, rewards, and challenges.* Paper presented at Annual Ethnography in Education Research Forum. Philadelphia, PA.

Schwandt, T. A. (1996). Farewell to criteriology. *Qualitative Inquiry*, *2*, 58–72. doi:10.1177/107780049600200109

Schwandt, T. A. (2000). Three epistemological stances for qualitative inquiry. In N. K. Denzin & Y. S. Lincoln (Eds.), *Handbook of Qualitative Inquiry* (2nd ed., pp. 189–213). Thousand Oaks, CA: Sage.

Schwandt, T. A. (2005). On modeling our understanding of the practice fields. *Pedagogy, Culture & Society*, *13*, 313–332. doi:10.1080/14681360500200231

Schwartz, B., & Sharpe, K. (2010). *Practical wisdom*. New York, NY: Riverhead Books.

Shulman, L. (2007). Practical wisdom in the service of professional practice. *Educational Researcher*, *36*, 560–563. doi:10.3102/0013189X07313150

Simmons, P. E., Emory, A., Carter, T., Coker, T., Finnegan, B., & Crockett, D. et al. (1999). Beginning teachers: Beliefs and classroom actions. *Journal of Research in Science Teaching*, *36*, 930–954. doi:10.1002/(SICI)1098-2736(199910)36:8<930::AID-TEA3>3.0.CO;2-N

Smith, L., & Southerland, S. A. (2007). Reforming practice or modifying reforms? Elementary teachers' response to the tools of reform. *Journal of Research in Science Teaching*, *43*, 396–423. doi:10.1002/tea.20165

Southerland, S., Sinatra, G. M., & Mathews, M. (2001). Beliefs, knowledge, and science education. *Educational Psychology Review*, *13*, 325–351. doi:10.1023/A:1011913813847

Southerland, S. A., Gess-Newsome, J., & Johnson, A. (2003). Portraying science in the classroom: The manifestation of scientists' beliefs in classroom practice. *Journal of Research in Science Teaching*, *40*, 669–691. doi:10.1002/tea.10104

Tobin, K., & Roth, W.-M. (2005). Implementing coteaching and cogenerative dialoguing in urban science education. *School Science and Mathematics*, *105*, 313–322. doi:10.1111/j.1949-8594.2005.tb18132.x

Tsai, C.-C. (2002). Nested epistemologies: Science teachers' beliefs of teaching, learning, and science. *International Journal of Science Education*, *24*, 771–783. doi:10.1080/09500690110049132

van Driel, J. H., Beijaard, D., & Verloop, N. (2001). Professional development and reform in science education: The role of teachers' practical knowledge. *Journal of Research in Science Teaching*, *38*, 137–158. doi:10.1002/1098-2736(200102)38:2<137::AID-TEA1001>3.0.CO;2-U

van Manen, M. (1995). On the epistemology of reflective practice. *Teachers and Teaching: Theory and Practice*, *1*, 33–50. doi:10.1080/1354060950010104

van Manen, M. (1999). The practice of practice. In M. Lange, J. Olson, & W. BŸnder, (Eds.), Changing Schools/Changing Practices: Perspectives on Educational Reform and Teacher Professionalism. Luvain, Belgium: Garant.

Vianna, E., & Stetsenko, A. (2006). Embracing history through transforming it. *Theory & Psychology*, *16*, 16–81.

Wildy, H., & Wallace, J. (1995). Understanding teaching or teaching for understanding: Alternative frameworks for science classrooms. *Journal of Research in Science Teaching*, *32*, 143–156. doi:10.1002/tea.3660320205

Windschitl, M. (1999). A vision educators can put into practice: Portraying the constructivist classroom as a cultural system. *School Science and Mathematics*, *99*, 189–196. doi:10.1111/j.1949-8594.1999.tb17473.x

Woolfolk, A. E., & Galloway, C. M. (1985). Nonverbal communication and the study of teaching. *Theory into Practice*, *24*, 77–85. doi:10.1080/00405848509543150

Zagzebski, L. T. (1996). *Virtues of the mind: An inquiry into the nature of virtue and the ethical foundations of knowledge*. Cambridge, UK: Cambridge University Press. doi:10.1017/CBO9781139174763

ENDNOTES

1 Teaching for tested algorithmic skills rather than conceptual understandings is well documented in science education (e.g., BouJaoude, Salloum, & Abd-El-Khalick, 2004).

2 Theoretical knowledge is an umbrella term for theoretical academic knowledge involved in science teaching such knowledge of science disciplines, psychological, and sociocultural theories of learning.

3 Carr (2007) stated that the perceived 'content-less' aspect has been put forth as a critique of virtue ethics in education.

4 All names are pseudonyms.

5 For example, a parent or teacher may tell a child: "Is this low grade worthy of you?" or "This is not what I expected from you," etc.

6 'Goods' that teachers perceive are usually subjective and related to their experiences, and not necessarily reflecting reform notions in science education.

7 Pseudonym.

This work was previously published in "Approaches and Strategies in Next Generation Science Learning" edited by Myint Swe Khine and Issa M. Saleh, pages 27-51, copyright 2013 by Information Science Reference (an imprint of IGI Global).

Chapter 31
Learning about the Different Dimensions of Sustainability by Applying the Product Test Method in Science Classes

Mareike Burmeister
Institute of Science Education, University of Bremen, Germany

Janine von Döhlen
Institute of Science Education, University of Bremen, Germany

Ingo Eilks
Institute of Science Education, University of Bremen, Germany

ABSTRACT

This chapter reflects upon the development of a new pedagogy for secondary level science education: the product test method. This method mimics the authentic societal evaluation practice performed by professional product-testing agencies. The design of this particular pedagogy is tied directly to the three most prominent dimensions occurring in contemporary sustainability models: ecological, economic and societal sustainability. Application of this method focuses on learning about the evaluation of competing industrial products with regard to sustainability criteria. This chapter reflects upon the underlying theoretical framework of the consumer test method. It also illustrates its application in the classroom based on a lesson plan for the evaluation of different sorts of plastics. Experiences and feedback from teachers and students are discussed, based on the cyclical development process by Participatory Action Research. They reveal that the lesson plan has great potential for contributing to higher levels of student motivation and learner perception regarding issues of sustainable development.

INTRODUCTION

Application of science and technology is essential for the contemporary development of every modern society (Bradley, 2005). This is why all modern concepts of scientific literacy include the idea that students must acquire skills which prepare them for active participation in societal debates and decision-making processes concerning scientific and technological issues (Hofstein, Eilks & Bybee,

DOI: 10.4018/978-1-4666-7363-2.ch031

2011; Holbrook & Rannikmae, 2009). Science education in this sense must therefore contribute to Education for Sustainable Development (ESD) by promoting skills which make younger generations able to act responsibly and help shaping both our future society and the entire world in a more sustainable fashion (Burmeister, Rauch & Eilks, 2012; De Haan, 2006).

From an ESD and scientific literacy point-of-view, it is absolutely essential supporting students to become responsible citizens and in the means of authentic pedagogy to acquire proper skills in order to be able to participate in societal decision-making in the future (Holbrook & Rannikmae, 2007). These skills include not just communication and evaluation abilities *in* science and technology, but also a competency to debate and decide *about* science and technology in a societal context (Eilks, Nielsen & Hofstein, 2014).

One specific skill set belonging to responsible citizenry is critical media literacy, especially the ability to critically review and deal with information presented in the news media (Hobbs, 1998). Among other things, this area concerns information related to both the domains of science and technology (Rundgren, Chang-Rundgren, Tseng, Lin & Chang, 2012). However, this aspect touches upon far more than an ability to critically examine and accordingly react to science and technology related news in the media, as discussed by McClune and Jarman (2012). Information related to science and technology also appears frequently in the form of advertisements, product test reports and everyday life communications (Bauer, 2009).

This paper describes how product testing for media purposes is normally conducted and focuses on how to use the idea of product testings in science classes. It therefore presents a method to help pupils specifically learn about this issue. Using the example of evaluating different sorts of plastics along sustainability criteria, this method will be evaluated according to its potential for contributing to an understanding of issues and aspects of sustainable development.

SETTING THE STAGE

A Theoretical Framework for the Product Test Method

In all probability, the majority of high school students will never embark upon careers in which they become scientists or engineers (Hofstein et al., 2011). Even if some of them do, these students will generally become experts in one very specific area of science or engineering, effectively making them non-experts in all other domains of science. Therefore, only a small minority of all high school students will ever encounter future situations in which they are called upon to deal with authentic, scientific information in a specific domain. The authentic science in that domain can only be found in the respective research institutes, scientific publications or conferences associated with that area (Eilks et al., 2014). Access to such information is regularly limited to scientists in the specific area of interest due to both reasons of limited access and to the use of formal scientific language, technical terms, and symbols among experts.

Nevertheless, there is also another kind of 'scientific' information. The science-related information with which the vast majority of our students eventually come into contact comes only indirectly from the core of the authentic, scientific endeavor (see Bauer, 2009). As soon as we leave the world of the scientists, we are no longer dealing with original scientific information. Starting from the field of popular science magazines or school textbooks, the information available to readers has already become "filtered information" (Hofstein et al., 2011) in numerous steps by many persons along the chain of information transfer (Eilks et al., 2014). Almost all of our everyday communication, debates in society, and media reports are based on information which no longer comes directly from the field of science. Relevant information is presented by journalists, politicians or pressure groups. Their information regularly has been derived from the areas of mass media and political communication. These fields have hardly any direct contact with authentic science. In most cases, such media offerings have not been prepared (or checked) by any scientific experts active in the specific area in question.

The indirect link between real science and society was aptly illustrated by the philosophical works of Ludwik Fleck in the 1930s. According to Fleck (1935/1979), any domain of science can be understood as an esoteric core surrounded by concentric spheres representing information resources. The core is formed by the collective of scientists active in a particular area of science. According to Fleck's theory, the esoteric core covers as yet unrefined and codified views of scientific discourse within the specific scientific community and is presented via scientific conferences and journals. This core is surrounded by exoteric domains. If we leave the esoteric core, we find a compilation of the accepted "essentials," which stem from science in the form of research handbooks (so-called handbook science). A further domain is composed of textbook science, which represents a distillation of the basic core knowledge taken from the handbooks and applied to educational purposes. Beyond handbook science, there are further spheres representing the public understanding of science and non-science-related areas of society. Each step outwards from one domain to another can be described by the processes of selection, simplification and re-interpretation, until scientific information eventually reaches the non-expert citizen. The simplified model in Figure 1 makes it clear that nearly all non-academic activity within society has no overlap with the esoteric core of science, possibly not even with the closely allied area of handbook science. Referring to Fleck, Bauer (2009) described the relationship between the various spheres to be one characterized by a gradient of simplification, iconicity, concreteness, certainty of judgment, and controversial reception. The mechanisms behind this gradient are clearly driven by both the author's purpose and by the target audience for which the information is prepared.

Figure 1. The science-society relationship with regard to information sources

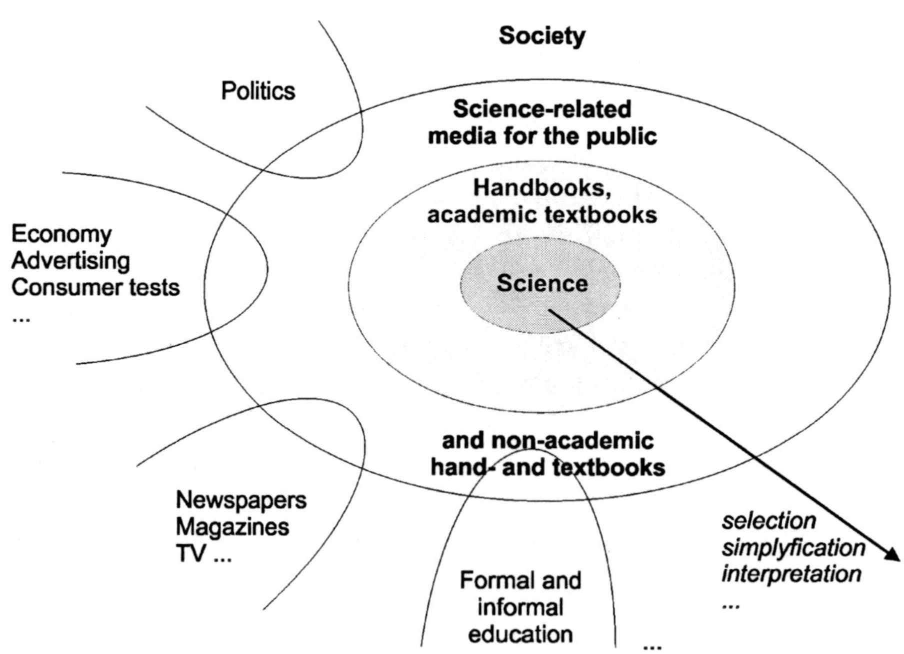

The process of filtering information described above is performed by individual persons in charge of producing and transferring the information (Eilks et al., 2014), e.g. information from and about testing procedures and their results which concern a specific technological product. Information gathering, selection, weighting, and interpretation are all based on the personal and individual foreknowledge, interests, and communication competencies of each person involved in the entire chain. With regard to product tests, people assume that experts must be in charge of performing the evaluations, but they know next to nothing about the experts' interests and levels of expertise. The same holds true for the journalists reporting on the topic. The further we move towards everyday life and society, e.g. the product information presented in a regional newspaper, the more likely it becomes that the person responsible for presenting the information does not possess the comprehensive, expert subject matter knowledge necessary to ensure the reliability of information transferred to the public from a specific domain. As a result, there is hardly any guarantee that either the producer of the underlying evaluation or the facilitator of information transfer possesses the necessary skills to evaluate the pertinent scientific facts and successfully route them onwards to inform consumers. Yet in order to critically evaluate product reports in the media, it is essential that readers have a minimum of basic knowledge about how such tests are performed and how they are disseminated to the public.

Sustainability Evaluations and Product Tests

In many countries, regularly-published product testing reports are available to the public in a variety of formats, including subject-specific magazines, daily newspapers, Internet pages, and radio and TV programs. In the case of Germany, test results for a large range of products and services are regularly made available by governmental (Stiftung Warentest) or private (Ökotest) product-testing agencies. In recent years the focus of such evaluation has taken a broader view, which reaches far beyond the mere technical quality or durability of a given product. New criteria include the ecological effects of a product and have quickly gained in importance. Consumers have a more critical view on ecological effects their products may cause and the demand of consumers for ecological product information has obviously increased considerably within the last years (Imkamp, 2000). Ecological criteria include factors such as item production, packaging, transport, disposal and potential for recycling. Of course, a product needs to be manufactured according to economically sound principles. Price also plays a large role in evaluating a product. The evaluation of socio-economic impacts surrounding an item's production has also increased in importance and covers - among other things - the social conditions under which a product is made. Today, consumers have a larger interest in those hidden qualities of a product which one can expose neither by subjecting it to a technical test nor through its use or consumption (Schoenheit, 2005). For example, it is important to know whether a guarantee exists that child workers are not involved in cheaply producing commodities or that textiles were not made under conditions that came to the public after the recent collapse of a factory in Bangladesh. From this discussion, we can recognize that modern product testing encompasses many of the dimensions addressed by modern concepts of sustainable development: ecological, economic and societal sustainability (Burmeister et al., 2012). Modern product testing and consumer policy includes not only reflection upon the technical quality of a given product, but also the ecological, economic and societal effects of its production, use and final disposal (Schoenheit, 2005; Reisch, 2004).

Many countries have similar sources of consumer information. One example is *Which?* magazine in the UK. But how do such reports, which often enjoy high levels of public trust and are often touted as measures of objective quality, actually arise? It is clear that in the initial phase of testing someone decides which product characteristics will be analyzed. Test dimensions and specific criteria must be set. Weighting factors must be defined, which allow the results in individual testing areas to affect the overall results. For example, should a decision be made to test an article for its packaging's ecological impact, if the packaging itself has absolutely no effect on the functionality or durability of the item? If so, which weighting factor should be applied? Decisions must somehow be reached about how various product characteristics like functionality, price, environmental effects or societal impact should stand in relationship to one another with regard to weighting purposes. These preliminary decisions may prove to be decisive in influencing the end results.

Despite many beliefs to the contrary, weighting of the various dimensions in product testing cannot be objectively achieved. It may be possible that the idea of a balanced view of ecological, economic and societal impacts may offer testers some guidance as is prevalent in most recent concepts of sustainable development (Burmeister et al., 2012). Nevertheless, even if the three most commonly used dimensions of sustainability are employed for evaluation purposes, the detailed criteria and concrete weightings must be set up, agreed upon, and commonality of understanding must be reached. This means that individuals actually decide which of the potential criteria are the most important (or irrelevant) for the test from their personal points-of-view. By doing so, these individuals act as filterers and interpreters of the evidence, which is to be revealed by technical measures during the actual testing procedures. This is an excellent example of the idea of filtered information as outlined in the previous section.

Logically, the weighting process can lead to widely differing end results for the same product, even when tested by multiple agencies or by various co-workers and experts within the same organization. Different teams within an agency evaluating a given item can reach different conclusions as compared to their colleagues. This means that a product can receive high marks from one testing agency and a lower grade from another, since the focus is different in the two testing runs. Evaluation by "normal" customers, such as those found on Internet portals like Amazon, frequently vary widely from the results reported by official testing agencies. Consumers often pay more attention to design, ease of use and price, instead factors such as the ecological or societal circumstances of an item's production. Nevertheless, a big study showed, that in 23 countries around the world, 40% of the 25,000 respondents have thought in the past year about punishing a specific company perceived as not socially responsible and at least 20% have avoided the product of a company or spoken out to others against the company (Environics International Ltd, 1999).

In the end, the consumer must decide upon the credibility of a given test and its source. This normally occurs as an unconscious process. But knowing that product ratings are always a compromise between different factors and competing criteria is crucial, if an individual is to develop a personal point-of-view with regard to the product testing results reported by the media. In this respect, consumer tests have the same characteristic as decisions based on sustainability criteria. Evaluations in the field of sustainability also present consumers with competing ecological, economic and societal effects. In order to reach a balanced decision we also need to negotiate the relative relationship between the weighting factors and their effects. Learning about this interrelationship by mimicking the authentic societal practice of product testing can aid us in better understanding the mechanisms in both domains.

CASE STUDY

Development of the Product Test Method by Participatory Action Research

The development of this pedagogy and the related lesson plan for evaluating plastics followed the Participatory Action Research (PAR) model in science education as described by Eilks and Ralle (2002). PAR is a specific interpretation of interactive Action Research for curriculum development and classroom investigation in domain-specific educational research (Eilks, 2014). PAR combines evidence-based knowledge taken from educational research and hands-on experience from classroom practitioners. It views them as the two ends of the knowledge spectrum of teaching and learning, since both are important and have their own strengths and weaknesses (McIntyre, 2005).

In PAR, teachers and researchers cooperatively design and investigate science teaching practices. At the start, evidence from educational research and the practical experience of the participating teachers are compared and contrasted in a group discussion process. The evidence from the two domains is reflected upon with respect to its relevance for innovation in teaching practices. A cyclical process is thereby initiated also reflecting didactical and methodological issues, the content, and explicating teachers' intuition and creativity. Lesson plans are developed, tested, evaluated, and refined based on evidence and joint reflection exercises. Innovative concepts and teaching materials are the end-products of this model. The accompanying research generates evidence and knowledge about teaching and learning, especially about the effects of changed teaching strategies and teachers' and students' perceptions of the new teaching approaches and pedagogies. The process also leads to developed practice, documentation of good practice, and contributes much to the Continuous Professional Development (CPD) in the practitioners (Mamlok-Naaman & Eilks, 2012; Eilks, 2014; Figure 2).

The project described in this paper started in December 2008 and was carried out by a group of experienced teachers which has worked together for a total of 10 years prior to this project (Eilks & Markic, 2011; Eilks, 2014). The group consists of ten teachers from various secondary schools. It meets once a month for about three hours to discuss and reflect upon the participants' classroom interventions. The group from the beginning is continuously accompanied by science educators from the University of Bremen.

The development of the overall lesson plan, which incorporated the pedagogy described here, required about one year of the group's time (Burmeister & Eilks, 2012). The focus of the project stemmed from an acknowledgement among the teachers that learning about sustainable development is underdeveloped in most German schools. Furthermore, Chemistry was identified as a subject which should largely contribute to learning about sustainability issues, as has been acknowledged in a study about German experienced Chemistry teachers (Burmeister, Schmidt-Jacob & Eilks, 2013). On this base the topic of plastics was selected as a suitable curriculum topic to introduce sustainability issues in the comparison of traditional and alternative plastic materials.

One of the accompanying science educators and the teachers then started first developing an overall structure for the lesson plan. Later they enriched the different steps by suitable media and innovative pedagogies. Single elements of the lesson plan then were pre-tested with student teachers in the university. Later the full lesson plan was applied by teachers from the PAR group in together five learning groups. The groups were selected on occasion, meaning teachers from the group volunteered to contribute to the testing in case they were teaching in a suitable learning group. The main challenge in the development of the product test method was to develop step by

Figure 2. Participatory action research in science education (Eilks & Ralle, 2002)

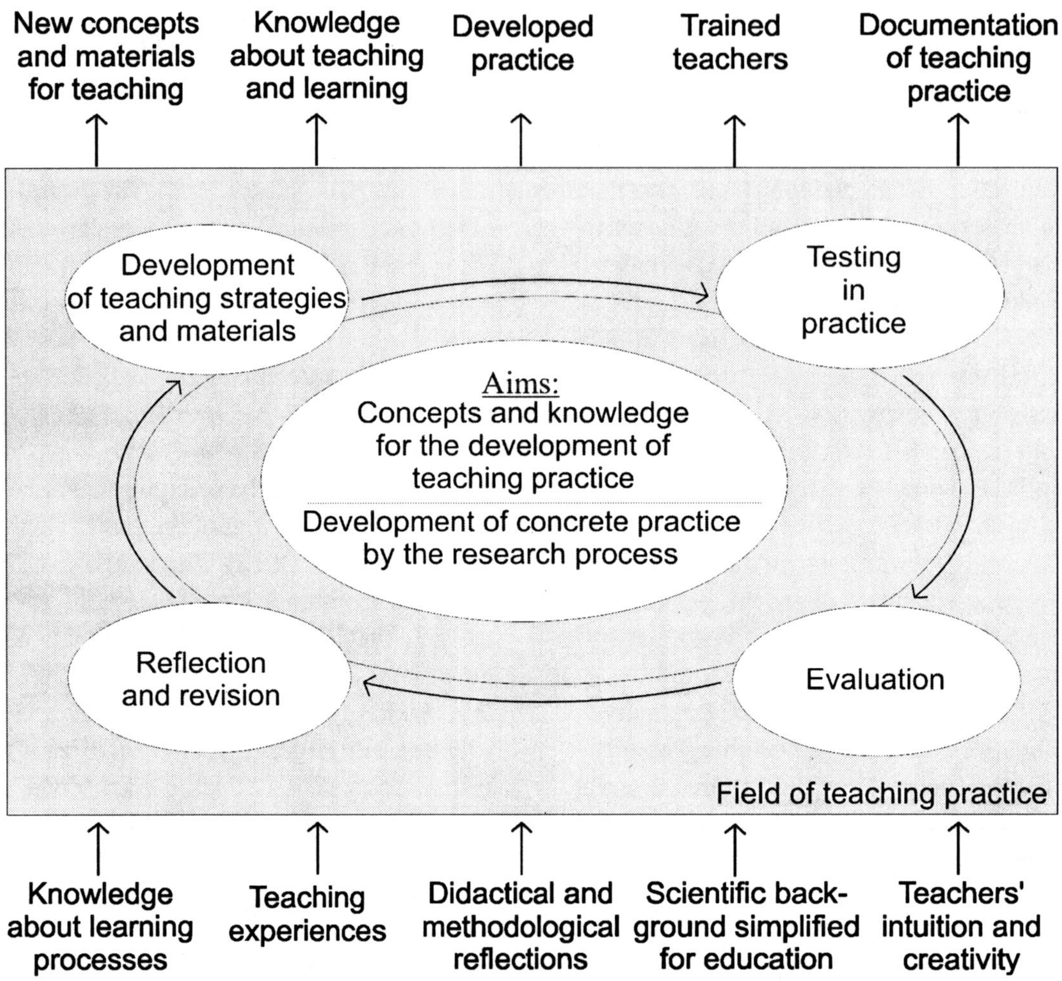

step a precise and comprehensive manual for how to operate the different steps. A set of worksheets and slides were developed to allow the teachers to introduce the method to their students. Also the selection and generation of feasible resources for evaluating the different sorts of plastics took several versions and intense discussions in the PAR teacher group.

The evaluation was based on the results gained by applying the lesson plan in its fully-developed form in different learning groups taught by the various PAR group members. For evaluation purposes, the lesson plan was tested in five learning groups representing varying levels of student knowledge with respect to content-based information in grade levels 9-12 (age range 15-18 years). A total of 95 students participated in this case study.

Data were collected through protocols recording the teachers' reflections during monthly PAR meetings. Student feedback was gathered with the help of open and Likert questionnaires, which were filled out directly after completing the lesson plan. The open-ended questions focused primarily on students' reflections about:

1. What they considered to be the most important things they had learned,

2. Their personal opinion about the lesson plan, and
3. What they now believed about the different types of plastics.

The twelve Likert items asked the students whether they had liked the topic, how they viewed the pedagogies which had been employed, and if they had experienced a change in their overall personal attitudes towards the issues of sustainability, consumer testing and plastics (Burmeister & Eilks, 2012). Parallel to this, the consumer testing pedagogy was also applied in several in-service science teacher training courses. It has been implemented in an undergraduate chemistry teacher education module concerning ESD at the University of Bremen. Feedback was also collected for these applications, based on protocols of teacher and student teacher feedback (Burmeister & Eilks, 2013). All sources of information were qualitatively analyzed and triangulated to construct meaning.

The Product Test Method in Class: Plastics as an Example

The product test method reveals many important aspects in the construction of consumer product reports as inspired by Germany's consumer protection group, *Stiftung Warentest*. It focuses on the points where negotiation and evaluation processes necessarily occur during the set-up phase of the testing procedure. By applying sustainability-related criteria for evaluation purposes, this method intends to reveal the competitive nature of the various sustainability dimensions, including the need to balance them fairly in order to arrive at an equitable and holistic evaluation for a given product.

A fictitious consumer testing process involving plastics was created in order to allow direct classroom access to a real debate centering around numerous sustainability issues involved with different plastics (see Wolf, Baitz, & Kreissig, 2010). This included the current problem of very large quantities of plastic waste which are found in the world's oceans (Barnes & Milner, 2005). The lesson plan was carried out using the idea of socio-scientific issues (SSI) based science education (Sadler, 2011). SSI education uses science-related societal issues which are controversial in nature. Issues are chosen from which a societal decision, which has an impact on the students' life or the development of society, must be made. The major objective is to learn how society is dealing with such controversial issues and how the individual can participate on societal debate and decision about them.

A specific SSI curriculum model was used: the socio-critical and problem-oriented approach to science teaching as developed by Marks and Eilks (2009). Lesson topics following the socio-critical and problem-oriented approach to science teaching are always authentic, controversial and will have direct impact on life in society. Topics are e.g. political decisions on taxes for renewable energy sources, restrictions in production and use of specific goods where the use may have an impact on the environment or health, or questions of technical innovations in comparison to standard technologies. All the lesson plans follow a five-step model of:

1. A textual approach by authentic media,
2. A phase of clarifying the science background information, in most cases by practical work,
3. Resuming the societal controversy and reflecting to which extend the science background helps for drawing a decision,
4. A discussion of the different perspectives and points of view available in the societal debate, e.g. by a role play or analysis of mass media, and
5. Finally a meta-reflection about societal discourse and decision making of the respective SSI.

Following the curriculum model, a clarification phase highlighting the scientific and technological background focusing the big advantages of modern plastics but also the issue follows the introduction of the controversial problem of plastic waste in the oceans (Burmeister & Eilks, 2012). A lab phase structured according to the learning-at-stations method (Eilks, 2002) then is employed in order to provide any necessary science background knowledge about plastics in general. Nine different stations offer various tasks for the learners to perform. Some stations are experimental in nature; others are theoretical. The stations cover not just the structures and production of polymers, but also the properties of different plastics and the processes of polymer synthesis. Thus the students start testing some relevant properties of different plastics in comparison themselves. A model-based explanation for differentiating between the basic types of plastics (elastomers, duroplasts and thermoplasts) is also included. The students are given two to three teaching periods of 45 minutes each to perform all of the tasks. Additional content material can be added depending on grade level. For example, specific reaction mechanisms for forming the most prominent examples of synthetic polymers were added for students in advanced courses at the upper secondary school level.

The product test method is then employed. This method aims to teach pupils about the different dimensions of sustainability-oriented evaluation. Three plastics are evaluated as examples of this: polyvinyl chloride (PVC), polyethylene terephthalate (PET) and thermoplastic starch (TPS). These particular plastics were chosen because of their significant (dis)advantages, which largely contradict one another when compared in a comprehensive analysis (see Table 1). PVC is a well-established, cheap and conventional technology. However, PVC remains quite controversial, despite its widespread use and diverse applications. Even though modern, high-quality PVC is hardly more problematic than other types of plastics and recycling it with modern waste treatment technology is safe, this fact is hardly ever heard in public discourse. The disadvantages of PVC, including the risks arising from unplanned combustion and those inherent in improper disposal, dominate the public perception. In contrast, the reputation of the alternative plastic TPS remains largely untarnished. It is a "bioplastic" and fulfills both overall criteria of bio-plastics. It is biologically degradable and is made (at least partially) from renewable resources. However, the uses of TPS are still very few and its price is relatively high compared to other alternatives. In the long run, such bioplastics may help us conserve our crude oil resources, but large-scale production of these substances might also increase the global risk of overly intensive agricultural land use. In the same vein, the public hardly ever hears about the negative side of PET, which is well-known for its use in drinking bottles. PET is largely ignored as a benign substance, even though recycling PET products successfully is a highly volatile societal problem. Until recently, old bottles were seldom recycled in Western countries. Instead, they were shipped off to developing countries or to China, where they were disposed or remanufactured into fleece pullovers and other clothing items under deplorable environmental and social conditions. Also the 'carbon footprint' of the transport is an often neglected factor in evaluating PET. With modern innovations in technology, however, PET can now be recycled under economically sound conditions in Western countries. But this technology is unfortunately not broadly applied in many countries.

The product testing phase of this teaching unit follows a sequence of eight steps:

1. Setting (or introducing) the relevant dimensions of an evaluation.
2. Individual weighting of the separate dimensions as a percentage.
3. Agreement on the weighting of the separate dimensions in small groups.

Table 1. An overview of TPS, PET and PVC

Thermoplastic Starches (TPS)	
• Largely composed of renewable resources • Biologically decomposable, which makes them especially suited for uses in agriculture, medicine and product packaging. • No dangers in their production, which occurs under relatively mild conditions and under minimal energy requirements in optimized production facilities. • Neither reactants nor products are dangerous for animals or humans. • The demand for bio plastics is constantly on the rise, causing a booming market for them.	• Unstable, therefore only useful in products requiring biodegradability, e.g. mulch foils, sutures or food packaging. • Biodegradable does not mean compostable, causing problems for regional disposal firms. • Intensive agricultural practices necessary to win raw materials, usually through gene-manipulated crops. • Up until now no recycling possible. • Composting leads to product loss without energy gain. • Production is currently non-competitive with respect to production facilities, production capacities, and market pricing.
Polyethene Terephthalate (PET)	
• Good physical properties: stabile, transparent, and flexible, especially good for packaging materials. • PET can be spun into fibers and is especially indispensable to the textile industry (Polyester). • PET can be easily recycled and the products of recycling evidence hardly any loss of quality. • PET is safe for humans and the environment, combusting to CO_2 and water, so that it can be burned in waste plants without problem. • PET manufacturing is well-established and has been optimized, so that cheap economic production is possible	• PET is good for many, but not all, uses. For example, it cannot be made into a rubberlike, elastic state. • PET is permeable to gases, so that carbon dioxide-containing drinks in PET bottles can lose their fizz and possibly pick up aromas from their storage environment. • PET is manufactured from fossil fuels. • PET is primarily recycled in China, where social and environmental conditions normally do not meet Western standards. Imported PET waste is typically recycled into fleece cloth, which is exported – frequently containing additives which are forbidden in Western lands.
Polyvinyl Chloride (PVC)	
• PVC is a well-researched plastic with many uses. • There are hard, soft and elastic forms of PVC, depending on the required use. • PVC is durable and especially fitted for outdoors use, e.g. windows, or locations under heavy traffic like flooring. • PVC is cheap in almost every area. • Chlorine, stemming from the synthesis of sodium hydroxide, is itself a reactant, which reduces disposal and extra production costs. • Pure PVC can be recycled.	• PVC creates carcinogenic substances in the intermediate production steps, thus increasing the risk for humans and environment in the case of industrial accidents. • Combustion of PVC frees poisonous gases (e.g. HCl), which causes additional problems in the case of unintended fires. • PVC can only be burned in modern waste disposal systems with advanced filtering systems. Otherwise, mixed waste must be stored (illegal in the EU), so that waste is often exported to foreign lands. • PVC is produced from fossil fuels.

4. Individual evaluation of one plastic using suitable sources (- to ++).
5. Exchange of considerations and comparison of the evaluations for one plastic in small groups.
6. Constructing new groups using the jigsaw classroom method, exchange of information about the results and the argumentation behind the evaluations.
7. Calculation of the weighted evaluation for the three different plastics.
8. Reflection on the results and the process of evaluation.

It begins by negotiating the individual evaluation dimensions. The utility of plastics, which has been discussed in previous lessons, reveals that plastic usage in the sense of technical properties or durability often becomes a logical factor in the evaluation. Availability and price are also easily derivable criteria. Environmental friendliness, decomposability and recycling suitability become additional touchstones drawing a large amount of student attention to themselves, thanks to introduction of the problem of plastic waste in the oceans in the beginning lessons. In order to lend more structure to the overall discussion, the pupils work on a worksheet dealing with a basic model of sustainability, driven by the interconnectedness of analyzing and balancing the ecological, economic and societal impacts of a given product. It is important to recognize that pragmatic criteria (a product's properties and utility) are combined with more economic and value-driven dimensions from the sustainability debate, namely economic (salability, cost-price-effectiveness, availability), ecological (environmental effects, health in production, disposal/recycling) and societal (production, recycling under appropriate social standards) impacts. These dimensions are decisive for developing the later product test.

During the actual process of product evaluation, the students must first decide which weighting factors will be attributed to the various dimensions. This step is introduced on a respective worksheet (Figure 3). The students are free to suggest each their own weighing based on their values, feelings, and preferences. The suggestions always cover an extremely wide spectrum. Some participants find environmental friendliness to be the most important factor; others view technical properties as more important. In the case of widely varying points-of-view, students can be asked to make compromises either in pairs or in small groups. The think-pair-share method (Kagan, 1992) can be applied to the joint weighting process. Groups can also decide to calculate mean weighting values. In all cases there are differences between different groups that can be operated into a discussion about potential reasons and justifications. (In the end a calculation can be done using different suggestions, so that the students will see how the weighing impacts the final results.) Our

Figure 3. Part of the working materials for the product testing method using the comparison of various plastics as an example

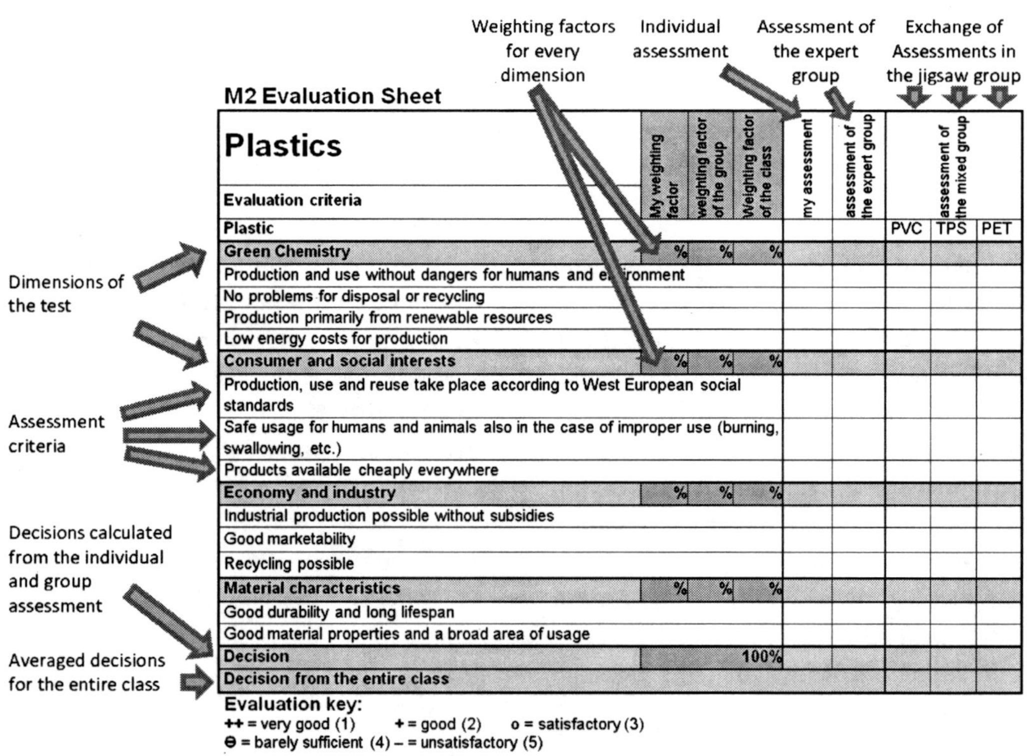

experience shows that in most cases all of these approaches eventually lead to an overall rounding off of extreme values. A weighting factor of roughly 20-30% for each category is generally the end result, but distributions vary by at least 10% in each of the categories.

These values and the following evaluations are noted by the pupils, either on a worksheet like that shown in Figure 3 or on a similar worksheet based on patterns found in product tests available in the country in question.

Evaluating the different plastics was embedded in a jigsaw-puzzle classroom scenario (Aronson, Stephan, Sikes, Blaney, & Snapp, 1978). Each student was assigned the task of individually rating one of the three plastics. Symbols taken from the *Stiftung Warentest* ranging from "++" to "-" were used, which parallel German school grades of 1-5 (see Figure 3). These are analogous to the grades A, B, C, D, and F in many English-speaking countries. Specially-designed texts were employed for younger secondary school pupils, in which information about each of the plastics was tailored to the various criteria. For older students it is also possible to use a variety of Internet pages as a resource. Pupils carry out the evaluation individually. Then they negotiate the values for their particular sort of plastic for each category in the expert round of the jigsaw classroom. The pupils then switch to the teaching phase of the jigsaw classroom, wherein the new learning groups are composed of at least one expert per plastic type. The learners discuss their various decisions with one another, including their reasons for positively or negatively evaluating a product. It quickly becomes clear while summarizing the evaluations that any advantages in one domain are often negated by disadvantages in other areas.

The sequence is ended with a meta-reflection discussing both the end decisions and the lessons which can be learned from the process. This reflection decisively focuses on both where and to what effect individual decisions can affect the end result. It must be made clear to the pupils that negotiation and decision-making processes between the different sustainability criteria, which also include the usefulness of the product, command the foreground in this respect. The goal of this exercise is to make the pupils think about the decisions made during the process of product evaluation. It also becomes abundantly clear that other school classes might come to different end results when testing the exact same products. The processes of weighting the criteria and decision-making can therefore easily lead to two equal overall values, despite different constellations in the chosen weighting factors and the selection of the final grades for all three plastics. It is clear that the end focus needs to be the question of how objective and balanced any evaluations reported to the public at large truly are. This includes the question of which role scientific evidence and/or sustainability criteria actually play in reaching such important decisions.

Experiences and Feedback

The objective of this case was to develop a feasible pedagogy to promote learning about the nature of sustainable development as well as to foster critical thinking skills and critical media literacy. The teachers in this study stated that the lesson plan was both highly feasible and intensely motivating for their students to engage themselves in issues of sustainable development. They described intensive debates during the pupils' product weighting decisions when the different sustainability dimensions were related to one another. Many students seemed to have great difficulty in assigning personal weighting factors to products and in jointly reaching agreements with regard to a given characteristic. The questions involved in assigning weighting factors to the various products incited intense debates regarding the specific role that the numerous evaluation dimensions should actually play when evaluating plastic products. In this way a critical reflection of a societal practice of argumentation, discourse and decision-making was mimicked. While applying the product test, students were plagued by the contradictions inherent between the different dimensions. This often led to productive confusion and an intense personal reflection about the overall dimensions of sustainability and the evaluation process itself. The teachers described many indicators that the students started becoming more critical against consumer test results and their presentation in the media. Some of the teachers stated that they themselves had never considered consumer testing activities and sustainability criteria in such an intense and critical manner. Some of the teachers concluded that their own view of such activities and criteria had shifted drastically.

As this case study was carried out with secondary school students, the results achieved for the three plastics commonly fell in the range between a low 2 and a high 3 (a grade of B- to C+ in many English-speaking countries). This revealed differences of single tenths of an evaluation point for all three plastics tested. As a generally negatively-viewed product overall in society, PVC is surprisingly evaluated as either equal to or even more highly rated than biologically degradable TPS. This occurs despite the risks present if PVC is accidentally burned. The unexpectedly poorer results of TPS are typically caused by its overall price and its relatively few useful applications in a very limited number of product areas. The latter is often an ignored aspect of TPS, especially when consumers are confronted by TPS advertisings as a bioplastic. In every case there were widely-strewn individual starting points, which led to very intense, very controversial discussions about the criteria themselves and the resulting weighting factors. The results proved to be unexpected, even on the part of the learners. In the test runs, there were generally no clear winners or losers at the end of the process. The pupils, teachers and student teachers had, however, expected the results to turn out differently. They normally assumed that the traditional, highly criticized product PVC would prove to be clearly inferior to the "bioplastic" TPS. Debate in the classroom showed that both the students and teachers tried to understand how the unexpected results emerged. This reflection showed good understanding of the often competing nature of the different dimensions of sustainability as well as on the influence on individuals in societal practices of consumer tests and their communication.

The evaluation accompanying classroom testing delivered positive assessments of the product test method in class. Roughly 65% of the learners stated in a 4-point Likert questionnaire that they now had a better understanding of how to judge reports on product testing results presented by the public media. A further 25% supported this statement at least partially. Only about 10% of the participants said that they didn't like the product test method because, in their opinion, they hadn't learned anything new with it. With respect to plastics, roughly 75% of the pupils fully believed that the unit had truly made them thoughtful about their choices of various plastic products. Another 20% thought the same was partially true. Nearly 65% of the total participants stated that they would judge plastics differently after having completed the lesson plan. A further 30% voiced partial agreement with this statement. All these results support the interpretation that the students after the lesson plan developed a better understanding of the idea of sustainability. These results also show indications for development in critical thinking skills and new insights contributing to critical media literacy. In the open-ended questionnaire, repeated remarks appeared that the teaching unit had been "useful." Students had the feeling they had learned something valuable for their future lives and now understood what the term 'sustainable' means. Overall, the lessons were rated as very positive and interesting. This evaluation coincided with that of the accompanying teachers.

CONCLUSION AND RECOMMENDATIONS

The product test method was able to rather unconventionally enrich the pedagogy of science teaching within the focus of this study. It provoked intense classroom debates and made students aware of science's potential to contribute to the evaluations of everyday products, including the potential role that the dimensions of sustainability might play in such choices. It also revealed the limitations of science when it attempts to achieve a holistic evaluation of products, e.g. in the foreground of the economic and societal dimensions of sustainability.

This pedagogy was developed for secondary school students in a society which is quite sensitive to sustainability issues. In other societies the pedagogy, if applied to a suitable context, might also be used to increase sustainability sensitization to other target groups, e.g. in informal, higher, or adult education. Nevertheless, even in mainly sustainability sensitized societies more mature learners can benefit from the activity. We implemented the activity into the chemistry teacher training program at the University of Bremen (Burmeister & Eilks, 2013). The idea was to give the student teachers a practical example of how to explicitly teach about sustainability issues in the secondary science classroom. By asking the student teachers to do the activity themself mainly the same processes and effects can be regularly observed as they were documented from the secondary science classroom. Also the feedback from the student teachers is almost the same as from the high school students.

In the follow up to this study, the product testing method was transferred to other examples by the PAR group. Two examples dealt with different types of fuels for car engines (based on conventional gasoline, bio-ethanol, or hydrogen technology) and with bicycle frames (made of steel, aluminum, and carbon fibers). The later example was even operated for grade 7 students of age range 12-14. In reflecting upon the parallels existing between the examples, the action research group attempted to localize criteria for developing product test lesson plans. In all the examples (see Table 2), the group found common support for a design principle that was already applied in the original lesson plan. There were always three principally different technologies selected. One technology was purposely chosen to represent a traditional, well established, cheap technology with some deficits either in its ecological impact or in its technical quality. A second alternative technology was new to the market, but not yet fully competitive, mainly because of its potential market volume or due to a much higher price bracket. The third technology was typically one with possibly large market potential for the future, but as yet incomplete development, yielding a very limited product range and extremely high material costs. In each case there were competitive advantages and disadvantages with respect to ecological, economic, and societal impacts. Additionally, all of these impacts compete with and partially overshadow the end price and usefulness of the product.

Table 2. Potential topics for the product test method when evaluating products according to sustainability criteria

	Example Plastics	Example Fuels	Example Bicycles
Traditional technology	PVC	conventional gasoline	steel
Alternative technology	TPS	bioethanol	aluminum
Future technology	PET	hydrogen	carbon fiber

The development of the new teaching units revealed to the researchers that transfer of the original method developed for the case study on plastics is definitely possible. New units also have similar potential to make students more contemplative, especially with respect to the use of science for individual and societal decision-making when it comes to science and technology applications in our everyday lives. This also includes learning about the competitive nature of criteria chosen for specific evaluations, especially when the different dimensions of sustainability are comprised.

ACKNOWLEDGMENT

We gratefully acknowledge the funding of the Deutsche Bundesstiftung Umwelt (DBU) and the German Funds of the Chemical Industry (FCI) for supporting the development of the lesson materials.

REFERENCES

Aronson, E., Stephan, C., Sikes, J., Blaney, N., & Snapp, M. (1978). *The jigsaw classroom*. Beverly Hills, CA: Sage.

Barnes, D. K. A., & Milner, P. (2005). Drifting plastic and its consequences for sessile organism dispersal in the Atlantic Ocean. *Marine Biology*, *146*, 815–825. doi:10.1007/s00227-004-1474-8

Bauer, M. W. (2009). The evolution of public understanding of science – Discourse and comparative evidence. *Science, Technology & Society*, *14*, 221–240. doi:10.1177/097172180901400202

Bradley, J. D. (2005). Chemistry education for development. *Chemical Education International, 7*.

Burmeister, M., & Eilks, I. (2012). An example of learning about plastics and their evaluation as a contribution to education for sustainable development in secondary school chemistry teaching. *Chemistry Education Research and Practice*, *13*, 93–102. doi:10.1039/c1rp90067f

Burmeister, M., & Eilks, I. (2013). Using participatory action research to develop a course module on education for sustainable development in pre-service chemistry teacher education. *Centre for Educational Policy Studies Journal*, *3*, 59–78.

Burmeister, M., Rauch, F., & Eilks, I. (2012). Education for sustainable development (ESD) and secondary chemistry education. *Chemistry Education Research and Practice*, *13*(2), 59–68. doi:10.1039/c1rp90060a

Burmeister, M., Schmidt-Jacob, S., & Eilks, I. (2013). German chemistry teachers' knowledge and PCK of green chemistry and education for sustainable development. *Chemistry Education Research and Practice*, *14*, 169–176. doi:10.1039/c2rp20137b

De Haan, G. (2006). The BLK '21' programme in Germany: A 'Gestaltungskompetenz' based model for Education for Sustainable Development. *Environmental Education Research*, *12*, 19–32. doi:10.1080/13504620500526362

Eilks, I. (2002). Learning at stations in secondary level chemistry lessons. *Science Education International*, *13*(1), 11–18.

Eilks, I. (2014). Action research in science education – From a general justification to a specific model in practice. In T. Stern, F. Rauch, A. Schuster, & A. Townsend (Eds.), *Action research, innovation and change* (pp. 156–176). London: Routledge.

Eilks, I., & Markic, S. (2011). Effects of a long-term participatory action research project on science teachers' professional development. *Eurasia Journal of Mathematics. Science and Technology Education*, *7*, 149–160.

Eilks, I., Nielsen, J. A., & Hofstein, A. (2014). Learning about the role of science in public debate as an essential component of scientific literacy. In C. Bruguière, P. Clément, A. Tiberghien (Eds.), *Book of selected presentations, ESERA Conference Lyon 2011*. Dordrecht, The Netherlands: Springer.

Eilks, I., & Ralle, B. (2002). Participatory action research in chemical education. In B. Ralle & I. Eilks (Eds.), *Research in chemical education - What does this mean?* (pp. 87–98). Aachen, Germany: Shaker.

Environics International Ltd. (1999). *The millennium poll on corporate social responsibility*. Retrieved July 01, 2013, from www.globescan.com/news_archives/MPExecBrief.pdf

Fleck, L. (1935/1979). *Genesis and development of a scientific fact (English translation 1979)*. Chicago, IL: University of Chicago.

Hobbs, R. (1998). Building citizenship skills through media literacy education. In M. Salvador & P. Sias (Eds.), *The public voice in a democracy at risk* (pp. 57–76). Westport, CT: Praeger.

Hofstein, A., Eilks, I., & Bybee, R. (2011). Societal issues and their importance for contemporary science education: A pedagogical justification and the state of the art in Israel, Germany and the USA. *International Journal of Science and Mathematics Education*, *9*, 1459–1483. doi:10.1007/s10763-010-9273-9

Holbrook, J., & Rannikmäe, M. (2007). The nature of science education for enhancing scientific literacy. *International Journal of Science Education*, *29*, 1347–1362. doi:10.1080/09500690601007549

Holbrook, J., & Rannikmäe, M. (2009). The meaning of scientific literacy. *International Journal of Science and Environmental Education*, *4*, 275–288.

Imkamp, H. (2000). The interest of consumers in ecological product information is growing – Evidence from two German surveys. *Journal of Consumer Policy*, *23*, 193–202. doi:10.1023/A:1006430601842

Kagan, S. (1992). *Cooperative learning*. San Juan Capistrano, CA: Kagan.

Mamlok-Naaman, R., & Eilks, I. (2012). Action research to promote chemistry teachers' professional development – Cases and experiences from Israel and Germany. *International Journal of Mathematics and Science Education*, *10*, 581–610. doi:10.1007/s10763-011-9306-z

Marks, R., & Eilks, I. (2009). Promoting scientific literacy using a socio-critical and problem-oriented approach in chemistry education: Concept, examples, experiences. *International Journal of Environmental and Science Education*, *4*, 131–145.

McClune, B., & Jarman, R. (2012). Encouraging and equipping students to engage critically with science in the news: what can we learn from the literature? *Studies in Science Education*, *48*, 1–49. doi:10.1080/03057267.2012.655036

McIntyre, D. (2005). Bridging the gap between research and practice. *Cambridge Journal of Education*, *35*, 357–382. doi:10.1080/03057640500319065

Reisch, L. A. (2004). Principles and visions of a new consumer policy. *Journal of Consumer Policy*, *27*, 1–42. doi:10.1023/B:COPO.0000014103.38175.df

Rundgren, C.-J., Chang Rundgren, S.-N., Tseng, Y.-H., Lin, P.-L., & Chang, C.-Y. (2012). Are you SLiM? Developing an instrument for civic scientific literacy measurement (SLiM) based on media coverage. *Public Understanding of Science*, *21*, 759–773. doi:10.1177/0963662510377562 PMID:23832159

Sadler, T. D. (2011). *Socio-scientific issues in the classroom*. Dordrecht, The Netherlands: Springer. doi:10.1007/978-94-007-1159-4

Schoenheit, I. (2005). Social quality of products – Assessment and signalling. In U. Oltersdorf, E. Claupein, C. Pfau, & J. Stiebel (Eds.), *Consumer & nutrition challenges and chances for research and society* (pp. 108–116). Karlsruhe: Bundesforschungsanstalt für Ernährung und Lebensmittel.

Wolf, M.-A., Baitz, M., & Kreissig, J. (2010). Assessing the sustainability of polymer products. In P. Eyerer (Ed.), *Polymers – Opportunities and risks II: Sustainability, product design and processing* (pp. 1–55). Dordrecht, The Netherlands: Springer.

This work was previously published in a "Handbook of Research on Pedagogical Innovations for Sustainable Development" edited by Ken D. Thomas and Helen E. Muga, pages 154-169, copyright 2014 by Information Science Reference (an imprint of IGI Global).

Chapter 32
Providing Elementary and Middle School Science Teachers with Content and Pedagogical Professional Development in an Online Environment

Mary V. Mawn
SUNY Empire State College, USA

Kathleen S. Davis
University of Massachusetts – Amherst, USA

ABSTRACT

There is a great need to provide science teachers with on-going and relevant professional development, but access to such opportunities can be challenging due to time, distance, and budget pressures. Online courses and programs can provide alternatives to address these challenges. This chapter presents approaches, findings, and recommendations for online professional development of elementary and middle school science teachers based on a case study of an online science education course and an online chemistry course. Three themes are discussed: the ability to incorporate inquiry-based teaching and learning in online environments, the importance of online discourse and reflection, and the role of linking theory with practice. Teacher participants reported increased experience exploring content via inquiry, felt actively engaged with their peers as they constructed their knowledge, and expected to adapt inquiry-based activities in their classrooms as a result of these online courses.

INTRODUCTION

Good teaching matters! According to the NAEd Education Policy White Papers Project, "there is persuasive evidence that students benefit from high quality instruction and that these benefits are cumulative for students who have good teachers for several years" (Wilson et al., 2009, p. 1). Thus, there is a great need to provide science teachers with on-going and relevant professional develop-

DOI: 10.4018/978-1-4666-7363-2.ch032

ment (PD). The Glenn Commission report, *Before It's Too Late* (Department of Education, 2000), states that better mathematics and science teaching is grounded in improving the quality of teacher preparation and making continuing PD available.

Having access to PD programs can be problematic. Teachers must deal with time and travel constraints and budget pressures, leaving little opportunity to pursue PD. However, online courses and programs can allow teachers to fit coursework into their schedules as they can be accessed at any time, from any place (Asbell-Clarke & Rowe, 2007). For some teachers, online coursework may be the only option for furthering their subject knowledge (McNall Krall, Straley, Shafer, & Osborn, 2009).

There is growing evidence supporting online PD (Clary & Wandersee, 2009; Davis & Snyder, 2012; McNall Krall et al., 2009), but developing an effective course involves more than putting notes and assignments online. This chapter will describe considerations and approaches for developing online PD courses for science teachers and how two online courses engaged teachers in inquiry, meaningful discourse, and making connections to their classroom practice.

BACKGROUND

The Importance of Inquiry-Based Professional Development

Science is not simply a collection of facts to be memorized and explained, but rather, it is a way of thinking and approaching real-world problems. Scientific inquiry is described as:

...a multifaceted activity that involves making observations; posing questions; examining books and other sources of information to see what is already known; planning investigations; reviewing what is already known in light of experimental evidence; using tools to gather, analyze, and interpret data; proposing answers, explanations, and predictions; and communicating the results. Inquiry requires identification of assumptions, use of critical and logical thinking, and consideration of alternative explanations. (National Research Council (NRC), 1996, p. 23)

Students who engage in scientific inquiry use many of the same activities and thinking processes as scientists, yet these activities and processes are not always familiar to teachers (Olson & Loucks-Horsley, 2000). Since teachers' knowledge, experiences, and beliefs greatly impact what takes place in the classroom, teachers should learn content and pedagogy through engagement in activities that mirrors the approaches it is hoped they will bring into their classrooms (NRC 1996; Loucks-Horsley, Stiles, Love, & Hewson, 2010).

PD should incorporate science practices that promote teachers' understanding of science content and inquiry-based approaches (Capps, Crawford, & Constas, 2012). As with students, teachers learn best by doing science, investigating and constructing their understandings. Teachers should have significant and substantial involvement in laboratory experiences where they actively investigate phenomena, devise research questions, design procedures, collect and analyze data, and report findings (NRC, 1996).

PD that engages teachers in authentic research experiences and provides opportunities to develop inquiry-based lessons, may be key in assisting teachers in bringing such approaches to their classrooms (Capps et al., 2012). There is a need for rigorous, research-based PD for science teachers that empowers them to utilize the most effective science teaching methods, including unstructured problem-solving and inquiry-based learning (National Science Board, 2010).

The Value of Communication in Online Professional Development Courses

Effective online courses should actively engage educators and follow the basic principles of good teaching: encourages student-faculty contact and collaboration among students; provides prompt feedback; emphasizes time on task; communicates high expectations; and respects diverse talents and ways of learning (Chickering & Gamson, 1999). Many of these principles rely on open and frequent communication between the instructor and students and among course participants. Discussion boards, chat rooms, whiteboards, email, and assignment drop-boxes (with options to provide feedback) allow for multiple modes of interaction and opportunities for online communication (Asbell-Clarke & Rowe, 2007).

In general online courses use predominantly asynchronous discussions, which can build a sense of community, keep participants on task, break-down difficult concepts, and encourage synthesis and analysis of materials (Offenholley, 2006). Discussions also provide unique opportunities for scientific inquiry due to their text-based and archivable nature (Asbell-Clarke & Rowe, 2007). These written exchanges make students' thinking visible and is available for review, reflection, and reference. Learners can articulate their ideas and respond to classmates at their own pace, share and clarify ideas or observations, give and receive comments based on differing experiences and interpretations, and review and reflect on multiple exchanges (Harlen & Doubler, 2004).

Finally, online communities allow participants to feel part of a group with similar interests (Asbell-Clarke & Rowe, 2007). This sense of community can contribute to teachers' overall success. Participants who feel connected often place a higher priority on, and devote more time to, course content and activities (Lee et al., 2004).

Connections of Online Professional Development to Classroom Practice

A key feature for effective PD includes deepening teachers' content knowledge, including understanding how students learn and the difficulties they may encounter in mastering key concepts (Wilson et al., 2009). When surveyed, teachers enrolled in online science courses positively rated the value of online PD for furthering their knowledge in the subject areas they teach (Clary & Wandersee, 2009).

Likewise, fostering teachers' use of instructional practices in the classroom is important (Desimone, 2011). Online courses should provide teachers with opportunities to develop learning materials that embed these practices and can be implemented in their classrooms.

Online Professional Development for Science Teachers

Based on the literature, we believe that a high-quality course for teachers (whether online or face-to-face) would require teacher engagement in inquiry-based activities, interaction between instructor-students and students-students, and links to classroom practice. As a result, in this case study we sought to describe the characteristics of one online science education (SE) course and one online chemistry course from the perspectives of elementary and middle school teachers. Specifically, in what ways did these online courses engage teachers in inquiry, meaningful discourse, and make connections to their classroom practice, and of what value was that to the course participants?

STUDY CONTEXT AND METHODS

Online Program and Study Site

Faculty collaborated to develop twelve online, graduate-level courses for in-service elementary and middle school science teachers. These courses are tied to state and national standards and incorporate inquiry-based, hands-on experiences. Aided by kits of materials, teachers engage in guided and open-ended inquiries as they develop their understanding of science concepts. Discussions, journaling, and course activities provide multiple opportunities for interaction.

The SE instructors aimed to: introduce teachers to the nature of science through inquiry; help teachers identify the elements of inquiry teaching and learning; and explore the topics of conceptual learning, teacher questioning, and assessment as they relate to learning in the inquiry-based science classroom.

The chemistry instructors sought to: introduce teachers to the chemical and physical properties of matter; allow teachers to reinforce and extend their understanding of chemicals and their interactions; enable teachers to problem-solve and be aware of variables in experimental design; and examine the importance of pattern awareness in the learning process.

Each course was divided into twelve sessions. Course activities included readings, hands-on experiments, online discussions, lab notebook/journal, and quizzes. Course materials included textbooks, laboratory kit (chemistry), and online resources. The lab kit included materials needed to conduct hands-on experiments, with teachers supplying a small number of household items (e.g., food coloring, toothpicks, paper clips).

Methods and Participants

Study participants included the two science education instructors and six teachers enrolled during one semester, and two chemistry instructors and nine teachers enrolled during one semester. Ten teachers were female; five were male; six were elementary; nine taught at the middle-level. Two teachers had been teaching five or fewer years; 10 had 6-15 years of experience; and three had been teaching 20 or more years. Teachers taught in a variety of settings: urban, rural, and suburban.

Case study methods were used (Yin, 2009) to describe the online courses and the teachers' participation in them. Case study methods are used when the researcher asks "how" or "why" research questions, seeking to explain or describe a present-day occurrence within a real-life setting that is bounded (Merriam, 1998) and where "relevant behaviors cannot be manipulated" (Yin, 2009, p. 11). This approach does not seek to report a causal relationship between a treatment and results. Rather, it is explanatory and descriptive of the occurrence studied (Merriam, 1998). Data sources included pre- and post-instructional surveys, interviews, and course artifacts.

Pre- and post-surveys were constructed to investigate participants' beliefs and attitudes about science, teaching and learning. In addition, surveys provided demographic information, previous skills and experience with computer use, and informants' experiences in course activities.

In-depth, semi-structured interviews (Merriam, 1998) provided insight into participants' beliefs and attitudes about teaching science and their students' learning, their experiences within the project, the significance and interest level in course activities, the benefits of the project, and the obstacles/limitations encountered. All interviews were approximately 60 minutes in length, tape-recorded, and later transcribed.

Course artifacts include students' postings during weekly discussions, students' journals documenting project work, course syllabi and weekly lessons and prompts.

Data sources were analyzed for evidence of inquiry, meaningful discourse, and connections to teachers' classroom practice. Relevant patterns and themes were drawn based on the research questions (Merriam, 1998) and triangulated across the data sources. The analysis includes direct quotes, tables, and interpretive commentary. Pseudonyms were used throughout the analysis to maintain the informants' privacy.

STUDY RESULTS AND FINDINGS

Key Themes

Three themes are discussed below, the importance of: incorporating inquiry-based, hands-on activities in online science courses; promoting online discourse and reflection; and bridging science content and teaching practice. Through teacher interviews, course artifacts, and pre- and post-surveys, these aspects are described, and teachers explained each element's value.

Engaging in Inquiry-Based, Hands-on Activities

The two courses in this study made extensive use of inquiry-based, hands-on activities. During a typical weekly session in the chemistry course, teachers engaged in two to four inquiry-based experiments. They conducted investigations individually and discussed their findings with their classmates. Three chemistry experiments were selected for in-depth analysis:

- The Fast, the Slow, the Big, and the Small: Teachers placed vanilla extract inside a balloon, inflated this balloon, and observed the states of matter;
- Taping the Charge: Teachers explored attractive/repulsive forces between adhesive tape placed under different conditions; and,
- Making "Cents" of Surface Tension: Teachers placed water on a penny and experimented with the cohesive forces of water under different conditions.

These experiments engaged teachers in various aspects of scientific inquiry (see Table 1). They explored, explained, and questioned the fundamental properties of solids, liquids, and gases. Teachers identified and tested variables while experimenting with attractive and repulsive forces. Finally, they developed investigations, made predictions,

Table 1. Teachers' demonstration of inquiry skills

Inquiry Descriptor \ Participant	SE1	SE2	SE3	SE4	SE5	SE6	CH1	CH2	CH3	CH4	CH5	CH6	CH7	CH8	CH9
Identifies questions that can be scientifically investigated	X	X	X	X	X	X	X	X	X	X	X	X	X	X	X
Designs and conducts a scientific investigation	X	X	X	X	X	X	X	X	X	X	X	X	X	X	X
Uses appropriate tools and techniques to gather, analyze, and interpret data	X	X	X	X	X	X	X	X	X	X	X	X	X	X	X
Uses data to construct a reasonable explanation	X	X	X	X	X	X	X	X	X	X	X	X	X	X	X
Communicates investigations and explanations	X	X	X	X	X	X	X	X	X	X	X	X	X	X	X
Makes critical and logical connections between evidence and explanations	X	X	X	X	X	X	X	X	X	X	X	X	X	X	X
Recognizes and analyzes alternative explanations and predictions	X	X	X	X	X	X	X	X	X	X	X	X	X	X	X

Note: *Inquiry Descriptors drawn from the NSES (NRC, 1996). SE1=Science Education Teacher #1; CH1=Chemistry Teacher #1; X = indicates skill present in teacher's investigation postings and/or journal.*

tested hypotheses and extended investigations as they explored cohesive forces. For example, one teacher recorded the details of the surface tension experiment in her journal as follows:

A clean, dry penny was placed on a paper towel.... Prediction:...40 drops will be the maximum before the water overflows off of the penny. 47 drops of the water were placed on the penny before the water overflowed....A clean dry toothpick was gently touched to the dome of water. It penetrated the dome of water without causing the water to overflow. Now the toothpick was dipped in liquid dish soap. When the toothpick touched the center of the water, the water flowed off of the penny.

Similarly, SE course participants engaged in a "rusty nail" activity--an open-ended, inquiry-based project. (see Table 1) The project opened with a question about one instructor's eight-year-old pickup truck. After recently moving to Massachusetts, she found a hole, crusted with rust about the size of a quarter, on the bed of her truck. She hadn't seen a sign of rust until that fall. As she showed them pictures of the hole in question, she questioned the group: "What caused this rusted hole? What is rust? What is the prognosis for my truck?"

After some discussion of what they knew about rust, the group was charged with making two nails, devoid of any protective covering, as rusty as possible. They designed their investigations, explored their questions around rust, submitted three Lab Reports based on their experiments, and recorded information about their investigations in a Science Journal. For example, Michelle noted in her journal her initial thoughts and questions about rust:

...road salt "eats" cars, fine sand and salty moist air is why we are advised to hose our cars off after a trip to the seacoast. What is the role of salt? How does it participate in the rust process?

I think if I scratch up one the nails to expose more surfaces to the salt water, that will encourage faster rusting. What am I going to scratch that small surface with? Another nail?

Why am I thinking about this "expose more surfaces of the nail?" What do I think rust is and why do I think smaller or thinner pieces of metal will get rustier, or rust faster than a denser piece of metal?

Why do I think that these nails are made of the same metal that cars are made of? What kind of metals are nails made of? And cars?

If salt water will rust metal faster than fresh water, why are ships and boats that are made of metal allowed on the oceans?

Michelle designed and conducted her investigations. Below are excerpts from her journal, where she describes her investigation, observations, and reflects on her results.

Lab 1: Rust is a Sign of Neglect

I put each nail in its own (expendable) saucer. One on the kitchen counter, one on the porch. Each nail sat in a puddle of water. In a few hours (If I knew rust happened so fast I would have measured the time then!), [the] inside nail was "bleeding rust," outside nail was encased in ice, not rusty.

Cold preserves nail. Warmth encourages evaporation, which seems to promote rust.

Have to redo [the] experiment for time measure.

...Outside nail is slightly rusted, but took longer because snow and ice preserved nail. Inside nail needs more water added constantly. It is very cruddy, flaky, big rust. The rust on outside nail is fine textured, not nearly as extensive.

Rust can be a sign of careful, deliberate care to cause rust. I want to think about "why salt" before I do salt. Also need more nails.

Lab 2: Only One Variable, and Only One Nail

Meantime, my experiment was limited to observations on the effect of temperature....

The problem is not having any way to measure, to explain how rusty the nails are or are not. This only: When cold tap water was used, the rust process was slower. (How slow?) The rust was finer. (How fine?) NOW: After the boiling water treatment, the rust is flakier. (How much flakier?) I BELIEVE the nail looks skinnier. (A caliper, do I need a caliper?)

Based on an analysis of discussion postings, online journals, and survey data, teachers painted a picture of learning science through "doing." They discussed the benefits of conducting experiments individually and being able to confer with their online classmates. Through their investigations, teachers made predictions, designed and redesigned experiments, made observations, and ultimately drew conclusions about their studies. The ways in which the teachers engaged in the processes of scientific inquiry were evident in their postings and in the reports they recorded in their journals.

In a post-course survey, teachers enrolled in the chemistry course strongly indicated that inquiry methods were used. One teacher commented:

Almost every lesson ... [was] inquiry based. Every week we had several labs or activities that involved concepts demonstrated through inquiry-based learning.

Another teacher wrote:

We collected data, analyzed it, [and] further investigated with questions of "wonder what would happen if.".. [I]n some cases we were asked to search for answer[s] to questions that could only be found through investigation.

Another noted:

There were questions posed, and we followed activity directions, hypothesized on what we believed would happen, carr[ied] out the experiments, gather[ed] data, [made] conclusions, and share[d] our results, and sometimes [went] back to...investigate...

A fourth stated: "Yes, [this] course definitely encouraged inquiry methods throughout all the lessons. As students, we were encouraged to ask questions, to develop an understanding and discuss our answers with our peers." She provided this example:

You often do not think about what is happening to the atomic particles when you begin to smell a substance from the other side of a room, such as a turkey cooking in the oven. Yet, we did question the atomic movements of particles through the wall of a balloon. We asked questions and tested liquids such as vanilla extract and lemon juice among others in which we enclosed inside the sealed balloon. We could not see the substances dripping out anywhere, yet we could begin to smell the liquids within a few minutes.. ...[W]e explored and analyzed the properties and behaviors of gases that we would then use as a foundation for new questions we asked throughout the course.

When asked how the activity-based nature of their course impacted the learning of science content, one teacher responded that "inquiry based activities really gave me a greater depth of understanding." Another teacher commented: "the activities made me slow down a bit and take a more hands-on versus theory-only approach to learning science content." A third teacher stated: "I have been able to remember most of the content because of the hands-on activities that allowed me to experience the learning."

Promoting Online Discourse and Reflection

Online discussions played an important role in supporting teachers' learning. Early in each course, teachers were asked to provide introductions and share information about their teaching contexts, course goals, and personal interests. Instructors fostered student interaction during each class session by asking teachers to respond to the posts of at least one or two other course participants. All postings and responses to posts were graded with a rubric.

As we analyzed their postings, we noted that teachers engaged with course concepts as they interacted online; participants' exchanges provided guidance and clarification as teachers explored course topics, experienced new approaches, and considered ways to teach these concepts to their students; and teachers shared perspectives and approaches as they built on and extended their knowledge.

Teachers noted the benefits of discussing experimental findings with their classmates. During these online conversations, the teachers provided a summary of their experiments, formulated explanations based on their observations, responded to specific questions, and commented on their classmates' postings. On a post-course survey one teacher commented:

The inquiry was all based on hands-on activities. It was an awesome experience to have everyone replicating the same experiment, and then actively discuss all of the viewpoints...[This experience gave] a real sense of not leaving any stone unturned.

In the excerpt below, four teachers in the chemistry course discussed what happened when vanilla was placed inside of an inflated balloon. This conversation highlights the constructive nature of online discussions as these teachers explored the states of matter:

Anna: *...Matter changes states. I put a liquid into the balloon. A gas was smelled outside the balloon. ...I can fill a balloon up with water and it doesn't leak, but when I put vanilla or lemon extracts in the balloon I could clearly smell them. The gas molecules of those substances were fitting through the gaps in the balloon material. I put liquid extracts into the balloons, but clearly a gas or vapor was outside the balloon as part of the air mass. This is evidence of multiple states of matter of the same substance.*

Brent: *...I liked your example. ...If the air molecules that were used to fill the balloon can't fit through the openings of the balloon that allowed gas molecules of the vanilla extract to come out of, then why is it that eventually a balloon will slightly deflate?...*

Anna: *While I was writing that I had the same question. Perhaps there is greater attraction between liquid molecules than gas molecules. They tend to stick together and don't slip through the holes while gas molecules are solitary creatures and can slip through?*

Chris: *I, too, was wondering the same thing... I am thinking that the air molecules must escape over time. Perhaps this all relates to pressure, temperature, etc?...*

Diana: *...I think air molecules do get out over time, because they find their way between all of the nooks and crannies eventually...*

These online discussions were an integral part of teachers' investigations. In a post-course survey, one teacher from the chemistry course described the learning approach as "...read, discuss, experiment, discuss, report." Another teacher commented on the benefit of online discussions:

The best part was the open discussions with other classmates. Someone always added a different perspective. It was a great way of learning, sharing ideas, and collaborating on our conclusions.

Based on the interview data, teachers in the SE course also noted that they were able to communicate ideas, feelings, and experiences and provide others with alternative ideas, critique, suggestions, questions, and encouragement. One teacher commented:

The actual back and forth was useful getting people's comments - that was interesting....There was a diversity of experience and background among the [teachers].. Some were perhaps asking more sophisticated questions...but they were all... interesting questions. Interacting back and forth... you make connections between them....They would bring up questions you might not had thought of - that was useful....By encouraging everybody to comment on everyone else's questions and statements, encouraged people to really look at other people's way of looking at things and that doesn't happen often among peers and which you wouldn't have necessarily done if [you] were in a regular (face-to-face) class.

Another stated:

I found myself asking questions of everybody..".. can you give me a really good idea" because I want to tap into the best ideas of everybody....I felt more grounded because some their stuff wasn't working, or they had questions and I said "Oh thank God," so now I can do something silly like that too and not feel bad. So I would say it was the inquiry part of that piece of it that was really important.

A third teacher noted:

I think everyone ran into problems...the encouragement made such a big difference. You don't necessarily find that in every course that you take, and the atmosphere that was created was one of cooperation more than competition....I felt that really strong.

Michelle's investigation experiences provided a good example of the problem-solving discourse. She described her struggle to quantify the rusting of the nails. In her journal, she noted how, missing this element, she found her investigation scientifically wanting. She stated:

All the teachers are designing their own experiments and doing them, and commenting on each other's variables, and controls, observations, conclusions, and questions. These are mine, so far. They are not what I would call the most scientific...some of the other teachers' experiments are more scientific than mine. Why? Because I didn't measure measurable things, like how much time, how much water, the exact temperature.

Course participants and instructors responded to Michelle's queries about how to measure the rust on her nails. Below are their comments:

Mary: *I, too, am looking for a way to quantify the degree of corrosion; a fine scale before and after? scraping off the rust and weighing that? I am open to ideas. I am currently estimating it crudely by the depth of the sediment on the bottom of the jars, but that is pretty rough.....I, too, have trouble trust-*

ing my experimental technique and design. I keep thinking of things I forgot to control...).

Carrie: *You say you have a problem, not being able to measure or describe how rusty a nail is... Is this a good next idea to focus on as you plan your further rusty nail investigations? Could you create a descriptive scale upon which you could rate your nails? An interesting idea!*

David: *Carrie's suggestion is a good one. How can we set up a scale of "rustiness."*

Michelle: *Thank you for your comments! It occurred to me the other day that weighing the nail would be a measure....I like Mary's idea of measuring the rust scrapings. I agree with you all that some kind of design has to be made to depict and communicate, to let us rusters know, as well as convey our beliefs to each other.*

Mary: *Re: rust weight, I don't know where to find a scale with fine enough measurement capability, though maybe they have one at the HS.*

In her journal, Michelle recorded how she followed through on several of the suggestions and some new ideas of her own. She drew pictures with the aid of computer graphics. She took photos, but found them misleading. She also devised a "verbal scale for rustiness" to define the types/stages of rust (e.g., "rust blood or powder: oozes from immersed nail, sinks to bottom of container; small flakes: float away, specks are visible; ash stage: nail can be broken or crumbled by bare hand").

Teachers ably questioned and self-critiqued their investigative process. This was a component of each participant's inquiry process whether it involved reassessing preconceived ideas, acknowledging the need for more study, or rethinking their investigative design. The interaction between teachers appeared to facilitate their inquiry process. Course participants related that, although initially they may have had some uneasiness posting their ideas for everyone else to see, they felt that the course format provided a safe place to share. One teacher pointed out:

[The course] was a little bit scary, then it was fun...sometimes I'm very critical [of] my own writings....Sometimes I would go...online and see, I think it was Mary, she was always first and always thorough. I would be like "errrrrr!" But what was good about it was that I thought she was very bold. I think what I'm trying to say is that I was scared and hesitant to post something not knowing how it was going to look. "Is this what they're looking for?" Some people, myself included, will inevitably sound better on paper after...hav[ing a] chance to edit, think it through.... So this way would be easier for me in terms of... [a] style of communicating.

Another teacher pointed out:

You know it was motivating to go and see...if anybody answered my thoughts...when I posted a message. I always went to see if anybody answered. Maybe now that I'm not so afraid of science, I might enjoy an adult science course, but I really don't know. I don't think I would ever sign up for one for fear that I would be the least knowledgeable person in class. Over the Internet, it was more comforting to know that I was just out there in Cyberspace, and nobody really knew me, and I could type in anything....The anonymity was comforting. (laughter) I could ask a silly question, or I could ask whatever and write it... and...I wouldn't have to see their faces in disbelief that [I] asked this question. And people wrote nice [things] back, "Oh, I know what you mean." That was a nice part of it.

One participant described how the online setting differed from a traditional course:

Frequently, you might interact with people you know in the class but you don't have to interact with everyone in a class...unless you're in a small seminar type.... It's usually...more the teacher has set-up the issues you're going to discuss....[There was] certainly more [peer communication] than a lecture course --even more than a lab.

In addition, the online format allowed more space to talk. A teacher said:

When you post your responses - it's different than sitting in a classroom setting where not everyone might get a chance to speak.... Some people in a group discussion want to talk to all the time, not that they don't want anyone else to talk. So you have that kind of group dynamics that the design set-up eliminated that which can be problem for all teachers and learners.

In the online setting, there were no interruptions from instructors and other students; there was no need to "share air time" with others. There was no limit to how much one could report about whatever, and all "voices" could be heard.

Bridging Science Content and Teaching Practice

In the online courses, teachers actively explored science content through experimentation and scientific inquiry and reflected on these experiences individually and with their peers. As teachers engaged as learners, they gained first-hand experiences in inquiry-based learning and teaching. Teachers reported how they could then draw upon these experiences when designing comparable learning opportunities for their students (see Table 2).

When asked whether the chemistry course influenced their teaching methods and/or if they used any of the experiments in their classrooms, one teacher commented: "I feel more secure in teaching chemistry concepts. I ask students to hypothesize about what would happen if...?" A second teacher responded: "...the course gave me a broader depth of understanding, which enhanced

Table 2. Course linkages to teachers' practices

Teaching Link \ Participant	SE1	SE2	SE3	SE4	SE5	SE6	CH1	CH2	CH3	CH4	CH5	CH6	CH7	CH8	CH9
Indicates increased confidence in teaching science concepts and/or inquiry processes.	X		X				X	X	X	X	X	X	X	X	X
Indicates increased content understanding, which enhances teaching.		X					X	X	X	X	X	X	X	X	X
Reports use of course activities in science classroom.	X	X	X		X			X				X	X		
Indicates incorporation of inquiry processes (making predictions, asking questions, designing investigations, etc.) in classroom teaching.	X	X	X	X	X			X	X				X		
Provides students with more autonomy in their science investigations.	X	X	X		X			X							
Incorporates more opportunities for students to write about their science activity.	X	X		X	X										
Incorporates more opportunities for students to talk about their science activity.	X				X										

Note: SE1=Science Education Teacher #1; CH1=Chemistry Teacher #1; X = indicates teacher's linkage of course to teaching practices; shaded boxes = indicates survey data not submitted.

my teaching" and, also, "The activities from the online course added several more quick lessons to my curriculum."

A third teacher shared the following:

One change that I have definitely added since taking the Chemistry course is having the students "make predictions." I often remind the students how important it is that everything should be constant in their experiments except for the testing condition, in which this chemistry course reinforced. ...[In addition] I have modified some of the activities. One activity we did during the course was to determine the properties and behavior of atoms and molecules using food coloring that we added to warm and cold water. ...I also use the activity in which two pieces of tape are temporarily taped to the same surface. Then they are removed and show how they repel each other. Students again make predictions before and during the various steps of the demonstration.

The following teacher commented on his recent classroom experience with "Making 'Sense' of Surface Tension":

I actually used this activity the other day as a kind of fun activity to settle in the class on a Friday. ... Many of them have done this experiment before, however most were quite surprised. We talked about surface tension and cohesive forces, and I mentioned to them how this will relate to physical properties when we study them in about one week.

Importantly, teachers came to see how active learning is critical to students' gleaning of new knowledge and understandings. Although the overall topic of study for the SE project (rust) was predetermined by the instructors, the teachers had a great deal of say and ownership within the project. Teachers decided what questions to explore, how to design for their experiments, what was important to observe, what data to collect and how to go about collecting it, how to resolve problems, and how to redesign. They also interpreted their data and constructed their understandings based on the results of their investigations.

One teacher noted that the investigation design aspect was the most challenging, and yet, it helped her see the openness of science and how she could be more open-ended in her teaching. She said:

I felt comfortable...to come in and say, okay, this is the question. What are all the different ways we can look at the question. Feeling like there were no tidy boxes or correct answers. The most important thing was the inquiry - to get kids interested and stimulate [their] thinking....I think I'm much more open ended. I was always open-ended - it's kind of my personality, but I think I'm much more open-ended now.

Another teacher described how the students in her fifth-grade class engaged in the rusty nail activity with her.

The fifth graders... rusted nails....did [Internet] searches on rust, and...learned that there were businesses that were very actively interested in preventing rust. We...[made] a list [of] where the children had observed rust through the weekend – that was their homework. At the very, very end...the fifth grade had... made up their own experiments – what they wanted to do was to see what substance would rust nails faster or retard rust and so on....

She explains:

They don't want to hear about variables. I partake in experimenting with them, explaining why I am using two bowls, one with salt and one w/ out salt. Someone adds a galvanized nail to each of my bowls...The galvanized never rusted--yet. Although we had talked about galvanization, and I showed info on a web site about it, one student was intrigued by these nails and asked," Why didn't those get rusty?"

One child showed me that he was going to rust a nail quicker by coating it with Vaseline. "What will the Vaseline do?" I asked. "Add moisture and keep the moisture in so more rust will happen." The next week he was SO EXCITED. "I made a discovery!" he crowed. "Vaseline prevents rust!!!!!"

This teacher indicated that, as a result of her experience, she:

... will be less rigid....Instead of dealing with vocabulary first, the vocabulary will come last. Instead of telling students what they will find, asking them what they found

Another teacher indicated that she had "a new way of looking at science...the importance of asking the right question at the right time... and [students] being able to write a lot about what they're learning and just the whole thinking component." She indicated that she had become "patient with kids and really let them come to their own meanings...come up with questions and their own kind of inquiries." This was evidenced in a plant unit that began with her asking her class:

...to collaborate in groups of four to come up with things they already knew about plants and what makes a plant grow, what they thought they knew about plants, and what they would like to find out about plants...The kids were excited to get going. This list became the focus of several weeks of inquiry as we began our study.

The SE course participants had autonomy in the rusty nail activity. Their questions, ideas, interests, and skills were acknowledged, addressed, and built upon; they were able to set goals and plan next steps in their investigations; and they explained and justified their work to themselves and others. Their experience influenced several of them to provide students with a voice in their science learning. As one teacher noted, it was important to "[let] the children have the reigns more."

DISCUSSION

Scientific Inquiry Online

At the forefront of the *Next Generation Science Standards (NGSS)* is the view that learners "cannot fully understand scientific ideas without engaging in the practices of inquiry and the discourses by which these ideas are developed and refined" (NGSS, 2013, p. 48). Thus, it is critical to not only 'know' science content but to be able to use "understanding to investigate the natural world through the practices of science inquiry" (NGSS, 2013, p. 48).

Evidence from this study shows that hands-on, scientific inquiry can be designed for the online environment. Teachers experimented with locally-purchased household materials and with additional equipment and materials acquired from a science supply company. During their investigations, teachers explored research questions, designed procedures, collected and analyzed data, and reported findings.

Meaningful and autonomous science inquiry was key to teachers' participation. They had much to report regarding the questions asked; the design of their investigations; and the critique of their exploratory processes and thinking. This provided them with a science that was authentic and constructed. Teachers also came to appreciate the value of scientific practices such as making predictions and observations, engaging in open-ended investigations, and synthesizing and communicating findings.

As a result of these experiences, several teachers reported being more receptive to using such approaches in their classrooms (see Table 2). Several noted the value of asking the "right" questions, engaging students in making predictions and allowing them more autonomy with their science. Many described taking course activities directly to their students and others reported engaging students in the same research explorations they were investigating in their course.

Scientific Discourse Online

The online environment provides an ideal forum for engaging teachers in science content and teaching practice and providing time and space for thought, discussion, and reflection. In addition, in these courses, the online interactions between teachers were an integral part of their scientific investigations.

The course structure provided participants with more space for talk than the traditional classroom context and, for some, a more comfortable space. All voices were heard since learners could equally contribute to the discussions. In addition, the anonymity of the online environment, created, for many teachers, a safer setting to ask questions; to share ideas, experiences, and difficulties; and to provide support, critique, and suggestions.

These online settings were rich in science talk and were cooperative and constructive. This provided an inviting setting for participants to share their inquiry process, read others' science doings, and reflect upon their peers' explorations and their own. Participants found the setting to be supportive and critical to their learning process.

Several approaches can be used when designing online forums based on size (whole class, small group, or individual journals) and focus (science content, teaching practice, or both). The instructor's efforts are key to enabling course participants to build social connections; requiring teachers to respond to classmates' posts appears important. As teachers come to recognize the support of their peers, they often become self-motivated to review, analyze, and discuss their peers' contributions.

Engaging, productive questions (Eltsgeest, 2001) are key to launching learners into reflection about natural phenomena and observations of investigations that they or others have initiated. These can begin with the simplest questions (e.g. "What do you notice? What happens?") to those that involve measurement and comparison. "What would happen if" questions (e.g. "What would happen if you placed the nails in boiling hot water?") are of "great value...to explore the properties of unfamiliar materials, living or nonliving, of forces at work, and of small events taking place" (p. 39). Problem-posing questions set up real situations that, after a bit of exploration with materials or phenomena, learners can engage in enthusiastically, much like asking the SE teachers to make a nail as rusty as possible. Lastly, reasoning questions, such as "Why do you think…?," engage learners in looking at evidence to make sense of an event and/or experience. Instead of focusing on the "right answer," the emphasis is on building on ideas that individual students bring to the course. Such a questioning atmosphere is counter to the traditional science classroom where there is a right answer, usually held by the instructor. In contrast, learners engage in sharing ideas and questioning and, together, build knowledge and understanding.

Learners, whether children or their teachers, need to focus not only on "what we know" but also '"how do we know what we know," "why do we believe what we know," and "what should we do to find out?" (Sampson, Grooms, & Walker, 2010, p. 253). Having learners construct an explanation for what they observe and/or experience requires them to "clarify their thinking, to generate examples, to recognize the need for additional information, and to monitor and repair the gaps in their understanding" (p. 253). Like Michelle's experience, "[i]t also requires [them] to learn and use the criteria by which these explanations...will be judged or evaluated" (p. 253).

Furthermore, student-student interchange is vital as "socially meaningful activity is prerequisite to cognitive development" and critical thought (Waltonen-Moore, Stuart, & Newton (2006), p. 305). Course instructors play an important role in designing questions and structures that launch students' inquiries; encourage them to share their learning processes, observations, and questions; and foster critical dialogue that leads to student understanding.

Teachers valued these in-depth, online discussion opportunities. Several noted that they provided increased opportunities for student science talk. However, none discussed ways to incorporate online discourse in their classrooms. A key question to explore would be how teachers might incorporate approaches and technologies in their classrooms that increase students' science dialogue?

Limitations of the Study

In this study we sought to describe the characteristics of two online courses from the perspectives of elementary and middle school teachers, who were participants. Specifically, how did these online courses engage teachers in inquiry, meaningful discourse, and make connections to their classroom practice, and of what value was that to the course participants? We did not seek to report a causal relationship between a particular "treatment" in each course and subsequent "results." External methods were not employed to determine participants' learning.

The study did not include observations of the participants' classrooms. Researchers relied on teachers' descriptions of methodologies and lessons they implemented as reported in discussions and through surveys and interviews.

The second author of this study was one of the instructors for the SE course. To address potential effect, interviews were conducted after the course was completed. A researcher, not associated with the course, conducted one round of interviews. A graduate student, who assisted with the course, conducted a second round.

Finally, given the distributed nature of online courses, it was difficult for the researchers to "observe" teachers engaged in certain course activities. For example, as they conducted experiments and reported their findings, we analyzed teachers' descriptions of their participation in these activities as reported in discussion forums, journals, surveys, and interviews. While a teacher might not have described a specific inquiry process, this does not mean that it did not occur; the reverse should also be considered. Ways to address this in the future might be to ask teachers to video-record themselves as they conduct experiments, or to explicitly ask them if they engaged in specific inquiry processes and to request examples.

FUTURE RESEARCH DIRECTIONS

As we reflect upon these study results and the existing scholarship in online PD, future research would prove useful in supporting the growth and development of rigorous online PD for science teachers.

One research direction could focus on impacts on teachers' practices. To what extent do teachers, who have engaged in online PD, implement instructional changes? How do these changes impact student learning? As noted by Desimone (2011), PD should provide opportunities for teachers to improve the content of their instruction and their pedagogical methods. Building on this theme, what types of online PD activities are most beneficial to teachers' learning of subject matter and pedagogical approaches? The online environment provides multiple ways for teachers to interact with content and each other as they explore concepts, approaches, and resources. Identifying "best practices" for the online PD of science teachers would be greatly beneficial.

A second research direction could explore online instructional approaches. What discourse practices, initiated by instructors, are most beneficial to promoting teachers' subject matter and pedagogical learning in online science PD? Are open-ended simulations or the exploration of data-sets effective approaches to engage teachers in inquiry and support their learning of subject matter and pedagogical learning? By exploring the interactions between instructor-teacher, teacher-teacher, and teacher-content, these findings would greatly inform online PD instructors and developers as they design course assignments, discussions, and activities.

Finally, a third research direction could focus on changes in teachers' attitudes and beliefs. How do teachers' confidence levels for teaching specific subject matter change following engagement in online PD? How do teachers' views of the nature of science and scientific inquiry change after engaging in science practices in online PD? How does online PD impact teachers' willingness to use collaborative instructional approaches and web-based technologies? Again, these findings would greatly inform online course developers and instructors, while also providing insights into how online PD impacts teachers' beliefs and practices.

Such research directions would greatly inform teacher educators, instructional designers, administrators, and key stakeholders, not just in science education but in other PD fields as well. These findings would also provide support for online PD, strengthen learning opportunities for teachers, and lead to new research directions in the field of science teacher PD.

RECOMMENDATIONS AND CONCLUSION

By overcoming time and distance barriers, online programs can provide teachers with access to PD who otherwise are unable to engage in such opportunities. In many cases, online coursework may be the only option for extending their subject and pedagogical knowledge.

In this study, we show that teachers benefit from engaging with course concepts as learners as they conducted experiments, discussed findings, and proposed explanations. This approach provided teachers with first-hand knowledge of what their students would experience in similar conceptual explorations. Also, since these teachers were working from a distance, they had to conduct their own experiments and could not rely on the "expertise" of a lab partner by their side. As a result, each teacher was fully engaged in the experimental process.

As educators design experiments for online science PD, they need to provide opportunities for "guided inquiry"--where participants conduct an experiment within certain parameters and design an extension based on their observations. Online PD should also engage teachers in open-ended inquiry--where participants identify questions and test variables from the start. Both forms of scientific inquiry would provide teachers with concrete, first-hand experiences learning science content and processes.

Although the teachers in this study experimented individually, they were part of a community of teachers who engaged in the same processes, and who served as a resource as they discussed findings, debated explanations, and proposed extensions. These teachers were supportive and readily offered suggestions when a classmate had questions. They formed strong communities as they explored scientific concepts and shared findings, questions, knowledge, and resources.

In working towards building communities, teacher educators should design online PD that includes multiple formats for interaction (whole class discussions, small group forums, individual online journals). These provide opportunities for participants to articulate and reflect on their learning. Discussions should begin with productive questions that encourage prediction, observation, reflection, thought and a range of responses. Journals should provide opportunities to reflect and expand on course activities, experiences, and interactions, and to apply their learning in the context of their classrooms. The teachers in this study greatly benefited from such exchanges; these interactions also provided the instructors with information regarding the teachers' explorations and perceptions.

Finally, as is the case with all PD, it is important to link teachers' content learning and experiences with instructional practices in their classrooms. The online courses studied for this chapter provided teachers with opportunities to explore new topics and consider relevant teaching approaches. The results, shared online, provided a rich set of resources that teachers could draw from when developing learning activities. Since these online interactions were text-based, this information could be readily saved and referenced at a later time.

In sum, this study's themes – inquiry-based teaching and learning, online discourse and reflection, and linking theory with practice – closely align with the standards for the PD of science teachers as outlined in the *NSES* (NRC, 1996). These standards state that PD programs should involve teachers in actively investigating phenomena that can be studied scientifically; build on the teacher's current science understanding, ability, and attitudes; incorporate reflection on the process and outcomes of understanding science through inquiry; encourage and support collaboration among teachers; make connections between science and science teaching; build on teachers' current knowledge of science content, teaching, and learning; model and guide science teaching practice; and provide opportunities for reflection, feedback, and support during PD activities.

The mix of individual investigation, group interactions, and individual reflection in these two online PD courses provided teachers with many opportunities to explore and engage in course concepts and reflect on their teaching practices. Upon completion of their studies, teachers reported that these online courses were applicable to their classroom curriculum; they gained new insights into the teaching and learning of science that will improve their teaching; and they expected to use or adapt inquiry-based activities from one or more of their online courses.

AUTHOR NOTE

This research was funded in part by a grant from the National Science Foundation 0243536.

REFERENCES

Asbell-Clarke, J., & Rowe, E. (2007). Learning science online: A descriptive study of online science courses for teachers. *Journal of Asynchronous Learning Networks*, *11*(3), 95–121.

Capps, D., Crawford, B., & Constas, M. (2012). A review of empirical literature on inquiry professional development: Alignment with best practices and a critique of the findings. *Journal of Science Teacher Education*, *23*(3), 291–318. doi:10.1007/s10972-012-9275-2

Chickering, A. W., & Gamson, Z. F. (1999). Development and adaptations of the seven principles for good practice in undergraduate education. *New Directions for Teaching and Learning*, *1999*(80), 75–81. doi:10.1002/tl.8006

Clary, R. M., & Wandersee, J. H. (2009). Can teachers learn in an online environment? *Kappa Delta Pi Record*, *46*(1), 34–38. doi:10.1080/00228958.2009.10516689

Davis, K. S., & Snyder, W. (2012). Fostering science education in an online environment: Are we there yet? *Journal of College Science Teaching*, *42*(2), 24–31.

Department of Education. (2000). *Before it's too late: A report to the nation from the National Commission on Mathematics and Science Teaching for the 21st century*. Author.

Desimone, L. M. (2011). A primer on effective professional development. *Phi Delta Kappan*, *92*(6), 68–71. doi:10.1177/003172171109200616

Elstgeest, J. (2001). The right question at the right time. In W. Harlen (Ed.), *Primary science...taking the plunge. How to teach science for effectively for ages 5 to 12* (pp. 37–46). Portsmouth, NH: Heinemann.

Harlen, W., & Doubler, S. J. (2004). Can teachers learn through enquiry on-line? Studying professional development in science delivered on-line and on-campus. *International Journal of Science Education*, *26*(10), 1247–1267. doi:10.1080/0950069042000177253

Lee, J., Bray, M., Carter-Wells, J., Glaeser, B., Ivers, K., & Street, C. (2004). *Discovering the meaning of community in an online master's degree program*. Paper presented at the meeting of the Association for Educational Communications and Technology, Chicago, IL.

Loucks-Horsley, S., Stiles, K. E., Love, N., & Hewson, P. W. (2010). *Designing professional development for teachers of science and mathematics* (3rd ed.). Thousand Oaks, CA: Corwin Press.

McNall Krall, R., Straley, J. P., Shafer, S. A., & Osborn, J. L. (2009). Hands-on at a distance: Evaluation of a temperature and heat distance learning course. *Journal of Science Education and Technology*, *18*(2), 173–186. doi:10.1007/s10956-008-9142-2

Merriam, S. B. (1998). *Qualitative research and case study applications in education*. San Francisco, CA: Jossey-Bass Publishers.

National Research Council. (1996). *National Science Education Standards*. Washington, D.C.: National Academies Press.

National Science Board. (2010). *Preparing the next generation of STEM innovators: Identifying and developing our nation's human capital*. Washington, D.C.: National Science Foundation.

Next Generation Science Standards. (2013). *Next Generation Science Standards: For states, by states*. Washington, D.C.: National Academies Press.

Offenholley, K. (2006). Successful discussions in online mathematics and science courses. *Online Classroom*, *7*, 1–3.

Olson, S., & Loucks-Horsley, S. (2000). *Inquiry and the National Science Education Standards: A guide for teaching and learning*. Washington, D.C.: National Academies Press.

Sampson, V., Grooms, J., & Walker, J. P. (2011). Argument-driven inquiry as a way to help students learn how to participate in scientific argumentation and craft written arguments: An exploratory study. *Science Education*, *95*(2), 217–257. doi:10.1002/sce.20421

Waltonen-Moore, S., Stuart, D., & Newton, E. (2006). From virtual strangers to a cohesive online learning community: The evolution of online group development in a professional development course. *Journal of Technology and Teacher Education*, *14*(2), 287–311.

Wilson, S., Ball, D. L., Bryk, A., Figlio, D., Grossman, P., Irvine, J., et al. (2009). *National Academy of Education Teacher Quality White Paper*. Washington, DC: National Academy of Education.

Yin, R. K. (2009). *Case study research: Design and methods*. Thousand Oaks, CA: SAGE Publications, Inc.

ADDITIONAL READING

Allen, I. E., & Seaman, J. (2013). *Changing course: Ten years of tracking online education in the United States*. Sloan Consortium, Babson Survey Research Group, Pearson.

Annetta, L., Klesath, M., & Meyer, J. (2009). Taking science online: Evaluating presence and immersion through a laboratory experience in a virtual learning environment for entomology students. *Journal of College Science Teaching*, *39*(1), 27–33.

Bannier, B. (2010). Motivating and assisting adult, online chemistry students: A review of the literature. *Journal of Science Education and Technology*, *19*(3), 215–236. doi:10.1007/s10956-009-9195-x

Baron, J. D., & McKay, M. M. (2001). Designing and delivering an online course for K-12 educators. *T.H.E. Journal*, *28*(9), 68–75.

Birman, B. F., Desimone, L. M., & Porter, A. C. (2000). Designing professional development that works. *Educational Leadership*, *57*(8), 28–33.

Brown, A., & Green, T. (2003). Showing up to class in pajamas (or less!): The fantasies and realities of on-line professional development courses for teachers. *Clearing House (Menasha, Wis.)*, *76*(3), 148–151. doi:10.1080/00098650309601992

Brown, R. E. (2001). The process of community building in distance learning classes. *Journal of Asynchronous Learning Networks*, *5*(2), 18–35.

Cady, J., & Rearden, K. (2009). Delivering online professional development in mathematics to rural educators. *Journal of Technology and Teacher Education*, *17*(3), 281–298.

Cancilla, D. A., & Albon, S. P. (2008). Reflections from the moving the laboratory online workshops: Emerging themes. *Journal of Asynchronous Learning Networks*, *12*(3-4), 53–59.

Cavanaugh, C., & Dawson, K. (2010). Design of online professional development in science content and pedagogy: A pilot study in Florida. *Journal of Science Education and Technology*, *19*(5), 438–446. doi:10.1007/s10956-010-9210-2

Chitanana, L. (2012). A constructivist approach to the design and delivery of an online professional development course: A case of the iEARN online course. *International Journal of Instruction*, *5*(1), 23–48.

Christopher, M. M., Thomas, J. A., & Tallent-Runnels, M. (2004). Raising the bar: Encouraging high level thinking in online discussion forums. *Roeper Review*, *26*(3), 166–171. doi:10.1080/02783190409554262

Clary, R. M., & Wandersee, J. H. (2010). Virtual field exercises in the online classroom: Practicing science teachers' perceptions of effectiveness, best practices, and implementation. *Journal of College Science Teaching*, *39*(4), 50–58.

Dede, C., Jass Ketelhut, D., Whitehouse, P., Breit, L., & McCloskey, E. M. (2008). A research agenda for online teacher professional development. *Journal of Teacher Education*, *60*(1), 8–19. doi:10.1177/0022487108327554

Desimone, L. M. (2009). Improving impact studies of teachers' professional development: Toward better conceptualizations and measures. *Educational Researcher*, *38*(3), 181–199. doi:10.3102/0013189X08331140

Desimone, L. M. (2011). A primer on effective professional development. *Phi Delta Kappan*, *92*(6), 68–71. doi:10.1177/003172171109200616

Desimone, L. M., Porter, A. C., Garet, M. S., Yoon, K. S., & Birman, B. F. (2002). Effects of professional development on teacher's instruction: Results from a three-year longitudinal study. *Educational Evaluation and Policy Analysis*, *24*(2), 81–112. doi:10.3102/01623737024002081

Evans, K., & Leinhardt, G. (2008). A cognitive framework for the analysis of online chemistry courses. *Journal of Science Education and Technology*, *17*(1), 100–120. doi:10.1007/s10956-007-9087-x

Feger, S., & Zibit, M. (2005). *The Role of Facilitation in Online Professional Development: Engendering Co-Construction of Knowledge*. Providence, RI: The Education Alliance at Brown University.

Fies, C. (2008). Online jigsaw science inquiry for preservice teachers. *Journal of Computing in Teacher Education*, *24*(3), 85–92.

Fishman, B., Konstantopoulos, S., Kubitskey, B. W., Vath, R., Park, G., Johnson, H., & Edelson, D. C. (2013). Comparing the impact of online and face-to-face professional development in the context of curriculum implementation. *Journal of Teacher Education*, *64*(5), 426–438. doi:10.1177/0022487113494413

Hudson, P. (2006). Examining preservice teachers' involvement in online science education. *Journal of Interactive Learning Research*, *17*(4), 367–383.

Kim, M., & Hannafin, M. (2004). Designing online learning environments to support scientific inquiry. *Quarterly Review of Distance Education*, *5*(1), 1–10.

Lock, J. V. (2006). A new image: Online communities to facilitate teacher professional development. *Journal of Technology and Teacher Education*, *14*(4), 663–678.

Lunsford, E. (2008). Guided inquiry and social collaboration in an online classroom. *Bioscene*, *34*(2), 12–21.

Mawn, M. V., Carrico, P., Charuk, K., Stote, K. S., & Lawrence, B. (2011). Hands-on and online: Scientific explorations through distance learning. *Open Learning*, *26*(2), 135–146.

Means, B., Toyama, Y., Murphy, R., & Baki, M. (2013). The effectiveness of online and blended learning: A meta-analysis of the empirical literature. *Teachers College Record*, *115*(3), 1–47.

National Science Foundation Task Force on Cyberlearning. (2008). *Fostering learning in the networked world: The cyberlearning opportunity and challenge; A 21st century agenda for the National Science Foundation*. Washington, D.C.: National Science Foundation.

Nicholas, H., & Ng, W. (2009). Fostering online social construction of science knowledge with primary pre-service teachers working in virtual teams. *Asia-Pacific Journal of Teacher Education*, *37*(4), 379–398. doi:10.1080/13598660903050336

Parsad, B., & Lewis, L. (2008). *Distance education at postsecondary institutions: 2006-07*. Washington, D.C.: National Center for Education Statistics, Institute of Education Sciences, U.S. Department of Education.

Reeves, J., & Kimbrough, D. (2004). Solving the laboratory dilemma in distance learning general chemistry. *Journal of Asynchronous Learning Networks*, *8*(3), 47–51.

Rowe, E., & Asbell-Clarke, J. (2008). Learning science online: What matters for science teachers? *Journal of Interactive Online Learning*, *7*(2), 75–104.

Sherman, G., Byers, A., & Rapp, S. (2008). Evaluation of online, on-demand science professional development material involving two different implementation models. *Journal of Science Education and Technology*, *17*(1), 19–31. doi:10.1007/s10956-007-9075-1

Sprague, D. (2006). Research agenda for online teacher professional development. *Journal of Technology and Teacher Education*, *14*(4), 657–661.

U.S. Department of Education, Office of Planning, Evaluation, and Policy Development. (2010). *Evaluation of evidence-based practices in online learning: A meta-analysis and review of online learning studies*. Washington, D.C.: U.S. Department of Education.

U.S. Department of Education, Office of the Under Secretary, Planning and Evaluation Service, Elementary and Secondary Education Division. (2000). *Does professional development change teaching practice? Results from a three-year study. Executive summary*. Washington, D.C.: U.S. Department of Education.

Waldrop, M. M. (2013). Education online: The virtual lab. *Nature*, *499*(7458), 268–270. doi:10.1038/499268a PMID:23868243

KEY TERMS AND DEFINITIONS

Online Course: A course delivered via the Internet, providing a means for content delivery, assignment completion, and synchronous and asynchronous communication for learners in different locations.

Online Discourse: Ways of talking in the online environment.

Professional Development: Continuing teacher education in content knowledge and instructional practice.

Reflection: The process of examining one's thoughts in relation to new observations, experiences, and information, and considering how this may shape and influence one's beliefs.

Science Discourse: Ways of talking in the field of science.

Science Education: Study of science teaching and learning.

Scientific Inquiry: Practices that scientists employ as they investigate the world.

This work was previously published in "Exploring the Effectiveness of Online Education in K-12 Environments" edited by Tina L. Heafner, Richard Hartshorne, and Teresa Petty, pages 228-249, copyright 2015 by Information Science Reference (an imprint of IGI Global).

Chapter 33
Implementing the Understanding by Design Framework in Higher Education

Judy Alhamisi
Marygrove College, USA

Blanche Jackson Glimps
Tennessee State University, USA

Chukwunyere E. Okezie
Marygrove College, USA

ABSTRACT

This chapter describes an organizational initiative to develop and implement the Understanding by Design (UbD) curriculum-planning framework to improve learning outcomes for teacher candidates and their students during clinical experiences and in their future classrooms. This case study explores a pedagogical approach that has met with success in working with teacher candidates. The focus is on a narrow range of knowledge, skills, and dispositions related to effective teaching in science education: the ability to design, plan, and implement curriculum. Curriculum design using the Understanding by Design (UbD) Framework is a high priority when moving from simply covering subject matter to ensuring deep understanding. Using "Backward Design" helped many teacher candidates develop skills to plan effective science curriculum, units, and lessons. The experiences of two teacher education programs in building teacher candidates' skills in planning and implementing science education curriculum using the UbD Framework are presented in this case study.

ORGANIZATION BACKGROUND

Colleges and universities, in responding to political, social, economic, and technological pressures, are becoming more responsive to teacher candidates' needs and are more concerned about how well they are being prepared to assume future roles as teachers. Faculty are feeling the pressure to change their teaching strategies by developing rigorous curriculum, lecturing less, making learning environments more interactive, integrating technology into the classroom, and us-

DOI: 10.4018/978-1-4666-7363-2.ch033

ing collaborative learning strategies. This chapter describes two institutions of higher education and their experiences in scaffolding teacher candidates regarding the use of UbD as a framework for unit and lesson planning, including assessment.

These urban institutions are identified as University A and University B. University A, a Historically Black College/University (HBCU), is located in the southern part of the United States and has a long tradition of educating students who have been historically underrepresented. The mission of University A is to build a cadre of graduates who are prepared to conduct scholarly inquiry and research, become life-long learners, and are committed to service. Currently, University A offers bachelor's degree programs, master's programs, and awards doctoral degrees in several disciplines. University A is comprised of eight colleges and schools. The program for preparing teachers, at University A, is located within the College of Education. The program prepares teachers for elementary and secondary classrooms in specific content areas including science. Education faculty teach courses in programs, such as: curriculum and instruction, special education, reading, science education, and math education. Content area specialization for majors in secondary area is provided by faculty in different colleges and schools, including: math, science, history, humanities, music, art, and physical education. Students enrolled in the teacher certification programs take traditional courses to prepare them to design curriculum, develop assessments, engage all learners, and become reflective practitioners. In terms of professional development, University A faculty in the education department do not have access to funds to support attendance at conferences or bring well-known educational researchers and experts to the campus.

University B is a four-year, co-educational private liberal arts college. The institution moved to its current location in 1927 to serve young women who would otherwise be unable to obtain a college degree. While University B has evolved to meet the changing needs of its students throughout its 86-year history in the city, the institution has remained committed to the city and to the education of those disadvantaged by gender, race, economic circumstances, or social limitations. The mission of University B is to educate each student to become intellectually and professionally competent; ensure career flexibility through grounding in the liberal arts; and develop active compassion and commitment. Through excellent teaching in its undergraduate, graduate, and continuing education programs, the university provides a personalized learning environment for its students. The Institution, in an open, caring, nurturing, and friendly environment, provides learning experiences and opportunities for students to demonstrate leadership and develop confidence and self-reliance. The individual and collective excellence for which University B stands will continue to be measured by the quality of its graduates and their successes in serving society.

The Education Department of University B offers undergraduate and graduate programs to prepare students for careers as certified teachers at both the elementary and secondary levels. Perhaps no program at University B has focused more directly on the commitment to enabling positive change than the teacher certification program. Like most small, private liberal arts colleges, teacher training has been a curricular centerpiece, graduating students who staffed and led schools in the city, throughout the metropolitan area, and in many other states. Teachers from University B continue the legacy of service and leadership. University B is committed to professional development, with faculty receiving financial support to attend conferences and workshops. In addition, the Education Department hosts an annual conference that brings nationally known researchers and educators to the campus.

This chapter presents the Understanding by Design (UbD) Framework and its use in preparing teacher candidates in science and other curricular areas. In this framework, learning is about the acquisition of factual knowledge and the fluent patterning of behavior, as well as about making meaning. While these are elements of learning, the "goal is to teach enduring understandings that should be the central focus of curriculum building" (Wang & Allen, 2003, p. 37).The chapter focuses on elaborating steps embedded in UbD that target deep understanding. The goal is to acquaint readers about the UbD framework as an effective tool for developing understanding, particularly in science education. The steps followed to introduce students and guide them through creating science planning using UbD are presented.

SETTING THE STAGE

The National Science Teachers Association (NSTA, 2006) supports strong performance-based science teacher education programs and state licensure standards for teacher candidates for both elementary and secondary schools. The NSTA recommended that teacher education programs should use a curriculum framework based on National Science Education Standards that provide teacher candidates with deep science knowledge and skills at the grade levels for which they are teaching.

UbD challenges teacher candidates to design, develop, and implement units and lessons to assist their students in acquiring science literacy. To achieve this objective, faculty must engage their teacher candidates in the learning process. Craven and Penick (2001) asserted that "students not engaged in the learning process leave with little more than shallow understandings, weak connections between big ideas, trivial knowledge, unchallenged naive conceptions of how the natural world operates, and an inability to apply knowledge in new settings" (para. 2). Although improvements have been found in student performance in science, a need exists to improve science education (National Science Foundation, 2006). According to Craven and Penick (2001), teacher educators must assist their teacher candidates move from planning units and lessons based on traditional curriculum models used to teach and learn science. UbD provides a tool to help achieve this objective.

Grant Wiggins and Jay McTighe (1998, 2013) developed UbD as a curriculum framework for improving student achievement through specific elements, including standards-driven curriculum, instructional design, assessment, and professional development. In practice, UbD uses a three-stage backward planning curriculum design process, a set of design standards with rubrics, and a comprehensive training program to help teachers design, edit, critique, peer-review, share, and improve their lessons and assessments. Teachers, using backward planning, identify essential questions that students must be able to answer by the end of the unit. Jones, Vermette, and Jones (2009) asserted that with the end in mind, "teachers then design the assessment of those understandings, followed by carefully crafted lessons to achieve this set of objectives" (p. 357). Wiggins and McTighe (1998, 2013) identified six facets of understanding within the UbD framework that should be evident in teachers' work with students, including:

- Explaining,
- Interpreting,
- Applying,
- Shifting perspective,
- Empathizing, and
- Self-assessing.

UbD details the three-stage backward planning curriculum design process.

Stage One – Desired Results: This stage considers which transfer and content goals need to be met and what big ideas students should understand. The essential questions that students will explore and address and what knowledge and skills with which students will leave are included in this stage. Table 1, an example of this stage, presents the title and brief lesson summary, produced by a teacher candidate at University B, including state content

Table 1. Title and brief lesson summary including established goals (state standards)

Life Cycle – Apple Trees is a two-week unit that teaches first graders to explore the stages in the life cycle of an apple tree. Students theorize, investigate, and discover how apple seeds sprout to become mature apple trees that reproduce after their own kind. Students engage in various activities and assignments across subject areas within an overarching apple tree cycle/apple theme. **English Language Arts** (ELA) Standards: Common Core State Standards (CCSS). ELA – Literacy.W.1.2; CCSS.ELA – Literacy.W.1.3; CCSS.ELA-Literacy.SL.1.2; CCSS.ELA-Literacy.SL.1.5 Science Standards: S.IP.E1; S.IA.E.1; SRS.E.1; L.OL.E.2 Art Standards: ART.VA.II.1.; ART.VA.III.1.1; ART.VA.V.1.4 Math Standards: N.ME.01.08; N.MR.01.11; D.RE.01.03

standards of a first grade integrated science unit on the Life Cycle of an Apple Tree.

The student teacher reviewed the science standards needed to address the science topic at the first grade level. One set of science standards had to do with the Life Cycle of Apple Trees. The student teacher selected the topic based upon the first grade curriculum, as well as her own interest and motivation to teach this particular topic. She also observed during her clinical experience that students naturally learned across subject areas. As a result, she researched the state and national standards pertaining to the related disciplines of English Language Arts, Literacy, Art, and Math.

Next, the student teacher translated the standards into transfer goals, indicating the student's ability to transfer learning from one context to others:

- Independently, students will be able to use their learning to...
  - Describe the life cycle of a plant – the apple tree. (L.OL.01.21)
  - Plan and conduct simple investigations. (S.IP.01.13)

She determined the Desired Results including what the students should know, understand, and be able to do at the end of the unit. The students will explore "big ideas" and related thought-provoking essential questions throughout the unit. Examples are as follows:

- **Understandings:** Students will understand that...
  - Apple seeds develop into trees.
  - Investigations entail coming up with questions, making predictions on outcomes, and doing an activity to see if the predictions were true and if the original questions were answered.
- **Essential Questions:** What essential questions will be considered?
  - Where do apple trees come from?
  - What can happen if you plant an apple seed?
  - Why are apple trees important?

Specific knowledge and skill statements linked to the standards, understanding and essential questions also are identified:

- **Knowledge:** Students will know...
  - The parts of the life cycle of an apple tree include seed form, sprout, sapling, blossoming tree, and apple producing tree.
  - Apple seeds grow into sprouts, which develop into fruit bearing trees.
- **Skill:** Students will be able to...
  - Construct a plant cycle diagram using the pars given (cut-outs used) or draw them.
  - Label pictures of the stages of the apple seed (a. Seed, b. Planted in dirt, c. Breaking through the soil, d. Small sprout, e. Tall sprout with long roots).

Stage Two – Determine Assessment Evidence: At this stage, the student teacher determined the extent to which students have achieved the desired results in Stage 1. The student teacher decided which performances and products could reveal evidence of understanding, as well as other evidence that could be collected to reflect desired results. When using UbD, assessment evidence must be aligned with the desired results detailed in the first stage. To accomplish this alignment, teachers and curriculum planners need to think like an assessor when designing specific units and lessons. In the Life-Cycle of Apple Trees Unit, the student teacher identified two transfer tasks that students need to complete that show their learning by doing:

- **Transfer Task(s):**
  - Draw a picture diagram of how an apple seed goes to a sprout, to a sapling, to a grown tree that produces apples, to apples falling off the tree, and back to a seed germinating in the ground after the apple has disintegrated, with each part labeled.
  - Plant their own apple seed in small containers of dirt or Miracle Gro, water them daily, and chart their growth over the course of two weeks. They will also write "How To" books on growing an apple seed.
- **Evaluative Criteria:** Performance is judged in terms of…
  - Accuracy of diagram according to the model.
  - Accuracy of Growth Chart according to the model and rubric
- **Sample Performance Tasks and Rubric:** Watering the Seed Charting Growth

3. Meets the standard of using 1 or 2 dropper full or no more than 1 Dixie cup of water to avoid spillage Accurately shows progress of the seed development using pictures and written details of observation according to the model
2. Uses the standard, but does not implement effectively (only using half a Dixie cup amount or not using the dropper's full capacity) Uses some aspect of the model to chart progress, with some aspects omitted
1. Ignores the standard by using other forms of water measurement or overflowing the plant with excess water Does not accurately chart progress or shows unrelated information

Stage Three – Learning Plan: During this stage, the student teacher considered which activities, experiences, and lessons could lead to achievement of desired results (Stage 1) and success with assessment (Stage 2; Wiggins & McTighe, 1998). Graff (2011) believed that "teachers can approach unit planning in any order as long as they aim for coherence of all three components" (p. 155). The acronym WHERETO summarizes the key elements the student teacher considered as she designed the learning plan (Wiggins & McTighe, 2013):

W: Where are we going? Why? What is expected? We are going to learn about where apples come from and how to get more of them. This is important to know because it is part of life science and explains where a source of nutritious food comes from. Students are expected to meet the state standard of being able to understand and describe the life cycle of a plant [apple tree] and produce an apple sprout from scientific experiment of planting and nurturing apple seeds.

H: How will we hook the students? Students will be hooked with lively videos and hands-on experiments, such as seed planting.

E: How will we equip students for expected performances? Students will be provided with appropriate scaffolds including many examples, visual aids, step-by-step instruction, and rubrics to guide them.

R: **How will we rethink or revise?** If planned lessons are not as grade level appropriate or conducive to the culture of the classroom due to limited schema as expected, then lessons and activities will be revised to be accessible to all students and meet them where they are (Zone of Proximal Development). As long as students come away with knowing more than they did before, the unit is successful.

E: **How will students self-evaluate and reflect their learning?** Students will self-evaluate by indicating their feelings and reflections of how they performed or how well they understood via communicating in writing/ written journal entries, showing a thumbs up or happy face for understanding/doing well or a thumbs down for confusion/not doing so well, etc.

T: **How will we tailor learning to varied needs, interests, and learning styles?** Some students are considered low students and need more scaffolding than others. Thus, their lessons can consist more of verbal and pictorial expression. They can also point to pictures when asked about sequence and for labeling assessment. Students who are more visual will benefit from the videos and actual plants being used instead of just earing about the apple tree cycle. The use of dirt and actively planting seeds may appeal to kinesthetic learners.

O: **How will we organize the sequence of learning?** Young students should always start with the simplest understanding (basic knowledge) and build from there. A read-aloud introduction and videos set the stage for the scientific experiment of planting and documenting growth.

During stage three, the student teacher designed lesson plans for each teaching event for the two-week unit. The individual lesson plans addressed the essential questions and performance tasks, and were connected to the goals and assessments for the unit.

The big picture of the UbD approach (Meier, n.d.) illustrated the key instructional design questions, design considerations, filters (design criteria), and what the final design accomplishes. At stage one, the key instructional design question is what is worthy and requiring of understanding. The answer to this question is determined by an examination of national and state standards, as well as student learning outcomes. At this stage, the design criteria are enduring authentic ideas and discipline-based work. Stage two focuses on the evidence of understanding. The design criteria are the six facets of understanding in combination with a continuum of assessment types. Stage three considers what learning experiences and teaching promote understanding, interest, and excellence.

UbD is responsive to the No Child Left Behind legislation that emphasizes research-based programs to influence student achievement positively. UbD is not a program of scope and sequence of skills, as no prescribed teaching activities are delineated. No direct, causal evidence exists on the influence of UbD on student achievement (McTighe & Seif, 2003); however, UbD principles and practices reflect views of learning in cognitive psychology. UbD shifts views on effective learning from drill and practice to understanding and application. Learning, using the UbD framework, is guided by generalized principles that are applicable for most students. Superficial coverage of topics may be ineffective in helping students develop competencies that prepare them for future learning and work (McTighe & Seif, 2003). Feedback is fundamental to learning, with such opportunities often scarce in classrooms. Most standardized assessments measure factual knowledge, and fail to ask if students know when, where, and why to use that knowledge. In designing units and lessons with UbD principles, students can develop deep knowledge of a topic that can be used as a scaffold for future learning.

Constructivism is an approach to learning where prior experiences and current knowledge are integrated to form new understanding. Constructivism has implications for teaching and learning, using discovery, hands-on, experiential, collaborative, project-based, and task-based learning as strategies to improve academic achievement. Backward design, when compared to the traditional way of teaching, is a more constructivist route to teaching. UbD is a form of guided discovery that focuses on authentic learning of content by starting with the finished product and working backward. Teachers' use of creative, innovative, and interesting teaching strategies that relate to real-life situations result in students developing deeper understanding of the subject matter being taught.

UbD is a tool that can be used to facilitate a quality education for all students and is a framework for improving student achievement through standards-driven curriculum development, instructional design, assessment, and professional development. McTighe and Thomas (2003) suggested that for "backward design to work, educators need to identify desired results, analyze multiple sources of data, and determine appropriate action plans" (p. 53). These components of the three stage backward planning process were introduced to students at both universities during the study of UbD.

CASE DESCRIPTION

The current case involves two urban universities and their attempt to introduce teacher candidates to UbD with the goal of stimulating innovation in preparation for their future classrooms. The National Science Teachers Association (NSTA) considers strong, performance-based science teacher education program and science teacher licensure standards to be essential for all science teachers (National Science Teacher Association, 2004). As they prepare teacher candidates for future science classrooms, faculty at both universities must be aware of continuously changing NSTA and state licensure standards.

University A and University B are terms used to distinguish the two institutions of higher education that comprise this case study. The universities each selected different routes for equipping teacher candidates with skills embedded in the UbD Framework. University A followed the traditional model of preparing teacher candidates for science and general education classroom. Consequently, at University A, only graduate students enrolled in a curriculum design course were exposed to UbD. Members of the course included students whose majors represented science education, English education, school psychology, and special education. This course introduced students to several models of curriculum design, including UbD. University A had no institutional support to follow any particular curriculum design paradigm. The choice of the paradigm to which students were introduced depended on the instructor. The faculty member responsible for the course at University A conducted a review of the literature (e.g., research articles, webinars, and podcasts) on different curriculum frameworks before deciding to learn about UbD in preparation for sharing the curriculum framework with students. Course evaluations, using the traditional student course evaluation system, were minimal as the majority of students did not complete the assessment.

At University B, UbD was embedded throughout the courses that students were required to complete in the teacher education sequence. A number of strategies for curriculum, instruction, and assessment had been used at the university for several years before the introduction of UbD as a curriculum framework. As the Education Department faculty began preparing for state and national accreditation, they began developing goals that included implementation of learning outcomes driven curricula and use of different approaches for engaging faculty and teacher candidates to achieve these goals. During the 2007-08 academic year, a university-wide accreditation review was completed by the Higher Learning Commission of the North Central Association. The results identified areas of strengths as well as challenges that needed to be addressed to improve program quality and academic excellence for all students, including teacher candidates.

The initial challenge was assessment, which led the Education Department (ED) faculty to begin a review of the curriculum. In an attempt to identify ways to improve the ED programs, the faculty began reviewing researched-based instructional models, approaches, and strategies. After examining several models, the education faculty decided to adopt the "Understanding by Design (UbD) Framework" in undergraduate and graduate teacher education programs. This case study focuses on the process that led to the integration of "Understanding by Design" as a framework to contextualize the teacher education programs. To provide a chronological review used to introduce, develop, and embed the UbD framework into the Teacher Education curriculum, the following steps were taken.

On March 14, 2011; the institution invited Grant Wiggins, an education expert, to deliver the keynote address at its two-day education conference, "Authentic Performance Assessment in Urban Education." During his presentation, Wiggins discussed the importance of assessment in developing units and lessons using the UbD curriculum framework. Understanding by Design (UbD) is an important, practical framework used locally, nationally, and internationally to improve student-learning outcomes in college programs and in K – 12 schools. Wiggins and McTighe's framework emphasizes teachers' roles as designers and planners of student learning experiences. Course planning using UbD also is called *backwards design* or *teaching with the end in mind.* Using this curriculum framework provides a way for teacher candidates to understand the big ideas by investigating, exploring, testing, and verifying important concepts. They are able to transfer knowledge using the previously learned concepts.

As a result of the 2011 Education Conference and the emphasis placed upon UbD, the ED faculty began integrating UbD as the curriculum framework for Teacher Education. The ED began a curriculum mapping process, incorporating principles of UbD in redesigning the programs and courses in the Teacher Certification Program. Faculty began to map curriculum with the Backwards Design process used in UbD by focusing on what teacher candidates need to know at the end of the program/course/unit. In addition, a course, Curriculum Instruction and Assessment, was designed to prepare teacher candidates to use Backward Design to create units and lessons for their students in elementary and secondary schools.

In July 2011, the elementary and secondary education coordinators participated in two national UbD training programs, conducted by Grant Wiggins and Associates. The programs were "UbD Train the Trainers" and "Designing from the Standards." Participants included elementary and secondary teachers, higher education faculty, and administrators from across the United States and other countries. The attendees were at various stages in learning about and implementing UbD in their educational agencies and institutions.

During the 2011-2012 academic year, the ED submitted its Elementary Education Program application to the state department of education to assure the program complied with state standards. In this application, the elementary teacher preparation curriculum was aligned to the elementary education standards, with the UbD framework included across all phases of the program. The ED also participated in the Teacher Education Accreditation Council (TEAC; now known as Council on Accreditation of Educator Preparation [CAEP]) national accreditation process of its teacher education programs. The UbD Framework was incorporated into the Inquiry Brief Proposal submitted to TEAC. Both the state and national reviews were positive, solidifying the Education Department's commitment to UbD as a viable curriculum framework.

University B hired two adjunct faculty members who were experienced in UbD to assist with instruction and implementation across the Teacher Certification Programs. In 2012, the education coordinators and one of the newly hired adjunct faculty designed, planned, and delivered a three-session workshop for college supervisors, cooperating teachers, as well as the education and liberal arts faculty. The evaluation feedback was positive and indicated that additional training and support were needed to infuse UbD into the education programs adequately.

During the 2012-13 academic year, education faculty began to embed the UbD Framework into the curriculum throughout the four phases of the Teacher Certification Program. In phase one, students became aware of the UbD Framework in an introductory course to the teaching profession. In phase two, pre-candidates began to develop and practice principles of basic unit and lesson planning using the framework of backward design in a curriculum, instruction, and assessment course. Pre-candidates also participated in microteaching that provided opportunities to practice teaching with friends and peers. The presentation was a 5 to 10 minute segment of a UbD lesson.

The teacher candidates complete their content major courses (integrated science, English language arts, social studies, and mathematics) that are offered through the liberal arts department. The methods courses for these subject areas are offered by the education department. While the liberal arts faculty attended some training on UbD, they had not fully adopted this curriculum framework into their courses.

In phase three, teacher candidates continue to develop and practice UbD units and lesson plans through a structured program of coordinated UbD theory, observation, and participation in field-based experiences. Teacher candidates apply ideas or procedures presented in their methods courses to K – 12 students in school settings, focusing on teaching integrated science, and using skills associated with classroom management, planning, and daily routines. They observe, participate in classroom activities, teach a science lesson using the UbD framework, and reflect on these experiences. During phase three, teacher candidates begin to develop an understanding of the daily routines that are part of the classroom environment. They develop a greater understanding of the need to know their students and learn to use this knowledge and understanding to inform the teaching-learning dyad, especially in a science classroom.

Student teaching is the college-supervised instructional experience that occurs during the fourth phase of the teacher education program. Student teaching is the culminating field experience in the teacher education program. Student teachers design a unit and develop lessons plans based on the UbD Framework. Each student teacher videotapes a continuous 30-minute segment of a planned teaching lesson that includes a self-evaluation and a reflective assessment of the teaching event. At the elementary level, integrated science is one of the content-area subjects that is included in the teaching event. The successful completion of requirements of this 15-week experience typically leads to a degree and teacher certification

The Education Faculty at University B meets on a regular basis to assess curricula, teaching, and assessment to ensure that key design issues are being addressed in course curriculum. During the current academic year, faculty will continue to update and revise curriculum to ensure that UbD is being implemented across disciplines, including science, in the education programs. The professional development team, including the elementary and secondary coordinators and adjunct professors, will continue to design training and consultation, along with follow-up activities to assist faculty in integrating UbD principles into their teaching and learning.

CURRENT CHALLENGES FACING THE ORGANIZATION

Understanding by Design helps teachers to "provide a broader focus on scientific concepts and processes in a 'big picture' sense and not overemphasize the parts of the scientific concepts and processes" (Designing Curriculum for In-depth Understanding in Science, n.d., Slide 4). The challenge is to equip education and liberal arts faculty with the skills and resources to incorporate UbD into their teacher preparation courses. Effective implementation of UbD requires institutional support because of the time needed to provide professional development to faculty who are involved with instruction in the teacher education programs (Newton, 2003). Faculty who coordinate the elementary and secondary teacher education programs at University B are also required to be involved in scholarship and service, as well as teaching full course loads. These coordinators have been responsible for the adoption of UbD by faculty in ED, but have had partial success with liberal arts faculty.

Teacher candidates, at University B, understand the need for planning, designing, and implementing UbD into their units and lessons during the third and fourth phase of their programs. Class size in the third and fourth phase of the teacher education programs are small, allowing instructors to provide frequent and prompt feedback as teacher candidates seek to implement UbD. Teacher candidates, working closely with faculty, incorporate UbD into their professional tools.

In this time of economic distress for higher education, faculty development monies are not available for University A. Some professional development activities are funded by University B, however, the monies are limited. To provide training for UbD, selected faculty attended training sessions and then returned to their institutions to become trainers for other faculty. Time and funds must be made available to provide professional development to faculty in education and liberal arts programs to ensure that the adoption and implementation of the UbD curriculum framework is provided for teacher candidates in the university.

SOLUTIONS AND RECOMMENDATIONS

Key personnel and shared vision were the overwhelmingly predominant supporting factors in adoption of UbD in teacher education programs. In the case of University A, the lead change agent was an individual faculty. However, the lack of program support limited the extent to which such change appeared within the program. The lead change agents at University B were two faculty from the Education Department. One faculty member from the department had a background in professional development and understood the level of professional development activities and ongoing support needed to produce organizational change.

Strategic vision is a predominant factor in supporting initiation of paradigm change in curriculum framework used in coursework for teacher candidates. As the courses are taught across disciplines (teacher education and liberal arts), a common vision is needed to adopt and implement curriculum to improve learning outcomes for teacher candidates and their students.

Sometimes change is not easy. Nevertheless, at University B, positive support for such a change was indicated by the numerous steps instituted to embed UbD across the teacher education program. At University A, there continues to be a lack of programmatic discussion of a need to introduce a common curricular planning tool. The lack of a common goal or vision for curriculum planning can be a major challenge in implementation.

Organizational structure was the predominant supporting factor in continuing the UbD curricular framework. Once the model was instituted at University B, faculty began to build the structure to support implementation and continuance of UbD.

Faculty met as a team and discussed the strengths and challenges involved in implementing UbD as the curriculum framework. At University A, the faculty member reported that she felt isolated in terms of exploring the use of UbD for curriculum design and development. Collaborative relationships and policy also can be challenges to the continuation of UbD at both universities. Personnel changes and expectations for preparing teachers are important challenges, at University B, in the continuing collaboration among faculty members and administration. Providing on-going training and support can ensure effective educational experiences in curriculum design and planning for teacher candidates.

In looking back at the activities involving both universities, assessment is an issue that needs to be addressed. Decisions about which curriculum framework should be presented to teacher candidates must be rooted in assessment. Data from program graduates who are practitioners and administrators need to be collected to determine the viability of using UbD as a curriculum-planning framework to improve student achievement. Data on teacher candidates' use of UbD and their students' achievement also need to be collected throughout the teacher preparation programs to examine progress in unit and lesson designing, planning, and implementing curriculum, especially in integrated science education.

REFERENCES

Craven, J., & Penck, J. (2001). Preparing new teachers to teach science: The role of the science teacher educator. *Electronic Journal of Science Education, 6*(1). Retrieved from http://ejse.southwestern.edu/article/view/7670

Designing Curriculum for In-depth Understanding in Science. (n.d.). *PowerPoint presentation.* Retrieved from www.fass.cecs.ucf.edu/.../Designing%20Curriculum%20-%20M.%20Hay

Hambrick, D., & Cannella, A. Jr. (1989). Strategy Implementation as substance and selling. *The Academy of Management Executive, 3*(4), 278–285. doi:10.5465/AME.1989.4277401

McTighe, J., & Seif, E. (2003). *A summary of underlying theory and research base for Understanding by Design.* Retrieved from assets.pearsonschool.com/asset_mgr/ubd_myworld_research.pdf

McTighe, J., & Thomas, R. (2003). Backward design for forward action. *Educational Leadership, 60*(5), 52–55.

National Science Foundation. (2006). *America's Pressing Challenge – Building a Stronger Foundation: Mathematics and Science Achievement is Critical.* Retrieved from http://www.nsf.gov/statistics/nsb0602/

National Science Teachers Association. (2004). *NSTA Position Statement: Science Teacher Preparation.* Retrieved from http://www.nsta.org/about/positions/preparation.aspx?print=true

Newton, J. (2003). Implementing an institution-wide learning and teaching strategy: Lessons in managing change. *Studies in Higher Education, 28*(4), 427–441. doi:10.1080/0307507032000122279

Prensky, M. (2006). Listen to the natives. *Educational Leadership, 63*(4), 8–13.

STEM Smart Brief. (n.d.). *Preparing and support STEM educators.* Retrieved from http://successfulstemeducation.org/sites/successfulstemeducation.org/files/Preparing%20Supporting%20STEM%20Educators_FINAL.pdf

Wang, D., & Allen, M. (2003). Understanding by Design Meets Integrated Science. *Science Teacher (Normal, Ill.), 70*(7), 37–41.

Wiggins, G., & McTighe, J. (1998). *Understanding by Design.* Alexandria, VA: ASCD.

Wiggins, G., & McTighe, J. (2013). *Understanding by Design* (2nd ed.). Alexandria, VA: ASCD.

ADDITIONAL READING

Alhamisi, J., Glimps, B., & Okezie, C. (2012). Understanding by design and culturally responsive teaching: Effective tools for instruction. *Tennessee Educational Leadership*, *39*(19), 21–28.

Authentic education's understanding by design – related products. (n.d.). Retrieved from http://www.authenticeducation.org

Graff, N. (2011). An "effective and agonizing way to learn": Backwards design and new teachers' preparation for planning curriculum. *Teacher Education Quarterly*, *38*(3), 151–168. Retrieved from http://www.teqjournal.org

Graffam, B. (2003). Constructivism and understanding: Implementing the teaching for understanding framework. *Journal of Secondary Gifted Education*, *15*(1), 13–22. doi:10.4219/jsge-2003-439

Jones, K., Vermette, P., & Jones, J. (2009). An integration of "Backwards Planning" unit design with the "Two-Step" lesson planning framework. *Education*, *130*(2), 357–360.

McTighe, J., & O'Connor, K. (2005). Seven practices for effective learning. *Educational Leadership*, *63*(3), 10–17.

Shumway, S., & Berrett, J. (2004). Standards-based curriculum development for pre- service and in-service: A "partnering" approach using modified backwards design. *Technology Teacher*, *64*(3), 26–29.

Staines, G. (2013). *The benefits of cloud computing in education are huge!* Retrieved from http://blogs.aspect.com/2013/06/27/the-benefits-of-cloud-computing-in-education-are-huge/

The flipped classroom. (n.d.). Retrieved from http://successfulstemeducation.org/sites/successfulstemeducation.org/files/Preparing%20Supporting%20STEM%20Educators_FINAL.pdf

KEY TERMS AND DEFINITIONS

Authentic Performance Assessment: These are curriculum based measures that require students to construct responses on real world tasks.

Backward Design: A form of guided discovery that focuses on truly understanding the content by starting with the finished product and working backward.

Curriculum: The educational term for what students experience in school.

Instructional Design: The systematic process of designing and delivering instructional materials.

Standards: Expectations for what students should know and be able to do.

Understanding by Design: Wiggins and McTighe indicate that UbD is a framework for improving student achievement through standards-driven curriculum development, instructional design, assessment, and professional development.

This work was previously published in "Cases on Research-Based Teaching Methods in Science Education" edited by Eugene de Silva, pages 64-79, copyright 2015 by Information Science Reference (an imprint of IGI Global).

Chapter 34
Designing and Teaching an Online Elementary Mathematics Methods Course:
Promises, Barriers, and Implications

Drew Polly
University of North Carolina - Charlotte, USA

ABSTRACT

This chapter discusses a longitudinal examination of a mathematics methods course for teacher candidates taught in hybrid and a 100% asynchronous online format. Using Guskey's (2000) framework for evaluating learning experiences for teachers, thematic analysis was conducted on teacher candidates' course feedback and two major course assignments. Data analysis indicated that teacher participants valued the amount of support provided by the instructor and communication with classmates, had mixed comments about having to take ownership of their learning, and disliked the amount of work in the course. Participants' work samples reflected the application of emphasized pedagogies in lesson plans and course projects, and participants also positively impacted student learning during their clinical project. Implications for future courses as well as the examination of online methods courses are shared.

INTRODUCTION

Online Learning Opportunities

During the last decade the demand to develop online methods courses in teacher education programs has increased dramatically (Ko & Rosen, 2010). While the benefits of online courses have been established (Tallent-Runnels et al., 2006), online courses in teacher education, especially mathematics education, provide a set of design issues that differ from other online courses outside of teacher education programs or other content areas (Delfino & Persico, 2007; Tallent-Runnels et al., 2006). To address these challenges, researchers have identified specific tools for online teaching that expand learning opportunities. These include: accessibility to the Internet and Web 2.0 technologies, a collaborative framework, and other course structures.

DOI: 10.4018/978-1-4666-7363-2.ch034

Higher education has embraced the possibilities of using the Internet as a medium to teach college courses. Nelson, Christopher, and Mims (2009) suggested that "the Internet and Web 2.0 technologies afford teachers ready access to collaborative, authentic opportunities for students to engage in meaningful experiences related to the curriculum" (p. 85). Oliver (2010) noted that the use of Web 2.0 technologies advances online learning opportunities that were not readily available before. There are numerous technologies to support teaching mathematics online (Hodges & Hunger, 2011). Blogs and wikis have been recognized by faculty as effective tools, especially in mathematics courses (Ajjan & Hartshorne, 2008; Carter, 2009; Peterson, 2009).

These technologies increase opportunities for student interaction and discussion; an essential component of online courses (Levin et al., 2001; Tallent-Runnels et al., 2006). In their comparison study of discussion formats, Im and Lee (2004) found that synchronous discussion was more effective at social interaction, while asynchronous discussion was better suited for task-oriented interaction. Regardless, the presence of peer collaboration and discussions in online courses is empirically-linked to students' perceptions of the course (Schlager & Fusco, 2004).

Further, specific course structures have been found to be effective in online courses. Greene and Land (2000) found that both guiding questions and frequent feedback from instructors helped them understand assignments and led to higher quality work products. Experts have highlighted the importance of relevant and challenging assignments (Levin, Waddoups, Levin, & Buell, 2001), clear expectations and evaluation methods (Moallem, 2003), and opportunities to reflect on assignments (Levin et al., 2001). Furthermore, online courses create new learning spaces for teachers and engender an environment that is learner-centered and align to characteristics of effective learning environments.

Online Learning Opportunities for Teachers

The research base regarding online learning opportunities for teachers includes studies regarding online professional development opportunities for teachers who are currently in classrooms and either working on advanced degrees or pursuing a deeper understanding of either content or pedagogy (e.g., Russell, Carey, Klieman, & Venable, 2009; Signer, 2008). Researchers that have compared online professional development to face-to-face models for mathematics teachers noted that there was no significant difference between the influence of the learning experiences on teachers' beliefs, knowledge or skills (Russell et al., 2009). Online learning experiences have found that the increase in written communication helps to promote more reflective inquiry about course content (Spicer, 2002; Treacy, Kleiman, & Peterson, 2002). In their five-year longitudinal study, Delfino and Persico (2007) found that teachers' written work in the online courses contained frequent instances of critical thinking and in-depth reflection. Further, on the whole teachers enjoy online learning opportunities and are willing to take more in the future (Russell et al., 2009; Signer, 2008).

However, while research on pre-service courses for pre-service teachers with limited or no teaching experience is scant. In a hybrid mathematics education methods course, Schwartz (in press) posed a task to a group of students in an online asynchronous format and compared their strategies and reactions to those in a face-to-face format. While pre-service teachers approached and solved the task in the same way, there was a lack of opportunities for the instructor and pre-service teachers to discuss the pedagogy and in-the-moment teacher decision making. O'Connor (2011) found that while pre-service teachers learned technology and formed effective collaborative relationships with their colleagues, videos of teaching experiences reflected a lack of student-centered pedagogies that were emphasized in the course.

Clearly, there is a need for further research on online learning opportunities for pre-service teachers, specifically in the area of mathematics education. This chapter provides an overview of the development of an online mathematics education methods course for elementary education students, including the theoretical foundation and course activities. The purpose of this study is to present a learner-centered framework for an online mathematics education course for elementary school pre-service teachers. That is followed by a case study that includes data collected and analyzed from course offerings. Lastly, implications are presented that inform current and future work related to this course.

OVERVIEW

Course Overview

The course described in this chapter is the second of two mathematics education courses in our Elementary Education Graduate Certificate in Teaching Program at the University of North Carolina at Charlotte. The course, ELED 5301: Assessing, Modifying, and Integrating Mathematics Instruction, includes numerous projects to prepare teacher candidates to assess students, plan instruction based on data, and then reassess students using formative and summative processes to monitor students' learning. ELED 5301 also includes content and pedagogy tasks related to fractions, algebraic thinking, and geometry.

Students in the Graduate Certificate in Elementary Education Program hold a bachelor's degree in another field, and are working on obtaining their initial teacher license in K-6 education. While some students have volunteered or worked in schools as teacher assistants or tutors, most students have little to no experience in elementary school classrooms. Students complete ELED 5301 when they are one semester away from student teaching, so the course includes a heavy focus on designing mathematics lessons, assessing students, and differentiating instruction.

Motivation to develop an online version of this course started in the 2007-2008 year, through discussions about alternative ways to provide teacher education courses to meet the scheduling needs of our students in the Graduate Certificate in Teaching Program.

While the entire Graduate Certificate in Teaching program is offered in a face-to-face format, there was interest from both College and Department leadership in exploring online courses. One reason for students' overwhelmed feeling is that they work full time during the day and typically take 3 3-hour courses each semester, and every course is offered in the evening. Students often report being overwhelmed with 9 seat hours of course work weekly in addition to 15 hours of clinical assignments per course. An online course provides students with flexibility to complete modules during the week and decrease the number of seat hours that students have, while maintaining the amount of clinical assignments.

The development of online modules in ELED 5301 was supported through a UNC Charlotte Curriculum Improvement and Development grant, which provided resources to purchase video vignettes of elementary school classroom teaching from the Developing Mathematical Ideas professional development materials (Schifter, Bastable & Russell, 2008). While these video vignettes were also used in face-to-face weeks, these videos were effectively used as part of online modules, in which teacher candidates viewed videos and analyzed how the teachers supported students' mathematical learning with pedagogies that were emphasized during the course.

Theoretical Foundation of the Course

The design and activities in this course are driven by two theoretical constructs: learner-centered instruction and mathematical knowledge for teaching. In this section, I provide an overview of each and then describe how each construct aligns to the course activities (Table 1).

Learner-Centered Instruction

ELED 5301 has been designed based on the theoretical framework of learner-centered instruction for teacher learning (Polly & Hannafin, 2010; National Partnership for Educational Accountability and Teaching [NPEAT], 2000).The American Psychological Associations' *Learner-centered Principles* (APA Work Group, 1997), provide the empirical basis for designing learner-centered environments for both students and adults. The NPEAT adopted these *Principles* for teachers' learning. Recently, the *Learner-Centered Principles* were synthesized with empirical research on teacher learning to derive a set of learner-centered principles for the design of learning experiences for teachers and teacher candidates (Polly & Hannafin, 2010). Specifically, these experiences should:

- Prepare teachers to address student learning issues (Heck, Banilower, Weiss, Rosenberg, 2008),
- Ensure ownership of their learning experiences (Garet et al., 2001),
- Promote collaboration among teachers or teacher candidates (Glazer & Hannafin, 2006),
- Emphasize comprehensive change processes through ongoing support (Fishman, Marx, Best, & Tal, 2003; Orrill, 2001)
- Develop knowledge and proficiency related to specific pedagogies, content and the intersection of content and pedagogy (Heck et al., 2008; Garet, et al., 2001); and
- Support reflection on work samples and artifacts from students and classrooms (Cohen, 2005; Loucks-Horsley et al., 2009).

Table 1. Alignment between characteristics of effective teacher learning environments and course activities

Characteristics of Learner-Centered Environments for Teachers	Course Activities Teacher Candidates…	Aspects of Mathematical Knowledge for Teaching
Address student learning issues	Complete the Culminating Diagnostic Project, which involves pre-assessing, teaching and post-assessing students.	Knowledge of Content and Students (KCS), Knowledge of Content and Teaching (KCT)
Provide teachers with ownership	Select topics for Culminating Diagnostic Project, Curriculum Evaluation and other activities.	All aspects
Promote collaboration	Collaborate with classmates on blogs and with in-service teachers on all clinical projects.	All aspects
Provide ongoing support	Turn in the Culminating Diagnostic Project at various times during the semester for feedback and support.	All aspects
Develop knowledge of content and pedagogy	Complete cognitively-demanding mathematical tasks in each module, analyze curriculum, and examine video-based and text-based vignettes.	All aspects
Support the reflection process	Write on blogs throughout the semester and reflect on the impact on students during the Culminating Diagnostic Project.	All aspects

Mathematical Knowledge for Teaching

In addition to the *Learner-centered Principles*, which specify how to structure effective learner environments for teachers, ELED 5301 is also grounded in the construct of mathematical knowledge for teaching (MKT). MKT (Figure 1) is a research-based framework that describes the different aspects of knowledge involved with teaching mathematics to students (Ball, Thames & Phelps, 2008; Thames & Ball, 2010). MKT includes both mathematical content knowledge and pedagogical knowledge. Based on MKT, opportunities for teacher learning should address these different knowledge types.

Table 1 shows the alignment between characteristics of learner-centered experiences for teachers and course activities. These activities are described in more detail later in the chapter.

Structure of the Course

Course Format

For every online module in either the online versions or the hybrid versions online content was asynchronous. Students communicated and interacted with each other and the instructor on their own time without any online meetings scheduled.

Learning Management System

All of the course activities were assigned through Moodle (http://moodle.org), an open source Learning Management System (LMS). Within Moodle, students interacted with course content (articles, videos, assignments), submitted work, and received feedback and grades on their progress.

Figure 1. Mathematical knowledge for teaching (Thames & Ball, 2010)

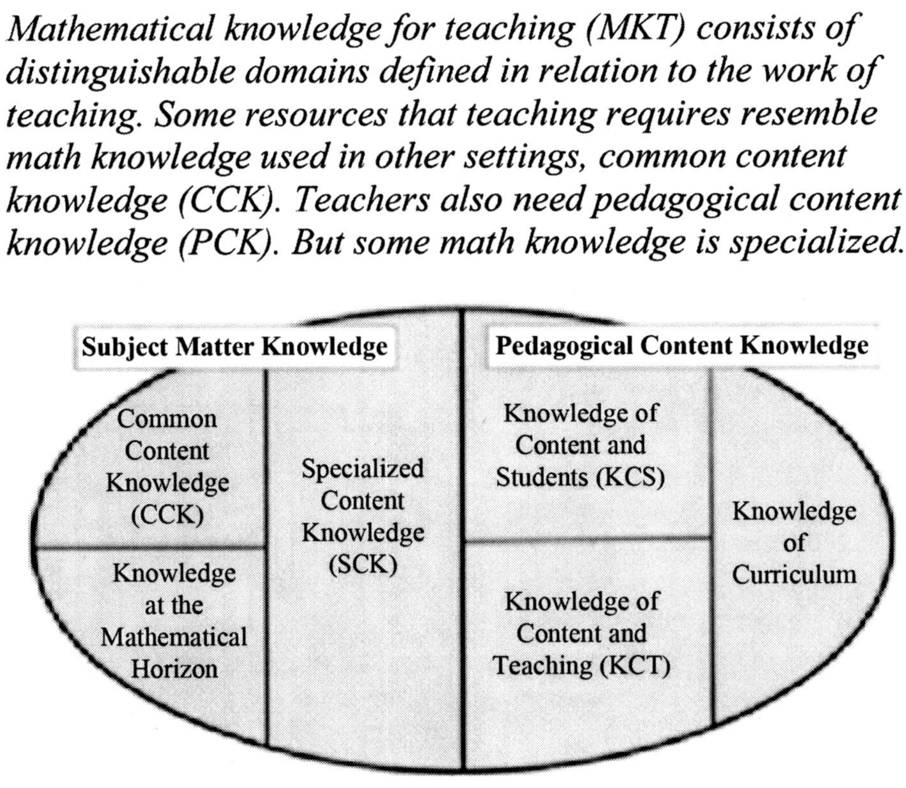

Modules

Assignments were bundled together in a module. Each module was organized as a series of tasks, which included a cognitively-demanding task, an analysis of classroom mathematics teaching, activities to complete based on their work with the state standards, curriculum, or elementary school students, and a reflection that connected the various activities. Modules were designed to last anywhere between 1-3 hours of work. Some modules also served as the foundation for major course projects, which are detailed.

Collaboration with Classmates

Each teacher candidate maintained their own blog, which was set up through either Word Press (http://wordpress.com) or Posterous (http://posterous.com). Teacher candidates posted to their blog each week, and responded to at least two of their classmates' postings. This communication between classmates provided opportunities for teacher candidates to share their thoughts and experiences with each other, and receive support from their peers. Teacher candidates also had access to a course wiki where they could go and receive support from each other and the instructor about course activities.

Course Activities

Grounded in the principles of learner-centered instruction, the course included various activities. Table 1 provides details about how course activities match these characteristics. These course activities are described in more detail later in the chapter.

Cognitively-Demanding Mathematical Tasks

During each module, teacher candidates completed mathematical tasks that were related to the Kindergarten through sixth grade state mathematics standards. Each of these tasks was a multi-step, complex problem that aligned with the construct of cognitively-demanding mathematical tasks (Stein, Grover, & Henningsen, 1997). These tasks varied in terms of the mathematical concepts that they dealt with, but most of them focused on number sense and algebraic thinking. In the face-to-face and hybrid formats, these tasks were solved during face-to-face course meetings.

Examining Artifacts of Mathematics Teaching

Teacher candidates examined examples of mathematics teaching through vignettes that were freely available on the Internet (http://learner.org) or in the course's text book, *My Kids Can* (Storeygard, 2010). Through viewing these video-based vignettes and focusing their attention on key aspects, such as teacher's questioning, student communication, or the structure of the lesson, teacher candidates developed a deeper sense of knowledge of content as it related to how to teach and how students learn content. These videos were incorporated into online modules during the hybrid versions of the course, and viewed during.

Assessing Students' Mathematical Learning

Teacher candidates conducted mathematical interviews with elementary school students. During these interviews, teacher candidates posed tasks, observed students as they work, and asked students to explain their processes and mathematical thinking. This assignment was a clinical assignment done in schools in all versions of the course.

Writing about Mathematics Teaching and Learning

Students completed a written reflection about activities in the module. These reflections focused on either their reaction to videos that they watched, articles they read, or their own experiences during their time observing and teaching mathematics to elementary school students. These were posted on blogs using either WordPress or Posterous. In all versions of the course, these blogs were used. However, students were required to post more frequently during hybrid and online sections of the course.

Curriculum Evaluation

During the middle of the semester, teacher candidates evaluated and analyzed a unit from a standards-based curriculum. They then read a few articles about the types of mathematics curriculum, the "Math Wars" and then reflected on their ideal math curriculum about being standards-based or more traditional. This assignment was completed in every version of the course. In the face-to-face section of the course, the background information was presented during a course meeting, and it was presented online during hybrid and online versions of the course.

Culminating Diagnostic Project

The culminating experience in the course was a diagnostic project that required teacher candidates to pre-assess, plan and teach five lessons, and post-assess a small group of students to examine students' growth. Teacher candidates were expected to base all of their instructional decisions on data that was collected during the pre-assessment and lessons. Teacher candidates also planned and taught a lesson to an entire class of students, in addition to planning a 10 lesson geometry unit that aligned to the state standards. Through these activities students developed pedagogical content knowledge, including knowledge related to teaching, students and curriculum. This assignment was completed in every version of the course.

METHODS OF EVALUATING THE COURSE

Evaluation Framework

Guskey (2000) provided a multi-level framework (Table 2) for evaluating the impact of learning experiences for teachers. While this framework was intended to evaluate the influence of professional development for practicing teachers, it can also be adapted for teacher candidates. In the remainder of this chapter, data is presented from each of these levels provides insight into teacher candidates' performance as well as the impact of the online course.

Research Questions

For this chapter, data was analyzed to examine:

- What are teacher candidates' reactions to participating in an online mathematics education course? (Level 1)

Table 2. Levels of evaluation (Adopted from Guskey, 2000)

Level	Name of Level	Questions Examined
Level 1	Teachers' reactions	How do teachers feel about their experiences?
Level 2	Teachers' knowledge and skills	What knowledge and skills do teachers learn during their experiences?
Level 3	Use of knowledge and skills	How do teachers use their new knowledge and skills in their own classroom?
Level 4	Impact on organization and program	How did the experience impact the program or organization responsible for providing the experience?
Level 5	Impact on student learning	How are PK-12 students impacted by teachers' participation in this experience?

- How do teacher candidates' demonstrate their understanding of course content? (Level 3)
- To what extent do teacher candidates' impact student learning on the courses' Culminating Diagnostic Project? (Level 5)
- How has data informed revisions and modifications of the course? (Level 4)

Context

Data in this chapter are shared from teacher candidates in ELED 5301 between 2008 and 2011. Table 3 describes the formats of the various courses. All students in the course were teacher candidates in the Graduate Certificate in Teaching Elementary Education program at UNC Charlotte, and held a bachelor's degree in a different field. Students' bachelor degrees vary, but common fields include Business, English, or Psychology.

Data Sources

Data sources for this study included: post-course feedback, online reflections, curriculum project reflections, and data from the Culminating Project.

Post-Course Feedback

Teacher candidates provided anonymous feedback using both the course evaluation form as well as a more detailed survey that I sent students. The survey asked them to share their opinions about: which assignments were the most beneficial at this stage in the teaching career, which assignments were not very beneficial, and what experiences would they like more of. Lastly, I asked participants to comment about the nature of the course and whether the online or hybrid design should be modified.

Online Reflections

Teacher candidates' online reflections on their blogs were analyzed to examine the topics of the blog posts. In each reflection, participants had freedom to focus their blogs topics of most interest to them. At the beginning of the semester, participants had a structured prompt, such as "what are the elements of an effective mathematics lesson?" or "what are your concerns related to planning a mathematics lesson?" Later in the semester, as participants are more involved with their clinical experiences, they are asked to write about their experiences and things that they are learning from teaching and interviewing students.

Curriculum Project

On their curriculum projects, teacher candidates analyzed standards-based curricula to look at the types of mathematical tasks, the way that content was presented, and the type of teacher support that was provided for them. After their analysis, candidates reflected on the assignment and discussed their perceptions of the curricula

Table 3. Formats of ELED 5301

Semester	Format	Number of Modules	Class Size
Fall, 2008	Online	10 online modules	28
	Hybrid	3 online modules, 12 face-to-face meetings	28
Fall, 2009	Hybrid	3 online modules, 12 face-to-face meetings	32
Fall, 2010	Hybrid	5 online modules, 10 face-to-face meetings	44
Fall, 2011	Online	10 online modules	39 students

that they examined. They also shared how the curricula aligns or conflicts with what they have learned in their coursework and their beliefs about teaching mathematics.

Culminating Diagnostic Project

Data sources from the diagnostic project included the types of mathematical tasks in the lesson plans as well as student growth between the pre- and post-assessments. Lesson plans were written using an indirect instruction format (http://coedpages.uncc.edu/abpolly/5301/coursedocs/diagnostic/lesson-plan-resources/diagnostic-6pt.doc), which focuses on designing the lesson around a cognitively demanding mathematical task, and supporting students through questions and follow-up tasks. Pre- and post-assessments were identical, and were either created by the teacher candidate, or modified from a curriculum.

DATA ANALYSIS

Course Feedback and Reflections

Course feedback and reflections from teacher candidates were examined using inductive, thematic analysis (Bogden & Biklen, 2003). All of the data was entered into a spreadsheet and labeled. During the data analysis process, data was coded using an open-coding process and then sorted all of the data by code. After coding and sorting, data was reread to confirm that the data matched the codes. The next step was to group the data by code in order to generate themes. Once themes were generated, data was reread again, to make sure that the themes truly reflected the data within that group. The themes were then matched up to the research question. Themes addressed either research questions 1 and 2. The Findings section of this paper describes the most common themes that were found during the data analysis process.

Types of Mathematical Tasks in Lesson Plans

Using the cognitively-demanding tasks framework (Stein et al., 1997), tasks from every lesson plan was coded and analyzed. The analysis was similar to the process detailed above, where tasks were coded, and once organized, reanalyzed to make sure that they were properly categorized. Once each task had been coded and checked, frequencies were determined. Table 4 provides descriptions and examples of the four types of mathematical tasks.

Table 4. Types of mathematical tasks

Cognitive Demand	Name of Task Type	Description	Example
High Cognitive Demand	Doing Mathematics	Students explore mathematical tasks that require them to choose an approach, complete the task, and explain their steps and decision-making.	There are 24 yards of fencing for the garden. If you want to make a rectangular garden with side lengths that are whole yards, what are the possible dimensions of the garden? Which garden is the largest?
	Procedures with Mathematical Connections	Students explore tasks that can be solved with an algorithm, but have to generate more than one representation.	There are 9 dozen cookies in the bag. If you eat 6 cookies how many are left? Show your picture using a picture and an equation.
Low Cognitive Demand	Procedures with Mathematical Connections	Students explore tasks that require only an algorithm and only one mathematical representation.	There are 9 dozen cookies in the bag. If you eat 6 cookies how many are left?
	Memorization	Students recall a fact that is expected to be known.	What is the product of 9x6?

Student Growth on Pre- and Post- Assessments

As part of the Culminating Diagnostic Project, teacher candidates gave a pre- and post-assessment and taught 5 lessons around a concept. Since the assessments varied in length, topic, and grade level, data analysis focused on whether or not students demonstrated growth between the pre- and post-assessments. Since each teacher candidate gave the same assessment as both a pre- and post-assessment, growth was noted anytime a students' score increased on the post-assessment.

FINDINGS

What are Teacher Candidates' Reactions to Participating in an Online Mathematics Education Course?

Data regarding teachers' reactions to participating in an online (or hybrid) course centered on three major themes: responsible for learning, amount of work, and communication with class members and instructors.

Responsible for Learning

Participants across all semesters reported on post-course feedback that they felt more responsible for their own learning during online activities. This was true for the online courses as well as the online modules in hybrid courses. Participants expressed various emotions regarding this theme.

In some cases, students enjoyed the responsibility that they perceived. "It was really hard to concentrate more and learn the material better without having regular meetings with an instructor. I was completely in control of my own learning, which was nice, but I want more regular contact with my instructors." (Student, Fall, 2010 semester).

However, many students reported discontent and dissatisfaction with the amount of responsibility that they had to take on.

I feel like I am teaching myself how to teach math, and I don't feel confident that I am learning what I need to. The instructor provides modules and helps us if we get stuck, but we basically are on our own to figure things out. I don't think that's fair. (Student, Fall, 2008 semester).

While feedback from each course included both positive and negative reactions about students' responsibility for learning, students in the hybrid courses compared the face-to-face and online learning. Many commented about the benefit of having a blend of both face-to-face and online work.

I feel like a hybrid model best met my needs. The weeks in class allowed us to see and work with our colleagues, but we were also able to have the flexibility of working on our own during online weeks. (Student, Fall, 2009 semester).

Amount of Work

Data from teacher candidates' reactions also focused on the amount of work in the courses. Nearly every comment compared the amount of work in an online course to a face-to-face course.

My friend is taking the face-to-face section, and my online section has much more work to do. The concepts are the same, but it seems as if we have a lot more reading and activities to work on. I would rather spend my evenings in a classroom sitting there instead of working on these modules. (Student, Fall, 2008 semester)

Some students, however, reported that they thought there was much less work in a 100% asynchronous online course and that they were able to better differentiate their attention on concepts and activities that they were more interested in and needed to learn more about.

The amount of work takes up much less time since it is an online class. I think that I still learned a lot, but was able to focus on things that I individually found interesting or needed, instead of sitting in a classroom and having to learn things that are not as relevant to me. (Student, Fall, 2011, semester)

Teacher candidates reported in the online only courses reported mixed reactions about the amount of work. While a face-to-face class is designated as a 3 hour weekly course meeting followed by approximately 3 hours of weekly activities and assignments, students were not comfortable with the idea that an online course may vary in terms of the length of the modules and the amount of work each week. While the amount of work did not change between versions of the online course, more recent teacher candidates reported less frustration with the amount of work involved in the course.

It was hard to predict how long a module would take. There were some where I knew the math concept and I flew through it, and some like algebra, that took me an entire Saturday. I feel like a face-to-face class would have been better since they are typically more consistent about the amount of time that you have to spend on projects and homework. (Student, Fall, 2008 semester).

Over the past few years, the number of comments about the inconsistent lengths of the online modules has dissipated.

Communication with Class Members and Instructors

Data from participants' reactions also frequently mentioned communication with class members and instructors. Many students commented about the surprisingly high amount of communication that they had with students via their weekly online reflections.

I did not expect to interact with classmates at all in this course. It was nice to be able to get responses from my weekly reflections, and be able to hear what others thought about the course activities. The interaction helped me feel like there were others in the same situation as myself. (Student, Fall, 2011 semester).

Other teacher candidates shared disappointment with the lack of communication in the asynchronous online course. In the course feedback, they cited the amount of group work that goes on in face-to-face courses in the program.

I just felt awkward and disappointed in this course. We do group activities in every face-to-face course in our program, including the first math course, but we were very isolated here. Commenting on reflections is definitely not enough communication for a course like this.

One change between the 2008 online and 2011 online course was a more frequent use of the class wiki to share problem solving strategies and ideas. Most students in the Fall, 2011 course commented about the benefit of the wiki and the opportunity to frequently communicate with each other.

I felt supported, especially when we had to solve those math tasks. Having the wiki where the instructor and classmates put ideas about how to start working on those tasks was a huge help. It lessened the anxiety I had trying to solve hard math problems. (Student, Fall, 2011 semester).

The data showed that participants were fairly satisfied with the amount and frequency of communication between them and the instructor in all versions of the course. However, some students did admit that in the online course, it was difficult to generate questions that they said would most likely be answered during face-to-face course meetings. A majority of the comments related to communication focused on the amount of support they received on drafts of assignments, including the major Culminating Diagnostic Project.

I felt like this instructor lived on e-mail. He was constantly available to give feedback and answer questions. The fact that he gave detailed feedback on the Diagnostic Project draft was a huge amount of help. (Student, Fall, 2011 semester).

The data indicated that while some participants did not like the perceived large amount of work, and a few felt that they were overly responsibly for their own learning, the communication mechanisms established in the course were beneficial. On the whole, though, participants expressed mixed reactions about their experiences in the course. However, over time student responses seemed more favorable toward online course delivery.

How do Teacher Candidates Demonstrate their Understanding of Course Content?

In order to explore this question, participants' online reflections and their curriculum projects were analyzed. In both cases, analysis was constrained to topics related to standards-based instruction, specifically pedagogies, such as posing cognitively-demanding mathematical tasks, asking higher-level questions, and the use of manipulatives.

Mathematical Tasks in Lesson Plans

In the Culminating Diagnostic Project, teacher candidates were encouraged to use cognitively-demanding mathematical tasks, specifically either Procedures with Mathematical Connections tasks or Doing Mathematics tasks (Table 4). The analysis of the mathematical tasks found that there were a substantial amount of Procedures with Mathematical Connections tasks in all lesson plans across all courses (Table 5). Further, there was no difference noted between the online and hybrid courses; teacher candidates in both courses used cognitively-demanding tasks regardless of the course format.

Based on the data in Table 5, there was little difference between the number of cognitively-demanding tasks labeled Procedures with Connections or Doing Mathematics. A one-way t-test was conducted to compare the types of mathematical tasks between online and hybrid courses. In SPSS 17.0, tasks were coded in the following way:

Table 5. Types of mathematical tasks in lesson plans

			M	P w/out C	P w/C	DM
Fall, 2008	Online	28 students	0	3 (1.08%)	245 (88.45%)	29 (10.47%)
	Hybrid	28 students	0	5 (1.67%)	262 (87.33%)	33 (11%)
Fall, 2009	Hybrid	32 students	0	6 (1.63%)	325 (88.08%)	38 (10.3%)
Fall, 2010	Hybrid	44 students	0	5 (1.05%)	422 (88.47%)	50 (10.48%)
Fall, 2011	Online	39 students	0	7 (2.21%)	286 (90.22%)	24 (7.57%)
	Total	171 students	0	26 (1.49%)	1540 (88.51%)	174 (10.0%)

M: Memorization
P w/out C: Procedures without Connections
P w/C: Procedures with Connections
DM: Doing Mathematics

1. Memorization,
2. Procedures without Connections,
3. Procedures with Connections,
4. Doing Mathematics.

The t-test results showed no statistical significant difference between course format and the types of mathematical tasks in lesson plans, $t\,(1738) = -1.159$, $p = 0.247$.

Online Reflections

Data from online reflections focused largely on teacher candidates' responses to the various modules and course activities. Most of these activities included watching a video or reading a classroom vignette and responding about both the teacher's and students' actions.

Data across all courses showed that participants reported favorable impressions of the use of cognitively-demanding mathematical tasks and higher-level questions during the vignettes. Participants shared that the vignettes provided opportunities to see these concepts in action in classrooms, which gave a clearer picture of what they look like.

The video clips were a huge help. In the first course and this one, we were taught about the value of posing good math tasks and asking high level questions. I haven't seen that though in my clinical classrooms, so I didn't know what it looked like until I saw these videos. (Student, Fall, 2011 semester)

Many students shared how the use of video vignettes coupled with clinical experiences helped increase their buy-in about how the type of tasks and questions influence students' learning.

Until I saw the videos, I didn't really believe math could be taught this way. The teacher in the first video posed a hard task, let students struggle, and then asked questions to help them. The approach actually worked. I was surprised, but it was nice to see. (Student, Fall, 2011 semester).

In some cases the online reflections provided an opportunity for teacher candidates to spend time on more broad issues, such as classroom management issues.

In the [vignette] that we read, I was mostly interested in how the teacher managed classroom behavior and her system of disciplining students. It was interesting to see how she dealt with students who acted out during math class. (Student, Fall, 2008 semester).

Another student commented, "I was surprised at how wild the students were on the video. I am not sure I could even start a math lesson if a class were behaving that way." (Student, Fall, 2011 semester).

The online reflections provided an opportunity for teacher candidates to process what they were gleaned from these video and text-based vignettes with topics that they were learning. Still, as seen above, some comments were not focused on the mathematics, but about much more broad issues, such as classroom management.

Curriculum Reflections

The goal of the curriculum reflection was to have students share their views on the standards-based mathematics textbook that they analyzed in light of what they have read and their beliefs about how students learn mathematics. The analysis of teacher candidates' curriculum reflections from the 2008 semester indicated that teacher candidates acquired a value for standards-based curriculum, but their views of curriculum were much more focused on the supplemental materials and teacher support than the types of tasks included in the curriculum.

I have learned to value this type of mathematics teaching with lots of hands-on activities and engaging activities for students. However, the curriculum that I looked at did not have as many manipulatives or books as the textbook that my school has. (Student, Fall, 2008 semester).

All of the good tasks in the world don't help, if the textbook doesn't give me a detailed picture of what the lesson should look like. I feel like I need a textbook that has a lot of different resources so I can choose which ones to use with my students. (Student, Fall, 2008 semester).

During the 2011 online course, teacher candidates reported that their preferences were largely influenced by their 15 hours of clinical experiences and by observing their clinical teachers' work with curriculum.

My teacher uses standards-based curriculum in third grade, and it is amazing to hear how the students are talking about mathematics. At first, I thought that I would favor a traditional book with more problems and just hands-on activities, but I really like the idea of standards-based materials that have rich tasks and a lot of opportunities for students to share their ideas and strategies.

The influence of clinical teachers also negatively influenced teacher candidates' feelings about curriculum. This was especially true in Grades 3 through 5, which have a high-stakes state-wide assessment to take later in the year. "My grade level was forced to use standards-based books and it is too hard for these students. They don't have any freedom and I would hate being in that position."

The reflections from students in the hybrid courses during 2009 and 2010 focused more on the quality of mathematical tasks. For example, "I think that my ideal textbook is one with demanding tasks, and opportunities for students to use a lot of hands-on manipulatives. I think teacher support is important, but not as important as a book with good mathematical tasks." (Student, Fall, 2009 semester).

Another commented, "A textbook is just a resource. If it has high-level tasks which are hard to find, then that is a huge plus for teachers."

In general, reflections from online courses focused much more on the use of manipulatives in 2008, and their observations from clinical experiences in 2011. Responses from teacher candidates in the hybrid sections spent much more time talking about how a curriculum is a set of resources and the quality of mathematical tasks is the top factor in its effectiveness.

To what Extent do Teacher Candidates' Impact Student Learning on the Courses' Culminating Diagnostic Project?

On the Culminating Diagnostic Project, teacher candidates taught a small group of 3-5 students after conducting a pre-assessment. After the lessons, a post-assessment was also given. Table 6 shows the percentage of students showing growth between the pre- and post-assessments. Over 97 percent of students showed growth between assessments in all courses.

As shown in the table nearly every elementary school student (97.84%) made growth between their pre-test and post-test. A t-test comparing online to hybrid models showed no statistically significant difference between course formats, $t(555) = -0.103$, $p = 0.918$.

How Has Data Informed Revisions and Modifications of the Course?

The modifications of ELED 5301, including the development of an online course has been largely based on data collected from teacher candidates during the course, course evaluations, and through conversations with past students after they had completed the course and their student teaching

Table 6. Percent of students showing growth

Semester	Format	Teacher Candidates	Total Students	Students Showing Growth	Percent of Students Showing Growth
Fall, 2008	Online	28	111	108	97.30
	Hybrid	28	110	107	97.27
Fall, 2009	Hybrid	32	98	96	97.96
Fall, 2010	Hybrid	44	125	123	98.40
Fall, 2011	Online	39	113	111	98.23
	Total	171	557	545	97.84%

semester. The revisions mainly fall into three categories: mathematics content, pedagogies, and preparation to teach.

Mathematics Content

This course was originally designed to address geometry, data and non-number sense concepts. The first course focuses heavily on number sense and the use of manipulatives to facilitate computation. However, as I continued to teach the course, especially in online and hybrid formats, I noticed that students' understanding of mathematical content in many areas, especially fractions and algebra were incredibly weak. This is incredibly troubling, as a large portion of content emphasized in the upper elementary school grades is fractions and algebra. During student teaching, teacher candidates anecdotally report a desire to spend more time working on fractions and algebra. Therefore, the course content since 2008 has focused more on fractions and algebra. Geometry is still taught, but a majority of the content is fractions and algebra.

Pedagogies

One of the goals of the course is to continue to develop teacher candidates' understanding of standards-based mathematics with a focus on cognitively-demanding tasks and high-level questions that address students' mathematical thinking. The online modules in this course have provided a venue to use more video vignettes in the course. Currently, teacher candidates view videos from *My Kids Can* (Storeygard, 2009) and the Annenberg Mathematics Library (http://www.learner.org/resources/series32.html). Prior to the 2010 semester, the Annenburg videos were the only resource used. Teacher candidates responded favorably to the use of videos in online modules, so *My Kids Can* was adopted as the course textbook. It includes video and text-based vignettes.

The focus on mathematical tasks and questions was emphasized in the videos, but also as students observe in classrooms, and plan and teach lessons. As teacher candidates wrote lesson plans and reflections from their experiences, emphasis was heavily placed on the types of mathematical tasks and the questions that were posed.

Teaching Experiences

During the 2008 semester, the number of clinical projects that teacher candidates completed in elementary school classrooms was increased. Based on program data about the lack of time that our teacher candidates spend in classrooms, the current course stresses course experiences with elementary school students (Table 7).

Based on the data collected during the past few years, teacher candidates have expressed a higher level of preparedness to teach mathematics after student teaching. Part of that is likely contributed to the amount of time that they spend

Table 7. Current clinical experiences in the course

Experience	Estimated Time
Culminating Diagnostic Project	7 hours
Observing Mathematics Lessons	3 hours
Teaching a Whole Group Lesson	1-2 hours
Student Interviews	1-3 hours

in mathematics-related activities during their clinical experiences.

The potential downside of the intensive amount of clinical experiences is the quality of the classroom where teacher candidates are placed. With the courses' emphasis on standards-based mathematics instruction, teacher candidates who are placed in classrooms where the teacher does not use cognitively-demanding tasks or high-level questions express frustration and a disconnect between what they are learning in the course and seeing in the field.

DISCUSSION AND IMPLICATIONS

The evaluation data presented above describe the influence of an online mathematics education methods course for elementary school teacher candidates. The findings above extend the knowledge base on online courses and have relevant implications for future teaching of online mathematics education methods courses.

Clearly Stating Course Expectations to Teacher Candidates

Students' feedback of the online course focused heavily on their perceptions that they had to take a lot of responsibility for their own learning and there was more work involved than a face-to-face course. This finding extends the work of prior studies that found that online courses were more successful when they included relevant and challenging assignments (Levin et al., 2001) and opportunities for critical thinking (Delfino & Persico, 2007).

Consistent with Moallem's (2003) work, the data indicated that the gradual refinement of instructions led to less student frustration about the clarity of assignments. At the start of the 2011 semester the instructor e-mailed students to clearly explain the nature of the course, time demands, and the philosophy behind the course's focus on projects and self-paced learning. Providing clarity helped students to have a better understanding of course expectations. Online instructors of methods courses might also lean more towards a face-to-face orientation meeting prior to the beginning of the semester, or incorporating synchronous experiences to increase the social nature of courses (Im & Lee, 2004; Tallent-Runnels et al., 2006). Future studies should examine ways that course instructors can clearly communicate expectations to students, and the influence that those initial and during-course messages have on the quality of students' work.

Ensuring Effective Clinical Activities

All versions of this course described leveraged numerous clinical activities to provide teacher candidates with experiences in elementary school classrooms. The most recent version of the National Council for Accreditation in Teacher Education (NCATE) Standards (2010) heavily emphasizes the need for teacher candidates to participate in worthwhile clinical activities throughout their program. In a 100% online course, clinical activities are more critical since teacher candidates do not have face-to-face course meetings to see the instructor model methods or to discuss these pedagogies. In this study, some teacher candidates completed their clinical requirements with teachers who used standards-based pedagogies that aligned to course objectives. These students reported a high degree of buy-in that these strategies are effective. Schwartz (in press) noted that online courses are limited in that they cannot easily model pedagogies as well as the teacher action of in-the-moment decision making that occurs frequently in classrooms. Instructors of online methods courses for teacher candidates need to ensure that students are placed in classrooms that embody the types of pedagogies emphasized during the course.

Future studies should more closely examine the influence of clinical activities on teacher candidates' knowledge and skills. As O'Connor (2011) noted, the online courses did not effectively prepare pre-service teachers to implement standards-based pedagogies that were emphasized during the course. In an online course, these pedagogies and skills related to in-the-moment decision making must be included in clinical activities and in vignettes that are included in the courses.

Creating a Feeling of Support

Teacher candidates frequently reported comments about the amount of support that they received during the course. In this study, the courses' heavy focus on projects and the application of course content in elementary school classrooms created a high level of stress and the perception that there was a lot of work in the course. For many of the teacher candidates, this was the first time they taught the same group of students multiple lessons. The course instructor provided feedback by helping them to revise their assessments and gave them feedback about their lesson plans. By providing feedback and giving suggestions about lesson plans, teacher candidates reported feeling supported while working on this major assignment. This aligns with Moallem's (2003) findings that students in online courses need to have clear expectations, including information about rubrics and evaluation.

Greene and Land (2000) found that students had clearer expectations of assignments and produced higher quality work products when instructors provided guiding questions and frequent amounts of feedback. This study extends the results of their work. In future studies, there is a need to examine how the different types and manners in which feedback is given influence the quality of students' work products.

Providing Opportunities for Communication

In this study, teacher candidates communicated with each other via their online reflections on either WordPress or Posterous. These required reflections allowed teacher candidates to communicate their thoughts about vignettes that they had examined, their clinical experiences, and other issues related to mathematics teaching and learning. By reading and responding to their classmates' reflections, students were able to interact with their classmates and have shared experiences. Based on the feedback from teacher candidates, this helped reduce the feeling of isolation that is common in online courses (Tallent-Runnels et al., 2006).

The requirement to participate in these online discussions is critical; very few students participated in the optional activities of contributing to and viewing the class wiki. If posting and responding to online reflections was optional, some students would likely not participate at all. This extends the work of Im and Lee (2003/2004), who found that creating asynchronous communication effectively supported discussion of content, but not social conversations. The addition of synchronous communication might have increased participants' willingness to also participate in the optional course activities on the wiki. Future studies should examine types of synchronous and asynchronous communication and compare which types may best support pre-service teachers in a methods course.

CONCLUDING THOUGHTS

Impact of Course

The online mathematics methods course provided an opportunity for teacher candidates to develop the knowledge and skills related to teaching elementary school mathematics. Based on the evaluation data, teacher candidates demonstrated knowledge about mathematics, an understanding of planning cognitively-demanding mathematical tasks, gains in student learning outcomes, and relatively positive perceptions of their experience. While this study was not a direct comparison between the impact of online and face-to-face versions of the same course, the data indicate that the course experience successfully developed aspects of teacher candidates' mathematical knowledge for teaching (MKT) and positively impacted their students.

As program faculty continue to decide whether or not this course is worthwhile to teach in a 100% online format, the evaluation data suggests that it has potential, under the conditions that the course supports communication between students, and the course instructor readily and frequently provides feedback on assignments.

Limitations

While this study provided data about the benefit of an online methods course for pre-service teachers, this study includes some limitations, making it difficult to generalize the findings across various contexts. While the data analysis process included multiple iterations of reading, coding, and reanalyzing the data, the data collected provides a description of the course on a few levels of Guskey's evaluation model. Further, the impact on student learning is limited, as nearly every student showed growth, but it is difficult to acknowledge how much growth was made without a group of control students to compare.

While the findings from this chapter come with limitations, there is enough data regarding the effectiveness to continue to pursue the refinement

Table 8. Research questions for examining online methods courses for teacher candidates

Level		Questions Examined	Data Sources
Level 1	Teachers' reactions	How do teachers feel about their experience? How do they feel about course structures (e.g., communication, course feedback)? How do they feel about course activities (e.g., projects, assignments)?	Course evaluation forms Surveys Interviews Focus Groups
Level 2	Teachers' knowledge and skills	What knowledge and skills do teachers learn during their experiences? What knowledge related to content do teachers learn? What knowledge related to pedagogy do teachers learn? What knowledge related to assessment do teachers learn?	Course assignments Lesson plans Exams
Level 3	Use of knowledge and skills	How do teachers use their new knowledge and skills in their own classroom? To what extent and how do teachers apply emphasized pedagogies? To what extent and how do teachers modify instruction while using emphasized pedagogies?	Student work samples Videos of classroom teaching
Level 4	Impact on organization and program	How did the experience impact the program or organization responsible for providing the experience? What revisions are made to course structures between semesters? What revisions are made to course assignments between semesters?	Course documents, including syllabi, assignments, rubrics Interviews Surveys
Level 5	Impact on student learning	How are PK-12 students impacted by the course? How well do PK-12 students perform on assessments compared to those not influenced by the course?	Student work samples Assessments Student interviews

of an online mathematics education course and continue to research the impact of the course. To that end, future studies should continue to evaluate the impact of the course on teacher candidates as well as K-6 students.

Directions for Future Research

Guskey's (2000) framework for evaluating opportunities for teacher learning provided a research-based foundation to examine how the course influenced various aspects of teacher candidates, the institution and students. I close this chapter with a set of possible questions worth examining for those interested in researching online methods courses for teacher candidates (Table 8).

Due to the cyclical nature of teaching, where instructors teach a semester, and then revise their course based on data, including feedback from their students as well as their own reflections, design-based research ([DBR]; Edelson, 2002; Reeves, Herrington, & Oliver, 2005) offers a potential way of framing research studies regarding online teaching. DBR is based off a preliminary theory that is grounded in a combination of empirical and anecdotal information. The theory leads to a proposed intervention, which is then carried out. Based on formative data that is collected and analyzed during the study, researchers make decisions regarding both the implementation of their intervention as well as make conclusions from the data.

In the case of online teaching, course instructors' structures and content is based off a combination of empirical and anecdotal information. As data is collected during and after the semester, the instructor makes empirically-based decisions, which influence the course, as well as provide information about the influence of the course on their students. In the case of teacher candidates, these courses influence not only teacher candidates, but also the students that they work with during their clinical experiences. DBR provides a systematic way for instructors and researchers involved with online teaching to analyze the impact of the course on students and others involved in the teaching and learning process.

REFERENCES

Ajjan, H., & Hartshorne, R. (2008). Investigating faculty decisions to adopt Web 2.0 technologies: Theory and empirical tests. *The Internet and Higher Education*, *11*(2), 71–80. doi:10.1016/j.iheduc.2008.05.002

Ball, D. L., Thames, M. H., & Phelps, G. (2008). Content knowledge for teaching: What makes it special? *Journal of Teacher Education*, *59*(5), 389–407. doi:10.1177/0022487108324554

Bogden, R. R., & Biklen, S. K. (2003). *Qualitative research in education: An introduction to theories and methods* (4th ed.). Boston, MA: Allyn & Bacon.

Carter, J. F. (2009). Lines of communication: Using a wiki in a mathematics course. *Problems, Resources, and Issues in Mathematics Undergraduate Studies*, *19*(1), 1–17.

Cohen, S. (2005). *Teachers' professional development and the elementary mathematics classroom: Bringing understandings to light*. Mahwaw, NJ: Lawrence Erlbaum Associates, Inc.

Delfino, M., & Persico, D. (2007). Online or face-to-face? Experimenting with different techniques in teacher training. *Journal of Computer Assisted Learning*, *23*, 351–365. doi:10.1111/j.1365-2729.2007.00220.x

Edelson, D. C. (2002). Design research: What we learn when we engage in design. *Journal of the Learning Sciences*, *11*(1), 105–121. doi:10.1207/S15327809JLS1101_4

Fishman, B. J., Marx, R. W., Best, S., & Tal, R. T. (2003). Linking teachers and student learning to improve professional development in systemic reform. *Teaching and Teacher Education*, *19*(6), 643–658. doi:10.1016/S0742-051X(03)00059-3

Garet, M., Porter, A., Desimone, L., Briman, B., & Yoon, K. (2001). What makes professional development effective? Analysis of a national sample of teachers. *American Educational Research Journal*, *38*(4), 915–945. doi:10.3102/00028312038004915

Glazer, E. M., & Hannafin, M. J. (2006). The collaborative apprenticeship model: Situated professional development within school settings. *Teaching and Teacher Education*, *22*(2), 179–193. doi:10.1016/j.tate.2005.09.004

Greene, B. A., & Land, S. M. (2000). A qualitative analysis of scaffolding use in a resource-based learning environment involving the World Wide Web. *Journal of Educational Computing Research*, *23*(2), 151–180. doi:10.2190/1GUB-8UE9-NW80-CQAD

Guskey, T. R. (2000). *Evaluating professional development*. Thousand Oaks, CA: Corwin Press.

Heck, D. J., Banilower, E. R., Weiss, I. R., & Rosenberg, S. L. (2008). Studying the effects of professional development: The case of the NSF's local systemic change through teacher enhancement initiative. *Journal for Research in Mathematics Education*, *39*(2), 113–152.

Hodges, C. B., & Hunger, C. M. (2011). Communicating mathematics on the Internet: Synchronous and asynchronous tools. *TechTrends*, *55*(5), 39–44. doi:10.1007/s11528-011-0526-4

Im, Y., & Lee, 0. (2004). Pedagogical implications of online discussion for preservice teacher training. *Journal of Research on Technology in Education*, *36*, 155–170.

Ko, S., & Rosen, T. (2010). *Teaching online: A practical guide*. New York, NY: Routledge.

Levin, S. R., Waddoups, G. L., Levin, J., & Buell, J. (2001). Highly interactive and effective online learning environments for teacher professional development (electronic version). *International Journal of Educational Technology*, *2*. Retrieved from http://smi.curtin.edu.au/ijet/v2n2/slevin/index.html

Loucks-Horsley, S., Love, N., Stiles, K. E., Mundry, S., & Hewson, P. W. (2009). *Designing professional development for teachers of science and mathematics* (3rd ed.). Thousand Oaks, CA: Corwin Press.

Moallem, M. (2003). An interactive online course: a collaborative design model. *Educational Technology Research and Development*, *51*(4), 85–103. doi:10.1007/BF02504545

National Council for Accreditation of Teacher Education. (2010). *NCATE unit standards*. Retrieved from http://www.ncate.org/standards/ncateunitstandards/unitstandardsineffect2008/tabid/476/default.aspx

National Partnership for Excellence and Accountability in Teaching (NPEAT). (2000). *Revisioning professional development: What learner-centered professional development looks like*. Oxford, OH: Author. Retrieved from http://www.nsdc.org/library/policy/npeat213.pdf

Nelson, J., Christopher, A., & Mims, C. (2009). TPACK and web 2.0: Transformation of teaching and learning. *TechTrends*, *53*(5), 80–85. doi:10.1007/s11528-009-0329-z

O'Connor, E. A. (2011). The effect on learning, communication, and assessment when student-centered Youtubes of microteaching were used in an online teacher-education course. *Journal of Educational Technology Systems*, *39*(2), 135–154. doi:10.2190/ET.39.2.d

Oliver, K. (2010). Integrating Web 2.0 across the curriculum. *TechTrends*, *54*(2), 50–60. doi:10.1007/s11528-010-0382-7

Orrill, C. H. (2001). Building learner-centered classrooms: A professional development framework for supporting critical thinking. *Educational Technology Research and Development*, *49*(1), 15–34. doi:10.1007/BF02504504

Peterson, E. (2009). Using a wiki to enhance cooperative learning in a real analysis course. *PRIMUS (Terre Haute, Ind.)*, *19*(1), 18–28. doi:10.1080/10511970802475132

Polly, D., & Hannafin, M. J. (2010). Reexamining technology's role in learner-centered professional development. *Educational Technology Research and Development*, *58*(5), 557–571. doi:10.1007/s11423-009-9146-5

Reeves, T. C., Herrington, J., & Oliver, R. (2005). Design research: A socially responsible approach to instructional technology research in higher education. *Journal of Computing in Higher Education*, *16*(2), 97–116. doi:10.1007/BF02961476

Russell, M., Carey, R., Kleiman, G., & Venable, J. D. (2009). Face-to-face and online professional development for mathematics teachers: A comparative study. *Journal of Asynchronous Learning Networks*, *13*(2), 71–87.

Schifter, D., Bastable, V., & Russell, S. J. (2008). *Developing mathematical ideas*. Parsippany, NJ: Dale Seymour.

Schlager, M. S., & Fusco, J. (2004). Teacher professional development, technology, and communities of practice: Are we putting the cart before the horse? In S. Barab, R. Kling, & J. Gray (Eds.), *Designing virtual communities in the service of learning* (pp. 120–153). Cambridge, MA: Cambridge University Press. doi:10.1080/01972240309464

Schwartz, C. S. (2012). Counting to 20: Online implementation of a face-to-face, elementary mathematics methods problem-solving activity. *TechTrends*, *56*(1), 34–39. doi:10.1007/s11528-011-0551-3

Signer, B. (2008). Online professional development: Combining best practices from teacher, technology and distance education. *Journal of In-service Education*, *34*(2), 205–218. doi:10.1080/13674580801951079

Spicer, J. (2002). Even better than face-to-face? In A. Thorson (Ed.), *By your own design: A teacher's professional learning guide* (pp. 32–33). Columbus, OH: Eisenhower National Clearinghouse for Mathematics and Science Education.

Stein, M. K., Grover, B. W., & Henningsen, M. (1996). Building student capacity for mathematical thinking and reasoning: An analysis of mathematical tasks used in reform classrooms. *American Educational Research Journal*, *33*, 455–488.

Storeygard, J. (2009). *My kids can*. Boston, MA: Heinemann.

Tallent-Runnels, M. K., Thomas, J. A., Lan, W. Y., Cooper, S., Ahern, T. C., Shaw, S. M., & Liu, X. (2006). Teaching courses online: A review of the research. *Review of Educational Research*, *76*(1), 93–135. doi:10.3102/00346543076001093

Thames, M. H., & Ball, D. L. (2010). What mathematical knowledge does teaching require? Knowing mathematics in and for teaching. *Teaching Children Mathematics*, *17*(4), 220–225.

Treacy, B., Kleiman, G., & Peterson, K. (2002). Successful online professional development. *Leading & Learning with Technology*, *30*(1), 42–47.

Work, A. P. A. Group of the Board of Educational Affairs. (1997). Learner-centered psychological principles: A framework for school reform and redesign. Washington, DC: Author.

ADDITIONAL READING

Ball, D. L., Thames, M. H., & Phelps, G. (2008). Content knowledge for teaching: What makes it special? *Journal of Teacher Education*, *59*(5), 389–407. doi:10.1177/0022487108324554

Ko, S., & Rosen, T. (2010). *Teaching online: A practical guide*. New York, NY: Routledge.

Moallem, M. (2003). An interactive online course: A collaborative design model. *Educational Technology Research and Development*, *51*(4), 85–103. doi:10.1007/BF02504545

Tallent-Runnels, M. K., Thomas, J. A., Lan, W. Y., Cooper, S., Ahern, T. C., Shaw, S. M., & Liu, X. (2006). Teaching courses online: A review of the research. *Review of Educational Research*, *76*(1), 93–135. doi:10.3102/00346543076001093

KEY TERMS AND DEFINITIONS

Asynchronous: A format of an online course in which the students and course instructor communicate electronically but not in real-time.

Blog: A shortened phrase for weblog, an online system in which users can include text, audio, or video, and others can comment on their postings.

Clinical Experience: Activities that are completed by teacher candidates in PK-12 classrooms.

Hybrid: A format of an online course in which students have a combination of face-to-face meetings and online modules.

Mathematical Task: Any mathematical problem or activity completed by a student.

Methods Course: A course in a teacher education program that specifically focuses on content and pedagogy related to a topic (e.g., mathematics, science, social studies).

Synchronous: A format of an online course in which the students and course instructor communicate in real-time, typically in a computer system that has audio and/or a chat room interface.

This work was previously published in "Teacher Education Programs and Online Learning Tools" edited by Richard Hartshorne, Tina L. Heafner, and Teresa Petty, pages 335-356, copyright 2013 by Information Science Reference (an imprint of IGI Global).

Chapter 35
Sustainability in Higher Education through Basic Science Research:
Strategies for Corporate Bodies in Pharmaceuticals

P. Yogeeswari
Birla Institute of Technology and Science-Pilani, Hyderabad Campus, India

D. Sriram
Birla Institute of Technology and Science-Pilani, Hyderabad Campus, India

ABSTRACT

Basic research in Universities is essential for a sustainable development. Recent developments in higher education have seen the inclusion of curiculum redesigned to serve the concept of developming young minds in the intrepretation, execution, and use of basic science research. The challenge for educators is to "demystify" research and teach in ways that are professionally meaningful as well as intellectually acceptable. The objective of this chapter is to bring in various case studies to prove the essentiality of basic research in higher education with specific concern over pharmaceutical industrial growth plans. Data on R&D in higher education can be broken down by field of science (natural sciences, engineering, medical sciences, agricultural sciences, social sciences, and humanities), by type of costs (current expenditures, capital expenditures), and by source of funds (business enterprise, government, higher education, private non-profit, and funds from abroad). Measures of R&D performance in the higher education sector are often estimated by national authorities, and evaluation methods are periodically revised. It is necessary to review the design and conduct of higher education R&D surveys to improve the comparability of these indicators.

INTRODUCTION

Especially in the field of Pharmaceuticals, there are perceptions that the pace of new drug development has slowed down and the pharmaceutical industries with high profit have sparked concerns of the future drug development. Pharmaceutical manufacturing lags far behind the manufacturing techniques of the 21st Century and the science required to transform a new discovery to a market-

DOI: 10.4018/978-1-4666-7363-2.ch035

able new drug has definitely not kept pace with advances on the drug discovery side. This is mainly attributed to lack of basic scientific research training. Universities like Birla Institute of Technology & Science, Pilani have taken initiatives way back in 1970s to involve corporate leaders in their curriculum design and implementation.

The present chapter would include various strategic thoughts and implementation plans based on various university initiatives and some ideas on executive development programs in the areas of pharmaceutical education.

This book chapter would raise question among the readers to what the mission and value of basic research in higher education means in the field of pharmaceutical sciences and how corporates could get an impact or value-addition in the light of changing biomedical research and healthcare environments globally. The questions, and the views presented in this chapter are intended to promote discussion among all of the many stakeholders in pharmacy education. In the history of education, contribution by various eminents including Prof. Schroff in the harmonisation of pharmacy education in the global arena is worth mentioning. Prof. Schroff on the call of Pundit Madan Mohan Malavya, the then Vice-Chancellor of the prestigious Banaras Hindu University started the regular B.Pharm course of three years duration in the year 1937. These graduates were expected to be efficient in quality control and standardization of drugs for pharmaceutical industries. Consistent with the changing society and development of health care system, there is an incessant redesigning of the quality of undergraduate-level courses (Commission to Implement Change in Pharmaceutical Education, 1993). In the year 1944, Dr. Khem Singh Grewal introduced pharmacology as a subject in the curriculum and a mandatory three months. Since then the pharmacy education is making progressive strides in India. Pharmacy education is in the cross roads of chemistry and biology and now integrated with information technology. Later on research dimension is also added constituting an important instrument for the creation of new knowledge.

From handful of 50 degree institutions 50 years ago, we now have more 300 degree colleges training more than 10-20 thousands of students. The number of M.Pharm and Ph. D. aspirants has also quite significantly increased in the past 2 decades. Although, graduate education and research programs have made significant contributions to the profession of pharmacy, to science, and to society, changes in today's research, education, and health care environments are stimulating the need for evaluation of whether some of these programs have outlived their usefulness. To understand the effectiveness of our need to revolutionize higher education in India, we need to understand the current status and changing scenario with regard to industrial expectations.

In various nature India forums the problems of basic science research in India has been discussed and debated and the root problem has been indicated to the basic science education system in India. With regard to the courses in pharmacy, these are not well defined and mostly directed to the conventional needs of the industry (Etzkowitz & Leydesdorff, 2000). The elements of the course have become outdated and provide no or little interaction with the status and growth of the pharmaceutical industries. It becomes necessary to follow the rapid and continuous changes in pharmaceutical industry, which makes academia-industry interaction a necessity. In the current curriculum, interaction with the pharmaceutical industries is hard to witness, which ultimately creates a gap between the academia and industries.

CURRENT STATUS OF HIGHER EDUCATION AND BASIC RESEARCH IN INDIA

A disquieting tendency in India and in many of the developing countries is the increasing disinterest in science among the younger generation.

Many colleges are closing down science departments because students prefer to take courses in management, commerce, and related areas. And there are fewer bright students studying for careers in science and engineering research or higher education. This situation has to be remedied by promoting talent and by offering incentives. It is likely that the vast population of India and other developing countries harbors more than a few geniuses, possibly future Faradays and Newtons. But we have to find them and encourage them to pursue science.

From the school days a child opting science stream is destined to take utmost pressure and struggle to attain a seat in medicine, engineering or information technology. Today these three streams are considered by most of the children and parents to be the most sought for as if these only form the whole tree of science. This is because there is a strong gut feeling among school children that it is easy to get jobs with engineering, medicine, computer science and allied fields. Although some are able to make into these fields, a majority of good budding science students count themselves as failures. And by chance some students end up in basic science education or pharmacy with lot of frustrations. And when such frustrated and demoralized failure cases enter into higher education and then research, how would we expect India to compete in Modern Science. This is extremely debate-provoking scenario. One of the major problems facing youth of our country which dissuades them to take up science or pharmacy as a career is the requirement to understand the concerned subject at deepest possible level.

The proportion of the population with a 10th (high school) and 12th (higher secondary) degree has increased significantly, from 8.2% (69.7 million) in 1991 to 23% (246.9 million) in 2004. Those with graduate degrees and above have risen from 2.4% (20.5 million) of the population in 1991 to around 4.5% (48.7 million) today. The proportion of diplomas has risen more than ten times and is currently around 0.4% (3.9 million) of the population. In 2004, about a fourth of those qualified to the level of graduate and above had a background of science education. There are 39.2 million graduates in all (22.3% of whom are from the science stream), 9.3 million postgraduates (19.4% of whom are from the science stream), and 0.3 million doctorates (one-third from the science stream). While both the central government and the state governments spend around four per cent of GDP on education each year, there has been a sharp hike in private spending on education. Between 35% and 40% of government expenditure gets spent on elementary education, another fourth on secondary education while just a tenth goes to university and higher education. Though private spending is higher in the richer states, where government spending also tends to be high, private spending as a proportion of the total spending on education in the state tends to be higher in poorer areas. For instance, in Punjab, the government spent Rs 845 per person in 2000–01 and the state's citizens reciprocated by spending 30% less. In Bihar, where the government spent only Rs 44 per person in 2000–01, the average citizen spent Rs 168 in 2001–02 on education. Interestingly, it is states like West Bengal and Punjab, not Kerala, which emerge as states with the highest per capital expenditure on education.

Science and technology (S&T) are the drivers of economic growth and science education the backbone of all S&T efforts in any country. It is being realised that further improvement in the nation's competitiveness is possible by having a better-educated population. Within the broad umbrella of human capital, the role of scientific manpower is critical, and there is a close relationship between human resource in science and technology and economic growth. For better socio-economic growth, it is important to know how this pool of skilled manpower is being utilized (Stone, 2012).

As per Report of the Higher education in India, Issues Related to Expansion, Inclusiveness, Quality and Finance, the access to higher education measured in term of gross enrolment ratio increased from 0.7% in 1950/51 to 1.4% in 1960–61. By 2006/7 the GER increased to about 11 percent. By 2012, (the end of 11th plan objective) is to increase it to 15%. The current growth in education is clearly depicted in numbers as shown in Table 1.

India's higher education system is the third largest in the world, after China and the United States. The main governing body at the tertiary level is the University Grants Commission (UGC) (India), which enforces its standards, advises the government, and helps coordinate between the centre and the state. Accreditation for higher learning is overseen by 12 autonomous institutions established by the University Grants Commission.

PHARMACEUTICAL GLOBALIZATION AND DRUG DISCOVERY RESEARCH

Pharmaceutical drug discovery is considered to be a major key for any country's national security, public health and economic development. Domestic drug development is very challenging and depends on the status of basic research in our country. United States has the largest world market for drug market due to the fact that the country has the world's strongest patent protection, absolute expenditures on R&D and due to no price control by the government. If our country does not produce drugs on its own, then it will have to purchase essential medicines from foreign countries at a higher cost (Wadwa et al., 2008). This could be prevented by domestic production by inculcating training at the level of higher education. The major advantage of developing drugs on our own would drive tax revenues and would yield foreign currency from exports. India adopts agreement on trade-related aspects of intellectual property rights (TRIPS) to protect product patents and which had many failure cases including that of Novaritis's product Gleevac which made the CEO to comment as follows,

Table 1. Growth of higher education based on UGC report published in March 2012

S. No.	Sector	Fold Increase
1.	Universities	18.8
2.	Teaching staff	34.69
3.	Student enrollments	42.76
4.	Colleges	47.52

Unfavourable patent ruling is not an invitation to invest in Indian R&D. Company will redirect hundreds of millions of dollars in investment to countries where it has greater IP protection.

These are a major concern for our country to revisit issues related to drug discovery and production domestically to become an economically well-driven country. In 2006, revolutionary improvements started taking place in the drug innovation sector. About 5.5 percent of all global pharmaceutical patent applications (WIPO PCT applications) named one inventor or more located in India, and 8.4 percent named one or more located in China. This increase accounts for a four-fold from 1995. Through interviews with executives of 16 pharmaceutical firms in China and India on their business models, value-chain activities, partnerships and technology capabilities, the researchers found that:

- Indian and Chinese companies are making strides in the most lucrative segments of global value chains. In less lucrative segments, such as preclinical testing, animal experimentation and manufacturing, Chinese firms appear to be more prevalent.
- India is regarded as a more mature venue for chemistry and drug-discovery activities than China.
- Domestic Indian and Chinese firms rarely have the capital and the regulatory expertise to develop a drug beyond phase II clinical trials. Their commercial develop-

ment of new intellectual property therefore necessitates relationships with major multinational corporations.

- Pharmaceutical multinationals have begun partnering with Indian firms to conduct fixed dose-eruption skin-reaction tests. Competition has now significantly reduced the cost of this work, from USD 80,000 to approximately USD 30,000.

According to the study, because Indian drug companies have the most experience in selling generic drugs that meet FDA standards, India is playing a more strategic role in early discovery. Companies such as Ranbaxy, Aurigene, Advinus, Nicholas Piramal and Jubilant have negotiated long-term deals with Western pharmaceutical companies to discover and develop new chemical entities.

PHARMACY EDUCATION AND IMPACT ON PHARMACEUTICAL INDUSTRIES

The history of pharmacy education in India is as old as when the country was being gravitated towards British Emperorship. As mentioned earlier, the epicentre of pharmacy education started with Banaras Hindu University at Varanasi at the heart of India. Globalisation has revolutionised the educational process across the globe. Pharmaceutical education plays a very prominent role in attaining sustainable and equitable development of a country. The curriculum of the degree in some developed countries (B. Pharm.) usually requires 5 academic years of study. In most of the European countries successful completion of a university degree leads to a one-year internship. The formal pharmacy education in India (3 year degree in BHU) dates back 1932, and since then, there has been a continuous growth in number of such institutions. As per Pharmacy Council of India (PCI) 2005 diary calendar, the total numbers of recognized degree institutions are 220 with intake of 12506 students. And as per All India Council of Technical education (AICTE), the total number of degree colleges is 445 with the intake of 24672 students as well 30 institutions for the post graduation in various fields (Suresh, 2012).

There is no doubt that currently there is enormous gap existing between education and practice of pharmacy. Most of the academic institutions providing education in pharmacy are away from practice environment. The overall basis of pharmacy education is still extrabiological synthesis, physicochemical studies, analysis, and manufacturing aspects of drug. It is a common feeling that the medical practitioner is better placed for pharmacists' job than the pharmacists themselves. The dispensing services are poor. The syllabus and duration of the two-year diploma course in pharmacy education in India is completely outdated and irrelevant in the present industry context. Pharmacy as a nascent science developed like this in the last century (Yeole & Puranik, 2005). During 1940s and 50s, hospitals and industries were established in large numbers in India. Consequently, pharmacists and pharmaceutical chemists were required in huge numbers. Hence pharmacy education was developed in such a way to satisfy the requirement of industry and hospital. Short-term compounders and or D. Pharm. course to satisfy the needs of hospital and medical shops and B. Pharm. course for the industry were started. This is proved by the fact that in the last few decades D. Pharm. holders are not employed by the industry and B. Pharm. holders are not in many numbers in hospitals or medical shops. In the West, pharmacy education is patient-oriented and is responsible for Healthcare Management, while in India pharmacy education is industry-oriented. Nearly 55 per cent of the jobs are available in the industry sector while 30 per cent in education. The approach of carrying both the aspects of pharmacy practice and industrial pharmacy together hasn't shown success as the course structure for practicing pharmacy is not relevant to the requirements of an industry and vice-versa. Hence, it is pertinent and relevant to have separate degree courses of education in practicing pharmacy and industrial pharmacy.

NEED FOR INDUSTRY-ORIENTED TRAINING FOR PHARMACIST

Pharmacy courses are not well defined and not even updated in most of the universities in our country. With the evolving scenario of more multinationals setting up their research facility in India, pharmacy education in India need to redesign curriculum to introduce course more relevant to the industries. Moreover like an internship program for doctors, pharmacy students should also have an internship in industry to understand the application of the knowledge relevant to the need of the industry (Fischli, 1999). The elements of the course have become outdated and provide no or little interaction with the status and growth of the pharmaceutical industries. It becomes necessary to follow the rapid and continuous changes in pharmaceutical industry, which makes academia-industry interaction a necessity. This could be facilitated by more industry-institute interactions and collaborations. This definitely strengthens the ties for organizing campus placement services for the graduates and to generate resources for research and development also through consultancy services. In the current curriculum, interaction with the pharmaceutical industries is hard to witness, which ultimately creates a gap between the academia and industries. Our academic area is that end which has been at a fixed place for a long time while the industry on the other end has continuously made rapid advancements in technology. If this gap keeps on widening then industries will no longer prefer pharmacy students in their companies as they will not be able to fulfill their current requirements.

Academic institutes with a focus on industry-oriented curriculum will be a better place to generate expertise in patentable ideas and industries would be benefitted by collaborating with such institutes. In turn, institutes are benefitted by sending students and faculty for internship program to acquire the latest technological know-how from the industry. Benefits of these mutual exchanges of ideas and symbiotic relationship are multifold. To achieve these, regular upgradation of curriculum will generate manpower in ready to-use mode that could be utilized and absorbed by industries (Friedman, 2010). More value should be given to projects/assignments, conceptual and practical approaches of teaching so that students can spend more time on understanding the subject and develop specific skills. Involvement of industry experts in framing the syllabus is yet another way to get updated with industrial developments. In the new millennium due to technological innovation and improved communication, drastic changes are taking place in every field. The field of pharmacy education cannot keep away from these changes. Harmonisation of pharmaceutical education has to be made a global agenda that will encompass the development that have taken place in basic, medical, and pharmaceutical sciences in serving the needs and expectations of industry and universities.

DEVELOPMENTS IN THE BASIC SCIENCE RESEARCH IN THE CONTEXT OF HIGHER EDUCATION

In our rapidly expanding global scientific research enterprise, good science anywhere is good for science everywhere, provided that there exists an open flow of information with transparent processes to promote rigorous peer review and scientific integrity. Basic sciences in the area of pharmacy include the traditional areas like pharmacology, medicinal chemistry and pharmaceutics. The end of the Cold War, the rapid growth of international competition in technology-based industries, and a variety of constraints on federal research spending have altered the market for PhD-trained scientists. Although the demand for scientists has remained strong (as evidenced by low and stable unemployment rates), the three areas of primary employment for PhD scientists–universities and colleges, industry, and government–are undergoing simultaneous change. Two compelling indicators of problems for the science work force, particularly in academia, are the growing postdoc population and the declining grant approval rate.

Recognizing and promoting excellence, so as to discover and access new frontiers of knowledge, is an imperative which should be possible for all countries whatever their level of economic development. Yet these frontiers are often in the fields of science, technology and engineering, health care, agriculture and economics where highly-educated and skilled human capital, along with large-scale investment, is essential to appropriate context of enquiry. The research function of academia remains a prime source of knowledge and innovation at national, regional and international levels. Yet, over the past decade, most industrialized states have been obliged to address the double challenge of providing wider access to postsecondary education and training and ensuring adequate investment in high-level research. This is proving to be a delicate balancing act, which hinges on visionary policies and a more diversified funding base. Governments pursue reforms to build world-class systems of higher education, which assure quality in both research and teaching. In contrast, the term "World-Class University" tends to denote research-oriented institutions, although this should also recognize those who achieve excellence through innovative approaches to learning. For universities wishing to enhance their research reputations, the challenges continue to grow. Today, some twenty-two of the world's elite twenty-five research universities (known as "Super RUs") are located in one country, the United States of America (USA). While American higher education deserves full credit for the breadth and resourcing of this sector, this monopoly cannot be expected to meet global needs in terms of research (Williard, 1996). For this reason support for research universities, notably those with science, technology and innovation strengths, has become an important priority in OECD (Organisation of Economic Co-operation Development) Member countries. The rise in status and influence of various ranking systems (aiming to evaluate excellence in academic research) has influenced this situation. Investment in research is increasing in emerging economies including India (Shukla, 2005). Postgraduate education and training has assumed new importance as an underpinning to this policy approach, and a dual agenda must be adopted: resources should be made available at this level, even where countries currently struggle with the provision of basic and secondary education. Even the poorest nations require research capacity, or access to research findings, to progress; and so it could be argued that support for the principle of a research university in these contexts is more urgent than ever before. Reaching this goal, and maintaining the quality and relevance of these essential institutions, requires national commitment and must remain a major objective for international cooperation in the years ahead. Current issues facing the research function and its environment include equity; quality; relevance; ownership; and international networking. An ever-growing number of nations of varying size have now given priority to developing their knowledge base through higher education, research and innovation, and to commit the necessary resources to this goal. Success stories are becoming more common in all regions, and they are characterized by specific indicators (Meek e al., 2009):

- Innovative policies in higher education and research and in Science, Technology and Innovation (STI).
- A will to improve and profile the necessary infrastructure, including universities.
- Efforts to train and retain and attract highly-skilled human capital.
- Increased levels of investment in research and higher education.

The mandate of the UNESCO Forum is to chart these important processes and help promote their replication and adaptation worldwide, in order to render the global knowledge society a more level playing field.

It seems reasonable to assert that graduate education, research, and/or scholarly activity of faculty is necessary for the maintenance of a high level of excellence in professional education in pharmacy. The essential question seems to be what types of graduate education; research and/or scholarly activity in the pharmaceutical sciences are likely to promote a high level of excellence in professional pharmacy education today and in the future? Although graduate education cannot exist without research, research can exist without graduate education. Pharmaceutical education is responsible to the profession and to society for generating new knowledge about drugs, drug products, drug therapy, and drug use through the conduct of basic and applied research. This would promote the pharmaceutical sciences by fostering graduate education and research within its schools and colleges. Pharmaceutical education is responsible for both professional education and graduate education for research.

At the recent Indian Science Congress held in January 2012, our Prime Minister, Dr. Manmohan Singh pledged to hike R&D expenditures from around $3 billion last year to $8 billion in 2017 (Jogalekar, 2012). The windfall is meant to turbocharge initiatives to create elite research institutions, bring expatriate Indian scientists home, enrich science education, and equip smart new laboratories. Indian scientists in Universities would be cheering the government's new initiative.

ADVANTAGES OF INTEGRATING BASIC SCIENCE RESEARCH PROGRAM

Science and Technology have played a very decisive role in the overall development of mankind. Scientific knowledge has led to remarkable innovations that have been of great benefit to humankind. Rapid advancements in science and technology have brought the subject to a centre-stage. The global scientific knowledge and information are increasing at a very fast rate. The doubling time of knowledge has drastically reduced from decades to years and from years to a few months (Khar et al., 2011). The economic growth critically hinges on inputs from science & technology of the highest kind and this gets linked to the doctoral programme ion Universities.

What are the contributions of research programs in basic sciences to professional education? How significant are these contributions? Are the contributions essential and unique or are there other means of achieving them? There are primarily two ways in which basic science research departments are seen to contribute to the mission of the pharmacy profession. The first is that the faculty in these departments provides instruction in basic science coursework to the professional students (Meek et al., 2009). The second is that the graduate education and research enterprises in basic sciences create directions and opportunities for the professional enterprise. In addition to contributions to the profession, basic science research programs make contributions to the overall national research enterprise.

SCOPE FOR IMPROVEMENT AND FUTURE PERSPECTIVES

Universities should realistically evaluate the strengths and weaknesses of their basic science research programs, the ways in which basic science programs are contributing to and detracting from the professional mission, and new organizational paradigms for professional education that best serve the evolving needs of the professional mission. The issues associated with evaluating whether basic science research programs should be strengthened, downsized, or eliminated are different for each institution and need to be considered on an individual basis (Rao, 1999). Universities should re-examine what course work in basic sciences is critical and explicitly relevant to professional training and practice today. The organizational framework in which such course work is offered should be re-evaluated to determine to what extent instruction needs could be met by faculty and departments outside departments of pharmacy or by alternative organizational structures (Report of the Task Force for Basic Scientific Research in Universities, 2005). Next, the quality and competitiveness of research programs in each university should be evaluated based on criteria such as amount of research funding received and number of publications in peer-reviewed journals (both inside and outside pharmacy).

Science research should be made the key activity of the Universities with greater emphasis on research leading to PhD degree without which quality teaching is not sustainable. Access to scientific research journals and related resources must be strengthened and its scope should be widened through Information and Communication Technology (ICT). Science research should mean research in Basic Sciences, engineering, medicine and applied interdisciplinary sciences. The system needs to be made more flexible to strengthen research in interdisciplinary needs. Improvements in the education system could be achieved by the some of the following steps. There are a large number of sanctioned faculty positions in Universities that have been lying vacant for a prolonged period. These positions should be expeditiously filled in by inducting talented faculty with a view to strengthening the research base. There is a need to create networking centres in Basic Sciences in leading Departments of Universities in different parts of the country to promote collaborative research, access to advanced facilities and imparting training in frontier areas. These centres should be supported on a long term basis in a substantial manner to enable them to realize internationally competitive status. Formal linkage between the Universities and national level institutions including CSIR laboratories be promoted through joint research project and training. Linkages of Universities and their faculty with industries and user organizations should be encouraged. There should be a provision for joint appointments/ visiting professorship to stimulate collaborative research. The system of both Winter and Summer Schools must be supported. Each subject area of Basic Sciences may offer up to ten programmes a year. There should be a provision for visiting Fellowships for faculty within the country.

In order to promote quality scientific research in Universities, infrastructural facilities ought to be upgraded by providing development grants for modern laboratories, physical infrastructure including uninterrupted power supply, water, safety requirements of international standards, high level equipment with built-in maintenance mechanism, exclusive research students' hostel and a research starter fund etc. The number of PhDs from Indian Universities should increase five-fold within a span of ten years with proper standards. This should be seen in the context of extremely low fraction of PhDs in India in relation to B.Pharm/M.Pharm., as compared to what it is in USA, UK, Germany, Japan etc. The emphasis for research will clearly emerge if we have Universities with only departments and separately Universities having affiliated colleges.

CONCLUSION

The funding scenario for science in India is better than ever, with lots of infrastructure, startup funds and facilities provided to young aspiring professors. A major recent initiative has been the creation of the Indian Institutes of Science Education and Research (IISERs) whose primary goal has been to meld undergraduate education with research, something that's traditionally been almost completely missing in India. Expensive new centers including a neutrino observatory, a clean nanotechnology lab and most impressively, an open-source tuberculosis drug discovery project are highlighting India's ambition and capabilities to focus on relevant and cutting edge fields. For a developing country India has always been pretty big on science. Nobel Prize winners like H. G. Khorana (who died a few months back) and C. V. Raman emerged from the old education system which was modeled after the British system. In some fields like statistics India has consistently produced world-class researchers. After India became independent in 1947, science was one of the top priorities for the country's first Prime Minister, Jawaharlal Nehru, a Cambridge-educated lawyer. Even today Nehru remains India's most scientifically literate leader and he was instrumental in establishing both the well-known Indian Institutes of Technology and several leading federal research laboratories.

To draw a meaningful conclusion, opinion from leaders from academia, industries, government funding bodies and of course the student feedback would be fruitful in implimentation of research policies to facilitate basic research in universities offering pharmacy courses to attract industries for various activities including placement opportunities.

REFERENCES

Commission to Implement Change in Pharmaceutical Education. (1993). The papers of the Commission to Implement Change in Pharmaceutical Education. American Journal of Pharmaceutical Education, 57, 366–398.

Etzkowitz, H., & Leydesdorff, L. (2000). The dynamics of innovation: From national system and mode 2 to a triple helix of university industry government relations. Research Policy, 29, 109–123. doi:10.1016/S0048-7333(99)00055-4 doi:10.1016/S0048-7333(99)00055-4

Fischli, A. E. (1999). Chemical education and the pharmaceutical industry. Pure and Applied Chemistry, 71(5), 865–879. doi:10.1351/pac199971050865 doi:10.1351/pac199971050865

Friedman, Y. (2010). Location of pharmaceutical innovation: 2000-2009. Nature Reviews. Drug Discovery, 9, 835–836. doi:10.1038/nrd3298 doi:10.1038/nrd3298 PMID:21030994

Government of India. (2005). Report of the Task Force for Basic Scientific Research in Universities. Ministry of Human Resource Development Department of Secondary & Higher Education Government of India, New Delhi.

Jogalekar, A. (2012, March 1). Science in India: Moon shot or bust? *Critical Twenties, Science & Technology.* Retrieved from http://www.criticaltwenties.in/sciencetechnology

Khar, R. K., Ahmed, F. J., & Jain, G. K. (2011) Pharmacy education: Gearing up to meet industrial needs. *Pharmacology Times, 43*(3).

Meek, V. L., Teichler, U., & Kearney, M. L. (2009). *Higher education research and innovation: Changing dynamics*. UNESCO Forum Report, INCHER-Kassel.

Rao, C. N. R. (1999). Science in the future of India. Science, 286, 1295. PubMed doi:10.1126/science.286.5443.1295 doi:10.1126/science.286.5443.1295 PMID:19589967

Shukla, R. (2005). India Science Report. National Council of Applied Economic Research.

Stone, R. (2012). India Rising. Science, 335, 904–910. PubMed doi:10.1126/science.335.6071.904 doi:10.1126/science.335.6071.904 PMID:22362982

Suresh, S. (2012). Cultivating Global science. Science, 336, 959. PubMed doi:10.1126/science.1224580 doi:10.1126/science.1224580 PMID:22628620

UGC. (2012). Report on higher education in India. Retrieved from UGC.ac.in

Wadwa, V., Rissing, B., Gereffi, G., Trumpbour, J., & Engardio, P. (2008). Globalization of innovation [Can India and China Cure the Global Pharmaceutical Market? Kauffman- The Foundation for Entrepreneurship.]. *Pharmaceuticals (Ottawa)*, 3–5.

Williard, R. (1996). The changing biomedical research and health care environments: Implications for basic science graduate education and research in pharmacy. American Journal of Pharmaceutical Education, 60, 308–314.

Yeole, P. G., & Puranik, M. P. (2005). Pharmacy in India: At crossroads. Express Pharma Pulse.

This work was previously published in "Evolving Corporate Education Strategies for Developing Countries" edited by B. PanduRanga Narasimharao, S. Rangappa Kanchugarakoppal, and Tukaram U. Fulzele, pages 56-66, copyright 2013 by Information Science Reference (an imprint of IGI Global).

Chapter 36
cSELF (Computer Science Education from Life): Broadening Participation through Design Agency

Audrey Bennett
Rensselaer Polytechnic Institute, USA

Ron Eglash
Rensselaer Polytechnic Institute, USA

ABSTRACT

The phrase "broadening participation" is often used to describe efforts to decrease the race and gender gap in science and engineering education, and in this paper the authors describe an educational program focused on addressing the lower achievement rates and career interests of underrepresented ethnic groups (African American, Native American, and Latino students). However "broadening participation" can also describe the more general problem of a narrow, decontextualized form of education that can alienate all demographics. Broadening the scope of computing education can not only help address disparities in different social groups, but also make technical education more attractive to all individuals, and help us create a generation of science and engineering professionals who can better incorporate an understanding of the world into their technical work. The program the authors report on, Computer Science Education from Life (cSELF) takes a modest step in this direction. Using the concept of "design agency" the authors describe how this merging of abstract formal structures, material creative practice, and cultural knowledge can improve underrepresented student engagement, and foster learning practices in computing that offer broader forms of social expression for all students.

1. THE NEED FOR BROADENING PARTICIPATION

Underrepresented ethnic groups in the US consist of three groups: African American, Latino, and Native (which includes Native Alaskan, American Indians, and Pacific Islanders). Although they constitute about 45% of the college-age population, underrepresented ethnic groups comprised only 12% of engineering bachelor's degrees (NACME, 2008). This is not a problem that is simply resolving itself over time: the shares to black and Native

DOI: 10.4018/978-1-4666-7363-2.ch036

students have remained flat since 2000. Graduate computer science is particularly troubling: students from under-represented ethnicities comprised only 3% of the total number of degrees granted in 2008 (Ladner, 2012).

The low representation of underrepresented ethnic groups (African American, Native American, and Latino students) in the STEM workforce in general, and computing specifically, has two root causes. One is career interest: as noted by Simard (2009) narrow perceptions of career paths and stereotypes about what is an "appropriate" profession contributes to the lack of diversity. But more problematic is the lower academic achievement, especially for low-income students. High school drop-out rates for African American and Latino students are double that of white students, and triple for Native Americans (Stillwell & Sable, 2013).

These lower levels of STEM achievement and interest are detrimental to these populations, resulting in lower income levels and even contributing to health disparities (academic achievement is correlated with lower rates for HIV infection and substance abuse, higher rates for vaccination, etc... (Bridges & Alford, 2010; Fields et al. (2007).

As noted in the introduction, the concept of "broadening participation" can also be applied to students in general. In computer science, the percentage of high school students taking computing courses has surprisingly dropped from 25% to 19% (Nord, et al. (2011). In recent years, the computer science advanced placement (CS AP) test has sustained the lowest participation rate in comparison with other STEM disciplines. The CS AP exam also shows a strong gender gap: only 19% of girls compared to 81% of boys comprise the CS AP test-takers (NCWIT 2012).

A broader approach to computing education could also improve the ability of the STEM workforce to address critical humanitarian and sustainability issues. The fiscal meltdown of 2008, for example, was a destructive force in much of the US economy, and precipitated a global recession. Many scholars attribute the incorporation of computational models of risk--for example the Gaussian copula function--as a key ingredient (Salmon, 2009). Similar issues in the role of computational risk modeling arise in environmental disasters; for example the engineering professionals in the 2010 Gulf Oil disaster (Deep Water Horizon Study Group 2011). Narrow conceptions of what it means to be a computational scientist are inculcated in our classrooms; it is this narrowness that allows these professionals to say "it's not my place to think about consequences, I'm just here to crunch the numbers." Broadening the *forms* of participation–educating students in the use of computation as an expressive medium with deep connections to the social world–can serve as a powerful counter-balance to this tendency to abdicate responsibility.

2. THE ROLE OF CULTURALLY SITUATED EDUCATION

Barriers to participation and achievement in STEM disciplines for underrepresented youth can be framed in three categories. The first concerns the barriers due to economic conditions, which are correlated with underrepresented ethnic groups, include lower quality schools, health care, and other aspects of the learning environment and experience. A 2010 study of California schools, for example, found that African American students were six times more likely than white students to attend one of the schools at the bottom third of the state ranking (Education Trust-West, 2010). The second category encompasses myths of genetic determinism: the belief that a "math gene" or some similar genetic construct prevents certain racial groups from STEM success. There is no evidence for such a phenomenon, but the myth itself can have strong negative consequences, discouraging students and diminishing their confidence (e.g. Geary, 1994) The third category covers myths of cultural determinism: conflicts with stereotypes of authenticity, a perceived lack of social relevance, accusations of "acting white" (Ogbu, 1998; Downey & Lucena, 1997; Eglash, 2002). Culturally situated education can be an important resource for combating the second and third categories. Like any pedagogy, this can be done well or poorly. Poor versions include attempts to paste a thin veneer of culture onto standard lessons–replacing Dick and Jane counting marbles with Tatuk and Estaban counting coconuts. A more promising path can be found in disciplines such as ethnomathematics and ethnocomputing, which "translates" from the STEM concepts and practices embedded in indigenous cultural practices and contemporary vernacular activities to their contemporary equivalents (Eglash, et al. (2006; Lipka & Adams, 2004). Like many scholars investigating STEM education for underrepresented ethnic groups (Hammond, 2001; Eisenhart, 2001; Rennie, et al. (2003), we agree that "learning is authentic when it takes as its starting point the interests, perspectives, desires, and needs of the students" (Buxton, 2006).

3. COMPUTER SCIENCE EDUCATION FROM LIFE (cSELF)

In our prior work (e.g. Eglash, et al. (2006; Eglash & Bennett, 2009; Boyce, et al. (2011), our team developed a suite of applets for simulating traditional cultural arts: Culturally Situated Design Tools (CSDTs), which are freely accessible via the internet at www.csdt.rpi.edu. As noted above, CSDTs do not impose math and computing ideas from outside the culture; rather they make use of the mathematical and computational ideas that are already present, whether explicit or implicit, in the cultural practices they simulate. Native American beadwork, for example, makes use of iterative algorithms on a Cartesian grid; African American cornrow hairstyles show recursive geometric transformations; urban graffiti includes polar coordinate curves, etc.

In the process of designing the software, we begin with an investigation of the original cultural context of the artisans, through established literature as well as our own interviews and ethnographic investigations. Native American culture, for example, has the Cartesian layout of orthogonal two axes, or four-fold symmetry, as a strong underlying geometric theme throughout its designs, concepts and practices. This "translation" works in both directions, as deeper mathematical understanding can help in futher understanding of the cultural material. For example, in the famous oral history of Iroquois peace-maker Hiawatha, we hear that phrase "he will split the sky" – a curious wording unless you know about four-fold symmetry: he meant that he would travel the north-south axis, since the Iroquois know that sun, moon and stars travel in "the sky's direction" of east-west. Other examples of four-fold symmetry include the Shoshone prayers that begin with "the four winds," the Navajo orientation to four sacred mountains, the four poles of the teepee, and so on (Eglash 2010). Thus a simulation which is modeling the beadloom's arrangement of rows and columns as a Cartesian grid, as seen in Figure 1, is not imposing western concepts on unrelated forms; to the contrary it is "translating" a deep cosmological principle to its western equivalent.

The suite of online applets, CSDTs, provide this cultural background as well as applets (flash, java, and javascript) which allow students to utilize

Figure 1. Comparison of Native American beadwork and its simulation by student

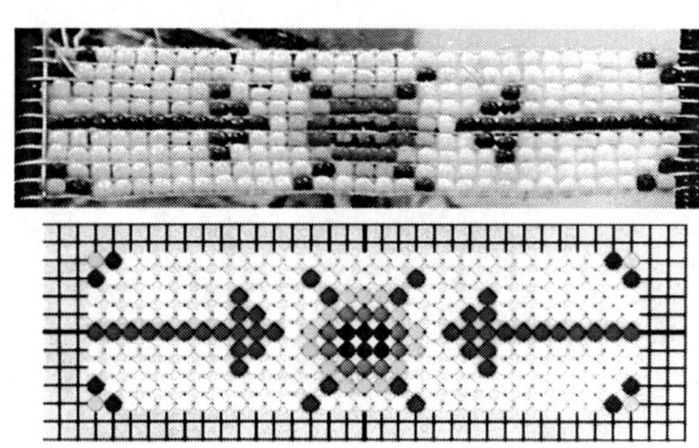

these indigenous math and computing concepts in simulations. There are two types of CSDT interface. The parametric interface is focused primarily on math concepts. For example in beadloom tools (Figure 2) the drop-down interface offers built-in algorithms for create a line, rectangle, and triangle as geometric figures in which the parameters are the vertices, as well as iterative patterns in which the parameters are the number of iterative cycles and the initial conditions. This allows students to understand the parameters in a more narrative form; for example "starting with a row of 7 beads, subtract 1 from each side as you iterate in the +y direction." The algorithm is fixed, but each of the underlined parameters can be changed by the user. The second interface is "programmable," using drag and drop code blocks, so that students can creatively invent their own algorithms, after learning the mathematical representation in the first interface (Babbitt et al., 2011).

The cSELF program takes this process full circle, allowing students to take the culture-based simulations they have created, and render them as physical art objects. The goal is to create a learning environment in which computational thinking (CT) fundamentals are gained through a synergistic encounter between creativity, culture and computing; thus generating new avenues for broadening participation.

The cSELF pilot program begins with summer workshops for training art teachers, who subsequently pass that on to their underrepresented art students in high school art classes during the academic year. In the initial pilots our assessment examines the extent to which this is sufficiently engaging from an arts education perspective, and its ability to teach CT fundamentals. If successful, long-term studies will be used to examine the impact on selection of elective computing courses in high school, selection of major (computing vs. non-computing related) in college, and retention in those majors for students in cSELF in comparison to their peers.

In early workshops with CSDTs we observed that they offer a flexible format that allows underrepresented students to engage in both structured

Figure 2. The virtual beadloom tool

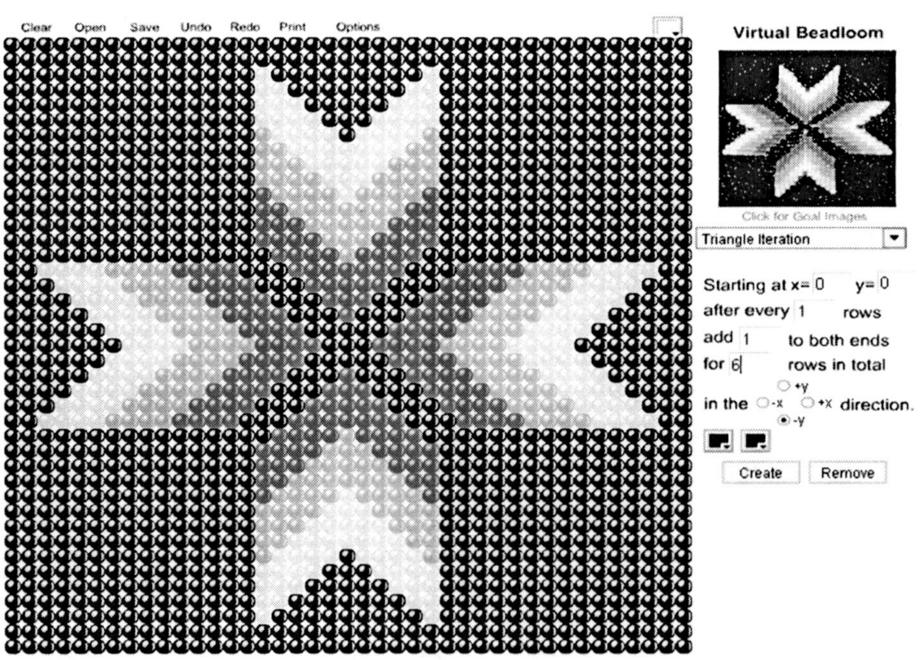

learning and exploratory learning where they openly create their patterns. This connection between the computational skills and understanding required to create the simulations, and open, unrestrained creativity can be particularly important for students from under-represented groups who may think of themselves as lacking technical inclination but willing to explore artistic practices. Students' use of physical arts media–painting, ceramics, etc.–to render their virtual simulations adds an important experiential component that is critical to the concept.

4. THE cSELF PROGRAM

cSELF began with a one-day, professional development workshop in Summer 2012 for local art teachers from the NY capital region: teachers from 3 school districts attended (Schenectady, Albany and Troy high schools). The art teachers were introduced to 3 CSDTs: African-American Cornrow Curves, the Native American Beadloom, and African Fractals (which offers fractal simulations for a variety of African architectural and artistic designs as well as fractal models in nature). The workshop activities modeled many of the classroom activities that teachers would later repeat with their students, as seen in Figure 3. These included:

1. Group discussion of the cultural background offered in the CSDT website;
2. Step-by-step instruction on creating basic simulations, using the website tutorials. At this point the participants are only duplicating patterns created by the original artisans;
3. Reflection on how basic CT concepts--iterative loops, variables, algorithms, geometric transforms, etc.–are present both informally in the original artistic practice and formally in the simulation (Table 1);
4. A "performance art" activity which reinforces the computational/mathematical concepts through the creative use of body motion;
5. Creative exploration of the tools; participants now move from duplicating previous patterns to creating new patterns of their own design;
6. Group discussion of the artistic patterns in which participants provide each other with supportive comments and constructive criticism.

Figure 3. A conceptual drawing of the cSELF program

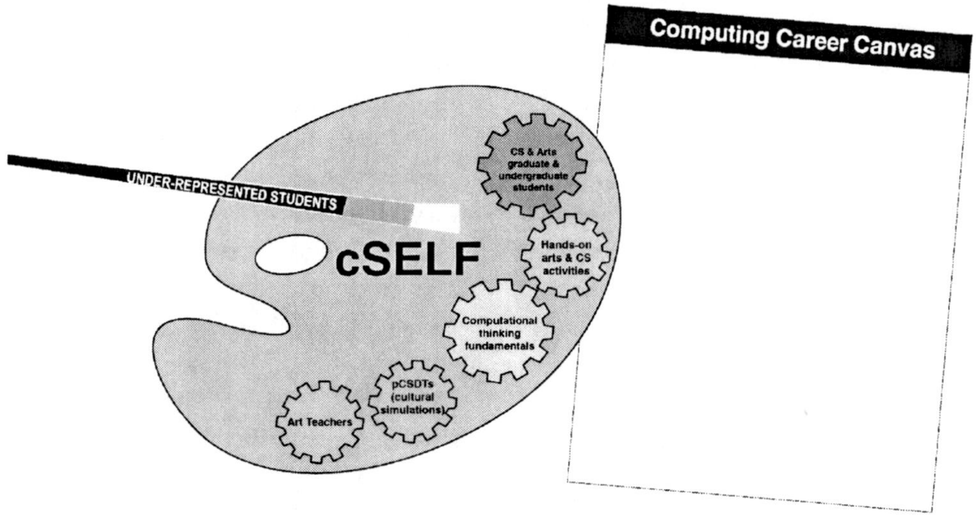

Table 1. Computational concepts taught through CSDTs

Native American Virtual Bead Loom	African-American Cornrow Curves
• Geometric shapes • Line of symmetry • Reflection symmetry • Four-fold symmetry • Cartesian coordinate system • Plotting points on a Cartesian coordinate system • Plotting points on a Cartesian coordinate system to create lines. • Plotting points on a Cartesian coordinate system to create 2D shapes • Plotting points on a Cartesian coordinate system using iteration algorithms	• Iterating copies of a seed shape • Changing copies with transformational geometry parameters: dilation, rotation, translation, reflection • Parameters as variables • Same algorithm with different initial values for variables produces different outcomes • Differentiating variables inside and outside of iterative loop

During the school year, art teachers add a final phase to the cSELF learning process, using these virtual simulations in the design of physical artistic counterparts. This component offered an opportunity for the art teachers to bring their own skills and creative insights into the process.

5. DESIGN AGENCY

cSELF builds on the concept of ethnocomputing–the idea that important computational concepts are already present in the heritage culture and vernacular culture of underrepresented students. Evidence that the ethnocomputing approach, even in its purely virtual form, can enhance learning comes from both qualitative and quantitative sources. For example, in this unsolicited communication, a teacher at a Lakota Nation school using the Virtual Bead Loom tool reported:

You might be interested to hear that one of the students, who is an IT major, an artist, a very traditional beadworker and fluent Lakota speaker, was so delighted with the software that he decided to go ahead and develop his own algorithms independently. He was really inspired. He said it was the first time that math/graphing seemed to really make sense or "click" for him. I haven't seen how far he got with computer algorithms, but his final project for our math class was full of linear models that described his most recent beadwork creations.

Such anecdotal evidence is complimented by our statistical studies. One published study was carried out in two high school computing classes in New York City (Eglash, et al. (2011). Students in the study used one of two websites, both with java applets, in two classes taught by the same instructor: one class used a popular (non-cultural) site for fractals education, the other our African Fractals site. The results were surprisingly strong: both the pre/post differences in skills and the pre/post differences in attitudes toward computing careers show statistically significant improvement (.001 confidence level) in the class using the CSDT website.

The main hypothesis of the cSELF program is that these advantages of ethnocomputing can be enhanced by including a physical component. At the core of cSELF lays the concept of "design agency." Agency has been defined differently in various disciplines. In philosophy, agency is usually defined as the capacity to take action on one's mental decisions; it is considered separately from "free will" since someone might express their agency on the basis of motivations that are caused by the will of someone else. Philosophers often use agency in this sense as something which distinguishes the consciousness of humans from the actions of machines or other non-conscious entities (Johnson, 2006). Sociologists on the other hand often focus on the contrast between the agency of individuals (thus more closely associated with free will) and the social structure in which they are embedded; for example the barriers we describe in the case of underrepresented students who do not feel free to engage in computer science education because would cast them as "nerds" or violates other social expectations would be a case of blocked agency. "Design agency" makes use of this sociological sense of human agency in relation to free will, but allows its mixture or synthesis with non-human agency as described by Pickering (1995). According to Pickering, non-human agency in nature or machines lacks intentionality, but as it mixes with human agency the resulting "mangle" can shift both human and non-human sides in unexpected ways.

In Pickering's work, the focus is on scientific inquiry. Pickering notes that in the typical description of science, the lab apparatus is described as a sort of transparent window that passively delivers the facts about the world. But he points out that if you actually follow the scientific process in detail, there is a kind of "resistance" that is often encountered: nature won't respond in the exactly the ways the scientist was hoping it would. So you make adjustments–"tune" the machine in various ways, change chemical mixtures, and sometimes even change your goals. Rather than a transparent window, it is a gradual "negotiation" between people, nature and machines; success occurs when they arrive together at some mutually acceptable conclusion.

In similar ways that Pickering describes the scientific process, we posit that "design agency" is a mangle between the human and non-human elements of a design process. While it might include a similar "negotiation" with nature–the physics of clay, for example, will both facilitate and constrain the sculpting process–culture can play a similar role here. Consider, for example, the case of a student who is trying to create a straight line with the beadloom tool (Figure 4). As long as the line is oriented at a multiple of 45 degrees–0, 45, 90, etc--the line is smooth, as we see with the line from (-14, -14) to (14, 14). But at other angles it becomes jagged (as we see with the other line). That is not a bug; it's simply the fact that any Cartesian grid with intersections on integer values must have a kind of "staircase" appearance when it is fitting lines at those angles. Students respond to such discoveries of "resistance" in various ways. Some restrict their design to angles with multiples of 45 degrees. Others make use of the "jaggedness," reflecting the same jagged line over the Y axis to form a symmetric pair. Culture is present prior to the design process because the applet is based on the native American beadloom–had it been based on other beadwork techniques there would have been a different set of constraining/enabling features. But it can also play a role in the on-going design process when the student begins to see that traditional beadwork has been influenced by the same "resistance"–some traditional designs restrict themselves to multiples of 45 degrees, and some make use of symmetric jaggedness. And culture can guide the design process in other directions as well. One student, for example, decided that he wanted to replicate the Puerto Rican flag in the beadloom tool (Eglash et al., 2006). He found that the flag included an equilateral triangle with strong symbolic significance. The "resistance" he encountered to creating an equilateral triangle on a Cartesian grid (you can't simply count the beads on each side because the beads across the diagonal are farther apart than those on the axis) resulted in his investigation of triangle geometry; eventually he discovered he could use two 30-60-90 triangles to create the 60-60-60 required by the flag. Resistance is not simply negative, it can positively guide as well as restrict, and that guidance creates new affordances and learning opportunities in the process.

Figure 4. A student who encounters the contrast between smooth and jagged lines in the virtual beadloom can respond to this "resistance" in at least two different ways. Both can be found in ordinary beadwork.

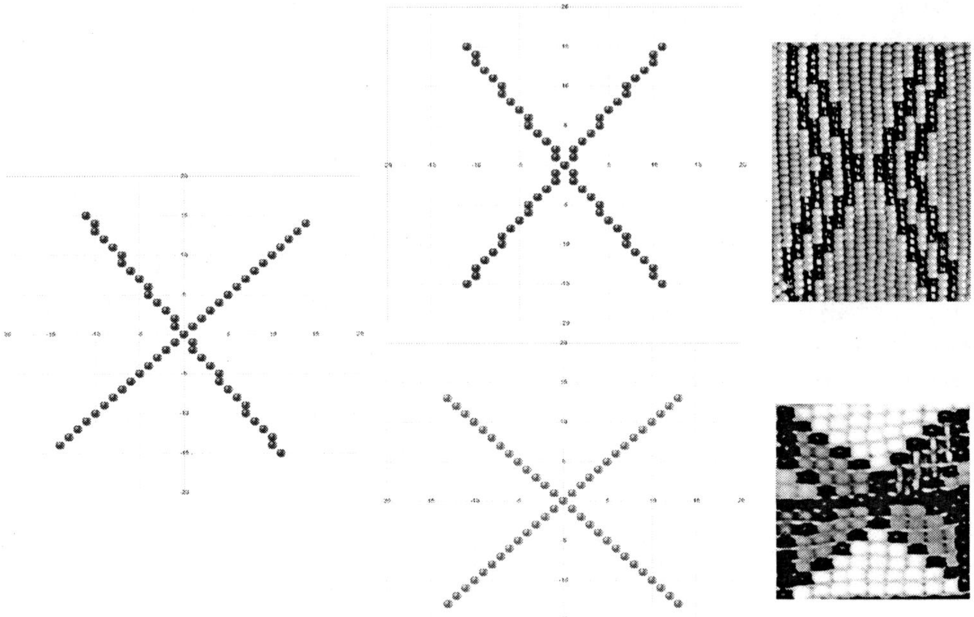

Creativity is sometimes thought of as an exploration in the space of possible designs (e.g. Gero & Kumar, 1993). But like real-world explorers, the places you visit can both constrain and enable your travel, and even "mangle" your goals: your car breaks down but you manage to buy a horse; the horse tends to look for green grass which is by a creek, the creek has a raft, and so on. Just as it may have never occurred to the traveler to explore the waterways, it never occurred to the student of Puerto Rican heritage that he would be eagerly engaged in a geometry problem. The framework of design agency helps us understand how the "mangle" between human intentions and non-human facilitation in creates new bridges (or rafts) between social and technical worlds. Thus one way to look at the cSELF program is that it is creates a learning environment in which cultural and artistic resources enter into a productive negotiation with computational design agency.

Culture is not the only source of these enabling and constraining mangles in cSELF; the physical arts become another component; a means to bring together head, heart and hand. The students and art teachers were highly creative in their process of devising physical renderings for the virtual designs. Of course this is the norm for an art class, but it was striking that even in cases where there was little algorithmic change, the physical medium became an additional opportunity for creative agency. In the case of designs in the virtual beadloom, for example, the Cartesian grid design remained largely the same, but its medium was transformed from virtual image to ceramics, mixed media, and even electrical optics (jars of color water lit from below), as seen in Figure 5.

In other cases the physical renderings became more interpretive. One African American student, for example, noted that one of his simulation for scaling sequences in cornrow braids looked like spiders (Figure 6). The topic of spiders also came up during a presentation on African traditional design, and perhaps that connection inspired him to focus on the spider theme when he rendered the

Figure 5. Physical renderings offer an additional opportunity for creative agency, even in the case of purely abstract patterns

final physical version (Figure 7). Other students carried the interpretive license even farther. One student re-interpreted the scaling sequence of geometric figures as a series of cigarettes that had burned to decreasing lengths; her final piece featured a scaling sequence of physical cigarette butts glued to a painted figure with a death's head rising from her chest. Starting with similar scaling sequences from the fractal CSDT using triangles (Figure 8), a different student created the paper sculpture of Figure 9, along with a more upbeat narrative:

My 3D fractal piece was inspired by my daily encounters with teenagers in my school. When I got to high school I noticed that more and more people judge others based on their personalities, style of clothing, their hobbies, who they're friends with and much more. Because of this, bullying often occurs and kids feel pressured to change their persona to fit in and stop the bullying they experience. This false personality is something that hides some one's true self from the world around them, but will never truly be who they really are. My project is of a boy hiding behind a window. This red window is his ticket to popularity and acceptance. However, the 3D triangle fractals represent the boy's true personality. The triangles create a mask that represents his personality coming through. As the triangles get bigger and bigger, the personality gets stronger and stronger, eventually coming up onto the window like it's breaking through his false persona. Everybody will always have their true being inside of them which is what makes them happy. Who we are is who we are and it will always prevail in the end.

Figure 6. Using the cornrows tool, a student is inspired to title this "spiders"

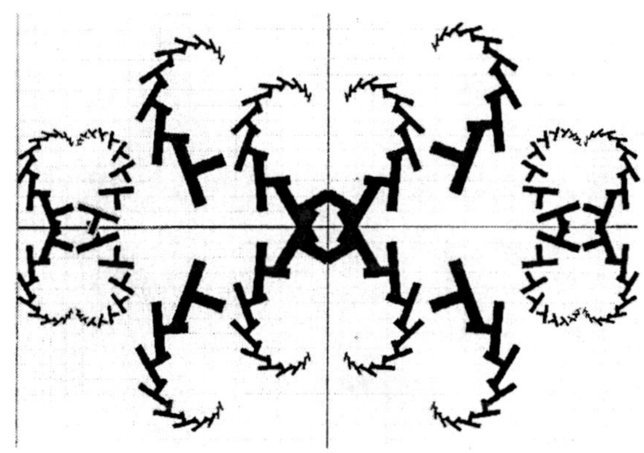

Figure 7. The same student emphasizes the spiders theme in his physical rendering

Figure 8. A fractal virtual design, and paper rendering of the pattern, provided by the teacher as an example

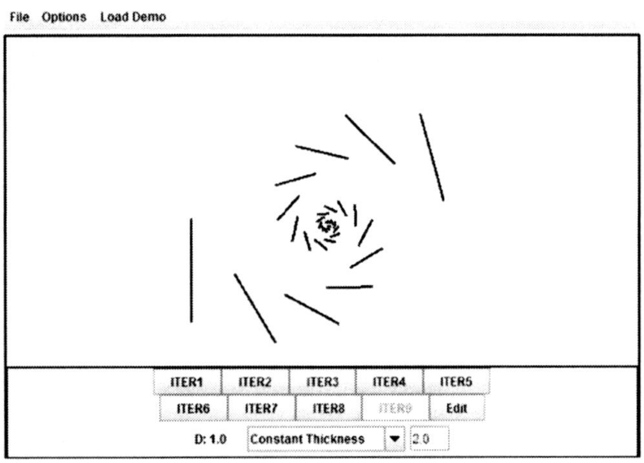

Figure 9. After modifying the teacher's virtual example, the student added an artistic rendering of the design

One concern that often arises in discussions of this program with math and computing instructors is the fear that students who are not technically inclined will "ditch" the technical content and focus on artistic interpretations. Pre/post test scores on math and computing concepts showed clear learning gains for many of the students. There was no correlation between scores on pre/post tests and the decision about how closely the physical form would duplicate its virtual inspiration, indicating that creative agency was not in competition with technical learning.

6. THE RELEVANCE OF cSELF'S CURRICULUM TO COMPUTER SCIENCE EDUCATION STANDARDS

cSELF's curricula provides high school youth with training in computational thinking that correlates with the College Board's seven core principles for computer science education (Table 2) (see complete descriptions of core principles at collegeboard.com/html/computerscience).

7. THE ARTS-COMPUTING DISCONNECT: WHY WE NEED THE cSELF PROGRAM

At first glance, it might seem that the arts connection is already well-covered in computing

Table 2. The College Board's seven core principles in correlation with computer science education

College Board's Principle	cSELF's Pedagogical Activity or Component
1. Computing fosters creative expression and the creation of artifacts. Computing is a creative act.	cSELF engages youth in a design process explicitly for creative expression.
2. People use models and simulations to generate new understanding and knowledge.	In cSELF, students are simulating culture-based arts, essentially "reverse-engineering" the original patterns and uncovering the algorithms that were at work.
3. Insight and knowledge can result from translating and transforming digitally represented information.	As noted in the discussion of "resistance" and design agency, the cSELF gain insight into the enabling and constraining aspects of the different simulations. Looking back on the traditional patterns, it becomes more clear why certain structures are present. In addition, the final step allows them to transform their digital works into physical renderings, which allows them further insight into the ways that our world can be seen in terms of an underlying computational structure.
4. Sequencing, selection, iteration, and recursion are building blocks for an algorithm which is a precise sequence of instructions for a process that can be executed by a computer.	Iteration, recursion and algorithms are concepts that students must explicitly utilize in the CSDT tools--The Virtual Beadloom and Cornrow Curves--as well as artistically with hands-on, arts activities. The aesthetic components naturally lend themselves to concepts such as selection (colors of beads for example) and sequencing (for instance difference cornrow braids are associated with different iterative sequences).
5. Programming is an iterative process that requires an understanding of how instructions are processed.	The CSDT interface offers immediate visual feedback; students are constantly adjusting parameters and checking results as they "hone in" on a particular design.
6. The internet and the systems built on it facilitate collaboration.	The CSDT community website is freely accessible via http://community.csdt.rpi.edu/and allows users to upload simulations and share them in open-source fashion, encouraging collaborations.
7. Computing is situated within economic, social, and cultural contexts.	cSELF begins with cultural art forms that embody both a social background and the computational thinking that is embedded in its designs. The process of physical rendering allows students to move away from the screen and interact with each other and the art studio environment, further enhancing the social and material components of the experience. The facilitation of design agency offers the opportunity for further social commentary.

education. After all, many computing classes include graphics and animation components. However our initial investigation has revealed three significant gaps:

1. Art in computing education is generally limited to a role as motivator or "content provider;" something to attract student interest. While a project to create an online gallery might be inspiring for some students, it still keeps a conceptual barrier in place between the artistic and technical process. Allowing students to see computational thinking in the artistic works themselves changes the status of both art and computing in the mind of the student; it makes computation itself available as a medium of artistic expression;
2. The arts simulated by CSDTs–Native American beadwork, African American cornrow hairstyles, etc.– are distinctly associated with the cultures of underrepresented students. By using physical algorithms in culture-based arts, and matching these with computer simulations of the artifacts, students from these background can learn computational thinking through their own heritage and vernacular culture, and other students can learn greater cross-cultural appreciation;
3. The high school courses that offer computational thinking are, not surprisingly, mostly limited to computing classes. However, cSELF offers computational thinking through an arts curriculum. This allows three potential advantages: first, it brings a new pool of students into the computing education pipeline. Second, it potentially increases the number of underrepresented students in the computing education pipeline. Finally, it increases the number of creatively inclined students in college computing disciplines, adding intellectual diversity as well as ethnic diversity.

One potential objection against this last point might be that students entering into the computing education pipeline through cSELF will tend towards interdisciplinary programs such as gaming, electronic arts, animation etc. rather than a pure computer science major. That is an empirical question we intend to investigate in future studies. But we would argue that even if that does turn out to be the case, cSELF is merely allowing pre-college computing education to match a process that is already occurring at the college level: the dramatic increase in computer science/artistic practice synthesis that is increasingly featured in the aforementioned college programs, and beyond college, for computing careers in industries such as gaming and media production.

Another potential objection is that this ethnocomputing concept is already present in computer education in the form of computational origami. It is true that the computation work on origami does exactly what we intend: computer scientists take seriously the concept that origami is algorithmic (Bern and Hayes 1996; Lang 1994; Huffman 1976); they have software that allows movement between virtual and physical instantiations (Balkcom & Mason, 2004; Kasem & Ida, 2008; Mitani, 2009); and they use it in the classroom (cf. Goadrich, 2010). But the association of computational thinking with origami does not challenge the stereotypes that hold back underrepresented students: indeed the assumption that Asian students are "inherently" better suited for computing careers is exactly the kind of damaging myth that discourages underrepresented students (Wu, 2002). The broader ethnocomputing approach offered by the cSELF program, in contrast, can show this same resource–physical instantiations of computational thinking in an artistic process–based on the cultural heritage and vernacular innovations of underrepresented students. Rather than confirm the stereotypes that discourage their participation (Ogbu, 1998; Fordham, 1991; Steele, et al., 2002), cSELF's approach is seeks to empower underrepresented students to see their cultural "self" as having a computational heritage.

8. CONCLUSION

cSELF is primarily designed for students who may be loath to think of themselves as "computing geeks"–students who are more inclined towards the arts and humanities. Data on under-represented students suggest that they may find this approach particularly helpful. By engaging this under-served student population, cSELF aims to contribute--in the long run–to diversifying the discipline of computing. But it would be a mistake to say that the problem in the system is completely on the side of the children. The discipline of computing is itself far too decontextualize and narrowly conceived to effectively serve human needs in an increasingly interconnected world. Developing forms of computing education that can offer new opportunities for social expression and creative engagement is a critical first step: not just in better serving up human resources for the computing industry, but for improving the ability of computing as a resource for human needs.

REFERENCES

Babbitt, B., Lyles, D., & Eglash, R. (2012). From ethnomathematics to ethnocomputing: indigenous algorithms in traditional context. In S. Mukhopadhyay & W.-M. Roth (Eds.), *Alternative forms of knowing (in) mathematics* (pp. 71–89). Rotterdam, The Netherlands: Sense Publishers. doi:10.1007/978-94-6091-921-3_10

Balkcom, D. J., & Mason, M. T. (2004, April 26, May 1). Introducing robotic origami folding. In *Proceedings of the 2004 IEEE International Conference on Robotics and Automation (ICRA '04).*

Bern, M., & Hayes, B. (1996). On the complexity of flat origami. In *Proceedings of the 7th ACM-SIAM Symposium on Discrete Algorithms*, Atlanta, GA (pp. 175-183).

Boyce, A., Campbell, A., Pickford, S., Culler, D., & Barnes, T. (2011). Experimental evaluation of BeadLoom game: How adding game elements to an educational tool improves motivation and learning. In *Proceedings of the ACM Innovation and Technology in Computer Science Education Conference,* Germany (pp. 243-247).

Bridges, E., & Alford, S. (2013). *Comprehensive sex education and academic success*. Advocates for Youth.

Buxton, C. A. (2006). Creating contextually authentic science in a 'low-performing' urban elementary school. *Journal of Research in Science Teaching*, *43*(7), 695–721. doi:10.1002/tea.20105

Deep Water Horizon Study Group. (2013). *Final report on the investigation of the Macondo well blowout*. Center for Catastrophic Risk Management UC Berkeley.

Downey, G. L., & Lucena, J. (1997). Weeding out and hiring in: How engineers succeed. In G. L. Downey & J. Dumit (Eds.), *Cyborgs & citadels: Anthropological interventions in emerging sciences and technologies*. Santa Fe, NM: School of American Research Press.

Education Trust-West. (2010). *Access denied: 2009 API rankings reveal unequal access to California's best schools.*

Eglash, R. (2002). Race, sex and nerds: From black geeks to Asian-American hipsters. *Social Text*, *20*(2), 49–64. doi:10.1215/01642472-20-2_71-49

Eglash, R. (2009). Native-American analogues to the Cartesian coordinate system. In B. Greer, S. Mukhopadhyay, A. Powell, & S. Nelson-Barber (Eds.), *Culturally responsive mathematics education* (pp. 281–294). New York, NY: Routledge.

Eglash, R., & Bennett, A. (2009). Teaching with hidden capital: Agency in children's computational explorations of cornrow hairstyles. *Children, Youth and Environments*, *19*(1), 58–73.

Eglash, R., Bennett, A., O'Donnell, C., Jennings, S., & Cintorino, M. (2006). Culturally situated design tools: Ethnocomputing from field site to classroom. *American Anthropologist*, *108*(2), 347–362. doi:10.1525/aa.2006.108.2.347

Eglash, R., Krishnamoorthy, M., Sanchez, J., & Woodbridge, A. (2011). Fractal simulations of African design in pre-college computing education. *ACM Transactions on Computing Education*, *11*(3).

Eisenhart, M. (2001). Educational ethnography past, present, and future: Ideas to think with. *Educational Researcher*, *30*, 16–27. doi:10.3102/0013189X030008016

Fields, V., Sumpter, C., Seagraves, L., Laney, A., Lott, A., & McBride, K. (2013). *Determining factors affecting parental non-compliance with vaccination schedules of children ages 6 months to 2 years*. LaGrange College Department of Nursing.

Fordham, S. (1991). Peer-proofing academic competition among black adolescents: "Acting White" Black American style. In C. Sleeter (Ed.), *Empowerment through multicultural education* (pp. 69–94). Albany, NY: State University of New York Press.

Geary, D. C. (1994). *Children's mathematical development: Research and practical applications*. Washington, DC: American Psychological Association. doi:10.1037/10163-000

Gero, J. S., & Kumar, B. (1993). Expanding design spaces through new design variables. *Design Studies*, *14*(2), 210–221. doi:10.1016/0142-694X(93)80048-H

Goadrich, M. (2010, March 25-26). TopSpin and communication/origami. In *Proceedings of the Mid South Conference,* Searcy, AR.

Hammond, L. (2001). Notes from California: An anthropological approach to urban science education for language minority families. *Journal of Research in Science Teaching*, *38*(9), 983–999. doi:10.1002/tea.1043

Huffman, D. A. (1976). Curvatures and creases: A primer on paper. *IEEE Transactions on Computers*, *C-25*, 1010–1019. doi:10.1109/TC.1976.1674542

Johnson, D. G. (2006). Computer systems: Moral entities but not moral agents. *Ethics and Information Technology*, *8*(4), 195–204. doi:10.1007/s10676-006-9111-5

Kasem, A., & Ida, T. (2008). Computational origami environment on the web. *Frontiers of Computer Science in China*, *2*(1), 39–54. doi:10.1007/s11704-008-0009-8

Ladner, R. (2012). Broadening participation: Data and resources. In *Proceedings of the NCWIT*.

Lang, R. J. (1994). Mathematical algorithms for origami design. *Symmetry: Culture and Science*, *5*(2), 115–152.

Lipka, J., & Adams, B. (2004). Culturally based math education as a way to improve Alaskan Native students' math performance. *ACCLAIM Working Papers #20.*

Mitani, J. (2009). A design method for 3D origami based on rotational sweep. *Computer-Aided Design & Applications*, *6*(1), 69–79. doi:10.3722/cadaps.2009.69-79

National Action Council for Minorities in Engineering, Inc. (NACME). (2008). *NACME 2008 data book: Confronting the "new" American dilemma.*

NCWIT. (2012). *Fact sheet.*

Ogbu, J. (1998). Voluntary and involuntary minorities: A cultural-ecological theory of school performance. *Anthropology & Education Quarterly*, *29*(2), 155–188. doi:10.1525/aeq.1998.29.2.155

Pickering, A. (1995). *The mangle of practice: Time, agency, and science*. Chicago, IL: The University of Chicago Press. doi:10.7208/chicago/9780226668253.001.0001

Rennie, L., Feher, E., Dierking, L., & Falk, J. (2003). Toward an agenda for advancing research on science learning in out-of-school settings. *Journal of Research in Science Teaching*, *40*, 112–120. doi:10.1002/tea.10067

Salmon, F. (2009). Recipe for disaster: The formula that killed wall street. *Wired*, 23 February 2009.

Simard, C. (2009). *Obstacles and solutions for underrepresented minorities in technology*. Anita Borg Institute for Women and Technology.

Steele, C., Spencer, S., & Aronson, J. (2002). *Contending with group image: The psychology of stereotype and social identity threat* (M. Zanna, Ed.). Advances in experimental social psychology New York, NY: Academic Press.

Stillwell, R., & Sable, J. (2013). *Public school graduates and dropouts from the common core of data: School year 2009-10*. Institute of Education Sciences.

Wu, F. H. (2002). *Yellow: Race in America beyond black and white* (pp. 39–77). New York, NY: Basic Books.

This work was previously published in the International Journal of Web-Based Learning and Teaching Technologies (IJWLTT), 8(4); edited by Mahesh S. Raisinghani, pages 34-49, copyright 2013 by IGI Publishing (an imprint of IGI Global).

Chapter 37
TPACK Pathways that Facilitate CCSS Implementation for Secondary Mathematics Teacher Candidates

Nathan Borchelt
Western Carolina University, USA

Kathy Jaqua
Western Carolina University, USA

Axelle Faughn
Western Carolina University, USA

Kate Best
Western Carolina University, USA

ABSTRACT

Implementation of the Common Core State Standards in Mathematics has provided teacher educators a great opportunity to reexamine whether teacher preparation programs adequately provide the experiences to develop the base of knowledge and 21st century skills necessary to be effective teachers. The Mathematics TPACK Framework provides a roadmap for a series of pathways to integrate three knowledge components that are essential in teacher development: content knowledge, pedagogical knowledge, and technological knowledge. In this chapter, the authors examine how a teacher preparation program has evolved to integrate meaningful uses of digital technologies in content and pedagogy that are relevant to the teaching and learning of mathematics through the lens of implementing the Common Core State Standards.

INTRODUCTION

When examining the components of a teacher preparation program, one of the most pressing implications for practice is a need for meaningful experiences for teacher candidates similar to those provided through in-service professional development. Teacher candidates must be prepared to establish learning environments that emphasize 21st century skills.

People in the 21st century live in a technology and media-suffused environment, marked by various characteristics, including: 1) access to

DOI: 10.4018/978-1-4666-7363-2.ch037

an abundance of information, 2) rapid changes in technology tools, and 3) the ability to collaborate and make individual contributions on an unprecedented scale. To be effective in the 21st century, citizens and workers must be able to exhibit a range of functional and critical thinking skills related to information, media, and technology (Partnership for 21st Century Skills, Information, Media, and Technology Skills, 2011).

To that effect, technology should be an integral part of teacher candidates' experiences learning mathematics, or else they may not value the benefits of technology integration. They must also understand what the Common Core State Standards in Mathematics say about students' experiences with digital technologies in order to meet professional standards used to evaluate teachers.

Technology is valuable in leveraging and enhancing classroom experiences for students which promote mathematical reasoning and sense making (Dick & Hollebrands, 2011). In this chapter, we examine how a teacher preparation program has evolved to integrate meaningful uses of digital technologies in content and pedagogy which are relevant to the teaching and learning of mathematics. The Mathematics Technological Pedagogical and Content Knowledge (TPACK) Framework (Mishra & Koehler, 2006; Koehler & Mishra, 2008; 2009; Association of Mathematics Teacher Educators (AMTE), 2009) provides a roadmap for a series of pathways to integrate three knowledge components, which are essential in teacher development: content knowledge, pedagogical knowledge, and technological knowledge. We argue the necessity of moving teacher candidates through these pathways to help them promote in their students the types of mathematical practices described in the Common Core documents (Common Core State Standards Initiative (CCSSI), 2010).

BACKGROUND

Program Description

Our undergraduate pre-service mathematics education program includes extensive work both in mathematics and in education. Because of the comprehensive nature of our program, each of our teacher candidate graduates receives two degrees–a Bachelor of Science in Education, Mathematics and a Bachelor of Science, Mathematics. As part of this program teacher candidates complete 15 hours of education and mathematics methods courses, 42 hours of mathematics content courses, and at least 12 hours of full-time field experience. These degrees reflect the North Carolina Professional Teaching Standards (North Carolina Professional Teaching Standards Commission, 2006) and the 21st century knowledge, skills, and dispositions embedded in them.

Early classroom experience is a key component of our program. School partners are actively and continuously involved in providing multiple field experiences for our teacher candidates in the form of field observations and pre-service lesson study cycles. During internship and student teaching, school partners help evaluate teacher candidates' performance through observation protocols. As part of a prior evaluation of the program, a group of 58 teachers from local middle schools and high schools attended a presentation on our mathematics education program. Specific topics of discussion addressed what teacher candidates need to know and be able to do in the classroom upon graduation from a teacher education program. As a result of this evaluation, we designed a new course to enhance the connection of content, technology, and pedagogy. The new course, *Math 414: Introduction to Secondary Mathematics Teaching Methods*, has an emphasis on connecting technology and pedagogy. This course serves as a springboard to the second methods course,

Math 415: Methods and Materials for Teaching Mathematics in Secondary Schools. Now *Math 415* takes a more holistic approach to instructional design in which teacher candidates make instructional choices that integrate technology, pedagogy, and content.

TPACK Framework

As a guiding framework for promoting teacher candidates' acquisition of 21st Century skills and dispositions with respect to digital technologies, we use TPACK constructs that emerged from research in recent years. In this section, we provide a brief introduction to the TPACK Knowledge Bases, Standards, and Stages of Development.

As described by Koehler (2011), there are three Knowledge Bases:

Technology Knowledge is knowledge about standard technologies such as books and chalk and blackboard, as well as more advanced technologies such as the Internet and digital video. This would involve the skills required to operate particular technologies. In the case of digital technologies this would include knowledge of operating systems, and computer hardware, as well as the ability to use standard set of software tools such as word processors, spreadsheets, browsers, email etc. (TK)

Pedagogical Knowledge is deep knowledge about the processes and practices or methods of teaching and learning and how it encompasses (among other things) overall educational purposes, values and aims. This is a generic form of knowledge that is involved in all issues of student learning, classroom management, lesson plan development and implementation, and student evaluation. Pedagogical knowledge requires an understanding of cognitive, social and developmental theories of learning and how they apply to students in their classroom. (PK)

Content Knowledge is knowledge about the actual subject matter that is to be learned or taught. Clearly, teachers must know and understand the subjects they teach, including: knowledge of central facts, concepts, theories and procedures within a given field; knowledge of explanatory frameworks that organize and connect ideas; and knowledge of the rules of evidence and proof. (CK)

The TPACK framework focuses on essential knowledge created in the intersections of these three knowledge bases as illustrated in Figure 1 (Mishra & Koehler, 2006):

In addition to the framework, which provides a way to examine and categorize our course sequence in the light of overlapping sections of the diagram in Figure 1, the TPACK Standards listed provide a set of goals that proficient teacher candidates should strive to achieve by the end of their program. The following four TPACK Standards are suggested by AMTE (2009):

1. Design and develop technology enhanced mathematics learning environments and experiences.
2. Facilitate mathematics instruction with technology as an integrated tool.
3. Assess and evaluate technology enriched mathematics teaching and learning.
4. Engage in ongoing professional development to enhance technological pedagogical content knowledge (Information, Media, and Technology Skills).

In practice, these standards permeate the classroom culture and the everyday work of teachers who use technology to enhance students' comprehension of the subject matter through carefully designed "instruction, organization, and classroom management specific to the application of technology in the mathematics classroom." In particular, these teachers "understand their con-

Figure 1. TPACK Image (rights free image (http://tpack.org/))

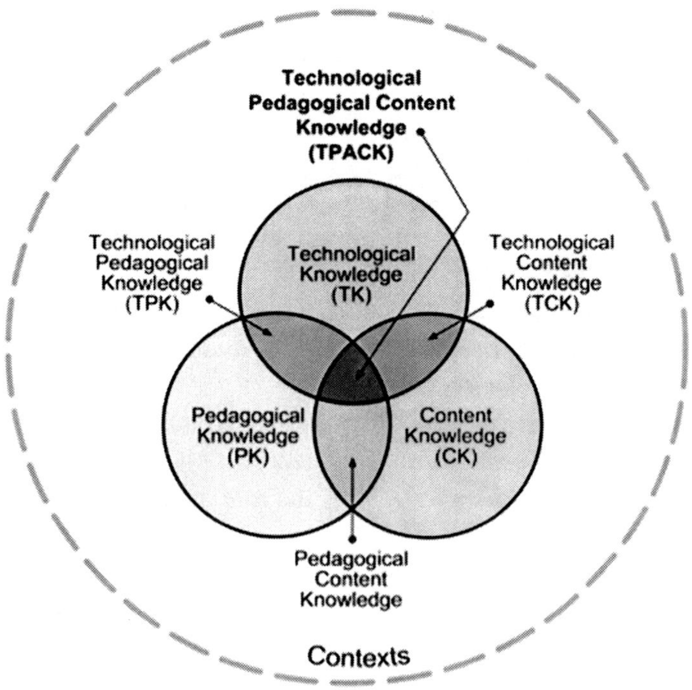

tent areas with both breadth and depth as a result of using technology as part of their instruction" (Guerrero, 2010).

Finally, the TPACK Stages of Development allow us to identify strategic places in the program that present opportunities for teacher candidates to transition from one stage to the next when using a specific technology in the teaching of mathematics. It is our goal as Mathematics Educators to facilitate these transitions through a carefully designed sequence of courses and experiences. In order for teacher candidates to demonstrate proficiency in the previously mentioned standards by the time they transition to the classroom, teacher educators need to select learning opportunities that promote TPACK acquisition in teacher preparation courses. In turn, the framework can be the basis for an in-depth examination of areas of growth for teacher candidates in relation to the use of technology as they move through the following stages of TPACK development when learning to use specific technology in the mathematics classroom (Niess et al., 2009):

- **Recognizing (Knowledge):** Where teachers are able to use the technology and recognize the alignment of the technology with mathematics content yet do not integrate the technology in teaching and learning of mathematics.
- **Accepting (Persuasion):** Where teachers form a favorable or unfavorable attitude toward teaching and learning mathematics with an appropriate technology.
- **Adapting (Decision):** Where teachers engage in activities that lead to a choice to adopt or reject teaching and learning mathematics with an appropriate technology.
- **Exploring (Implementation):** Where teachers actively integrate teaching and learning of mathematics with an appropriate technology.

- **Advancing (Confirmation):** Where teachers evaluate the results of the decision to integrate teaching and learning mathematics with an appropriate technology.

Mishra argues that teachers must "learn technology not by learning specific computer programs, but rather by designing technological solutions to pedagogical problems." (Mishra, 2012, May 12, para. 5). Likewise, the learning of a technology for technology's sake has been well documented as providing little support for students to attain reasoning and sense-making in the mathematics classroom (Jones, 2002). Furthermore, exposure to the use of technology through the modeling and observation of exemplary practice in the field will mold teacher candidates' beliefs and attitudes regarding the use of instructional technologies to enhance mathematical inquiry-based learning tasks (Meagher, et al., 2011a). Therefore successful TPACK acquisition requires moving teacher candidates along pathways through intersecting sections of the TPACK diagram, as well as through the Stages of Development as they engage in technology-intensive mathematics classes, in mathematics teaching methods classes, and in appropriate field experiences that model instruction specifically catered to the needs of 21st century learners.

CCSS-M Implementation and Technology (Secondary Education Focus)

The Common Core State Standards in Mathematics (CCSS-M) was developed as a state-led initiative out of concerns that existing sets of different standards across the country were not adequately providing students with knowledge and skills essential to compete in an increasingly mobile society.

The Standards for Mathematical Practices provide descriptions of the proficiencies with process that CCSS-M advocates for all students. Students need to be able to do the following:

1. Make sense of problems and persevere in solving them.
2. Reason abstractly and quantitatively.
3. Construct viable arguments and critique the reasoning of others.
4. Model with mathematics.
5. Use appropriate tools strategically.
6. Attend to precision.
7. Look for and make use of structure.
8. Look for and express regularity in repeated reasoning (CCSSI, 2010).

While the CCSS-M focuses heavily on the content proficiencies of students, it does acknowledge the role of technology as a tool for mathematics learning and constructing knowledge. "Strategic use of technology is expected in all work. This may include employing technological tools to assist students in forming and testing conjectures, creating graphs and data displays and determining and assessing lines of fit for data" (CCSSI, 2010, Appendix A).

Implementation of the Common Core State Standards in mathematics requires that teachers of mathematics must have deep content knowledge to teach mathematics effectively and that content preparation needs to be tied closely with pedagogical training. But where do teacher candidates get the practice and experience that they need to develop strong pedagogy using technology? Preparing future teacher for 21st century classrooms makes it essential for all educators to work together in providing adequate experiences and opportunities for future teachers throughout their program of study. This needs to be evident in content courses as well as methods courses. This chapter proposes a model for teacher preparation that seeks to develop strong content knowledge, pedagogical knowledge, technological knowledge, and intersections of these knowledge bases necessary for teacher candidates to be successful as secondary mathematics teachers.

MOVING TEACHER CANDIDATES ALONG TPACK PATHWAYS TO SUPPORT CCSS-M IMPLEMENTATION

The TPACK framework serves as a foundation for analyzing our program and identifying places where we provide teacher candidates with opportunities to advance their understanding of each intersection of Knowledge Bases. At the same time, we recognize different people will progress through the TPACK Stages of Development differently and will respond to a specific technology in various ways. Our goal is to provide learning experiences that allow exploration of all Knowledge Bases and their intersections with an emphasis on early and ongoing experiences in the center of the framework. These experiences help teacher candidates start on their journey of progressing towards advancing stages of TPACK. Figure 2 is an adaptation of the TPACK image that includes the most pertinent experiences in our program that help form the necessary pathways towards the center of the framework. The sections that follow will describe those pathways.

Beginning from Technology and Content Knowledge

In looking at the interplay of technological knowledge and content knowledge, it is clear that there are instances where these two areas are distinct. For instance, teacher candidates must have sufficient technological knowledge to access information on a particular network that may not involve any mathematical content knowledge. They must also have content knowledge that is primarily abstract in nature, such as what constitutes a group. There is evidence, however, to show that students can learn mathematics more deeply with appropriate use of technology (Jones, 2002; Ellington, 2003; Fey, Cuoco, Kieran, McMullin, & Zbiek, 2003). Therefore, it is very important that teacher candidates experience how to embrace the power of

Figure 2. Program components that help form TPACK pathways

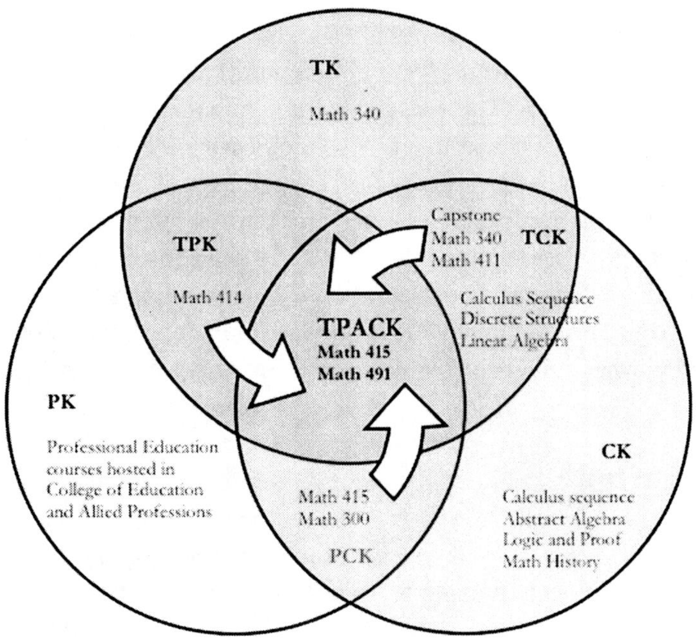

technology to better understand mathematics that challenges them during their teacher preparation program. Not only will teacher candidates experience how to use technology to do mathematics but also how to use technology to communicate mathematical thought and reasoning abilities.

In parallel with the CCSS-M Standards for Mathematical Practices #2 and #4, our teacher candidates are learning college level mathematics both abstractly and quantitatively with an emphasis on modeling. Our program requires several mathematics courses that integrate technology as a tool for learning. While instructors of each course determine exactly how much technology is used, it is a common tool. For example, graphing calculators are required for the three courses in the calculus sequence, and these courses tend to incorporate graphical and numerical techniques in conjunction with theoretical ones. Some instructors choose to use computer algebra systems in these courses to enhance the various ways of understanding. As teacher candidates move into higher-level mathematics classes, their ability to use a variety of technologies becomes important.

To ensure that teacher candidates have a sufficient background and proficiency with a variety of technologies, our program requires all of our majors to complete a course entitled *Math 340: Introduction to Scientific Computing*. The main goal of this course is to provide teacher candidates with a basis for using technology appropriately to aid in their own learning of higher-level mathematical content. In this way, they are moving into the intersection of content knowledge and technological knowledge within the TPACK framework from a learner's point of view as evidenced by its learning outcomes:

Math 340 (Introduction to Scientific Computing)

By the end of the course we expect students to be able to:

- Model problems mathematically and use mathematical software to solve or simulate these problems.
- Develop algorithms and implement them in the appropriate software or programming language.
- Present algorithms and solutions to problems in a mathematically sophisticated manner using a scientific documentation environment.
- Know the benefits and drawbacks of each of the computational tools used during the semester.

After successfully completing this course, teacher candidates are expected to be able to use a variety of mathematical software in other upper level mathematics courses without explicit instruction. In this way, teacher candidates in numerical analysis, linear algebra, discrete mathematics, and other courses are able to choose appropriate technologies to aid them in understanding the mathematical content. They also have developed the skills to present their understandings in mathematical ways. Through this integration, we encourage all of our teacher candidates to move into the zone that allows technology and content knowledge to interact seamlessly from the student point of view. They are trying to enhance their own understanding of the content through a technological lens.

Further integration of technology and content knowledge from the teacher perspective occurs as teacher candidates complete Capstone projects and prepare presentations of their work using technology, a part of the required course *Math 479: Capstone Seminar*. In this course, each teacher candidate chooses problems to explore and/or to solve. Often this work entails use of statistical analysis packages, computer algebra systems, dynamical graphing programs, or other digital technologies. What is common to all teacher candidates is the requirement to present their findings in print and as presentations using appropriate software. As each project is specific to the individual, each teacher candidate must demonstrate the integration of choosing appropriate technologies to aid in his or her understanding and in the presentation of that understanding to others. In this way, they begin to consider the intersection of technological and content knowledge from a teacher view.

In the required geometry class, *Math 411: Foundations in Geometry,* a more advanced level course, teacher candidates are introduced to *Geometer's Sketchpad*, a dynamical graphical software package, through an exploratory unit on transformational geometry. As teacher candidates explore particular examples and topics, they are asked to develop theorems that they then prove formally based on their experiences with the technology.

This allows students to play with shapes and form, making it easier to construct standard geometry proofs. In this regard, the software program merely emulates what was done earlier when learning geometry. However, the computer program does more than that. By allowing students to "play" with geometrical constructions, it also changes the nature of learning geometry itself – proofs by construction are a form of representation in mathematics that was not available prior to this technology (Koehler, 2011, TCK).

In this instance, teacher candidates experience how they can learn particular content with technological assistance and how that interaction allows for a different type of understanding. Through comparison of their own work in earlier parts of this course, teacher candidates are encouraged to consider the differences in the ways of understanding that are possible.

These examples demonstrate the incorporation of particular technologies for specific purposes in mathematics courses. Through these experiences, teacher candidates move along the early pathways from content knowledge and technological knowledge into the intersection of these two zones from both a student point of view and a teacher point of view. Furthermore, their continued work with content and technology aid our teacher candidates in moving from the early TPACK stage of Recognizing towards the next stage of Accepting in their TPACK development. Thus, additional steps on the pathway of TPACK development also help lead teacher candidates towards more sophisticated stages, Adapting and Exploring, as they begin to choose the software that would be most useful to their own learning and teaching.

Beginning from Technology and Pedagogical Knowledge

Pedagogical knowledge is the information we gather from research and experience of expert educators that helps us understand connections between teaching and learning. Effective teachers will use pedagogical knowledge to make decisions regarding how to convey and reveal subject matter to students. They are aware of the preconceptions and background knowledge that students typically bring to each subject and of strategies and instructional materials that can be of assistance. In addition, they understand and solve the possible difficulties likely to arise in the classroom and modify their practice accordingly. Their instructional repertoire allows them to create multiple paths to knowledge, in general, and to the subjects they teach, in particular. Obviously, this is not possible without specifically addressing the area in which the use of technology and pedagogical knowledge converge. Since the presence of technology is unavoidable in 21st century classrooms, what do teacher candidates need in order to be prepared for integrating technology appropriately into their teaching practices? As teacher candidates move into their methods courses, it is essential for them to develop a solid Technological Pedagogical Knowledge (TPK) which is described as "knowledge of the existence, components and capabilities of various technologies as they are used in teaching and learning settings, and conversely, knowing how teaching might change as the result of using particular technologies" (Koehler, 2011, TPK).

The CCSS Standard for Mathematical Practice #5 emphasizes that all students must be able to consider available tools and use them appropriately in their problem solving. This must include knowing how and when to embrace technological tools such as graphing calculators, dynamic geometry software, computer algebra systems, etc. If students are to be proficient in using the tools appropriately, then it is of the utmost importance that their teachers feel comfortable doing the same. Technology should be introduced and utilized to deepen understanding or reduce reliance on routine tasks so that students engage in higher-level mathematical reasoning. Teacher candidates need to spend significant time becoming familiar with the standards, considering multiple ways of presenting content addressed in the standards, and building technological knowledge necessary to support development of connections between and across those standards. Our teacher candidates begin a two-course sequence with *Math 414.* In this course, we want our teacher candidates to achieve four major learning outcomes:

Math 414 (Introduction to Secondary Mathematics Teaching Methods)

The future teacher will be able to:

- Understand and be familiar with the Common Core State Standards for secondary mathematics as well as the NCTM Principles and Standards.
- Identify mathematical connections among various topics and can apply content, processes, and reasoning to real world situations.
- Identify mathematics curricular goals and plan instruction to meet these goals, and identify appropriate uses of technology.
- Identify appropriate uses of technology and plan instruction that uses technology to enhance teaching and learning, and use technology to assess learning.

This course is taught based on a learning community model. Wenger and Lave (1991) proposed a model of situated learning that involves engagement in a community of practice. They argue that communities of practice are everywhere and that as individuals we are included in many of them – whether at work, school, home, or leisure activities. These communities evolve around things that matter to the people that are involved (Wenger, 1998). In keeping with this idea, we believe the shared experiences both in and out of the classroom can serve the development of a learning community in which teacher candidates and professor work collaboratively to explore technologies and develop strong TPK. The teacher candidates are all responsible for presenting and sharing their thoughts and ideas related to the Common Core State Standards.

This course is the first time during their program that teacher candidates are introduced to the Common Core State Standards in grades 9-12 and their implications for teaching. At the same time, teacher candidates are exposed to various forms of technologies that may be used to teach mathematics. Teacher candidates are given opportunities through presentations and written assignments to use technology productivity tools to complete required professional tasks. While not yet using the tools themselves in authentic teaching experiences, teacher candidates do begin to make meaningful connections between their own technological knowledge and how to teach. Through practice and demonstration using technology, they begin to align the content with the integration of technology. Teacher candidates are asked to develop lesson plans based on standards. While doing so, there is a focus on differentiating between appropriate and inappropriate uses of technology for teaching and learning as well as using electronic resources to design and implement learning activities.

Through purposeful class activities in *Math 414* involving various technologies as a tool for mathematical sense making, it is hoped that the teacher candidates move to a level of acceptance in which decisions are made regarding the benefits or risks involved with using technology for instruction. It is important that teacher candidates begin to realize the capabilities of using technology as a cognitive tool to maximize student learning and to facilitate higher order thinking skills. To do so requires time to research and evaluate the accuracy, relevance, and appropriateness of these technologies. Teacher candidates also need time to explore electronic information resources that complement their developing pedagogical styles. The community of learning encourages valuable debate and open discussions that help each individual develop positive attitudes toward teaching and learning mathematics with an appropriate technology.

The course provides several experiences where technology is used to move between and within multiple representations of functions (numerical, graphical, verbal, and symbolic). For example, one activity involves using TI-Nspire technology to explore geometric relationships between the height of a vase and the volume of water in the vase by changing the values dynamically. The shape of the vase can be manipulated as well as the height of the water in the vase so that the relationships can be explored. The teacher candidates also get experience using data collection devices to collect information such as distances and temperatures for the purposes of developing mathematical models based on the data.

Additionally, teacher candidates are expected to explore and review numerous virtual manipulative and electronic resources to analyze whether they might be appropriate tools to use with students to address Common Core State Standards in grades 9-12. As the manipulatives are explored, the teacher candidates have a chance to investigate the use of interactive whiteboards for active teaching and learning. Ample time is allowed for discussion of how, when, and why to use the technologies to advance student thinking. As the course progresses, the teacher candidates have their first experiences developing lesson plans that address specific standards in the Common Core for which technology can play an important role. This aids in their development of recognizing the role of technology and adapting activities that can engage their students.

Strong technological knowledge is not enough when it comes to teaching mathematics effectively. Teachers must apply their skills in using technology with understanding of teaching strategies which engage students in mathematics. "The value of technology to the teacher lies not so much in the answers technology provides but rather in the questions it affords" (Dick & Hollebrands, 2011, p. xvi). It is the teaching pedagogy employed by the teacher that defines whether or not technology opens doors to understanding or becomes an instrument of distraction that diminishes mathematical sense-making. The experiences students have in *Math 414* create another pathway that aids our teacher candidates in moving toward the intersection of pedagogy, technology, and content. As teacher candidates progress in their program of study, they will begin to apply teaching strategies using technology that support the construction of content knowledge.

Beginning from Content and Pedagogical Knowledge

The notion of Pedagogical Content Knowledge was first described by Shulman (1986) as a way to consider pedagogy and content as inclusive components of a teacher's knowledge that inform the teaching decisions that best fit a specific subject matter. This includes knowing the different representations commonly used to present and make sense of curricular material, understanding the connections necessary to unveil so that students may overcome difficulties in learning a concept due to prior misconceptions, and making pedagogical decisions that promote students' mathematical proficiency.

Because of the strong emphasis on content inherent in most secondary education teacher preparation courses, there are often many factors that make it difficult for teacher candidates to gain access to authentic classroom experiences prior to their field placements. In order to provide additional opportunities for teacher candidates to plan lessons, Mathematics and Mathematics Education faculty work together to personalize end of course projects in several of our mathematics content courses. While applied mathematics students might be focused on a modeling project, teacher candidates will often be asked to develop an activity that integrates the course material into a secondary mathematics lesson plan. These recurring connections between higher level mathematics courses such as Discrete Structures or Linear Algebra, and the mathematics they will be expected to teach in high school, help teacher candidates understand the value of deeper content understanding that serves the purpose of sense making for appropriately challenging task design. This is a connective pathway from viewing mathematics as learners to constructing new knowledge of mathematics as teachers.

Another course specifically geared towards future mathematics teachers is *Math 300,* a problem-solving course in which teacher candidates are actively engaged in collaborative group work that helps them improve their problem solving abilities, while encouraging them to argue solutions with their peers in a constant open-communication atmosphere. In *Math 300*, teacher candidates have to simultaneously assume the roles of doers of mathematics, trusting members of a learning community where they learn to listen and to understand others' ways of thinking, and at times, teachers for their peers. This course specifically addresses the CCSS-M Standard for Mathematical Practice #1 where students are expected to "make sense of problems and persevere in solving them," as well as Standard #3 focused on constructing "viable arguments and critique the reasoning of others." Content standards addressed in *Math 300* cover a wide range of secondary mathematics topics across all major content strands of the CCSS-M, therefore providing teacher candidates with a collection of problem solving situations that they may immediately pull from once in the classroom. When exploring a variety of problem solving approaches, teacher candidates are also exposed to Standards for Mathematical Practices #7 and #8, which permeate all the mathematics content courses in the program. Learning outcomes for *Math 300* include:

Math 300 (Methods and Techniques for Problem Solving in Mathematics)

The future teachers will be able to:

- Identify and apply methods and techniques that are helpful in solving problems.
- Solve problems using problem solving steps.
- Explain and discuss problem solving strategies.
- Create problems that may be solved using a variety of strategies.

Using the TPACK framework to analyze our program has helped us identify areas where technology can be better integrated. For example, *Math 300* has previously been taught in ways that provided teacher candidates with limited opportunity to integrate technology in their reasoning and solutions. However it is an obvious place for us to move teacher candidates along a major pathway from pedagogical content knowledge experiences to more integrated TPACK experiences, therefore placing the course in the center of the TPACK diagram. In particular the use of Computer Algebra software can reinforce the nature of mathematical inquiry and investigation, conjecture building, conjecture testing, and developing logical arguments.

To build on the introductory methods course *Math 414*, seniors in our program take a second methods class, *Math 415*, which can be taken concurrent to a first semester internship, and directly precedes the full time student teaching semester. Learning outcomes for *Math 415* are as follows:

Math 415 (Methods for Teaching Mathematics in the Secondary Schools)

The future teachers will be able to:

- Reflect upon and discuss contemporary issues in teaching mathematics at the high school level.
- Identify mathematics curricular goals.
- Plan instruction to meet these goals.
- Implement lesson presentation skills.
- Evaluate mathematics instruction.

Following a similar community-based model as *Math 414*, students in *Math 415* routinely engage in lesson planning and teaching activities following discussions of important components of teaching, such as classroom management, questioning and student engagement, technology integration, assessment, curriculum, and mathematics. The community-modeling approach of the course assumes teacher candidates are actively involved in designing rubrics for the purpose of course assessments, leading and facilitating discussions, and taking turns teaching their peers. The students who are simultaneously enrolled in an internship experience often bring to class discussion topics which allow the community to provide a "just-in-time" model of trouble shooting professional development that is one of the highlights of communities of practice when implemented in the schools. A variety of teaching methods are enacted and discussed in the light of benefits they may bring to teaching, learning, and curricular concerns; these include interactive lecturing, inquiry-based practice, and collaborative group work.

As an overall course project, students are asked to design a full unit plan that will be one of the evidences required by the North Carolina Department of Public Instruction for obtaining a teacher credential. Within this unit, curricular goals must be clearly identified, alignment to the Common Core State Standards is required, and a variety of teaching and assessment strategies must be used, including integration of technology in instruction. Through this unit planning, teacher candidates begin to comprehend the bigger picture of curricular alignment, incorporation of multiple representations in instruction, and purposeful planning for promoting their future students' reasoning and sense-making. For teacher candidates, these tasks go beyond their own knowledge of mathematics and how to solve a particular problem, to include identifying student learning objectives, researching best ways to present a particular concept, becoming knowledgeable of high school students' common misconceptions, questioning for inquiry into students' prior knowledge, and students' understanding of new material. Regarding technology, the work started in *Math 414* expands to the necessity for reflections on when the use of a particular technology is appropriate depending on the content presented and student learning objectives. Teacher candidates are now asked to make choices that will have consequences for their instructional practices. Through these activities, teacher candidates are actively engaged in the center of the TPACK diagram with opportunities to reach Adapting, and perhaps Exploring, Stages of Development when deciding to incorporate specific technologies in their plans. Open sharing of these explorations into lesson planning and task design positions teacher candidates as active members of a 21st Century learning community.

When researching pre-service teachers' acquisition of TPACK in a teacher preparation program, Meagher, Özgün-Koca, and Edwards (2011) found that "if pre-service teachers are to develop a positive attitude to the use of advanced digital technologies in their instructional practice, they require more than a methods class to develop TPACK" (p. 259). They emphasize the role of field experiences in shaping teacher candidates' attitudes and dispositions towards using technology. With such findings in mind, we recently decided to address a need for increased collaboration between practicing teachers and teacher candidates earlier in the program through an adaptation of lesson study cycles into *Math 415* that will help teacher candidates better understand the importance of technology and curriculum integration.

Research on models of professional development interventions that best promote transfer of new teaching knowledge into the classroom shows that a lesson study approach is most efficient in ensuring teachers will actually implement the new ideas and materials in their teaching (Faughn, Brown, Kent, & Tuba, 2011). At the teacher preparation level, one way to adapt lesson study is to have teacher candidates commonly plan lessons during class time and receive feedback on those lesson plans from a practicing master teacher through a back and forth discussion until the lesson is ready to be implemented. Video recordings and conference calls can then be used in order for teacher candidates to view and reflect upon the teaching of their lesson by this experienced master teacher, who will also provide feedback on the lesson once implemented in the classroom. Teacher candidates can then revise the lesson once again in light of what went well and what needs to be changed (Meagher et al., 2011b).

We hope that requiring the use of technology in the lesson planning stage of the lesson study cycle and the teacher candidate-Master teacher feedback loops will move some of our teacher candidates to the Exploring Stage of Development with that particular piece of technology, while the post-lesson reflection with feedback from the mathematics education faculty teaching the course will facilitate reaching this stage with the technology used during the lesson. This is the first activity that our teacher candidates will engage in prior to their student teaching experience, where they may begin to observe and understand the different learning approaches of high school students when faced with having to use technology. It will be a crucial experience in reinforcing CCSS-M Standard for Mathematics Practice #5.

Reaching the Center: Technological, Pedagogical, and Content Knowledge

We have now argued the necessity of moving teacher candidates through multiple TPACK pathways to help them promote students' mathematical practices described in the Common Core documents.

Finally, at the intersection of all three elements is Technological Pedagogical Content Knowledge (TPACK). True technology integration is understanding and negotiating the relationships between these three components of knowledge. A teacher capable of negotiating these relationships represents a form of expertise different from, and greater than, the knowledge of a disciplinary expert (say a mathematician or a historian), a technology expert (a computer scientist) and a pedagogical expert (an experienced educator). Effective technology integration for pedagogy around specific subject matter requires developing sensitivity to the dynamic, [transactional] relationship between all three components (Koehler, 2011, TPACK).

As discussed in earlier sections, examples of early experiences in the center are included in courses throughout the program. Early opportunities for moving along pathways towards the center will position teacher candidates in our program to embrace digital technologies and the type of inquiry-based practices that accompany their use in the classroom. Table 1 summarizes activities

Table 1. Pre-field placement experiences that move teacher candidates along pathways to the center

Pathways	Courses	CCSS Experiences	Technology Examples
Content and Pedagogy to Center	Math 300 re-visited with technology Math 415	Revisiting Algebra Lesson study and unit plan	CAS, Graphing Calculators Student choice
Technology and Content to Center	Math 340 Capstone Math 411	Projects Projects Revisiting Geometry	Variety Variety Dynamic Geometry Software
Technology and Pedagogy to Center	Math 414	Lesson planning and curricular study	Graphing Calculators, Software, Interactive Whiteboards

that we identified as places where teacher candidates are introduced to the center with specific technological tools that support Common Core Mathematical Practices.

Based on the experiences summarized in Table 1, by the time teacher candidates go out for field placement, our goal is for them to demonstrate readiness to work in the TPACK center. During their last semester, all teacher candidates are enrolled in internship/student teaching, which serves as a culminating experience with the following outcomes:

Math 491 (Supervised Student Teaching – Mathematics 9-12)

By the end of the course we expect students to be able to:

- Demonstrate proficiency in teaching mathematics at the high school level.
- Be able to identify mathematics curricular goals, plan instruction to meet these goals, implement lesson presentation skills, and evaluate mathematics instruction.
- Reflect on teaching experiences addressing the following dimensions of teaching:
    1. Classroom Management;
    2. Planning;
    3. Questioning and Student Engagement;
    4. Technology;
    5. Assessment and Grading;
    6. Curricular Choices and Pacing;
    7. Knowledge of Mathematics for Teaching.

Once in the classroom, teacher candidates are able to directly experiment with the integration of technology in working with their own students. As they develop into reflective practitioners and define their identity as educators, it is expected that they will move toward the TPACK center and advance to higher Stages of Development.

IMPLICATIONS, CHALLENGES, AND FUTURE DIRECTIONS

Technology is an integral part of society, and it is essential that we embrace and explore opportunities for developing higher-order reasoning and thinking skills for secondary mathematics teacher candidates. All stakeholders have a responsibility to encourage appropriate uses of technology to engage students and develop mathematical understanding. We would like to offer a word of caution, however, that unequal access may exist from one placement school to another, which in turn will shape opportunities and development in the use of digital technologies. While the model presented here is a step in the right direction, there is much more work to be done. Wilson (2008) suggests that teacher education programs must serve as the conduit for connecting research and innovations in technology with practice and curriculum in mathematics (p. 425). Additional questions to be answered include:

1. Do our teacher candidates actually progress through the TPACK Stages of Development as we intend?
2. How do we assess teacher candidates' experiences with digital technologies as a means to enhance their students' experiences in support of the Common Core State Standards?
3. What means of support are necessary to encourage continued use of technology to further develop 21st century skills teaching and learning after graduation in classrooms with differing technology resources?

Addressing question #1 can take place within our program. There are several TPACK surveys (Koehler, 2011, TPACK surveys) that can be used as assessment instruments for evaluating teacher candidates' level of TPACK acquisition at various points in our program. This will help us gauge teacher candidates' changes in attitudes towards technology and towards its use for teaching and learning mathematics. These same instruments can help us to begin answering Question #2. According to Niess et al. (2009), the Stages of Development can be targeted to four specific themes to help us further unpack the levels of TPACK acquisition our teacher candidates may demonstrate: Curriculum and Assessment, Teaching, Learning, and Access. Indicators for these themes can serve as a tool for evaluating how accomplished our teacher candidates have become in using the technologies they were introduced to in their courses for the purpose of TPACK implementation in their own classrooms. In particular, analysis of teacher candidates' levels of TPACK acquisition before graduation can become part of the student teaching/internship review process. Continuing that assessment and addressing question #3 will be more difficult because of issues of access and control of external factors once teacher candidates have graduated and moved into the teacher role fulltime. Longitudinal research that follows these teacher candidates through their early years of teaching is clearly indicated. We can begin this process through surveys and other self-reporting instruments, but a more focused research design should be developed. It is only through following our teacher candidates' into their classrooms that we will be able to assess the long-term efficacy of our efforts to guide them along the pathways of TPACK acquisition.

CONCLUSION

The Conference Board of Mathematical Sciences insists that teacher candidates of mathematics be exposed to specific mathematical topics and progressions as well as applications of mathematics, modeling, and connections with other disciplines such as engineering and computational biology. This necessitates building strong 21st century skills to enhance mathematics teaching and learning. In addition, teacher preparation programs should immerse teacher candidates in various experiences throughout their program of study that challenge mathematical habits of mind, incorporate sound mathematical practices, and place a greater emphasis on field and clinical experiences. To ensure that teacher candidates are provided with these type of experiences, professional learning communities must develop including teachers at all levels, mathematicians (at two- and four-year institutions), and mathematics educators (Conference Board of Mathematical Sciences (CBMS), 2011, p. 15).

We have described a model for teacher preparation that emphasizes the importance of developing 21st century skills. The TPACK Framework provides a foundation to develop a series of pathways to integrate content knowledge, pedagogical knowledge, and technological knowledge. We argue the necessity of moving teacher candidates through these pathways to help them promote in their students the types of mathematical practices described in the Common Core documents.

REFERENCES

Association of Mathematics Teacher Educators. (2009). *Mathematics TPACK (technological pedagogical content knowledge) framework*. Retrieved from http://www.amte.net/sites/all/themes/amte/resources/MathTPACKFramework.pdf

Common Core State Standards Initiative (CCSSI). (2010). *Common core state standards for mathematics*. Washington, DC: National Governors Association Center for Best Practices and the Council of Chief State Officers. Retrieved from http://www.corestandards.org

Conference Board of Mathematical Sciences. (2011). *Common standards and the mathematical education of teachers* (White Paper). Retrieved from http://www.cbmsWeb.org/Forum3/CBMS_Forum_White_Paper.pdf

Dick, T. P., & Hollebrands, K. F. (2011). Introduction to focus in high school mathematics: Technology to support reasoning and sense making. In T. P. Dick & K. F. Hollebrands (Eds.), *Focus in high school mathematics: Technology to support reasoning and sense making* (pp. xi–xvii). Reston, VA: National Council of Teachers of Mathematics.

Ellington, A. J. (2003). A meta-analysis of the effects of calculators on students' achievement and attitude: Levels in precollege mathematics classes. *Journal for Research in Mathematics Education*, *34*, 433–463. doi:10.2307/30034795

Faughn, A. P., Brown, K., Kent, N., & Tuba, I. (2011). *Supporting beginning mathematics teachers with technology-based professional development*. Paper presented at the 2011 meeting of the American Educational Research Association. New Orleans, LA.

Fey, J. T., Cuoco, A., Kieran, C., McMullin, L., & Zbiek, R. M. (Eds.). (2003). *Computer algebra systems in secondary school mathematics education*. Reston, VA: National Council of Teachers of Mathematics.

Guerrero, S. (2010). Technological pedagogical content knowledge in the mathematics classroom. *Journal of Digital Learning in Teacher Education*, *26*(4), 132–139.

Jones, K. (2002). Research on the use of dynamic geometry software: Implications for the classroom. *MicroMath*, *18*(3), 18–20.

Koehler, M. J. (2011). TPACK – Technological *pedagogical and content knowledge*. Retrieved from http://tpck.org/

Meagher, M., Edwards, M. T., & Koca, A. (2011b). *Project CRAFTeD: An adapted lesson study partnering preservice mathematics teachers with a master teacher*. Paper presented at the 33rd annual meeting of the North American Chapter of the International Group for the Psychology of Mathematics Education. Reno, NV.

Meagher, M., Özgün-Koca, S. A., & Edwards, M. T. (2011a). Preservice teachers' experiences with advanced digital technologies: The interplay between technology in a preservice classroom and in field placements. *Contemporary Issues in Technology & Teacher Education*, *11*(3), 243–270.

Mishra, P. (2012). *Design and pedagogy*. Retrieved from http://punya.educ.msu.edu/research/design-pedagogy/

Mishra, P., & Koehler, M. J. (2006). Technological pedagogical content knowledge: A framework for teacher knowledge. *Teachers College Record, 108*(6), 1017–1054. doi:10.1111/j.1467-9620.2006.00684.x

Niess, M. L., Ronau, R. N., Shafer, K. G., Driskell, S. O., Harper, S. R., & Johnston, C. et al. (2009). Mathematics teacher TPACK standards and development model. *Contemporary Issues in Technology & Teacher Education, 9*(1), 4–24.

North Carolina Professional Teaching Standards Commission. (2006). *North Carolina professional teaching standards*. Retrieved from http://www.ncpublicschools.org/docs/profdev/standards/teachingstandards.pdf

Partnership for 21st Century Skills. (2011). Information, media and technology skills. Retrieved from http://www.p21.org/overview/skills-framework/61

Shulman, L. S. (1986). Those who understand: Knowledge growth in teaching. *Educational Researcher, 15*(2), 4–31. doi:10.3102/0013189X015002004

Wilson, P. S. (2008). Teacher education: A conduit to the classroom. In G.W. Blume & M.K. Heid (Eds.), Research on Technology and the Teaching and Learning of Mathematics: Volume 2: Cases and Perspectives (pp. 415-426). Charlotte, NC: Information Age Publishing, Inc.

ADDITIONAL READING

(2008). InBlume, G. W., & Heid, M. K. (Eds.). Research on technology and the teaching and learning of mathematics: Vol. 2. *Cases and perspectives*. Charlotte, NC: Information Age Publishing, Inc.

Conference Board of Mathematical Sciences. (2011). *Common standards and the mathematical education of teachers* (White Paper). Retrieved from http://www.cbmsWeb.org/Forum3/CBMS_Forum_White_Paper.pdf

(2008). InHeid, M. K., & Blume, G. W. (Eds.). Research on technology and the teaching and learning of mathematics: Vol. 1. *Research synthesis*. Charlotte, NC: Information Age Publishing, Inc.

Koehler, M. J. (2011). *TPACK – Technological pedagogical and content knowledge*. Retrieved from http://www.tpck.org/

Partnership for 21st Century Skills. (2011). Information, media and technology skills. Retrieved from http://www.p21.org/overview/skills-framework/61

KEY TERMS AND DEFINITIONS

21st Century Skills: Learning a profession in the 21st Century requires a specific set of skills that reflect an ability to solve problems collaboratively while communicating with and navigating evolving digital technologies. Examples of such skills include: Sharing and critically adopting knowledge, ideas and resources; addressing challenges as members of a team; and using social networks and other 21st century technologies to make individual contributions. Higher order and critical thinking skills will be demonstrated in a classroom where 21st Century skills are encouraged and developed.

Content Knowledge (CK): Or mathematical content knowledge in our case, refers to a teacher's understanding of mathematical ideas including concepts, procedures, and reasoning skills, as well as an understanding of the connections that serve as the underlying structure of the discipline.

Pedagogical Content Knowledge (PCK): Is more than just the intersection of knowledge about particular content and knowledge about particular pedagogies. It is the interaction of these two knowledge bases to create a separate type of knowledge that encompasses how to choose, organize, and present specific content so that learning occurs.

Pedagogical Knowledge (PK): Is familiarity with techniques, approaches, and strategies that may be used in a classroom to engage students and support the teaching and learning of mathematics.

Technological Content Knowledge (TCK): Develops when students have the opportunity to integrate technology with mathematical content in meaningful ways. Mathematics is not simply the backdrop in which technology is used but is an organizer and shaper of students' technological experiences. Likewise, the constraints and affordances that are offered by a particular mathematical technological tool help define the boundaries of mathematical learning.

Technological Pedagogical Content Knowledge (TPACK): Is the blending of all three knowledge bases to create new knowledge that is more than the sum of the individual parts. This new type of knowledge integrates aspects of all three knowledge bases with particular emphasis on the relationships created by those interactions.

Technological Pedagogical Knowledge (TPK): Refers to the interplay between pedagogical and technological knowledge bases. This entails being able to negotiate all aspects of how and when to use certain technologies to advance student learning. Moreover, it involves a teacher's understanding about how pedagogical choices and preferences inform the use of technology in the classroom.

Technology Knowledge (TK): Includes skills and understanding necessary to use, operate, and demonstrate technologies such as communication tools, software, and hardware to accomplish a variety of tasks.

This work was previously published in "Common Core Mathematics Standards and Implementing Digital Technologies" edited by Drew Polly, pages 353-369, copyright 2013 by Information Science Reference (an imprint of IGI Global).

Chapter 38
Bridging the Academia-Industry Gap in Software Engineering:
A Client-Oriented Open Source Software Projects Course

Bonnie K. MacKellar
St. John's University, USA

Mihaela Sabin
University of New Hampshire, USA

Allen B. Tucker
Bowdoin College, USA

ABSTRACT

Too often, computer science programs offer a software engineering course that emphasizes concepts, principles, and practical techniques, but fails to engage students in real-world software experiences. The authors have developed an approach to teaching undergraduate software engineering courses that integrates client-oriented project development and open source development practice. They call this approach the Client-Oriented Open Source Software (CO-FOSS) model. The advantages of this approach are that students are involved directly with a client, nonprofits gain a useful software application, and the project is available as open source for other students or organizations to extend and adapt. This chapter describes the motivation, elaborates the approach, and presents the results in substantial detail. The process is agile and the development framework is transferrable to other one-semester software engineering courses in a wide range of institutions.

MOTIVATION

Most computer science programs offer a software engineering course and view it as a critical link in ensuring the career-readiness of computer science graduates. However, too often this course is taught in terms of abstract principles, failing to engage students in real-world software experiences. Many of the skills required in industry are best learned by hands-on practice, such as the need for effective

DOI: 10.4018/978-1-4666-7363-2.ch038

communication among developers, or the need to interact with a non-technical client. Thus, students who have never engaged in a hands-on project in software engineering enter the workforce with gaps in their skills.

It is, however, difficult to bring a significant software development experience into the confines of a one-semester course in academia. The most common approach has been to introduce a "toy project," which is a small project designed by the instructor, and have students work in teams to complete the project by the end of the semester. The advantage of this approach is that students will ideally learn to work in teams and share responsibility for developing a codebase. The disadvantages are that the project may be oversimplified, and students gain no experience interacting with clients or with code written by others.

Another approach is to work with local companies in the private sector who sponsor *proprietary client-oriented software projects*. This has been used successfully by a number of schools, especially larger programs that already have established linkages with companies (Judith, Bair, & Börstler, 2003; Tadayon, 2004; Tan & Jones, 2008). Another setting that favors this approach is an internship course with the projects being developed onsite at local companies. The advantage is that students gain experience with real clients with high stakes in real projects. However, these projects are often standalone, one-off projects since companies may be reluctant to have students work on their internal codebase, or to develop mission critical software. This means that it may be difficult to get enough time and attention from personnel at the company while the students work on the project. Also, the project will normally become the property of the company, meaning that it cannot be freely shared with other schools trying to adopt a similar approach.

A third approach is to engage students in Free and Open Source Software (FOSS) development by having them contribute to *a large and active open source project*, such as Linux or Mozilla (Marmorstein, 2011; Ellis, Morelli, DeLanerolle, & Hislop, 2007). The advantage of this approach is that instructors and students can gain from the mentoring achieved through communication with the project's professional developers, and in some cases they contribute marginally to the "live" code base or the user documentation. The disadvantages of this approach are that most ongoing projects are large and complex, their developers may not be accessible, and given the time it takes to come up to speed in the project, students may gain little practical experience in a one-semester course.

A fourth approach, which occupies a middle ground between the proprietary client-oriented project model and the full-scale FOSS project, is to engage students in FOSS development via a relatively small project that fits in a one semester course, with a local nonprofit organization as the client. Local nonprofits are often happy to collaborate on these projects since they may have needs for mission-critical software systems that are not well met by the commercial software industry, yet they have limited technology budgets. Thus, it is relatively easy for an instructor to locate and collaborate with a local nonprofit. However, many instructors may still be unsure of how to get started or how to organize such a course.

This chapter describes our collective experience with the fourth approach, which we call *client-oriented free and open source software development* (CO-FOSS). The big advantage of treating client-oriented open source projects is the very openness of the project. An open source project developed in the context of one course for one client can be reused, extended, and adapted for new clients by subsequent iterations of the same course, or even by courses at different institutions. By providing not just the codebase but the course organization itself as an open source project, a collection of such projects can be built up to be used as models at different institutions. In addition, the tools and practices of open source projects provide a readymade infrastructure for software project courses.

Service learning projects for nonprofits have been used in a number of other software engineering courses and have been reported in these papers (Olsen, 2008; Poger & Bailie, 2006; Venkatagiri, 2006); however, these projects have not taken advantage of the reusable and extendable capabilities of the CO-FOSS approach. An example of a service learning approach that has been turned into a model for other universities is EPICS (Coyle, Jamieson & Oakes, 2005). This was developed for engineering schools and thus is not specific to computer science. In the EPICS model, interdisciplinary teams of eight to twenty students are assembled to work with community organizations on engineering problems. The teams are vertically integrated, consisting of freshmen through seniors. This model has been expanded to a number of universities and an implementation of it in a software engineering context was reported in (Linos, Herman & Lally, 2003). EPICS is similar to our model in that it provides a structure for community-oriented service learning projects, providing advice on team formation, identifying clients, structuring communication, and so on. However, it differs from our approach because it does not leverage the power of open source development, meaning that it is more similar to the proprietary client-oriented model in many ways. In addition, the vertically integrated teams are a challenge to integrate into many computer science programs.

However, the CO-FOSS approach has some challenges. It may be difficult for an instructor who has never done this type of project before to develop the course, to manage the project, and to attend to the post-course activities that accompany placing the new software into production. Therefore, we will present a framework for developing this type of course. In the CO-FOSS approach, one of the most important goals is that the students be able to actually complete a working prototype within the boundaries of a semester. To this end, the approach is highly structured, with weekly goals for completion. The instructor (rather than the students) develops the requirements. Since CO-FOSS projects are open source, architectures and code can be reused and leveraged into new projects. And because open source projects are visible to the world, students are motivated to achieve a higher level of quality. The openness of the course artifacts also means that instructors in other institutions can easily adapt existing projects to meet their clients' needs. Allen Tucker first developed this framework as part of the Humanitarian FOSS project (Morelli, de Lanerolle, & Tucker, 2012) and has taught numerous iterations of a software engineering course this way. Bonnie MacKellar and Mihaela Sabin adapted this framework to courses at their universities by taking into account students' characteristics that are specific to their institutions (MacKellar, Sabin, & Tucker, 2013). The result is a model for teaching software engineering that brings real world experience to a wide variety of institutions. In this chapter, we will present this framework, our experiences with adaptation, and a set of guidelines and best practices that instructors can use to integrate CO-FOSS development into their courses.

INTRODUCTION TO THE CO-FOSS MODEL

The CO-FOSS model has two major elements: 1) a *process* and 2) a *product*. The process is agile, participatory, and open. The product is real, functional, and useful to the mission-critical operations of the nonprofit client. This section elaborates each of these elements in turn.

The *process* of CO-FOSS development that we advocate is carried out by a team of developers, which necessarily includes a client representative who understands the manual activities that the software will replace. Because the process is agile, the client has frequent (weekly) opportunities to interact with each element of the software as it comes on-line and to provide feedback to the developers on which features work well and which do not. In turn, the developers can take that feedback into account as they refine and extend the software during the following week.

The team leader is typically the course instructor, who evaluates the work of the student developers each week and takes client feedback into account while preparing the following week's assignment. All of this activity is facilitated by a weekly meeting, either in-person or through an interactive video conference, where the developers demonstrate and explain their work to the client, the client provides feedback, and the instructor takes notes that inform the project's next steps.

The *product* of a CO-FOSS project begins with a complete requirements document and an initial codebase. The instructor must complete the requirements step before the beginning of the semester. This is a departure from the usual software engineering course, where students often work on requirements as well as software development. An alternative is a two-semester software development course, in which requirements elicitation and engineering concepts and practices are taught in the first semester, prior to designing and implementing the software in the second. The initial codebase may be the result of a prior similar open source development project, or it may consist only of an initial set of domain classes springing out of the requirements document. The final product is a real and viable software artifact, which fully implements the requirements that had been laid out at the beginning of the project, and which may be extended and adapted by other students for different clients in the future.

Our experience with nonprofits as clients for open source projects like this suggests that many nonprofits exhibit similar software needs that are not particularly well met by commercially available software at a price that the nonprofits can afford. For example, many nonprofits use numerous volunteers to help realize their mission. These nonprofits need software that assists them with scheduling volunteers into calendar shifts using the idea of a master schedule. Volunteers typically like to use simple repeating patterns (like "every other Thursday in the 12-3 shift" or "the first Monday of each month from 8-12") but commercially-available calendaring tools do not support this sort of master scheduling in an easily customizable way. Thus, *Homebase* was initially developed in 2008 to support the volunteer scheduling needs of the Ronald McDonald House in Portland, ME. Since then, the *Homebase* design and code has been reused and adapted for other clients that have similar volunteer scheduling needs: the Ronald McDonald House in Wilmington, DE, the Second Helping food rescue organization in South Carolina, and the Mid-Coast Hunger Prevention Program in Brunswick, ME. Because it is open source, *Homebase* can evolve over time and can be easily adapted and reused for new projects with similar needs. We have found that reusing the software architecture and underlying code for a successful project can greatly simplify an instructor's task of developing a new project for a different client.

Moving to a more detailed level of granularity, we find that the process of developing and delivering a one-semester CO-FOSS course can be divided into an eight-step framework:

1. Pre-course activity
2. Curriculum design: syllabus and milestones
3. Structuring client communication
4. Team formation and task assignment
5. Developer communication and code sharing
6. Writing user and developer documentation
7. Evaluation of team members' contributions
8. Post-course activity

The following section explains each step in this framework.

STEPS TO DEVELOPING AND DELIVERING A CO-FOSS COURSE

Pre-Course Activity

Client Sponsorship

Preparing a software development course so that students can have a real-life collaborative experience with a client and develop a useful software product during the course of a single semester requires significant effort on the part of the instructor. First, the instructor must find a willing nonprofit client and identify a specific software project. Professional or personal associations with people who are familiar with the day-to-day operation of the nonprofit (such as the executive director, operations manager, or a particularly active board member) can be very helpful in this effort. If the college has a service learning office, personnel there will often have contacts with local nonprofits. The local reputation of the nonprofit in the community can also provide leads. If the instructor has already worked with a CO-FOSS project for another client or wants to work with an existing CO-FOSS project from another school, it is best to locate nonprofits with similar needs or in a similar sector. The existence of a similar successful project for a previous client often heightens interest within a targeted nonprofit client. Once a client is located, the instructor must identify a specific employee who is responsible for the operations which the software can enhance, excited about helping with the development of such software, and able to dedicate time (a few hours before the semester begins and an hour a week throughout the semester itself) to work with the instructor and the student development team as they design and develop the software itself.

It is very important that the project be designed to ensure success. That is, the most important goal of the course should be that the students actually complete a working prototype for the software within the 13-week boundaries of a semester. This goal usually means that the instructor and the client representative make some hard choices about what will and will not be included in the final product at the end of the semester before the project begins. These decisions should inform the next step of the pre-course activity, that is, requirements elicitation and documentation.

Requirements Document

The instructor must work with the client to elicit requirements and tailor a project that both serves the needs of the nonprofit and can be completed within the context of a normal semester course by a team of students. The result of this activity will be a requirements document that spells out in substantial detail the functional, technical, and user interface requirements of the software, as well as the eventual use to which it will be put when it is completed. The requirements document must provide enough detailed domain-specific information so that students can identify initial development tasks almost from the get-go. This information should include, for example, initial domain classes for which representative attributes have been identified; a selection of programming and database languages appropriate to the project; and an overall architecture for the software so that students will clearly understand where each of their modules will fit within the larger product. Links to examples of our requirements documents can be found in the Resources section at the end of the chapter. Typically, a layered architecture that separates the user interface from persistent data representation and manipulation provides a good starting point for many Web-based applications.

While the idea of providing the requirements document to the students may seem surprising since many traditional software engineering courses involve the students in writing requirements as well as developing the software, we find that undergraduate students typically do not have the experience necessary for writing adequate requirements for a client. They either become bogged down in the task, leaving too little time for development, or they produce a requirements document that ends up having little to do with the final system. This has also been noted by Cheng & Lin (2010), who have developed a guided approach to the software engineering class project that is similar to ours, but without the client and open source focus. In their approach, the instructor also develops specifications and an overall design.

On the other hand, the presentation of a requirements document at the beginning of the course does not limit student-client interaction or student participation in requirements development. In the first case, students in this course interact weekly with the client to gain clarity on the details of requirements so that they can develop appropriate code for the weekly assignment. In the second case, the fact that the whole process is agile ensures that requirements evolve alongside the coding. So students are in fact first-class participants in the refinement of requirements, even though they do not develop the initial version of the requirements document.

Curriculum Design: Syllabus and Milestones

The syllabus can be structured in a way similar to that of other computer science courses. That is, each week of the semester, students are expected to complete a specific set of learning objectives and demonstrate their learning by completing an assignment that reflects those objectives. But there are two main differences:

1. The learning objectives and assignment are drawn directly from the requirements document, and
2. Students normally must collaborate and share a common code base to complete the assignment.

The first week or two in the semester are dedicated to team formation and setting up for collaboration. Each team must initialize a shared code repository for the project using a designated version control system, such as Subversion, GIT or Mercurial. Each student must also set up his/her own computer with a development environment that supports the programming tools chosen for the project. That environment must also be interfaced with the shared code base, so that each student can integrate his/her own work with teammates' work. This start-up step is not trivial for the student team, since students are not generally familiar with code sharing from earlier courses. Depending on the particular version control system and development environment chosen, sufficient tutorial materials should be made available to students so that they can independently set up their own computers to work effectively in the course. Links to supporting materials for Mercurial and Eclipse, for instance, can be found in the Resources section.

During each subsequent week in the course, each student team member develops, unit tests, and commits a new piece of the software to the repository. These weekly assignments are determined by the instructor, who takes into account client feedback from the previous week's assignment. The layered architecture of the software design facilitates task sequencing. That is, when the domain model, user interface, and database controller layers are distinguished, they can be helpful in organizing the weekly assignments in a course. In this case, it is often preferable to design and unit test the domain classes in weeks 3-5, the database modules in weeks 6-8, and the user interface modules in weeks 9-11 of the semester. That leaves weeks 12-13 free for students to perform more detailed integration testing, develop user documentation, and address other special circumstances that inevitably arise during the semester. The final exam date for the course provides a good opportunity for the student team to deliver the completed software prototype to the client in the setting of an oral presentation. Often, other members of the client organization are invited to this presentation, such as the executive director, other staff members, or even some board members.

Structuring Client Communication

As should be evident from the above discussion, client feedback is essential as each weekly assignment is completed and demonstrated. Students need feedback to determine the extent to which their coding efforts successfully addressed the assigned task. The instructor needs feedback to help assess the project's overall progress and to determine in detail the shape of the following week's assignment.

It is not practical for the client to meet face-to-face with the student team members at each weekly class session. Moreover, it may not even be practical for all the student team members or the instructor to be physically present at each team meeting. On the other hand, "virtual attendance" at each team meeting should be required of all students and the instructor and the client representative. Virtual weekly attendance among all team members can be ensured if everyone uses a visual teleconferencing medium, such as Skype or Google Hangout, to facilitate it.

The second ingredient to ensure client feedback at weekly team meetings is access to a shared "sandbox" server that can be used to demonstrate the current state of student progress with the project. This server can be provided by either the university or an independent Internet service provider. The important point is that at each team meeting, the client can actively view and work with the partially-developed software under the verbal guidance of the students who are developing it.

The dynamics of the weekly team meeting among the students, the instructor, and the client representative are as follows. Each student on the team presents to the client a brief description of what he/she accomplished during the prior week, and shows the client how that works. The client tries it out using the "sandbox" server and provides feedback on the spot. The instructor takes notes and uses the presentation and the feedback as factors in designing the following week's assignment. Typically, a week's assignment has two parts – one part that cleans up issues identified by the client in the team meeting and the other part that makes progress developing some additional aspect of the software. At the end of the development period, a more or less complete prototype will thus emerge. Critically, the software prototype that results from this process is not the product of the student developers working in a vacuum with the original requirements document. Rather, it is the product of weekly client feedback and immediate adjustment to that feedback. This is the essence of the term "agile development," brought to life within the setting of a real software project.

Team Formation and Task Assignment

Team formation and task assignment constitute one of the key tasks when designing this type of course. The task of team formation involves some key decisions:

1. How large will each team be?
2. What tasks will be assigned to each team?
3. Who will be assigned to each team? Who makes that decision?

Team Size

There are many factors beyond the instructor's control that impact these decisions. For example, student preparedness and the size of the client project will impact the size of the teams. The number of enrolled students, as well as the number and scope of available projects will also have an impact. However, the framework for CO-FOSS projects is flexible enough to incorporate varying team structures. For example, at Bowdoin College, there were several projects, and well-prepared students, so teams were structured by project. At St John's University, on the other hand, the instructor had already been using a way of structuring the course so that all students work on one project. This works well in that environment because the students are less well-prepared and can specialize in areas where they feel most comfortable, such as writing the help system, developing the database, or working on the user interface. At UNH Manchester, components of the project are partially developed in other courses. For example, students learn database model, design, and implementation techniques or how to integrate a Web-based user interface with database services in the Database Design and Development and Advanced Web Authoring courses.

Task Assignment

The architecture of the system plays a big role in the structure of the teams and the tasks they are assigned. This is another decision point that is eased by following the practice of building upon an existing open source project from past semesters. The *Homebase* architecture consists of a database, domain objects, a help system, and user interface. Thus, it is clear that students will be allocated to each of these tasks. If smaller teams are working at multiple smaller projects, and are following the Bowdoin model of focusing on separate layers at different points in the semester, then each student will end up working on each layer. This means that all students will need to gain expertise in every aspect of the system. This works best when all students are very well prepared for the course, and have seen topics such as database design and SQL before taking this course.

An alternative structure, used at St John's and UNH Manchester is to use the entire class as a team if the project scope is large enough. In this organization, the large team is split into subgroups along system architecture lines. If students are not familiar in advance with many of the project technologies, this organization has the advantage that students do not have to learn many new technologies at once. Also, students can be grouped according to their strengths. For example, the group working on the database layer can be composed of students who have already taken the database course. The advantages of this type of team organization are detailed in (MacKellar 2011). The disadvantage is that some layers depend on other layers. This means that interfaces between layers must be carefully specified and adhered to, and stubs for testing purposes may need to be developed.

Allocating Students to Teams

The CO-FOSS framework does not impose many constraints on the task of assigning students to teams, so the instructor has several choices to make in forming the teams. There has been quite a bit of research, both within computer science and in other disciplines, on the best way to form teams for group projects. Richards (2009) detailed many of the considerations in a survey. For example, students may collaborate more effectively if they are allowed to choose their own teams (Grundy, 1996) but when this is permitted, the more capable students are likely to end up together in one group. If it is important to distribute students of different abilities across the groups, then the instructor should assign students to groups. Criteria such as GPA in past courses or surveys of student can be used to determine placement (McConnell, 2006). Other criteria, such as gender, ethnicity, or simple time availability may come into play as well. Even issue tickets have been used as a way to form teams (Coppit & Haddox-Schatz, 2005). If the teams will consist of students working on all aspects of their project, as detailed above, then either student-based or instructor-based assignment can be used, and any of the various considerations can be used.

If the students will be working in larger teams, and assigned to specific components in the system architecture, then the instructor should do the assignment, and take their skills and interests into consideration. At St John's, the students are surveyed at the beginning of the semester, and also submit resumes. The instructor then assigns students to task groups in a way that minimizes project risks, much as a project manager in industry does. Thus, a student who indicates experience with SQL will be assigned to work on the database layer, whereas a student who has used PHP in the past will be assigned to the user interface layer, and a student who is familiar with QA, perhaps from an internship, will be assigned to the testing group. At UNH Manchester, the instructor solicits input from students in the first class about their academic and professional experiences, self-reported computing strengths, and areas of interests in their future careers. Two other sources of information are taken into consideration: student transcripts and evaluation from faculty members who know the students from their classes. Gathering this information is possible in a small department with less than 100 majors and a climate with frequent and meaningful interactions among faculty. This method of assigning development roles to students by the instructor only emphasizes learning to work in teams on large software systems rather than learning an array of specific technologies, such as PHP and SQL. The method is also inclusive of various talents and interests and avoids having students distracted by inexperience with a specific technology.

Developer Communication and Code Sharing

One of the advantages of approaching the software engineering course as an open source project is that open source development comes with a set of standard practices and tools which tend to work very well for students. The tools themselves are open source, so they are affordable in an academic setting. Open source work practices, which evolved from the need to support a highly asynchronous, distributed group of developers, also work well for students because the practices do not require as much face-to-face development effort.

What are these tools and practices? In general, most open source projects use the following tools (Fogel, 2005):

- A project repository
- Mechanisms for team communications, usually mailing lists and real time chat channels
- A version control system
- An issue tracking system

Since there are so many open source projects, a number of standard sites and tools have appeared to meet these needs. These sites constitute a ready-made toolbox for software engineering educators, greatly easing their task since they are already set up and integrated.

The Project Repository

Open source projects, like any other software project, need to reside somewhere. Not just the code, but design documents, installation instructions, build sequences, and records of defects must be maintained. Open source projects, however, are not usually tied to any one organization and must be accessible to developers and other contributors on a worldwide basis. Thus, quite a bit of effort has gone into building sites that allow sharing of project artifacts. These sites are referred to as project repositories, and often contain a number of related features, such as version control, bug tracking, and wikis. Examples of currently popular project repositories include SourceForge, Google Code, and GitHub. These repositories are free for open source projects. Obviously, the fact that they are free is very appealing for cash strapped universities. But they also are convenient for a number of other reasons. Since version control and bug tracking are typically integrated, the instructor is freed from needing to install and maintain complex software. Since they are designed to be shareable among distributed contributors, repositories allow students who may not be able to attend face-to-face meetings to still collaborate. And since these repositories host many projects, including some very famous ones, students can browse the repository site, see lots of interesting projects, and see that people much like themselves are contributing. This can add to their sense of connectedness with the discipline of software engineering and the open source development community.

Version control systems are critical to projects that require more than one developer to work on code in a controlled fashion. There are a number of version control systems in common use. Interestingly, most of them are open source systems. Examples include CVS, Subversion, Mercurial, and Git. A project repository will offer one or more of these systems, and the instructor, when setting up the project, must choose one. Using a project repository forces the students to use the version control system since they must interact with it to place their code into the repository.

Another typical feature of open source projects is the use of an online issue tracking (or bug tracking) system. These are used extensively in industry as well. In the open source world, these systems are usually integrated into the open source project repositories. For example, the issue tracker on Google Code integrates with Google Groups, which is an online discussion system, so that when an issue is entered, a message will automatically go out to the discussion group for the project, and when the issue is resolved, another message will be generated. This is easy for an instructor to configure, and helps ensure that everyone in the class is aware of the current status of all bugs in the software.

Online Messaging Systems

A characteristic of open source projects is that developers are distributed geographically, and are not likely to work on the project during set business hours. This means that tools to support asynchronous, online communication are critical to the success of the project. Very commonly, mailing lists are used. As mentioned above, support for such mailing lists may be integrated into the project repository. Mailing lists allow developers who are both geographically and temporally separated to maintain a conversation. Mailing lists are superior to regular email because they are topic-specific and they allow all of the developers to stay in the loop. Another tool that is used to support synchronous conversations is Internet Relay Chat (IRC), although many projects also use free videoconferencing tools such as Skype and Google Hangouts.

There are some significant advantages to using online messaging systems in a software engineering course. First, using these tools means that students do not have to hold as many face to face meetings, which can be a huge problem for non-traditional students with family commitments, commuter students, and even residential students who are taking a heavy course load. Students also appreciate the ability to search the conversations on the mailing list, making it less necessary to take careful meeting notes. And finally, mailing lists are very useful from the instructor's point of view, because he or she can monitor the conversations, getting a better idea of where the students are having trouble, which groups are not communicating very well, and what the various project statuses are. It is even possible to mine the conversations to create more careful analyses of student conversations (MacKellar, 2013).

Writing Developer and User Documentation

There are two types of writing that software developers usually engage in: writing documents aimed at other developers, and writing documents aimed at users. Software engineering courses usually try to have students engage in both types of writing. A very common set of documents produced during a traditional course would include a requirements document, a system design, a user manual or help system, and code level documentation. This follows the needs of the standard waterfall method, but in a one-semester software engineering course, this also can be very rushed. It is not clear how well students learn to write a requirements document in the two or so weeks that are typically devoted to the process in a traditional one-semester course.

In the CO-FOSS model, the requirements are already developed before the course starts, and if the course project is an iteration of earlier work, the architecture is also already defined. Thus, the bulk of the writing in the course is concentrated on two sets of artifacts: technical documentation aimed at other developers, and user documentation. The technical documentation consists of system design documents in the form of use cases and class designs, as well as comments in the code itself. The user documentation consists of an online help system.

Generally, open source projects have had a reputation of being poorly documented both in terms of developer-oriented documentation and user-oriented help systems and manuals (Madsen & Nürnberg, 2005; Meneely, Williams & Gehringer, 2008). This is not a desirable outcome for a software engineering course or for a client-oriented project, particularly when the clients are small nonprofits where there may not be a lot of technical expertise. This is an area where the instructor's guidance is vital to establishing effective standards for documentation. The client's feedback is also very important to help students develop a good help system, since most students have never written about or even thought about their software from a client perspective before.

Because of the open source nature of the CO-FOSS model, all artifacts are publicly available. This means that other students as well as the client can access, comment on, and even improve the documentation as needed. It also encourages the students writing the documents to concentrate on quality, since they know that other people will see their work. The code repository consists of an introductory page explaining the point of the project. The repository can also have a project wiki, which contains information of use to developers and users, such as how to install the software, and possibly instructions on writing new code modules. The use cases and class design are posted on the code repository, either in the wiki or as a downloadable document. Most open source repositories also contain an issue tracker and a discussion forum, and even those can be seen as forms of documentation.

Since students have usually not done a lot of technical writing before this course, they require a lot of guidance, structure, and examples when completing this part of the project. Students can be directed to other CO-FOSS student projects to find examples of good writing. The instructor can discuss these projects with students and point out ways in which the writing is effective or not effective. Even more powerfully, as a given client oriented open source project evolves over time, students will be able to work with existing help systems and design documents, and to use these as templates for their current project. Once an instructor has established a consistent framework for the open source project, it can be reused again and again, and students will be able to see a wide variety of examples, discovering what works and what does not work. For an example of an on-line help system developed by students, see rmh.myopensoftware.org and login as Admin1112345678 (same password) and hit the "HELP" tab.

Evaluation of Team Members' Contributions

Evaluating student contributions in team projects is one of the thorniest issues encountered by instructors of project-based courses. There is a large body of literature, and many competing ideas, on how to do this task effectively. The most common approaches are:

1. Individual grades based on peer evaluations.
2. Individual grades based on instructor evaluations.
3. A group grade that is assigned to all group members based on project success.
4. Mixed approaches incorporating the previous three approaches in various ways.

One of the problems that make it difficult to evaluate group projects is that individual contributions need to be measured to avoid the "free-loader" effect: that is, students who fail to make any meaningful contribution towards the project but who end up passing because they share a group grade. Research has found that students prefer individual grades in order to prevent the free-loader effect (Farrell, Ravalli, and Farrell, 2012). Thus, it is common to assess students with a mixture of approaches; assigning a group grade as well as an individual grade, weighting each in some fashion. Farrell finds that the group grade is often determined objectively (how many project objectives were achieved while the individual grade is often determined subjectively (how much of a contribution did a particular student make?).

At Bowdoin College, the project grade measured the degree to which the team completed all the project milestones. Individual grades were assigned in relation to this project grade in order to recognize differences among different students' contributions. At St John's University, students received individualized grades that took into account the overall group success, the students contribution to the group effort as measured by proportion of code written and activity at meetings and on the message board, and the quality of the student's code. At UNH Manchester students also received individualized grades that measured the degree to which assigned tasks were successfully completed. These tasks included: work on the software system artifacts (whether code or documentation), documenting the development process (team meeting agendas and minutes, participation on the team's reporting on status of the team's artifacts, and writing self-evaluations to reflect on progress and challenges with each student's individual contribution.

Post-Course Activity

A significant challenge to CO-FOSS projects is to ensure that the resulting software artifact is delivered and supported in a timely way to the nonprofit organization that helped develop it. This is a challenge because, just as in the case of pre-course CO-FOSS activities, this activity falls outside the normal expectations of a university faculty member.

Our experience in meeting this challenge is varied. At Bowdoin, for example, the instructor has dedicated his own time to ensuring that the software is delivered and properly supported through the first several months of its use. The quality of the software artifacts developed so far has been so high (in the case of Homebase and Homeplate, for example – see discussion in the next section) that the need for ongoing software support, once it is put into use at the nonprofit, has been minimal. At St John's and UNH Manchester, instructors have also devoted significant time to bringing the system up to production standards.

In the long run, the need for ongoing support for a successful CO-FOSS project can be met by the establishment of a partnership between the non-profit and a local software firm that has the capability and interest to provide that support on a cost-effective basis. For each of the Homebase and Homeplate projects, the non-profit has partnered with a local software firm that provides support at a reasonable cost.

We recognize that sustainability is a major challenge for any community service project involving student developers, simply because at the end of the semester both the students and the instructor typically move on to other courses and priorities. On the other hand, it is mandatory to the integrity of the CO-FOSS model that a support structure be put into place that facilitates the creation of client-software firm partnerships once a semester project is completed and the software is ready for deployment.

To this end, the authors are participating in the establishment of a new organization called the Non-Profit FOSS Institute (NPFI). The purpose of the NPFI is to facilitate the planning, creation, execution, and ongoing support for new CO-FOSS projects like the ones discussed in this chapter. Its 13-person Advisory Board represents all three types of participants in such projects – non-profits, instructors, and software firms. The Board members all have significant experience with, and enthusiasm for, developing CO-FOSS at a national level.

Key to the success of the NPFI is to facilitate and support the formation of "triads," each triad having an instructor, a non-profit client, and a local software firm. Facilitation includes supplying support materials – code bases, requirements document examples and templates, and other teaching materials that will help an instructor get started with a new CO-FOSS course. Because NPFI is in an early stage of development, we cannot say much more about it at this writing. We can say that in the spring of 2014, the Board plans to launch the http://npfi.org Web site, which will provide many more details about the mission and organization of the NPFI. Readers who are interested in becoming associated with NPFI in the future are encouraged to visit this Web site and become a member. NPFI membership will be open to all instructors, non-profits, software firms and others who embrace the CO-FOSS model.

CASE STUDIES: THE EXPERIENCE AT THREE SCHOOLS

The three authors have significant experience in adapting the CO-FOSS framework to different institutional and nonprofit environments. Overall, we have found this framework to be robust in the sense that it is adaptable to a variety of academic and nonprofit settings. However, an instructor wishing to use this framework in his/her own course must take into account local differences in student population, institutional support, and nonprofit availability and willingness to participate in such a project. The case studies in this section detail how the three authors customized the model to meet their local needs.

Case Study 1: *Homebase*

Developing the Framework with a Client-Server Application at a Small Private College

Bowdoin College is a selective residential liberal arts college with a fully-developed Computer Science Department, averaging about 10 majors per class. Students are enrolled full-time at the College, and computer science majors typically take four courses per semester. Many majors also choose to double major with mathematics, economics, or one of the sciences. At the time students enroll in the software development course, they are typically juniors and have taken a significant number of computer science courses, including data structures and algorithms, and have done a good deal of programming in different languages.

In Spring, 2008 Allen Tucker developed and taught the software course using the CO-FOSS approach for the first time. Inspired by the H-FOSS model (Morelli, DeLanerolle, & Tucker, 2012), this course aimed to develop on-line volunteer scheduling software for the Ronald McDonald House in Portland, Maine. Four brave students enrolled in and completed the course, all seniors (three CS majors and one economics/math major who had a lot of programming experience). The software that they completed was dubbed "Homebase" by the RMH staff at some point during the semester.

Prior to the beginning of the semester, the instructor met with RMH staff to gather information that would contribute to a requirements document for the *Homebase* project. The requirements document included a description of the then-current manual volunteer calendar scheduling process, a description of the methodology and tools that would be used to develop the software, a few screen-shot sketches of the desired user interface, and a time-line of milestones that had to be met to complete the project by the end of the semester.

The goal was to develop a complete working prototype for *Homebase*, including on-line user help and a week of training. The development team included the four students who would do the major programming, the instructor who would manage the project by giving weekly assignments and overseeing the sandbox server, and a client representative who would test the partially-completed software and provide weekly feedback to the students. Throughout the semester, each milestone was adjusted as a result of the weekly team meeting and feedback from the client. By the end of the semester, the client knew exactly what she was receiving as a software tool, making this a truly agile process.

The software architecture for *Homebase* may be its most important characteristic as a model for other CO-FOSS projects that use a similar teaching strategy. The client-server architecture can be viewed as having layers – the domain classes, the database modules, the user interface modules, and the user help modules. Confined by the limits of a single semester, the course can naturally flow by developing these four layers in order as a series of 3-week chunks, beginning with students developing and unit-testing the domain classes identified in the design document. This initial chunk breeds a vocabulary of terms that can be shared unambiguously between the students and the client – in the case of *Homebase*, terms like "Shift," "Week," and "Volunteer Availability" take on specific meanings that facilitate communication throughout the remainder of the development process.

The outcome of this project was a fully functional online volunteer scheduling module that integrated with the RMH Website and replaced the manual scheduling system during the summer of 2008. The instructor spent significant effort during that summer making the software "bullet-proof" so that it ran reliably and correctly on a 24/7 basis. Volunteers and House staff uniformly praised the software for its ease of use, security, and accessibility from anywhere there was a Web browser. The students completed this course knowing that they had not only experienced a real-world software development task but also made a significant service learning contribution. Since its completion in 2008, *Homebase* has been used as a starting point for similar software development activities, both at Bowdoin and at other universities and has been updated and expanded by two more teams of Bowdoin students in Spring 2012 and Fall 2013. A link to the current version of *Homebase* and its requirements document is listed in the Resources section at the end of this chapter.

Case Study 2: *Homeplate*

Adapting the Framework for a Mobile Computing Application

The CO-FOSS framework discussed above was used again at Bowdoin College in 2011 to develop a room scheduling module called *Homeroom*, and again in 2012 to develop a module called *Homeplate* for volunteers to record pickups and drop-offs for a food rescue and distribution organization called Second Helpings, in South Carolina. The *Homeplate* project was significant because it used the CO-FOSS framework for developing the domain, database, and user interface layers for an easy-to-use Web-based client-server application. But it was also significant because it provided a server-side platform with which an independent Android application could later be developed and deployed.

The server side of *Homeplate* was developed by a team of three Bowdoin students during the Spring 2012 semester, and deployed soon after the end of the semester. The *Homeplate* software architecture was essentially the same as that of *Homebase*, and we reused several key modules of *Homebase* in this new project. During the summer, one student worked with the instructor and the client to develop the Android tablet app that could be carried on the Second Helpings trucks themselves. This app facilitated volunteers' recording of food weights at the time the food was being picked up and dropped off from the trucks. To attempt this Android app as part of the semester project alongside the *Homeplate* software would have created both conceptual and practical overload. In general, it is always important for an instructor launching a CO-FOSS project to assess what can be accomplished in a semester, and trim the project appropriately in a way that ensures student success with a high probability.

The Android app runs on a tablet and sends and receives scheduling and weights data from the *Homeplate* server via FTP when the tablet is in a free wi-fi zone. This strategy avoids requiring Second Helpings to purchase expensive data plans to accompany their tablets, which would be a deal-breaker for most charitable nonprofits. The Android app was developed on a Java-like platform, which is provided freely through developer.android.com. Lots of tutorials about Android development are freely available on the Web, so that the student, the instructor, and the client representative were able to fully develop and deploy this enhancement during an 8-week period in the summer of 2012.

The Android app was deployed by Second Helpings on all 5 of their trucks in September of 2012, and has been running successfully ever since then. Volunteers (most of whom are retirees) remark that this app is extremely intuitive and easy to use. The success of *Homeplate* and its accompanying Android app has recently been reported in local a local newspaper article (Bredeson 2013) which highlights the essential role that students played in their success. Instructors interested in more details about *Homeplate* design, development, and source code downloads can access the *Homeplate* link in the Resources section.

Case Study 3: *RMHRoomReservationMaker*

Adapting the Framework at an Urban University with Students of Diverse Backgrounds and Abilities

St John's University (SJU) is a large, urban university whose students are ethnically diverse; a large percentage are Pell-eligible and first in their family to attend college. Many students work off campus or have significant family responsibilities. Software engineering is a required course in the computer science major. Students enroll in the software engineering course with various backgrounds – some have done little programming and others know quite a bit. In order to meet the needs of this diverse group of students, we usually center the software engineering course around one larger project, on which the entire class collaborates. This organization, which is described in more detail in (MacKellar, 2011), is based on the idea that students work in smaller groups organized around a project role, such as testing, development of the database, or help system development. This is a common organization in real world software projects.

In previous years, the project for the course was always a "toy project," designed strictly for the class and without a real world client. We wanted to improve the course by working with a real world client and by bringing more open source process into the course. We had a client in mind, the Ronald McDonald House of Manhattan, where there were several potential projects available. We were aware of the CO-FOSS through the Humanitarian Free and Open Source project. One of the CO-FOSS projects at Bowdoin College, *Homeroom*, was very similar to a project request from RMH Manhattan. The framework seemed to be a good approach; the challenge for us was adapting it to work with our students and with our whole-class project approach.

The first phase, the pre-course activities, was heavily facilitated by working within the CO-FOSS framework and making use of the architecture and various components of *Homeroom*, a pre-existing open source project that had been developed at Bowdoin College. There were some major differences in requirements between the two projects, so our project became one of extending the *Homeroom* code rather than simply adapting it. Since we had never worked with a real client for our software engineering course before, we used the *Homeroom* project as a model in many ways. One important way in which *Homeroom* served as a model was simply as a guide for sizing our project. Choosing a project of an appropriate size and scope is critical to success with client-facing projects, so being able to compare our potential project to an already successful one was critical.

We also used the organization and resources from the Bowdoin College projects quite extensively. We used the extended project description in (Morelli, Tucker, & DeSilva, 2011), materials on the project Website, and discussions with Dr. Tucker to become familiar with the structure and the decision points. We followed the organization as much as possible since it provided a successful structure.

However, there were significant areas where adaptations had to be made. This became most apparent during the curriculum design phase. Our student body is quite different from the students at Bowdoin College. The class was larger, with 25 students. Few of them had ever written a program longer than a hundred lines, and none had ever collaborated with another student on a software project. The *Homeroom* software was written in PHP, a programming language with which most of the students had little experience. A central concern in this course was to fill in their missing skills at the same time they were developing a system design for the project. To teach students the skills they would need for the project, the first half of the semester provided in-class labs on the basics of PHP, version control, and databases. Thus, students were not able to begin project implementation until about halfway into the course.

The project involved the room scheduling process at the Ronald McDonald House of Manhattan. As mentioned before, significant differences in requirements meant that we could not simply modify *Homeroom*. In particular, RMH-Manhattan needed two interfaces, a more complex workflow for processing room reservations that involved an approval process and a lot of automated email, and an audit system that tracked all changes made to reservations. Therefore, we reused the overall architecture and the lowest level layers of *Homeroom* – the database and domain object layers – but had to do an entirely new business logic layer. The project ended up spanning two semesters. In the first iteration of the course, the students completed a skeletal prototype. During the second iteration, a new group of students worked to close out issue tickets, complete functionality, and bring the system to a point where it might be deployed.

The differing approach, both in terms of the whole-class project approach and the necessity to spend time teaching students the needed skills, meant that the syllabus structure and task assignment phases had to be modified. As mentioned earlier, students spent the first half of the semester learning the technology and tools. This was done via a series of labs, including code reading exercises working with the *Homeroom* codebase from Bowdoin. Students began development at midsemester. The project work was organized into

2 multiweek chunks, similar to the organization of the *Homebase* project. Since all students in the course were working on the same project, they were organized into teams based on functionality. Students were chosen for teams by the instructor, who used the criterion of "least risk' in making the assignments, based largely on information on resumes submitted by the students in the first week.

Since the students only had half the semester to devote to project work, the project required two iterations of the course to be finished. In the second iteration of the course, the same overall schedule was used, with students learning the tools and the codebase in the first half of the semester. In the second half, the students worked on the project in feature oriented teams again, with tickets from the issue tracker assigned to the various groups.

The Google Code open source software repository was used to host the project. This repository has version control built into it, as well as an issue tracker which we used extensively. Because the students were mainly commuter students who do not spend much time on campus, a project specific discussion board was set up, which was used extensively by all of the students. Most communication relevant to the project happened either in class or on the discussion board. All tickets from the issue tracker were automatically forwarded to the discussion board so that students would all be aware of any bugs or problems. All code commits were also forwarded to a second discussion board which all students subscribed to.

Although finishing the project required two course iterations rather than one, this was not unexpected due to the fact that the students had weaker skills and less time to spend on the project than was the case at Bowdoin College. The open source codebase for *Homebase* and *Homeroom* served as both a model and as a set of classes and modules that could be adapted to the differing requirements of RMH-Manhattan. Even more importantly, the CO-FOSS course organization and materials were critical to our success; it is not likely we would even have attempted a project of this scale with a real client without a "recipe" which could be adapted to our needs.

Case Study 4: *DONATE*

Adapting the Framework for a Course at an Urban, Commuter College with Transfer Students Constrained Financially and by Work and Family Commitments

Affordability has been a compelling reason for adopting FOSS in the computing curricula and for equipping the computing labs with support infrastructure in the Computing Technology program at University of New Hampshire at Manchester (UNH Manchester). Using FOSS systems and services, however, is just the first step in taking advantage of how FOSS development principles and practices can impact students learning. In this case study we describe the experience with adapting the CO-FOSS framework for an upper-level elective course in the B.S. Computer Information Systems (BS CIS) major at UNH Manchester. The major requirements are structured into a two-layer core (eight courses), integrative and professional experience (four courses, including internship and capstone project courses), a self-designed concentration in an application domain (four courses), and three computing electives.

An overarching challenge for teaching CO-FOSS development in the BS CIS program is a collection of hard constraints placed on students by their work, family obligations, and other commitments. When time on campus is reduced to class meetings only, students are forced to do project work once a week, typically the night before the class meeting, with almost no time to coordinate their work with other team members. These constraints are compounded by the students' uneven academic preparedness. A large majority of students (70%) have more than 50 credits transferred from local two-year colleges, where:

1. Projects were individual endeavors;
2. Student exposure to FOSS principles and practices was limited or non-existent, and
3. Prior programming experience did not include algorithm design and using abstraction to tackle more complex problems.

In Spring, 2012, the *Homebase* approach was used in the Web Application Development course, an upper-level elective course. Eleven students enrolled in the course: four undergraduate CIS majors, two graduate IT majors, and five continuing education students. The continuing education students were engineers from a local company in Manchester, who were interested in open source technologies and gaining experience developing Web-based services for their company's in-house applications. The client was YWCA of Manchester, with whom the UNH-Manchester program had collaborated since Fall 2008 on various projects and student service learning activities. The project addressed the client's need for an information system that tracks donations from individuals and organizations. Like the other nonprofits mentioned in this chapter, YWCA cannot afford to buy software and pay developers or consultants. Open source is the only feasible approach that they have for developing a donation tracking system.

At the beginning of the semester, the project's prototype had a code base with the same layered architecture as the *Homebase* software. The project was a combined result from two other courses with student projects of smaller and dedicated scope – one database course and one Web authoring course, both at sophomore level. The prototype's back-end had a functional MySQL database, although incomplete, with a well-designed schema, sample data, and scripts to install, populate, and query the database. The front-end had a single use case implemented, i.e., viewing donors and searching donors by name. The prototype was staged on a virtual machine that runs on UNH-Manchester's server. The code base is hosted by Google Code.

In the first half of the semester students learned how to develop an open source client-based project through a variety of activities:

1. Experimenting with the *Homebase* and *Homeroom* projects in the textbook (Tucker, Morelli, & de Silva, 2011);
2. Meeting with the YWCA business director to get clarifications on the system requirements;
3. Learning from three experienced FOSS developers who joined the class via Skype;
4. Receiving feedback and having their work reviewed by the software engineers who were auditing the class; and
5. Doing assignments that provided practice with: PHP and SQL, model-view-controller architectural pattern, techniques for specifying requirements and design decisions, XAMPP run-time environment configuration, and a comprehensive development toolkit (Eclipse PHP, Xdebug, Doxygen, SimpleTest, and Balsamiq).

In addition, a project forum, wiki, and hosted project version control and issue tracking supported teamwork and collaboration. The code base became the "common denominator" for all design and implementation decisions. That is why the assignments were grounded in the code base and allowed students to learn first-hand about the developer roles needed for the project. By the time the six week-long project sprints occurred in the second half of the semester, students understood well the roles they would assume to maximize their contributions to the project.

At UNH-Manchester, the *Homebase* approach yielded several good outcomes. First, having continuing education engineers as observers allowed creation of the role of configuration manager, who was the most active and involved student during class meetings. These observers also participated in client meetings, reviewed and gave feedback on use cases, database and UI design, and code quality. Students also learned from one of the invited speakers about the Asana service for project management activities, which they ended up using for issue tracking.

CONCLUSION

Trying to incorporate real world software practices and project experience into the typical one semester undergraduate software engineering course is always challenging. The CO-FOSS approach, which occupies a middle ground between proprietary software development for individual customers and working within a large, existing FOSS project, offers a number of advantages. Projects developed in conjunction with a nonprofit client can be tailored to be student friendly and fit within the constraints of the academic semester. At the same time, following FOSS practices means that the architecture, codebase, and documentation can be reused and adapted for future projects. As shown in this chapter, a CO-FOSS project does not have to remain resident within one college but can be adapted by other schools in true open source fashion. Our future directions involve efforts to bring more schools to this approach, creating ways to support maintenance of these projects, and investigating ways to foster communication among students working on similar projects at different schools.

In conclusion, this chapter illustrates how a CO-FOSS project course can be designed and delivered. Such a course embodies a novel curriculum design that teaches communication skills, teamwork, and writing skills. It prepares students by engaging them in real-world problems, introducing them to fresh technologies (such as agile programming, layered architecture, and mobile computing), and leveraging all the advantages of open source development and tool use. We hope that our work will inspire others to rethink their courses in these ways, so that their students will become better prepared to enter the software industry with both feet on the ground.

REFERENCES

Bredeson, A. (2013, June). Second helpings volunteers get much-needed technological boost from app. built by students. *The Island Packet*.

Cheng, Y., & Lin, J. (2010). A constrained and guided approach for managing software engineering course projects. *IEEE Transactions on Education*, *53*(3), 430–436. doi:10.1109/TE.2009.2026738

Coppit, D., & Haddox-Schatz, J. (2005). Large team projects in software engineering courses. In *Proceedings of the 36th SIGCSE Technical Symposium on Computer Science Education* (pp. 137–141). ACM Computer Society.

Coyle, E. J., Jamieson, L. H., & Oakes, W. C. (2005). EPICS: Engineering projects in community service. *International Journal of Engineering Education*, *21*(1), 139–150.

Ellis, H. J., Morelli, R. A., De Lanerolle, T. R., & Hislop, G. W. (2007). Holistic software engineering education based on a humanitarian open source project. In *Proceedings of the 20th Conference on Software Engineering Education & Training* (pp. 327–335). IEEE.

Farrell, V., Ravalli, G., & Farrell, G. (2012). Capstone project: Fair, just and accountable assessment. In *Proceedings of the 17th ACM Annual Conference on Innovation and Technology in Computer Science Education* (pp. 168–173). ACM.

Fogel, K. (2005). *Producing open source software*. Sebastopol, CA: O'Reilly.

Grundy, J. (1996). A comparative analysis of design principles for project-based IT courses. In *Proceedings of the 2nd Australasian Conference on Computer Science Education* (pp. 170–177). Australian Computer Society.

Hauge, O., Ayala, C., & Conradi, R. (2010). Adoption of open source software in software-intensive organizations – A systematic literature review. *Information and Software Technology*, *52*(11), 1133–1154. doi:10.1016/j.infsof.2010.05.008

Judith, W., Bair, B., & Börstler, J. (2003). Client sponsored projects in software engineering courses. In *Proceedings of the 34th SIGCSE Technical Symposium on Computer Science Education* (pp. 401–402). ACM.

Linos, P. K., Herman, S., & Lally, J. (2003). A service-learning program for computer science and software engineering. In *Proceedings of the 8th Annual Conference on Innovation and Technology in Computer Science Education* (pp. 30–34). ACM.

MacKellar, B. (2013). Analyzing coordination among students in a software engineering project course. In *Proceedings of the 26th IEEE Conference on Software Engineering Education and Training* (pp. 279–283). IEEE Press.

Mackellar, B. K. (2011). A software engineering course with a large-scale project and diverse roles for students. *Journal of Computing Sciences in Colleges*, *26*(6), 93–100.

MacKellar, B. K., Sabin, M., & Tucker, A. (2013). Scaling a framework for client-driven open source software projects: A report from three schools. *Journal of Computing Sciences in Colleges*, *28*(6), 140–147.

Madsen, F., & Nürnberg, P. (2005). Calliope: Supporting high-level documentation of open-source projects. In *Proceedings of the 2005 Symposia on Metainformatics* (pp. 10). ACM.

Marmorstein, R. (2011). Open source contribution as an effective software engineering class project. In *Proceedings of the 16th Annual Joint Conference on Innovation and Technology in Computer Science Education* (pp. 268–272). ACM.

McConnell, J. J. (2006). Active and cooperative learning. *ACM SIGCSE Bulletin*, *38*(2), 24. doi:10.1145/1138403.1138426

Meneely, A., Williams, L., & Gehringer, E. (2008). ROSE: A repository of education-friendly open-source projects. In *Proceedings of the 13th Annual Conference on Innovation and Technology in Computer Science Education* (pp. 7–11). ACM.

Morelli, R., de Lanerolle, T., & Tucker, A. (2012). The HFOSS project: Engaging students in service learning through building software. In *Service learning in the computer and information sciences*. New York: Wiley and IEEE Press. doi:10.1002/9781118319130.ch5

Olsen, A. L. (2008). A service learning project for a software engineering course. *Journal of Computing Sciences in Colleges*, *24*(2), 130–136.

Poger, S., & Bailie, F. (2006). Student perspectives on a real world project. *Journal of Computing Sciences in Colleges*, *21*(6), 69–75.

Richards, D. (2009). Designing project-based courses with a focus on group formation and assessment. *Transactions on Computer Education*, *9*(1), 1–40. doi:10.1145/1513593.1513595

Tadayon, N. (2004). Software engineering based on the team software process with a real world project. *Journal of Computing Sciences in Colleges*, *19*(4), 133–142.

Tan, J., & Jones, M. (2008). A case study of classroom experience with client-based team projects. *Journal of Computing Sciences in Colleges*, *23*(5), 150–159.

Tucker, A., Morelli, R., & de Silva, C. (2011). *Software development: An open source approach*. Boca Raton, FL: CRC Press.

Venkatagiri, S. (2006). Engineering the software requirements of nonprofits: A service-learning approach. In *Proceedings of the 28th IEEE International Conference on Software Engineering* (pp. 643–648). IEEE.

ADDITIONAL READING

Anewalt, K. (2009). Dynamic group management in a software projects course. *Journal of Computing Sciences in Colleges*, *25*(2), 146–151.

Beck, J. (2005). Using the CVS version management system in a software engineering course. *Journal of Computing Sciences in Colleges*, *20*(6), 57–65.

Beck, J., Almstrum, V. L., Ellis, H. J. C., & Towhidnejad, M. (2009). Best practices in software engineering project class management. *ACM SIGCSE Bulletin*, *41*(1), 201–202. doi:10.1145/1539024.1508939

Bernhart, M., Grechenig, T., Hetzl, J., & Zuser, W. (2006). Dimensions of software engineering course design. In *Proceedings of the 28th International Conference on Software Engineering* (pp. 667–672). ACM.

Ellis, H., Morelli, R., & Hislop, G. (2008). Work in progress-challenges to educating students within the Community of Open Source Software for Humanity. In *Proceedings of Frontiers in Education Conference* (pp. 7–8). IEEE Press.

Ellis, H. J. C., & Mitchell, R. (2004). Self-grading in a project-based software engineering course. In *Proceedings of the 17th Conference on Software Engineering Education and Training* (pp. 138–143). IEEE Press.

Ellis, H. J. C., Purcell, M., & Hislop, G. W. (2012). An approach for evaluating FOSS projects for student participation. In *Proceedings of the 43rd ACM Technical Symposium on Computer Science Education*. (pp. 415). ACM.

Filho, W. P. (2006). A software process for time-constrained course projects. In *Proceedings of the 28th International Conference on Software Engineering* (pp. 707–710). ACM.

Gehringer, E. F. (2011). From the manager's perspective: Classroom contributions to open-source projects. In *Proceedings of Frontiers in Education Conference* (pp. 1–5). IEEE Press.

Gehrke, M., Giese, H., Nickel, U. A., Niere, J., Tichy, M., Wadsack, J. P., & Zündorf, A. (2002). Reporting about industrial strength software engineering courses for undergraduates. In *Proceedings of the 24th International Conference on Software Engineering* (pp. 395–405). ACM.

Hart, D. (2009). A survey of source code management tools for programming courses. *Journal of Computing Sciences in Colleges*, *24*(6), 113–114.

Hayes, J. H., Lethbridge, T. C., & Port, D. (2003). Evaluating individual contribution toward group software engineering projects. In *Proceedings of the 25th International Conference on Software Engineering* (pp. 622–627). IEEE.

Ikonen, M., & Kurhila, J. (2009). Discovering high-impact success factors in capstone software projects. In *Proceedings of the 10th ACM Conference on Information Technology Education* (pp. 235–244). ACM.

Lancor, L. (2008). Collaboration tools in a one-semester software engineering course: What worked? what didn't? *Journal of Computing Sciences in Colleges*, *23*(5), 160–168.

Lancor, L., & Katha, S. (2013). Analyzing PHP frameworks for use in a project-based software engineering course. In *Proceedings of the 44th ACM Technical Symposium on Computer Science Education* (pp. 519–524), ACM.

Larman, C. (2004). *Applying UML and patterns: An introduction to object-oriented analysis and design and iterative development*. Prentice Hall PTR.

Liu, C. (2005). Using issue tracking tools to facilitate student learning of communication skills in software engineering courses. In *Proceedings of the 18th Conference on Software Engineering Education & Training* (pp. 61–68). IEEE.

Liu, C. (2005). Enriching software engineering courses with service-learning projects and the open-source approach. In *Proceedings of the 27th International Conference on Software Engineering* (pp. 613–614). ACM.

Morelli, R., Tucker, A., Danner, N., De Lanerolle, T., Ellis, H., & Izmirli, O. et al. (2009). Revitalizing computing education through free and open source software for humanity. *Communications of the ACM*, *52*(8), 67–75. doi:10.1145/1536616.1536635

Nita-Rotaru, C., Dark, M., & Popescu, V. (2007). A Multi-Expertise Application-Driven Class. In *Proceedings of the 38th SIGCSE Technical Symposium on Computer Science Education* (pp. 119–123). ACM.

Pedroni, M., Bay, T., Oriol, M., & Pedroni, A. (2007). Open source projects in programming courses. *ACM SIGCSE Bulletin*, *39*(1), 454. doi:10.1145/1227504.1227465

Reichlmayr, T. (2003). The agile approach in an undergraduate software engineering course project. In *Proceedings of Frontiers in Education Conference* (pp. 13–18). IEEE Press.

Sanders, D. (2007). Using Scrum to manage student projects. *Journal of Computing Sciences in Colleges*, *23*(1), 79–79.

Sherrell, L. B., & Robertson, J. J. (2006). Pair programming and agile software development: Experiences in a college setting. *Journal of Computing Sciences in Colleges*, *22*(2), 145–153.

Stein, M. V. (2002). Using large vs. small group projects in capstone and software engineering courses. *Journal of Computing Sciences in Colleges*, *17*(4), 1–6.

Tan, J., & Jones, M. (2008). An evaluation of tools supporting enhanced student collaboration. In *Proceedings of the 38th Annual Frontiers in Education Conference* (pp. 7–12). IEEE Press.

Tucker, A. (2009). Teaching client-driven software development. *Journal of Computing Sciences in Colleges*, *24*(4), 29–39.

Way, T. (2005). A company-based framework for a software engineering course. In *Proceedings of the 36th SIGCSE Technical Symposium on Computer Science Education* (pp. 132–136). ACM.

Wilkins, D. E., & Lawhead, P. B. (2000). Evaluating individuals in team projects. In *Proceedings of the 31st SIGCSE Technical Symposium on Computer Science Education* (Vol. 32, pp. 172–175). ACM.

KEY TERMS AND DEFINITIONS

Agile Development: Methodology to develop software in short iterations, called sprints, in which existing code is first refactored, tests for a new requirement are written, and test-driven code is developed.

Code Base: All the source code files stored in a source control repository that handles various versions and tracks implementation issues.

Domain Model: Representation of conceptual classes or real-situation objects in the domain of interest.

Free and Open Source Software: Software licensed and freely distributed along with its underlying code.

Layered Architecture: Organization of the software classes into layers such that higher layers call upon services of lower layers.

Model-View-Controller Architecture: A three-layer architecture in which domain knowledge is maintained by the domain model objects, displayed by the view (user interface) objects, and manipulated by control (application logic) objects.

Nonprofit Organization: Usually a charity or service organization that uses any surplus revenues to support its mission and achieve its goals.

Real-World Projects: Software engineering projects that are sponsored by real clients and result in production-grade software systems.

Requirements: Functional capabilities (what the system will do) to which the software system must confirm.

This work was previously published in "Overcoming Challenges in Software Engineering Education" edited by Liguo Yu, pages 373-396, copyright 2014 by Engineering Science Reference (an imprint of IGI Global).

APPENDIX: CO-FOSS RESOURCES

- **Examples of the *Homebase* and *Homeplate* requirements documents:**
  - http://code.google.com/p/rmh-homebase
  - http://code.google.com/p/sh-homeplate
- **The current version of *Homebase* and its requirements document:**
  - http://code.google.com/p/rmh-homebase
- **The *Homeplate* project:**
  - http://code.google.com/p/sh-homeplate.
- **Student oriented Mercurial and Eclipse tutorials:**
  - http://myopensoftware.org/content/supporting-materials
- **Example syllabi for CO-FOSS project courses:**
  - http://myopensoftware.org/content/extended-course-syllabus.
- **The *RMH-RoomReservationMaker* project:**
  - https://code.google.com/p/rmh-roomreservation-maker/
- **Examples of currently popular project repositories:**
  - SourceForge http://sourceforge.net/
  - Google Code https://code.google.com/
  - GitHub https://github.com/.
- **Examples of currently popular version control systems:**
  - CVS http://savannah.nongnu.org/projects/cvs
  - Subversion http://subversion.apache.org/
  - Mercurial http://mercurial.selenic.com/
  - Git http://git-scm.com/

Chapter 39
Improving Learning Strategies for Mathematics through E-Learning

Cristina Bardelle
Università del Piemonte Orientale A. Avogadro, Italy

ABSTRACT

This chapter describes an experience concerning a mathematics course offered at university for first-year science students. The course, realized in a blended format, is aimed at supporting students in the critical stage of transition from high school to university. Beside standard materials and resources, students were provided with online tools for the achievement of learning strategies mainly based on semiotic and pragmatic aspects of mathematical language. The online activities are an effective tool from the cognitive and metacognitive perspective, promoting a shift from passive learning (listening to lectures and taking notes) to more active modalities of learning where the students were engaged in student-instructor or student-computer interaction activities. Overall, more than half of the students participated in the optional online activities, and a positive relationship between this participation and the results in the final written test serve to verify the potential effectiveness of this form of study.

INTRODUCTION

The learning of mathematics is a major challenge for a number of freshman students, and failure induces some of them to quit. An increasing number of students start their undergraduate courses without an adequate cultural background, and with problems both at the cognitive and metacognitive level. On the one hand, freshmen come from different kinds of schools, and on the other hand, they have to deal with a new educational environment. Our students' learning strategies are often based solely on listening to lectures, taking notes, and repeating by heart what they have written. The standard instructional setting at the university is usually not enough to support the development of learning strategies, particularly in the critical stage of transition from high school to university.

At the University of Eastern Piedmont Amedeo Avogadro in Italy, just prior to the beginning of the academic year, an intensive introductory mathematics course is held in order to give students all of the cultural background needed to face first year mathematics courses. In recent

DOI: 10.4018/978-1-4666-7363-2.ch039

years the standard course was supported by online activities introduced in a special online course implemented in the e-learning platform Moodle, in order to help students to face learning obstacles. The introduction of the online course did not require many organizational changes. Indeed, a part of the first face-to-face lesson was dedicated to introduce students the platform Moodle and to explain how to use it.

The online course had been realized to support the math bridging course and had been aimed at improving mathematics learning by bringing together the results of research in mathematics education and the opportunities provided by information and communication technology (ICT). The idea was to promote a shift from passive learning (listening to lectures and taking notes) to more active modalities of learning where the students were engaged in student-instructor or student-computer interaction activities. In particular, the online course was introduced to foster students' autonomy and to increase their ability to select appropriate resources and activities.

Special care was paid to language, according to the main principle that language is not only the carrier of pre-existing thinking but that it also fosters the development of thinking itself (Sfard, 2001). In particular, activities were designed in order to promote the coordination of different semiotic systems involved in the learning of mathematics (e.g., verbal language, symbolic expressions, visual components), thereby exploiting the opportunities provided by the e-learning platform.

E-learning platforms provided tools that allowed students to engage in active methods of learning. Several activities such as 'quiz,' 'lesson,' 'task,' 'glossary,' etc. were implemented in a structured way. Some of these activities provided self-assessment and self-reflection opportunities that were realized through an automated feedback system (quiz, lesson), as well as student-instructor interactions (quiz, lesson, task).

Both the face-to-face and the online parts of the course were designed, implemented, and managed by one instructor and one tutor only. The students that were involved attended undergraduate programs (biology, chemistry, computer science, mathematics, environmental sciences) offered by the Faculty of Mathematical, Physical and Natural Sciences.

BACKGROUND

Many studies have highlighted the need for further research into the secondary-tertiary transition, and with the emergence of technological tools, many questions about the role of technology in teaching and learning are arising. For example, Geudet (2008) wrote "The question of the effective and possible uses of technology in the secondary-tertiary transition has not been researched yet, as far as I know... Could technology be helpful to foster novice students' autonomy, for example by using appropriate online resources?" (p. 252). On the other hand, several studies, which investigated what role online environments should have in university education, drew attention to blended modalities (see for example Calvani, 2005). According to their findings, the potential of course management systems in the learning of mathematics was experienced as part of the critical secondary-tertiary transition. Moreover, the experiment was realized following the main idea that e-learning activities, in order to be effective, must be conceived taking into account pedagogical theories.

Potential benefits and constraints of online platforms have been analyzed in order to combine suggestions arising from research on mathematics education and that focusing on the tools available within e-learning environments (Albano & Ferrari, 2008; Bardelle & Ferrari, 2010). In particular, the online activities were designed taking into account the main theory that language and communication play a central role in the learning of mathematics. The terms 'language' and 'communication' are used here in a very broad sense. 'Language' refers, on the one hand, to all semiotic systems available in the communication of mathematics such as verbal language, images, graphs, symbolic expressions, etc., and, on the other hand, to all of the less explicit aspects such as functions, relationship with the context, etc. The word 'communication' is not confined to interactions between individuals only, but it also includes also thinking that is one's communication with oneself. This is what is called the 'communicational approach to the study of human cognition' presented in Sfard (2001) and Ferrarri (2004), that has its roots in Vygotskian writings. In this perspective, language heavily influences thinking and, moreover, the idea is adopted that the ability in switching from one representation to another is fundamental for the learning of mathematics (Duval, 2005).

Contemporary e-learning platforms provide the opportunity to engage students in interactive activities. This interaction can be of different kinds: student-computer type, student-instructor type and peer-to-peer interactions. Due to the constraints of the educational setting, the online course was designed using mainly student-computer interaction activities. Such activities allowed students to engage in an active method of learning: students were continuously required to solve problems, and to answer questions posed in a structured way, using the automated feedback system that fostered self-reflection and self-assessment on learning topics.

AN EXPERIENCE OF A BLENDED COURSE

The introductory mathematics course was organized using a blended modality. The face-to-face part consisted of nine daily lessons of three hours each. Each lesson had two hours of theory-focused teaching with the instructor in charge of the course, followed by an hour-long tutoring session wherein students were shown how to solve exercises. Due to time constraints and the presence of only one instructor and one tutor, with a large number of students (between one or three hundred), lessons were organized in a more traditional way, i.e., students just listened and took notes. The online part was realized and managed by the instructor and the tutor themselves.

Design of the Course

As a first step, math contents and math skills were identified according to the syllabus of the Italian Conference of Faculty of Science's directors. In particular, the course was developed around eight major topics: Arithmetic, Rational and Real Numbers, Geometry, Functions and Graphs, Equations and Inequalities, Exponentials and Logarithms, and, Trigonometry. Moreover, an analysis of the potential benefits and constraints of the platform was conducted in order to integrate learning theories, in particular for mathematics, with the online tools available. For further information on this topic see Albano and Ferrari (2008), and Bardelle and Ferrari (2010).

Carrying out the Course

All of the students were enrolled in the e-learning course from the beginning, and part of the first face-to-face lesson was dedicated to explaining how to use the platform and the resources and activities available on it to the students. Students could access the online course during the two weeks of face-to-face lessons. Students were made aware that after each face-to-face lesson, they would have access to all of the activities (see next section) related to the last topics treated in class. Students were explained that they can use these activities for training, and that they were not compulsory. Originally, all of the activities were implemented in order to be taken just once, but based on the explicit request of students, the opportunity to use the activities repeatedly was made available.

Evaluation of Students' Results

At the end of the course, students had to take a written admission test in order to verify their achievements. Both the course and the test were not compulsory, but rather were highly recommended. Moreover, students could achieve one or two credits according to the results of the test. There were no negative consequences if students failed the test, but in that case, students were then strongly recommended to attend tutoring sessions in the first semester.

Collection of Data

Finally data about the participation to the online course and the results to the admission test were analyzed and compared. Data about user behaviour concerning online activities are recorded by the platform Moodle. The system itself provides statistical information about their usage. Data relating to the online course requests (e.g. a request to view a file, to solve a quiz, etc) are stored in log files, which are subsequently processed to provide statistical information about usage as download of files, traffic patterns, resolution of activities, assigned grades per activity or per student, etc.

Special Aspects of the Experiment

The online course was implemented using Moodle, the e-learning platform that the university had provided in order to promote courses developed in a blended modality. Students were provided with resources such as notes, news about the course, examples of previously written examinations, and, online interactive activities in order to engage students in a more active form of learning. Besides the introductory section (including the description of the aims of the course, some tutorials about how to use the platform, the forum of the course, notes and the glossary), the course was characterized by eight macro-areas–one for each topic–that included several kinds of activities. The implemented activities relied mainly on student-computer interactions. This choice was due to the availability of only two instructors and to a lack of available time. The only student-instructor activities were eight assignments (tasks), one per topic. Since the instructor and the tutor could not give feedback to all submitted students' responses, examples of proper answers for each assignment were uploaded to the platform. All of the other activities (e.g., quizzes and interactive lessons), were implemented using the automated system of feedback provided by the platform. The course included three lessons on Arithmetic, Functions, and Equations, respectively. The e-learning course included eight quizzes: 1 opening quiz (18 questions) to help students to identify their major mathematical misconceptions, 3 short quizzes (5 questions), and 3 long quizzes (25 questions) on three main topics (Arithmetic, Geometry, and Rational and Real Numbers) for practicing math skills and 1 final quiz (25 questions). The quizzes fostered self-assessment, in particular the initial quiz which was designed to help students recognize their initial gaps, and the final quiz which was designed to help students observe improvements and to judge whether they were sufficiently prepared to pass the test, or to take the first year mathematics courses.

Strengths, Shortcomings, and Critical Issues

The main contribution of this experience is that students were engaged in a kind of active learning in which they were constantly required to solve problems. Moreover, a careful analysis of the learning problems inherent in the mathematical context, followed by a careful design of educational activities aimed at overcoming the difficulties of not only cognitive aspects but also metacognitive facets, was realized. Efforts have been made to try to integrate the tools offered by the platform with the studies on learning in general and mathematics education in particular. For example, e-learning environments give the opportunity to exploit different semiotic systems, among which are figures and symbolic expressions. This opportunity is very important in math education, since a good learning strategy must include the coordination of semiotic systems (Duval, 1995). In Figure 1, an example of a problem aimed at fostering the shift from the graphical system to the symbolic one is given. This problem is realized using a multiple choice question.

Data about the participation in the online course gives some information about students' satisfaction. In the last offering of the course, involving 169 students, the participation in the online activities was always below 57% and it decreased as time passed. Probably this behavior is heavily influenced by the overload of the entire course. The blended course is too intensive and requires students' considerable effort. Nevertheless, students seemed to appreciate the online course since most of the students who did use the online components used the activities several times until they managed to give the correct answers. In fact, they actually asked for this opportunity since the activities were initially set to be used only twice. In particular, the activities that were used most frequently were the quizzes.

Another critical issue was the availability of only a few staff for the implementation and management of the online environment. The implementation of online activities is very expensive in terms of time and the human resources involved. For this reason, the online activities implemented in the online course did not cover all of the topics. Also, the interaction between instructor and student through the platform was limited due to lack of time, and the fact that only one instructor and one tutor were involved in the experiment.

Figure 1. Example of a question about the coordination of semiotic systems

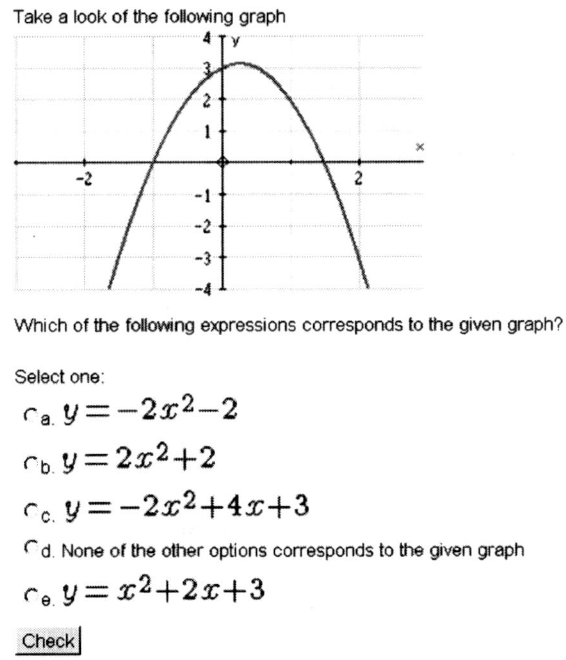

Solutions and Recommendations

Since the design and implementation of online activities requires a lot of time, when dealing with this kind of online environment one has to take into account that the number of activities will depend on the availability of human resources. In particular, for math education, in order to exploit all of the semiotic systems, such as figures and symbolic expression, which are available in such platforms, one has to acquire a certain degree of expertise. Moreover, for the management of online courses a suitable number of instructors or tutors must be available in order to also implement student-instructor activities and to foster peer-to-peer interactions. If there is not a sufficient number of instructors available for these important roles, the use on computer-student interaction activities is suggested, as was highlighted in the experience being presented here.

FUTURE RESEARCH DIRECTIONS

Even if technology platforms give the opportunity to exploit different semiotic systems, it is my opinion that additional innovations are needed in order to allow practitioners an easier access to them. On the one hand, an easier use of mathematical notation is needed, and, on the other hand, the integration of an e-learning platform with mathematical software, such as CAS (computer algebra systems), would lead to a significant improvement in the learning of mathematics through e-learning platforms. Indeed, our online course uses multiple representations, but they are almost always prearranged by the teacher (e.g., test items involving graphs) and do not fully exploit the opportunity of asking the student to build the representations him/herself.

To address the concern associated with the development of suitable learning strategies by means of tutorial systems, more research on teacher education is also fundamental. Some research has already drawn attention to blended modalities in university education (see for example Calvani, 2005). In Llinares and Olivero (2008) it is stated that, "More research still needs to be done to help us to understand better how prospective teachers learn specific professional skills when using new communication tools as a complement of face-to-face instruction" (p. 194). Their study on online discussions in teacher education is useful in designing virtual learning environments and the kinds of tasks through which the understanding of mathematics teaching and learning skills can be more fully developed.

CONCLUSION

For many years, the use of technology, and in particular that of e-learning, in the universities was limited to the transfer of traditional materials to online platforms, that were simply a source of resources (e.g., course notes, lists of exercises, communication of technical information such as timetables, exam results). Even if this was coherent with the main aim universities had of facilitating students access to learning resources (Singh, 2003), today the potentialities of e-learning platforms have not yet been fully explored for learning in mathematics, in particular at university. Starting from the idea that e-learning in tertiary transition is a great opportunity, the experience described in this chapter is an attempt to work towards this direction, providing an example of the development of e-learning activities set in a blended course for the support of secondary-tertiary transition in the field of mathematics. More broadly, this experience presents an example of the integration of education and ICT research (see for example, Conole, 2004, who highlighted the need for developing specific studies on learning through ICT tools).

A first evaluation of the results is positive (even though there are also some aspects that need further consideration), for example, a positive relation between the results of the written exam at the end of the course and the participation in the online activities emerged: better grading in the test corresponded to large participation in the online activities. Nevertheless, there is a need to develop long-term studies which would analyze the effects of the course over a longer period of time (e.g., a student's entire university career).

The diffusion of online environments, in particular Course Management Systems (CMS) for learning, together with the growing use of Internet by people in everyday life, represent customary trends in modern-day life. The e-learning modality therefore has a great, but still much unexplored, potential. In conclusion, e-learning provided plenty of opportunities to engage students in more active forms of learning based on activities of construction, conversion, and treatment of semiotic representations; it also facilitated several new forms of interpersonal communication (regardless of problems relating to available time and distance); and, it fostered self-assessment and self-reflection processes.

REFERENCES

Albano, G., & Ferrari, P. L. (2008). Integrating technology and research in mathematics education: The case of e-learning. In F. J. García Peñalvo (Ed.), Advances in e-learning: Experiences and methodologies (pp. 132-148). Hershey, PA: Information Science Reference (IGI Global).

Bardelle, C., & Ferrari, P. L. (2010). The potential of e-learning platforms to communicate mathematics. *IMSCI-EISTA 2010 Proceedings* (Vol. 1).

Calvani, A. (2005, November). E-learning at university. Which direction? *Journal of e-Learning and Knowledge Society, 3*, 139-146.

Conole, G., Dyke, M., Oliver, M., & Seale, J. (2004). Mapping pedagogy and tools for effective learning design. *Computers & Education, 43*, 17–33. doi:10.1016/j.compedu.2003.12.018

Duval, R. (1995). *Sémiosis et pensée humaine.* Berne, Switzerland: Peter Lang.

Ferrari, P. L. (2004). Mathematical language and advanced mathematics learning. In M. Johnsen Høines & A. Fuglestad (Eds.), *Proceedings of the 28th Conference of the International Group for the Psychology of Mathematics Education* (Vol. 2, pp. 383-390). Bergen, Norway.

Gueudet, G. (2008). Investigating the secondary-tertiary transition. *Educational Studies in Mathematics, 67*, 237–254. doi:10.1007/s10649-007-9100-6

Llinares, S., & Olivero, F. (2008). Virtual communities and networks of prospective mathematics teachers. In K. Krainer & T. Wood (Eds.), *Participants in mathematics teacher education* (pp. 155–179). Sense Publisher.

Sfard, A. (2001). There is more to discourse than meets the ears: Learning from mathematical communication things that we have not known before. *Educational Studies in Mathematics, 46*(1-3), 13–57. doi:10.1023/A:1014097416157

Singh, H. (2003). Building effective blended learning programs. *Educational Technology, 43*(6), 51–54.

ADDITIONAL READING

Albano, G. (2006). A case study about mathematics and e-learning: First investigations. In *International Commission for the Study and Improvement of Mathematics Education 58, Congress, Changes in Society: A Challenge for Mathematics Education* (pp. 146–151). Plzeň, Czech Republic: University of West Bohemia Press.

Borba, M. C., & Villareal, M. (2005). *Humans-with-media and reorganization of mathematical thinking: Information and communication technologies, modeling, experimentation and visualization*. New York, NY: Springer.

Halliday, M. A. K. (1985). *An introduction to functional grammar*. London, UK: Arnold.

Soulier, J. S. (1988). The design and development of computer-based instruction. Boston, MA: Allyn and Bacon. (ED 301 456).

Vygotsky, L. S. (1986). *Thought and language*. Cambridge, MA: The MIT Press.

Wild, I. (2009). *Moodle 1.9 math*. Birmingham, UK: Packt Publishing.

KEY TERMS AND DEFINITIONS

Blended Course: A course that combines traditional face-to-face classroom methods with online activities.

E-Learning: Learning realized by means of internet technologies.

Learning By Doing: The acquisition of knowledge or skills through direct experience of carrying out a task.

Online Activities: Online learning activities that require students to interact and participate actively.

Online Resources: Materials, notes, timetables, notices, etc., available online for reading.

Student-Computer Interaction: Activities in which students are engaged in interactions with the automated system of the online course.

Student-Instructor Interaction: Activities in which students are engaged in interactions with an instructor or tutor of the online course.

This work was previously published in a "Handbook of Research on Didactic Strategies and Technologies for Education" edited by Paolo M. Pumilia-Gnarini, Elena Favaron, Elena Pacetti, Jonathan Bishop, and Luigi Guerra, pages 673-680, copyright 2013 by Information Science Reference (an imprint of IGI Global).

Chapter 40
Collaboration Not Competition:
International Education Expanding Perspectives on Learning and Workforce Articulation

Cynthia J. Benton
State University of New York College at Cortland, USA

Orvil L. White
State University of New York College at Cortland, USA

Susan K. Stratton
State University of New York College at Cortland, USA

ABSTRACT

This chapter documents the development of a program for college faculty, public school teachers, graduate, and undergraduate students to pursue international travel, study, and teaching in Thailand. The program features collaboration among institutions, faculty colleagues, and school personnel focused particularly on science and mathematics education. The chapter reflects on the current methodological perspectives used to compare educational systems and the concomitant outcomes in workforce articulation. Details of expansion, development, and measurement of local, individual, and program success are provided and analyzed using current competitive international models. An alternative view of international relationships as collaborative learning opportunities is proposed.

COLLABORATION NOT COMPETITION: INTERNATIONAL EDUCATION EXPANDING PERSPECTIVES ON LEARNING AND WORKFORCE ARTICULATION

Findings from a wide range of studies have established the influence of rapidly expanding communication and technological resources which have resulted in increased contact among systems and organizations on an international scale (OECD, 2012; Friedman, 2006), and for the purposes of this chapter, in the systems practices of higher education (Egron-Polak, 2012; Sawchuk, 2012). In the context of increasing systems interdependence, global expansion in the past half-century has created frequent comparisons of student school achievement in both post-industrial and currently

DOI: 10.4018/978-1-4666-7363-2.ch040

developing nations (Schleicher, in Hargreaves, Lieberman, Fullan, & Hopkins, 2010). National and international comparisons have seldom focused on the possible contributions of a collaborative rather than a competitive model for promoting workforce preparation. This chapter argues for the positive learning contributions of a mutually beneficial international study program which takes a broad approach to participation and purpose, involving multiple international institutions, college faculty, K-12 school personnel, college students and K-12 students.

The thrust of the report focuses on programmatic issues which indicate effective educational experiences, but which also infer a revision in the ways success--both programmatic and educational--might be assessed on a global scale. Taking a collaborative stance, the authors establish the benefits for all stakeholders in the program, and provide indicators which project improvements in both intellectual and workforce outcomes.

THEORETICAL FRAMEWORK

International Comparisons and STEM Education

Typically, U.S. students and many of their teachers have a narrow, if not provincial, view of international experience and of the history and culture of other nations (Sawchuck, 2012). Generally unschooled in a functional knowledge of geography and international issues and politics, and often limited in knowledge about influences of other cultures and contributions of other ethnic and national groups, U.S. schools continue to lag behind others in assuming a global perspective in K-20 schooling, especially where it relates to the global workforce and productivity competition (Benton, Stratton, & Stearns, 2010; Hart, 2008). This limited view of an internationalized curriculum may figure largely in current comparisons of education-to-workforce articulation and must be examined from national and international perspectives.

Trends in the US: Historically, the U.S. has used competitive language in discussing education initiatives to draw attention to the need for US students to fare better on international measures of knowledge and skills (Clarke, 2009; Feistritzer, 2005). In the 1980s, practice and policies included language from *A Nation at Risk* (US DOE, 1983) a report which called for elected officials, educators, parents, and students to reform the public school system. The authors ominously cautioned that the nation was at risk because of the inadequate quality of American education, and expressed grave concern that a "once unchallenged pre-eminence in commerce, industry, science, and technological innovation is being overtaken by competitors throughout the world" (US DOE, 1983, p. 7). The document indicated the hostile, cold-war competitive stance in U.S.-global relationships, which in the ensuing 30 years has evolved into more inclusive possibilities, despite surviving vestiges of that view.

The subsequent *No Child Left Behind Act* (NCLB) was designed to drive broad gains in student achievement and to hold states and schools more accountable for student progress (U.S. DOE, 2001). This represented significant change in the education landscape, emphasizing annual testing, academic progress, report cards, teacher qualifications, reading first and funding changes. More recently, the American Recovery and Reinvestment Act of 2009 initiated *Race to the Top* (RTTT), competitive awards created to spur innovation and reforms in state and local district K-12 education (US DOE, 2009). States were awarded points for satisfying specified educational policies, such as performance-based standards for teachers and principals, complying with nationwide standards, promoting charter schools and privatization of education, and computerization. States that won the awards must demonstrate that goals were achieved within the four-year window (US DOE, 2009).

Running parallel to these competitive national goals and policy initiatives has been U. S. participation in international assessments examining academic performance ratings in a much larger context. The use of data from these comparative international assessments has made small inroads of influence into the policy development process (Egron-Polak, 2012). International study and placements, while increasing for higher education institutions, does not yet appear to be making inroads into workforce articulation and policy change. Instead, the nomenclature of international education appears to be grounded in competitive models (Bremer, 2006).

As early as the 1950s and 1960s, UNESCO's International Institute of Education developed and administered a study of 13-year-olds in 12 countries to determine comparative achievement of children in subject areas. Subsequent to that effort, the International Association for the Evaluation of Educational Achievement (IEA) emerged and conducted a series if international assessments over several decades (Schleicher, in Hargreaves et al., 2010). Currently, the most widely used assessments are the Trends in Mathematics and Science Study (TIMSS) conducted every four years, and Progress in Reading Literacy Study (PIRLS) conducted every five years. Most recently, the international assessment Organisation for Economic Co-operation and Development (OECD) as part of the Programme for International Student Assessment (PISA) has been administered. PISA surveys have been conducted every three years since 2000 in key content areas of reading, mathematics and science, but they also cover cross-curricular domains such as problem solving and a range of non-cognitive outcomes. PISA is one of the most rigorous and comprehensive assessments, not least in terms of its geographic coverage, with the 2006 survey measuring over 400,000 students who represented more than 20 million 15-year-olds in 60 countries that represent approximately 90% of the world economy (Mullis, Martin, Foy, & Arora, 2012). Such global movements have well established the structure and data for making international comparisons of participating countries.

In the U.S., the Science, Technology, Engineering and Mathematics (STEM) Education Coalition represents a broad and unified voice in advocating for policies to improve STEM education at all educational levels (STEM, 2013). An alliance of more than 500 business, professional, and education organizations, the group seeks to raise political and social awareness of the critical role that STEM education plays in enabling the U.S. to retain its position as a global economic and technological leader (STEM, 2013). The TIMSS, PISA and PIRLS assessments are focused on science and math literacy, and both organizations support STEM programs for teachers and students through the U.S. Department of Education, the National Science Foundation and other agencies that offer STEM related programs.

Increasingly, data from international assessments are being used by the U.S. and other countries to provide broader perspectives on their national performance targets. They have accomplished this by embedding components of the PISA or TIMMS assessments into their own national assessment tools (Schleicher, in Hargreaves et al., 2010). There are efforts underway to examine not only the techniques used in the classroom to achieve high test scores but the cultural and social situations that also influence those outcomes (Schleicher, in Hargreaves et al., 2010).

To what end is the teaching of STEM principles, and the TIMSS, PISA's and PIRLS assessments moving toward global workforce articulation? Drucker's (1994) description has become increasingly important in summarizing national as well as global workplace our high school and college students are entering:

The great majority of new jobs require qualifications the industrial worker does not possess and is poorly equipped to acquire. They require a good deal of formal education and the ability to acquire and to apply theoretical and analytical knowledge. They require a different approach to work and a different mindset. Above all, they require a habit of continuous learning. Displaced industrial workers thus cannot simply move into knowledge work or services the way displaced farmers and domestic workers moved into industrial work (at the turn of the last century). At the very least, they have to change their basic attitudes, values and beliefs. (Drucker, 1994, p. 62)

The U.S. continues to grapple with the question of how to improve learning outcomes and to assure relevant employment preparation as it faces global participation on a wider scale. To inform our efforts, examining emerging and established national policies in Asia is timely and important. What are their efforts in educational preparation and achievement that serve as models? The perceived lackluster performance of most U.S. school systems often point to its methods and assumptions about the instruction and learning of STEM subjects. Such perceptions generated the program presented in this study.

Trends in Asia: Although Asia may be behind the U.S. in some aspects of STEM education and ahead in others, the collaborative and multifaceted approaches that are taking place appear to be making positive changes at an increasingly rapid pace (IIE, 2007). There are a variety of STEM related initiatives in Asia, notably, UNESCO and OECD Initiatives (OECD, 2012). Notably, there is also an effort in Southeast Asia to foster collaboration among (currently) ten nations to promote educational and economic progress in the region. The Association of Southeast Asian Nations or ASEAN Group is organized for the purposes of promoting economic development, social progress, cultural development as well as protection of regional peace and stability and creates opportunities for member countries to discuss differences in a peaceful context (ASEAN, 2012). Toward the end of reaching these large international goals, efforts toward improving educational activities at the national, regional and international levels are underway.

Wu (2012) indicates higher education globalization has been continuously growing worldwide. The number of college students studying outside of their own national borders has jumped from 0.8 million in 1975 to two million in 2001, and reached close to 3.7 million in 2009 (Wu, 2012). This number is expected to approximately triple to nearly eight million by 2025 (Wildavsky, 2010). Further, with increased competition in global higher education, the migration of work effort and talent is active, and the directions of this intellectual diaspora increasingly diverse (Hulstrand, 2008). Asia is at the forefront of many of these global developments.

The United States has long been a leading destination for Asian students for several reasons, but changes in trends toward intra-Asian regionalization are emerging (Wu, 2012). The dominance of the Western model of higher education and the use of English as the *lingua franca* of academic interactions has contributed to this pattern and it may continue indefinitely. However, in Asia globalization at the regional level has been a growing trend, and may become a competing force for highly reputable Western higher education institutions (IIE, 2007). Higher education institutions in Korea, China and Japan have fast-growing courses and degree programs taught entirely in English, and as a result, U.S. students will find educational opportunities in Asia more accessible (OECD, 2012). With increased financial incentives and with employment offered by Asian countries, more American students will consider Asian destinations, while the percentage of Asian students coming to the U.S. may slightly decrease (Wu, 2012).

Within this context, The Millennium Declaration of 2000 (United Nations, 2010), a milestone in international cooperation which inspired development efforts to improve the lives of hundreds of millions of people around the world, directly influenced the goals of our current program. The

Millennium Development Goals Report (United Nations, 2010) references lack of education as the major obstacle to accessing tools that could improve lives and accomplish workforce articulation. Poverty and unequal access to schooling perpetuate high adolescent birth rates that jeopardize the health of girls and diminishing their opportunities for social and economic advancement. Of particular interest to this project was the *Asia and Pacific Regional Framework for Action* for 2015 whose target is to…"Ensure that, by 2015, children everywhere, boys and girls alike, will be able to complete a full course of primary schooling" (APR, 2000, p. 15). With such seemingly modest prospects for education, the Framework Preamble denotes a fundamental change in the consideration of access to education and training of the workforce.

Education is a fundamental human right of all people - of value in and of itself, for improving the quality of life, and as an essential part of social and human development. The provision of basic education, whether it be formally or non-formally delivered, is a core responsibility of the state with active and genuine collaboration of parents, communities, and civil society. All people, especially those most disadvantaged and excluded, must be guaranteed access to a basic education of decent quality (APR, 2000, p. 18).

Ironically, while many Asian nations excel in STEM subjects, Southeast Asian countries, including Thailand, the focus of this study, does not. Much of the impetus for the program came out of a series of innovations aimed at improving STEM in those countries. The creation of the ASEAN was aimed at economic, cultural and social advances that led to greater interface between nations. ASEAN countries have promoted greater workforce collaboration and exchange among their countries, and subsequently, a broader, more internationalized set of educational and career goals for its citizens (Wu, 2012). Among those goals was seeking collaborations with U.S. higher education programs which could design and deliver international exchanges, workshops, and supported instructional improvement in STEM subjects.

DEVELOPING INTERNATIONAL PROGRAMS TO BENEFIT TEACHING AND LEARNING

The current organizational and educational goals of international study and teacher development were an impetus for developing a college program for study abroad to change perspectives and opinions on the ideas of global competition. Documenting the process of establishing such a program indicates there are much broader issues than simply competitive rankings. Rather, involving college-level pre-professionals, faculty, and their international collaborators in international work indicates possible long-term benefits for collaborative models to advance learning STEM subjects, and therefore the preparedness of the global workforce.

Recent higher education efforts to create an internationalized curriculum have concluded that the need is to not only adapt the curriculum, but to create a broader perspective that goes beyond just curricular content and hosting international students (Stewart, 2012). Universities must create workforce articulation and sophistication at the local, regional and global levels while contending with the challenges of social migration, environmental and geographic issues, diversity of students in knowledge and global awareness (Bremer, 2006). The new emphasis aims to provide international experiences to all university staff and students so that they will perform successfully--professionally, economically and socially--in diverse contexts, exhibiting broader, multi-perspective understandings about the world. Only with such consideration of others' cultural identities can graduates navigate through and negotiate the changing nature of work, the labor force, and information technologies in a rapidly changing world.

Program Development and Process: Thailand Collaboration

A seven year program development process is reported here, focused first on methods of securing funding sources for international education, on establishing coherent curriculum, on vital faculty development opportunities among institutions, and finally, on individual student participation. The institutional history for the comprehensive college involved in this study showed a very small percentage of students seeking education degrees experienced international study compared with their peers in arts and sciences majors. A major challenge to creating an international program for teacher candidates was scarcity--not only funding, but lack of time, effort, and workload issues. For instance, faculty may have been interested in program participation but were not free to participate as they might wish; students whose programs have very few options for elective study or alternative coursework are limiting factors. In addition, the evolution of this program paralleled the identification of methods to improve STEM instruction in local K-12 and college-level instruction.

In addition to the scarcity of resources, there were also challenging decisions regarding critical teaching and learning issues in the implementation of an international education program. Creating an international exchange program between a teacher education program in a comprehensive college in New York State and Thailand, included challenges of how to address the purposes and goals of the Thai national education system and its schools, teachers and K-12 students, while at the same time expanding the opportunities for U.S. faculty and preservice teachers.

Following the Millennium Declaration of 2000 (United Nations, 2010), Thai Ministry of Education mandates required science and math to be taught in English by native English speakers at all grade levels and including gifted and talented programs. The Office of the Basic Education Commission (OBEC) is a division of Thai Ministry of Education, a result of King Rama V's experimental measures in administrative and political reform in the late nineteenth century, which was revived at the turn of the twenty-first century (Efron-Polak, 2012). From 1977 to the present day, all efforts of the Ministry of Education have been geared toward providing educational services as a means for a better quality of life and the advancement of Thai culture.

The impetus for an international exchange program with Thai educators began in 2005, when a senior professor from Indiana University, Bloomington (IU) arranged for four doctorial students from Srinakharinwirot University (SWU) in Bangkok Thailand to study at IU for one year. The U.S. study program was sponsored by the Institute for the Promotion of Teaching Science and Technology (IPST). One of the original exchange students currently works for the Bureau of Academic Affairs and Educational Standards, Office of The Basic Education Commission (OBEC), Ministry of Education in Bangkok Thailand. Her sustained collaborative efforts with two institutions of higher education have allowed students to travel between Thailand and the U.S. to conduct a total of 15 teacher professional development workshops.

Following the original cohort's return to Thailand, three additional graduate students began exchange study at IU in the summer of 2006 and returned to Thailand in 2007. An assessment and development trip was conducted in 2007, and a series of public lectures and teacher professional development workshops were delivered by the second author who moved from Indiana University to the State University of New York at Cortland. A similar follow up for the second cohort of participants and continued public lectures and teacher professional development workshops was conducted in 2008 by the second author and a special education public school teacher (see Table 1).

In 2008, the third cohort of two Thai graduate students studied science education at The State University College at Cortland for one year, a

Table 1. Example of workshop format. OBEC Thailand 2010 professional development workshop: Science inquiry to enhance the 21st century skills.

Day	Time	Activity	Notes
July 15/21	9:00 – 10:00	Introduction to Science Inquiry	Power Point # 1
	10:15 – 11:30	Observations/Inferences	Power Point # 2 & Candy Melt, Volume observation
	11:30 – 1:00	Lunch and Informal Discussions	Location:
	1:00 – 2:30	Science Inquiry in Earth Science	Oil Recovery and Clean up
	2:45 – 3:30	Models in the science classroom	Power Point # 3 & Earth models
July 16/22	9:00 – 10:00	Science Process Skills	Classification, Identification and Measurements in Inquiry
	10:15 – 11:30	Classification	My Pet Rock
	11:30 – 1:00	Lunch and Informal Discussions	Location:
	1:00 – 2:30	Critical/Creative Thinking	Dino Hunt
	2:45 – 3:30	Assessing the Inquiry Lesson: How can I know they Know?	Power Point # 4 Developing a plan for assessment
July 17/23	9:00 – 11:30	Inquiry lesson Planning	Participants will plan for the development of one Guided, one Challenge and one Open Inquiry Lesson
	11:30 – 1:00	Lunch and Informal Discussions	Location:
	1:00 – 3:30	Inquiry lesson Development	Participants will work on the development of their Guided, Challenge and Open Inquiry Lessons

collaboration with the second author. Both participants completed their U.S. study in 2009, and upon receiving their doctorates, assumed strong teaching positions to promote science study in Thailand. One of the doctoral participants now coordinates programs for the Science Education Center at Srinakharinwirot University, a government funded university in Bangkok and the leading Thai university in the fields of Education and Fine Arts, and the first institution of higher education in Thailand to concentrate solely on teacher training. The other original participant is employed at The Institute for the Promotion of Teaching Science and Technology in Bangkok, Thailand. The Institute for the Promotion of Teaching Science and Technology is a Thai government agency with responsibilities that include the development of national science and mathematics curricula, and sponsorship of science education, as well as the promotion of science in general. The agency is also Thailand's coordinator for the International Science Olympiads. IPST is a learning organization that can develop and increase Thailand's quality of science, mathematics and technology education at the international level. Such successful placement of former program participants is one of the fortuitous reasons for the success of this program. A summary timeline for development of the international program is as follows:

2005: First cohort of four Thai students arrive at Indiana University

2006: Second cohort of three Thai students arrive and first cohort returns to Thailand

2007: Second cohort returns to Thailand and a follow up trip by Cortland faculty member to Thailand is made to check on progress of first cohort. Public lectures and teacher professional development workshops begin during this trip

2008: Follow up trip with second cohort of Thai students and Cortland faculty member continues to conduct public lectures and teacher professional development workshops. A third cohort of two Thai students arrive at SUNY Cortland

2009: Third Thai cohort returns to Thailand and Cortland faculty joins them in Thailand to continue professional development and workshops. First group of 12 Thai teachers attend on site workshop, tour college classes and visit local public schools while at SUNY Cortland

2010: Workshops and lectures continue in Thailand

2011: Second group of 11 Thai teachers tour college classes and visit local public schools while at SUNY Cortland

2012: First SUNY 5 Cortland students travel to Thailand to assist in providing professional development workshops, third group of 13 Thai teachers at SUNY Cortland

In 2009, faculty development workshops taught by the second author continued in Thailand, and a group of 12 Thai teachers traveled to SUNY Cortland to participate in workshops and faculty development opportunities with local schools and college instructors (see Table 2). A similar process was followed for 2010 and 2011, where workshops and lectures were conducted in Thailand during the summer, and Thai teachers traveled during fall 2011 to participate in campus and K-12 school faculty development collaborations and workshops.

In 2012, a third group of Thai teachers visited SUNY Cortland, and the first cohort of preservice teachers from SUNY Cortland traveled to Thailand to participate in instruction in Thai professional development workshops and to travel and observe Thai culture. The culmination of program development was the institutionalization of the travel abroad as a course/program of study at the college, which will be maintained for the foreseeable future.

Challenges for Sustaining International Programs to Advance Workforce Articulation

A number of challenges, both unique to the world area studied, and for international programs in general are being assessed in and addressed in this program (Wildavsky, 2012). There are growing disparities within countries, despite the initiatives to offset them. A particularly persistent urban/rural gap in resources and participation is evident, as are achievement and access deficiencies for girls in South Asia. In addition, there is a relative lack of emphasis on alternative, non-formal approaches to basic education and lack of interest in workplace education, which resonates with the theme of this text. While much emphasis is placed on getting children into school, not enough attention is paid to the retention rate nor to the completion of schooling and the ultimate preparation of students for work and workplace articulation of skills.

There are ongoing shortfalls in national education budgets, especially for countries in economic crisis and in transition, which is especially problematic in relation to school-age population growth. Reflecting the uniquely urban bias of early childhood programs, and many programs in general, rural youth may be systematically eliminated from consideration of appropriate workplace skills.

In terms of school-based solutions to vocational options, there are particular weaknesses in identifying, refining, and expanding best practices in basic education that match the needs of the workplace. Particularly in South Asia, there has been difficulty in re-casting curricula to address the new risks and challenges facing youth in the region (Stewart, 2012). With limited in-country resources to improve the system, Thailand has taken purposeful steps to maintain collaborative international exchanges which transform the curriculum and teacher development status quo.

Table 2. Example of two-week teacher exchange workshop schedule

Day/Date	AM Activity/Class	PM Activity/Class	Instructor
Sunday 9/18	Arrive and move into Motel		
Monday 9/19	Drive to Raquette Lake	Raquette Lake	
Tuesday 9/20	Raquette Lake		
Wednesday 9/21	Raquette Lake	Return to Cortland	
Thursday 9/22	Rest	**Session # 1** Introduction to the Nature of Science (NOS) What can we expect to teach and what should our students understand? Observations vs. Inferences	
Friday 9/23	**Session # 2** Exploring/Teaching the Nature of Science with the process skills Tentativeness	**Session # 3** Creativity & Subjectivity (vs. Objectivity) Trip to Niagara Falls AMY H-H 1:30	
9/24 9/25	Trip to Niagara Falls		
Monday 9/26	Session # 4 Identifying the major aspects of the NOS as described within the National Science Education Standards Investigating the philosophy of science as it pertains to the so-called "teachable" aspects of the NOS Identifying several episodes from the history of science that illustrate and challenge the "teachable" aspects of the NOS President's Lunch reception 12:00 to 1:30	**Session # 5** Functions and Relationships of Theory and Law Social and Cultural Context Courtland Student Teacher Supervision Program	Session # 4/5 Dr. Lisa Donnelly* Ms. Renee Potter
Tuesday 9/27	Tully High School 8:30 to 11:00	Parker Elementary School 12:30 to 2:00	
Wednesday 9/28	Ban DHMO: Di-Hydrogen Monoxide, A case for misconceptions in science Ethics in the science classroom	Final tour Cortland, SUNY Cortland, 1890 House	
Thursday 9/29	Final Evaluation	Pack up	
Friday 30	**Check out of Motel** **Back to Thailand**		

Notes:
Each session will cover a different aspect of the NOS and the process skills used to teach them will use different activities from the text.
*Text used will be *Teaching the Nature of Science through Process Skills*, by Randy Bell
**Kent State University One day schedule/topic pending: Lisa will teach for an AM session and will help with the other session for the day
- Identifying the major aspects of the NOS as described within the National Science Education Standards
- Investigating the philosophy of science as it pertains to the so-called "teachable" aspects of the NOS
- Identifying several episodes from the history of science that illustrate and challenge the "teachable" aspects of the NOS

http://evolution.berkeley.edu/evosite/nature/index.shtml
http://www.indiana.edu/~ensiweb/natsc.fs.html

In addition to curricular and teacher workforce considerations, there has been a seeming inability to implement the required management reforms for the education systems of countries in transition (Wildavsky, 2010). There has been a noted lack of broad participation of communities and local leadership in management and delivery of education, which has gradually been addressed by some aspects of the program under study.

Finally, a number of challenges in the form of assessment design and delivery are reflected in some of the goals and ongoing development of the program under study. Generally, a lack of reliable data distinguish the region, sometimes due to inadequate means for assessing learning performance and achievement, and other times due to the lack of capacity to assess social problems which contribute to educational limitations (APR, 2000). In addition to limited testing, assessment and evaluation processes for learning, the methods used often are isolated from learning experiences which would contribute to effective workforce participation. Despite reforms instituted in the past quarter-century, a still large number of citizens in the region are illiterate and the challenge of delivering meaningful and relevant literacy programs to people living in different social, economic and political circumstances is monumental. Finally, the disruption or cessation of basic education provisions, facilities and support as a result of national or sub-national armed conflict or emergencies have provided additional obstacles to solid new programs for workforce education.

OUTCOMES OF INTERNATIONAL COLLABORATION

A central question in the development of the program was the need for consistent formative assessment, both of the program, and of the established relationships necessary for an international program. Stakeholders in such programs often have inequitable benefits, with one partner being the beneficiary of instruction or travel and the other being a host. A distinction of this program is the mutual benefits for all stakeholders–the Thai national education system, higher education institutions and their faculty development opportunities, public school teachers and STEM professional development, K-12 student learning, college pre-service teachers' ability to teach, and faculty development and study abroad (see Figure 1 for a representation of program development).

Figure 1. Diagram of program expansion

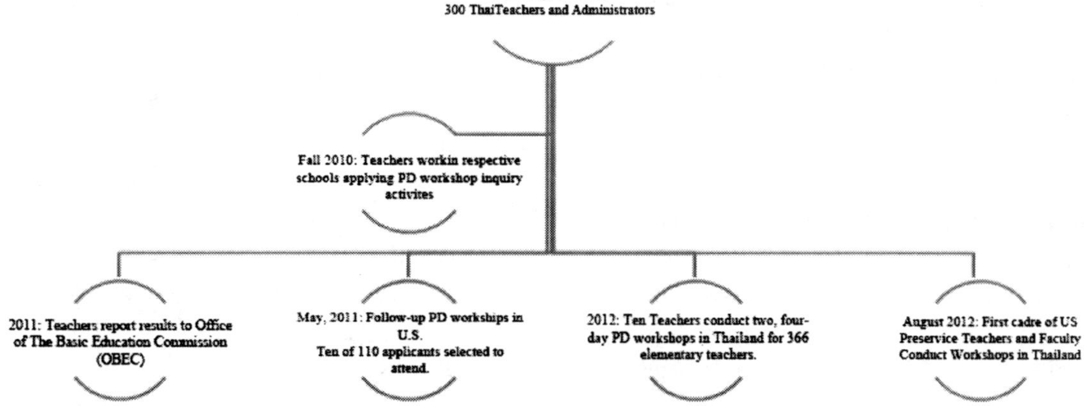

Seven Years of Program Development

Over the course of the program development under study, there has been an increase in primary school enrollment in the partner country. Thailand has seen an expansion of early childhood care and education programs aimed at providing an early start for children who previously may not have attended school at all (United Nations, 2010). A higher priority has been placed on quality of programs, curriculum and assessment for this region, and it has resulted in an increase in functional adult literacy. In addition, improvements in educational management information systems, and in national budgets for basic education have increased noticeably. More effective use of existing resources, increases in international assistance to basic education, and increases in the number of legislative measures, campaigns, projects and reforms in basic education have all become part of the education-to-work improvement characterizing this region.

Related to the present program, more innovative initiatives in basic education and more partnership between the private sector and civil society have contributed to the climate of collaboration which typified the current program (United Nations, 2010). While a seven-year development timeline may seem long to some, when considered in light of the mutual impact on teaching and learning by its respective participants, the gains are remarkable. The multiplicative nature of the program design recommends it as a trainer-of-trainers model. Science instruction in Thailand is improved by the participation of Thai teachers who do advanced study in the United States and who have participated in multiple workshops offered in Thailand (see Table 3). Individual faculty and school personnel in the U.S. have contact with visiting teachers, and faculty travel to Thailand to experience firsthand the curriculum and instruction of STEM subjects there. Pre-service and in-service teachers alike benefit from the in-

Table 3. Professional development workshops provided during seven years of project

Date	Workshop description	Attendees	Location
2007 (June)	Special 2 day Workshop **White, O**. *Science thru Art: Earth System Science in 2 & 3- Dimensions*	**60** staff members and teacher trainers from SWU & IPST	Srinakharinwirot University (SWU) Training Center, Bangkok, Thailand
	White, O., Guest lecturer *Using mathematics to connect and assess across curriculum areas.* Science education graduate course,	**35** graduate students	Mahidol University, Bangkok,
	Special Interactive 3 hour Public Lecture; **White, O.**, *Teach Me to Think: Using Thinking Skills and Problem Solving in Science and Mathematics.* Sponsored by the Institute for the Promotion of Science Teaching & Technology (IPST) and Srinakharinwirot University	Approximately **600** school science and mathematics teachers from the Bangkok metropolitan area and other provinces throughout Thailand. Additionally, the audience included teacher trainers from the Institute for the Promotion of Science Teaching & Technology (IPST) their university partners, educators and academic staff members from many of the departments under the Office of Basic Education Commission (OBEC)	Ambassador Hotel, Bangkok

continued on following page

Table 3. Continued

Date	Workshop description	Attendees	Location
2008 (March)	1 day Workshop **White, O.**, Waiyarod, S., (Ed.D) *Inquiry Based Teaching and the Learning Cycle.*	**20** teachers and administrators	Mahidol Wittayanusorn School in Salaya, Nakorn Pathom Province, Thailand.
	1 day Workshop **White, O.**, Punyain, K., (Ed. D) *Earth Science: The Global Curriculum and Links to other Curriculum areas.*	**45** graduate students	Science Education Center Srinakharinwirot University Bangkok
	Special 2 day Workshop: **White, O.,** *Earth Science: The Global Curriculum and Links to other Curriculum areas* *Assessments in the K – 12 setting: Standard assessments, Non-standard and alternative assessments.*	**25** staff and curriculum developers at the Institute for the Promotion of Teaching Science and Technology (IPST)	Bangkok
June-July 2008	1 day Workshop **White, O.**, *The Desirable Traits of the Science Educator: Providing Effective and Positive Science Education for the 21st Century*	**42** staff members, teacher trainers and graduate students	Science Education Center Srinakharinwirot University, Bangkok
	1 day Workshop **White, O.**, Teaching and Using the Science (life) Process Procedures, Ideas, and Activities (IPST)	**6** Thailand GLOBE program Director and Coordinators. Institute for the Promotion of Teaching Science and Technology	Bangkok
	2 day Workshop **White, O**. Earth System Science Institute	**35** staff and curriculum developers at the Institute for the Promotion of Teaching Science and Technology (IPST)	Bangkok
	1 day Workshop **White, O.,** Using Science Inquiry to Develop Critical and Creative Thinking in all Curriculum Areas	**47** teachers from 11 schools in North Central Thailand.	Thatakopittayakhom School Nakornsawan's province
	2 day workshop White, C., & **White, O**. Teaching All Learners, Easy Adaptations: *How to make lesson plans work for all grades and all learners,* Inquiry based teaching, and Assessments in Science (and other Curriculum Areas): Meeting the needs of all students. The 2 day workshop focused on inclusion strategies for teaching science.	**60** inclusion teachers' grades 1-6 from 40 schools in Eastern Thailand	Ratchabur Ratchaburi Province Thailand
January 2009	1 day Workshop **White, O.**, *Active Teaching in Science to Develop Science Process Skills*	**65** teachers from 11 schools in North Central Thailand	Thatakopittayakhom School Nakornsawan's province
	2 day Workshop **White, O.,** Punyain, K., (Ed. D). *Active teaching in the Science Classroom to Develop Creative and Critical Thinking,*	**45** staff members, teacher trainers and graduate students	Science Education Center Srinakharinwirot University Bangkok
	Guest lecturer, **White, O.,** Planning research; Data Collection and analysts	**15** graduate students	Science Education Center Srinakharinwirot University, Bangkok

continued on following page

Table 3. Continued

Date	Workshop description	Attendees	Location
2009 October	**White, O.,** *Education in Thailand: Moving into the 21st century; Building a stronger nation through reform*, Exemplary Models in Asian Education 2009 New York Conference on Asian Studies With the 12 secondary education teachers from Thailand	**35** attendees at a regional meeting of the Association for Asian Studies Theme: Asia Plural Cornell University,	Ithaca, New York
2009 Sept- October	2- week workshop **White, O.**, Pagano, A., Miller, W., Miller, D., Broyles, S., Anderson, S. Two week professional development workshop held on site at SUNY Cortland sponsored by the Office of The Basic Education Commission (OBEC), Bangkok Thailand.	**12** secondary education teachers from Thailand	State University Collage at Cortland
2010 July	July 15 – 17 Office of the Basic Education Commission, Ministry of Education Bangkok, Thailand Science Inquiry to Enhance the 21st Century Skills: A professional development program for science teachers.	**165** teachers and administrators from across Thailand	Windsor Hotel and conference center, Bangkok
	Guest lecturer **White, O.,** *Conducting the qualitative and qualitative research interview.* Science education graduate course, Guest of Dr. Sumalee Nakprada	**20** graduate students	Mahidol University, Salaya, Phutthamonthon District, Nakorn Pathom Province, Thailand.
	July 21 - 23 Office of the Basic Education Commission, Ministry of Education Bangkok, Thailand Science Inquiry to Enhance the 21st Century Skills: A professional development program for science teachers.	**150** teachers and administrators from across Thailand	Windsor Hotel and conference center, Bangkok
	July 29 – 30 SWU *Workshop Innovative Teaching strategies in Science for Higher Education: Using Innovative Teaching for Comprehensive Learning,*	**54** staff members, teacher trainers and graduate students from SWU	Science Education Center Srinakharinwirot University Bangkok
	Guest lecturer (July 28) **White, O.,** *Conducting the qualitative and qualitative research interview.* Science education graduate student workshop, Guest of Dr. Nason Phonphok	**12** graduate students	Science Education Center Srinakharinwirot University Bangkok
2011 Sept	2- week workshop **White, O.**, Borgerding, L., *Teaching the Nature of Science through the process skills,*	**11** administrators and teachers from Thailand	State University Collage at Cortland
August 2012	3 day Workshop August 15 – 17, ******Science inquiry to enhance 21st century skills*	**120** teachers from Northeast Thailand Given at the Hotel Rimpao	Kalasin, Thailand
	3 day Workshop August 20 – 22, ******Science inquiry to enhance 21st century skills*	**100** teachers from Northeast Thailand Given at the Avana Bangkok Hotel and Conference center	Bangkok
	1 day Workshop August 23 *Environmental Education, (Project WET & WILD), and Earth Science workshop	**43** Science Education doctorial students	Science Education Center Srinakharinwirot University Bangkok

continued on following page

Table 3. Continued

Date	Workshop description	Attendees	Location
	August 26, ***Invited Speaker*** STEM Education Workshop, National Conference: Upgrading Science Education	**85** Teachers and administrators from across Thailand	Ambassador Hotel, Bangkok
	Invited Speaker Round Table Discussion participant for 2 sessions	Round Table Discussion participants for both days: Office of Education Council, OBEC: Bureau of Educational Testing, Bureau of Academic Affairs and Standards, The National Institute of Educational Testing Service (NIETS), Office for National Education Standards and Quality Assessment(ONESQA), Institute for the Promotion of Teaching Science and Technology (IPST)	
2012 Sept	2- week workshop **White, O.**, Planned and coordinated *Teaching the Science, Technology, Engineering and Math (STEM) Education*, a two week professional development workshop held on site at SUNY Cortland	**13** administrators and teachers from Thailand	State University Collage at Cortland

* With Dr. Susan Stratton (C/ECE, SUNY Cortland)
Kayleigh Caswell, Courtney Cox SUNY Cortland Childhood Education students
Nicholas Decorse (History), Martina White (Fine Arts/Photography) students from TC_3

creased emphasis on science and math; those who travel gain transformative international experience which improves their professional lives.

Effective programs can only be based on a long-term vision with established, practical goals and realistic timelines. There must be willingness on the part of the program coordinators to work collaboratively across cultural and language barriers. This program has encompassed the call to action directions set out in multiple studies on contemporary educational and workforce articulation (Bremer, 2006; Hargreaves et al., 2010; IIE, 2007; Stewart, 2012). The level of program success experienced to date has been a result of realistic timelines, enthusiastic faculty involvement, and responsiveness to participant experiences and responses.

Participant Experiences and Responses

Informal and formal data from faculty, student and administrator feedback regarding the program were gained from all levels of participants. International programs have unique opportunities and singular challenges in terms of participant experience, assessment and data collection. The program featured in this report is no exception, and perhaps had additional challenges in terms of complexity of the program and the number of different constituencies served.

Comments from program participants indicate success in a number of areas. The first and most critical, participants gained life-altering perspectives by experiencing another culture. According to one preservice teacher…"I grew professionally and personally in so many ways… I learned how to work with a group of my peers to effectively communicate new concepts and ideas to a large group of people. I also learned how to work with teachers from a different culture which not only taught me about another country's views and beliefs on education but also how different the American culture is from places around the world... the appreciation and level of engagement that the Thai teachers showed us was inspiring." Another participant remarked, "Walking through ruins older than Columbus put new wonder and curiosity in me. It inspired me to want to see more...I want other people to experience the real world as it is and not as what we are taught it to be"

Second, the program created the opportunity to gain important skills in cultural context: "Working in an ELL setting made me more sensitive to children that come to our country and do not know English. Not only did I see the Thai teachers struggling to speak English to me but I also struggled to communicate with them. It was difficult when a group of Thai teachers were speaking Thai and I had no idea what they were saying at all. I have never been in a situation like that and it opened my eyes to just how frustrating being in that type of situation can be."

Another participant noted, "One of the first things I learned while in Thailand is to be incredibly flexible...because it is an inevitability that plans will change, locations, and ideas may change along the way, and our group had to learn to adapt and accept this... Since we were working in a setting with an interpreter, I learned how to simplify explanations, and allow time for the teachers to comprehend what was being explained to them. While there were times when the meaning of something was lost in translation, we were still able to communicate and get our ideas across to the teachers. Experiencing this was chilling; since I could see the impact I made on the teachers, and could see it in their eyes when comprehension dawned."

A final area of success noted by participants was in the area of professionalization and connection to workforce participation. Participants responded "...Going to Thailand has allowed me to grow as a teacher, student and professional..." I also realized how important it is to go out of your comfort zone in order to do something that can truly benefit your future. The presentation and the interview for the article written about our trip were the first times I have ever really spoken professionally about something that I have done. Thailand has really helped me to take that next step in my professional career. Since we returned from Thailand I have been able to connect my experience to so many different aspects of my education. From my social welfare class where we discussed social policies in different countries to my education classes where I was able to compare the American education system to the Thai system, I have felt myself grow intellectually. Going to Thailand was truly a life changing experience and it is one that will continue to affect me for the rest of my life."

Multi-directional development was generated from this project. Both Thai teachers experiencing the U.S. and U.S. preservice teachers and professors experienced each other's culture. In addition, they shared together the challenges of creating meaningful changes in curriculum and instruction in cross-cultural context. An example is one Thai teacher's experience as she progressed from participant to leader over the course of four years in the project (see Figure 1 for how the participants changed roles over time).

Ms. Usa, a primary and secondary science teacher attended the one day workshop in her home province focused on using science inquiry to develop critical and creative thinking across the curriculum in July 2008. She also attended a professional development program for inquiry teaching and learning Bangkok in 2010. Through a competitive application process, Ms. Usa was also chosen to attend the two week workshop conducted at the State University College at Cortland in 2011. She was one of eleven participants who then planned and conducted a series of teacher professional development workshops in Thailand which reached 366 elementary teachers during 2012. Such a trainer-of-trainer model demonstrates the fruition of a program that has disseminated STEM knowledge and skills to a wide audience, both in the U.S. and Thailand, and which has begun to change the Thai national curriculum as well as providing opportunities for mutual international exchange and experiences.

FUTURE RESEARCH DIRECTIONS

There is limited empirical support for the presumed connections between internationalizing the curriculum, the success of international study, methods of STEM instruction and the assumptions of global competitiveness which are the ambitious goals of this program. While numerous studies have focused on particular aspects of these connections, a more integrated approach may be indicated. After seven years of experiences, some important questions persist. What are the long-term benefits of international program participation for each of its stakeholders? What should be the measures of success? Is it possible to track consistently the spiral of curricular and interpersonal effects of the program in either country?

Tracking positive outcomes of international study is complex and should be longitudinal. Unfortunately, most programs do not have systems to support long-term follow up of student participants and the effect of their travel and study abroad. In this program, the impact on the national Thai curriculum in terms of inquiry learning will be studied as part of the national efforts to advance STEM learning. The impact of improving STEM instruction and inquiry methods are projected to create improvements in creativity, productivity and workforce articulation in Thailand. Monitoring the collaborative exchange of teachers and preservice teachers and the effects on their professional lives should also be followed. How is the teaching workforce affected by participation in such cultural exchange? Are there as many gains in STEM instruction by U.S. teachers who experience the benefits of the professional workshops?

The reciprocal exchange of skills and experiences would indicate both strong professional and personal connections and collaborations over time. The travel distance between nations is significant, but the increasing ability to collaborate via electronic means may contribute to the success of this program over time. The long-term impact may be difficult to track, but immediate interpersonal and cultural changes can be seen in the participants in this program. In the words of one student participant, "Thailand was awe inspiring and productive for my future. I have new ideas and things I want to try...Thailand left me longing for more, as to what that more is, it is another journey to find out."

CONCLUSION

Global competition, while an inevitable prospect in business models, must be reconsidered in the educational arena. If the intent is to produce global citizens prepared to collaborate and interact on a world stage, a dramatically different approach is necessary. While the U.S. needs to reconsider its approach to achievement in a number of areas, most notably STEM subjects, it must also endeavor to promote the positive application of collaborative models which lend themselves to cross-national, cross-cultural exchange. Only when educators recognize the importance of learning in global context and applying knowledge in collaborative ways will the U.S. be truly preparing its workforce for the future.

Establishing and promoting programs which can expand on our current models of international exchange, which define broader and more experiential teacher preparation, and which propose improving STEM instruction and learning is a model the authors advocate to promote success.

ACKNOWLEDGMENT

The authors wish to acknowledge our partners and collaborators, both in Thailand and the United States, who have made this work possible: Mary Schlarb, Director, International Programs Office, State University of New York College at Cortland, Chinnapat Bhumirat, Secretary General, Office of the Basic Education Commission, Thailand Ministry of Education and Kanchulee Punyain, Science Institute, Bureau of Academic Affairs and Educational Standards, Office of The Basic Education Commission; Nason Phonphok, Director, Science Education Center, Srinakharinwirot University Bangkok and Dr. Erik Bitterbaum, President of SUNY Cortland.

REFERENCES

Altbach, P. (2010). *Forward. Crossing borders in East Asian higher education*. Boston, MA: Center for International Higher Education.

ASEAN Foundation. (2012). *ASEAN Foundation comprehensive report*. Retrieved from http://www.aseanfoundation.org/

Asia and Pacific Regional Framework for Action. (2000). *Education for All, Guiding Principles, Specific Goals and Targets for 2015*. Adopted by the Asia Pacific Conference on EFA 2000 Assessment. Bangkok, Thailand.

Benton, C., Stratton, S., & Stearns, K. (2010). Action research empowering teacher development: Connecting teacher reflection, teaching effectiveness, and program change. *Excelsior: Leadership in Teaching & Learning*, *5*(1), 35–55.

Bremer, D. (2006). Wanted: Global workers. *International Educator, May-June*, 40-45.

Clarke, A. (2009). Teachers in comparative perspective. In J. L. Saha & A. G. Dworkin (Eds), International Handbook of Research on Teachers and Teaching, Springer International Handbooks of Education, 21(7), 527-538. New York: Springer Science & Business Media.

Coalition, S. T. E. M. (2013). *The Science, Technology, Engineering, and Mathematics (STEM) Education Coalition*. Washington, DC: National Science Foundation. Retrieved from http://www.stemedcoalition.org/membership/

Drucker, P. (1994). The age of social transformation. *Atlantic Monthly*, *62*, 53–80.

Egron-Polak, E. (2012). *Higher education internationalization: Seeking a new balance of values. Trends and Insights for International Education Leaders. Report to the International Association of Universities*. Washington, DC: National Center for Education Information.

Feistritzer, C. E. (2005). *National survey of K-12 public school teachers*. Washington, DC: National Center for Education Information.

Friedman, T. L. (2006). *The world is flat: A brief history of the 21st century*. New York: Farrar, Straus and Giroux.

Hargreaves, A., Lieberman, A., Fullan, M., & Hopkins, D. (Eds.). (2010). *Second international handbook of educational change*. Dordrecht, Netherlands: Springer.

Hart, P. D. (2008). *How should colleges assess and improve student learning? Employers' views on the accountability challenge*. Washington, DC: Association of American Colleges and Universities.

Hulstrand, J. (2008). Supply and demand: Building a fluent workforce. *International Education, September-October*, 24-31.

Institute of International Education (IIE). (2007). *Meeting America's global education challenge: Current trends in US study abroad & the impact of strategic diversity initiatives*. New York: Institute of International Education.

Mullis, I. V. S., Martin, M. O., & Foy, P. (2012). Chestnut Hill, MA: TIMSS & PIRLS International Study Center, Boston College.

OECD. (2012). *Education at a Glance 2012: Highlights, OECD Publishing*. Retrieved from.

Sawchuk, S. (2012). Teacher quality, status entwined among top-performing nations. *Education Week*. Retrieved from http://www.edweek.org/ew/articles/2012/01/12/16teachers.h31.html

Stewart, V. (2012). *A world-class education: Learning from international models of excellence and innovation*. Alexandria, VA: Association for Supervision and Curriculum Development.

Trends in International Mathematics and Science Study. (2011). *TIMSS 2011 Report*. Washington, DC: National Center for Education Statistics & Institute of Education Sciences.

United Nations. (2010). *Millennium Development Goals Report 2010*. New York: United Nations.

U.S. Department of Education. (1983). A nation at risk: The imperative for educational reform. Washington, DC: US D.O.E.

U.S. Department of Education. (2001). No child left behind. Washington, DC: US D.O.E.

U.S. Department of Education. (2008). *The National Science and Mathematics Access to Retain Talent Grant (National SMART Grant)*. Retrieved from http://studentaid.ed.gov/PORTALSWebApp/students/english/SmartGrants.jsp

Wildavsky, B. (2010). University globalization is here to stay. *The Chronicle of Higher Education*. Retrieved from http://chronicle.com/article/University-Globalization-Is/124148/

Wu, Y. (2012). Regional globalism and international higher education in Asia. *Trends and Insights for International Education Leaders: NAFSA*. Retrieved from http://www.nafsa.org/Explore_International_Education/Trends/TI/Regional_Globalism_and_International_Higher_Education_in_Asia/

This work was previously published in "International Education and the Next-Generation Workforce" edited by Victor C. X. Wang, pages 64-82, copyright 2014 by Information Science Reference (an imprint of IGI Global).

Chapter 41
Active Learning, Mentoring, and Mobile Technology:
Meeting Needs across Levels in One Place

Dianna L. Newman
University at Albany (SUNY), USA

Meghan Morris Deyoe
University at Albany (SUNY), USA

Jessica M. Lamendola
University at Albany (SUNY), USA

Kenneth A. Connor
Rensselaer Polytechnic Institute, USA

ABSTRACT

Educators are creating authentic settings that utilize active learning, mobile technology, and mentoring in efforts to promote students' success in developing 21st Century skills, motivation, and interest in STEM domains and STEM careers. Each of these approaches has been found to promote and transfer knowledge, as well as to develop problem-solving and communication skills in STEM. Little information, however, is available about the interactive effect of mobile technology and active learning in promoting learning in settings that use a hierarchical model of mentoring to promote the transfer of skills and knowledge. This chapter presents findings of a program that used mobile technology in active learning environments for five interrelated levels of an active, authentic environment, facilitated by mobile technology and hierarchical mentoring. Positive outcomes were documented at each level of participation; use of the mobile technology integrated within active learning settings supported by hierarchical mentoring increased learning in STEM content, skills, and affect.

INTRODUCTION

A major goal for the implementation of technology within educational settings is to provide an environment that engages students in opportunities to actively participate in the learning process (Wangler & Ziliak, 2013). When used appropriately, technology not only meets this goal, but also extends the engagement process, allowing students to reach beyond the onsite context by interacting with information as it is used in everyday life. Mobile technology is particularly useful in this setting in that it promotes applied learning by helping students gather information anytime and anywhere (Fisher & Baird, 2006). Research has found that when learning is supported by on-demand access to information it not only fosters short-term goals, but also promotes development

DOI: 10.4018/978-1-4666-7363-2.ch041

of lifelong learners (Naismith, Sharples, Vavoula, & Lonsdale, 2004).

The implementation of mobile technologies as a support for learning is becoming more popular in educational settings (Kim, Mims, & Holmes, 2006). In 2012, over 50% of high school students and 40% of middle school students owned or had access to smartphones or tablets (Nonono, 2012) and 82% of college freshman had access to online resources via handheld devices before they even arrived on campus (UCAS Media, 2013). As a result, instructors are looking for ways to not only use mobile devices as instructional resources, but also are attempting to find ways to blend them into other educational goals. One specific goal which can be supported via the use of mobile technology is mentoring, especially hierarchical mentoring. This model of mentoring, which involves cyclical, structured authentic interactions between a series of mentors and mentees across both formal and informal settings is now being used by many colleges and universities to help attract and retain STEM undergraduates (George & Neale, 2006; Olson & Riordan, 2012; Wilson, et al., 2012).

This chapter discusses the use and outcomes of experiential STEM learning using mobile technology embedded within a hierarchical mentoring setting. By using the mobile technology within an authentic educational setting, program planners were able to support active learning across five levels of mentor/mentee relations–higher education STEM faculty, STEM graduate students, STEM undergraduate students, secondary STEM educators, and STEM secondary education students. The chapter will describe the technology, the settings of implementation, the cognitive and affective outcomes of the participants, and the influence of hierarchical support for learning across levels.

OVERVIEW OF THE LITERATURE

Olson and Riordan (2012) note that during the 20th century, the interactive status of science, technology, and higher education had a major effect on the U.S. economy. During that latter part of the 20th century, more students were graduating from colleges and universities with STEM degrees than ever before, creating and supporting an economy based on new technologies and industries. Now, in the second decade of the 21st Century, societal demands require that we continue to reinforce and develop skills that are directly related to the use of technology such as information, media, and technology literacy; concurrently we must also develop skills related to career content and professional activities such as innovation, self-direction, adaptability, productivity, critical thinking, problem-solving, collaboration and communication (U.S. Department of Education, 2010).

To encourage the development of these skills amongst our youth and emerging adults, it is imperative that technology be incorporated within all aspects of education, and that student learning and pedagogy reflect technology integration, as well as technology content (U.S. Department of Education, 2010). The domain of STEM education is highly touted as a leader in this change. As of 2010, 46 of the 50 states in the United States now require that teachers implement technology within their instruction (National Science Board, 2010) in hopes that aligning instruction and technology will promote a growing workforce in STEM-related fields and a population prepared to take on STEM-related and STEM-supported careers.

The need for this change is further supported by workforce projects. According to the National Science Board, "the number of workers in STEM occupations from 1950-2007 (182,000 to 5.5 million) represents an average annual growth rate of 6.2%, nearly 4 times the 1.6% growth rate for the total workforce during this period" (2010, p. 6). The societal demands for improved and newer technologies, however, are increasing at a disproportionate rate to the number of students now obtaining STEM degrees and it is estimated that it will take the addition of approximately one million more STEM professionals to the current workforce over the next decade to maintain the U.S. global economic position (Olson & Riordan, 2012).

To fulfill the need for STEM professionals that our future workforce will demand, we need to ensure that students are obtaining and developing 21st Century knowledge and skills, but equally important, that these experiences foster the motivation for long-term learning and involvement within STEM careers. Opportunities to learn these skills need to occur from early elementary through college levels; as a result, teachers are now actively integrating technology into their instruction as both a tool and as content (Hucks & Ragan, 2013; Newman, Coyle, & McKenna, 2013). Current research indicates that successful integration and use of technology in learning experiences must allow for first, the introduction of the technology, second, meaningful exploration opportunities using the technology (Hucks & Ragan, 2013), and third, if learning is to be sustained, use of strategies that promote engagement, motivation, and long-term retention of technology-supported concepts (Dori, Hult, Breslow, and Belcher, 2007).

As noted above, students entering today's classrooms are more adept with and used to having access to digital resources and real world experiences than were previous generations. Consequently, student engagement in learning in the 21st Century requires use of instructional strategies that promote visual stimulation, experiential and authentic learning, and community-based exploration (Brown, Hansen-Brown, & Conte, 2011); these strategies need to include hands-on activities, problem-solving, collaboration, and cooperative environments as part of the process (Prince, 2004). All of these factors are major components of the underlying pedagogical elements of authentic curriculum development that supports constructivist, constructionist, and studio-based learning (Brophy, Klein, Portsmore, & Rogers, 2008; Taylor, 2009; Wiggins & McTighe, 2005).

The Role of Active Learning

Key to the implementation of approaches that support authentic education is a philosophy of "active learning" in classrooms for every level of students (Michael, 2006). Active learning reflects constructivist learning theory by engaging students in the development of their knowledge by involving them in gathering information, problem-solving, reflecting, and self-assessing understanding (Barak, Harward, Kocur, & Lerman, 2007; Michael, 2006; Prince, 2004; Sirinterlikci, Zane, & Sirinterlikci, 2009; Varak, Harward, Kocur, & Lerman, 2007). Through the process of active learning, students generally go above and beyond memorizing and recalling facts; instead they are embedded in settings that help them to construct new meaning from their prior knowledge and their current experiences, and to apply that new meaning to an authentic situation; they then are encouraged to transfer and articulate their new understanding to novel or advanced scenarios (Michael, 2006; Sirinterlikci, et al., 2009).

Active learning is versatile in that it can be implemented in a variety of pedagogical environments such as individual, cooperative, collaborative, peer-supported, and problem-based settings (Michael, 2006). Active learning supports different types of knowledge construction by encouraging students to utilize all of their senses in defining, generating, and solving problems; active learning also can be used to support development of a team environment and communication that fosters thinking on a higher level (Sirinterlickci, et al., 2009).

Research (e.g., Samson, 2010; Sirinterlikci et al., 2009) has shown that active learning is appropriate for students across all ranges of abilities and ages. Undergraduate, middle school, and elementary students have participated successfully in individual and team-based active learning environments, and have gained hands-on experience, enhanced communication skills, better collaborative work skills, and higher-order thinking skills. In addition, they have evidenced changes that support acquisition and transfer of knowledge such as a more positive affect toward the content/subject, more comfort with and greater self-efficacy in using resources, and an increased motivation to learn. Within STEM education settings, Newman and Gullie (2009) found when teachers used active learning, their students had higher scores in STEM content, and Rodd and Newman (2009) found that students participating in hands-on constructivist learning via mobile technology had more interest in continued coursework and knowledge extension; Newman, Deyoe, Connor, & Lamendola (2014) found that students who collaborated in active, authentic problem-solving expressed more interest in later learning, and expressed more interest in later learning, and students participating in hands-on constructivist based technology using mobile technology had more interest in continued coursework and knowledge extension (Rodd & Newman, 2009).

The Role of Mentorship

Education and professional preparation is not just the transfer of knowledge; education also supports affective, contextual, and communication skills needed to support a global society. As a result, an important feature in current STEM education is the value of mentorship (Kardash, 2000; Eagan, Sharkness, Hurtado, Mosqueda, Chang, 2011). Mentoring not only plays a critical role in promoting student learning in STEM programs, it also prepares participants for the workforce by fostering skills involving collaboration, communication, and self-awareness of their potential to contribute to the field and the world (Rodd & Newman, 2009; Wilson et al., 2011).

A particular type of mentoring that has proven successful is that which uses a hierarchical approach. Hierarchical, or turnkey, mentoring facilitates the establishment of relationships reflecting the various levels of the educational and work-related hierarchy. In this setting, higher education faculty work with graduate students who in turn establish a relationship with undergraduate students who then form a mentorship relationship with high school students; similarly, industry managers mentor novice employees who mentor prospective students at the graduate and undergraduate level. This hierarchical mentorship results in a linked community that connects K-12 students with the higher education campus, and similarly, campus students become connected to research and industry (Otterstetter, et al., 2011). One positive method of creating, cultivating, and supporting these connections across the various levels is the promotion and use of active engagement involving all members of the hierarchy across common authentic experiences (Feldman, Divoll, & Rogan-Klyve, 2013; Kiersma, et al., 2012). In addition, by combining societal elements within this mentoring and active learning, students gain real world experience in collaboration and communication that improve knowledge generation, strengthen cognitive processes, and enhance problem-solving. This use also increases awareness of their knowledge of multiple uses of STEM tools and how to use that knowledge to solve global issues and how to pass their knowledge both up and down the network to others. Combining these experiences facilitates the generation of a more complete and complex learning process, which helps students achieve more comprehensive learning results that will foster a global society (Niemi, 2002).

Mobile Technologies

The integration of technology in the classroom is not new to the field of education; however, the implementation of mobile technologies has generated new levels of excitement and opportunities for combining active learning and mentorship (Fisher & Baird, 2006; Kim et al., 2006). Mobile technologies give students and instructors access to information at their convenience (Kim et al., 2006; Looi, Seow, Zhang, So, Chen, Wong, 2009) and allows students and instructors to share information synchronously through text, audio, video, and kinetic means (Fisher & Baird, 2006; Kim et al., 2006). Currently, some of the most popular mobile devices supporting this effort include mobile phones, laptops, and tablets, (Kim et al., 2006).

Active Learning and Mobile Technology

Barak, Harward, Kocur, and Lerman (2007) expand on this integration noting that a key purpose for moving technology into instruction is to facilitate students' knowledge acquisition, to support theoretical understanding of the content, and to enable them to apply what they have learned via real world practice. By making technology portable, access can become ubiquitous, and students can engage in learning across time, place, and context. These uses of mobile technologies, when supporting innovative curriculum and instructional practices, provide an authentic learning experience. Studies have shown that this use promotes greater acquisition of knowledge and is successful in supporting transfer of new knowledge both within and outside the classroom (Brown, Hansen-Brown, & Conte, 2011; Looi, et al., 2009; Newman, Clure, Deyoe, & Connor, 2013; Newman et al., 2014; Newman & Gullie, 2009). This is especially true within the STEM domain where several studies have shown that hands-on activities supported by mobile technologies replicating authentic situations successfully engaged students in active learning and promoted positive long-term learning outcomes (e.g., Newman et al., 2013; Newman et al., 2014; Rodd & Newman, 2009; Won Hur & Anderson, 2013).

Little information, however, is available about the interactive effect of mobile technology and active learning in settings that use a hierarchical model of mentoring to support the broader transfer of skills and knowledge. This chapter initiates that discussion, by presenting the processes and outcomes of a pilot setting that used mobile technology to support active learning across five inter-related levels of mentored use. Faculty at an institution of higher education mentored graduate students and undergraduate students, who in turn mentored high school students. Secondary education teachers who served as both mentors and mentees participated as co-instructors and co-learners to gain and share knowledge and skills in STEM related lessons appropriate to their levels of instruction, and help faculty and graduate students understand STEM education issues.

BACKGROUND OF THE STUDY

The implementation of mobile technologies to support active learning of graduate students, undergraduate students, secondary educators, and high school students through incorporation and active learning pedagogy were utilized over the course of three years as part of a larger project and centered around three summer workshops, referred to as Pilot, Replication I, and Replication II in this chapter. The project was sponsored at a private institution of higher education (IHE) as part of educational outreach for a major National Science Foundation sponsored research center[1]. A key goal of this educational outreach is to develop innovative curriculum and instructional practices for university and pre-college educational settings. A supporting action is to offer hands-on STEM experiences during a one-week summer workshop. The goal of this summer workshop is to increase undergraduate students' interest in graduate school, promote high school students' interest in STEM careers, and to help increase retention and diversity within the field of STEM. Because Wilson et al. (2012) posits that to increase interest in and diversity within STEM classrooms both mentoring and active learning experiences at the pre-college and undergraduate level are beneficial, a hierarchical form of mentoring was introduced into both the planning and delivery of the summer workshop experience.

The Participants and Their Roles

As part of the larger program, higher education faculty mentored graduate students in collaborative research. As part of this experience, faculty assisted graduate mentees in developing content focused, authentic research and teaching experiences and methods of supporting collaborative team efforts. The undergraduate students, with the assistance and oversight of these graduate and faculty participants, were trained to provide direct mentorship to high school students; undergraduates activities were to include demonstrations, assistance in hands-on activities and informal career discussions both with faculty and graduate students and then with the secondary participants. Secondary STEM educators were mentored in content by faculty and graduate students, but served as mentors to these participants and also to the undergraduates in matters pertaining to pedagogy and assessment (see Figure 1). The active engagement of all these participants was facilitated in authentic research through the use of mobile technologies that simulated expensive equipment needed to conduct experiential studies, access information, and communicate findings. Technologies used included laptops, tablets, and simulation platforms. A more in-depth discussion of these interactive, collaborative mentorship roles is presented in the following subsections.

Higher Education Faculty

Higher education faculty from electrical, computer, and systems engineering and supporting areas specifically served as mentors to graduate

Figure 1.

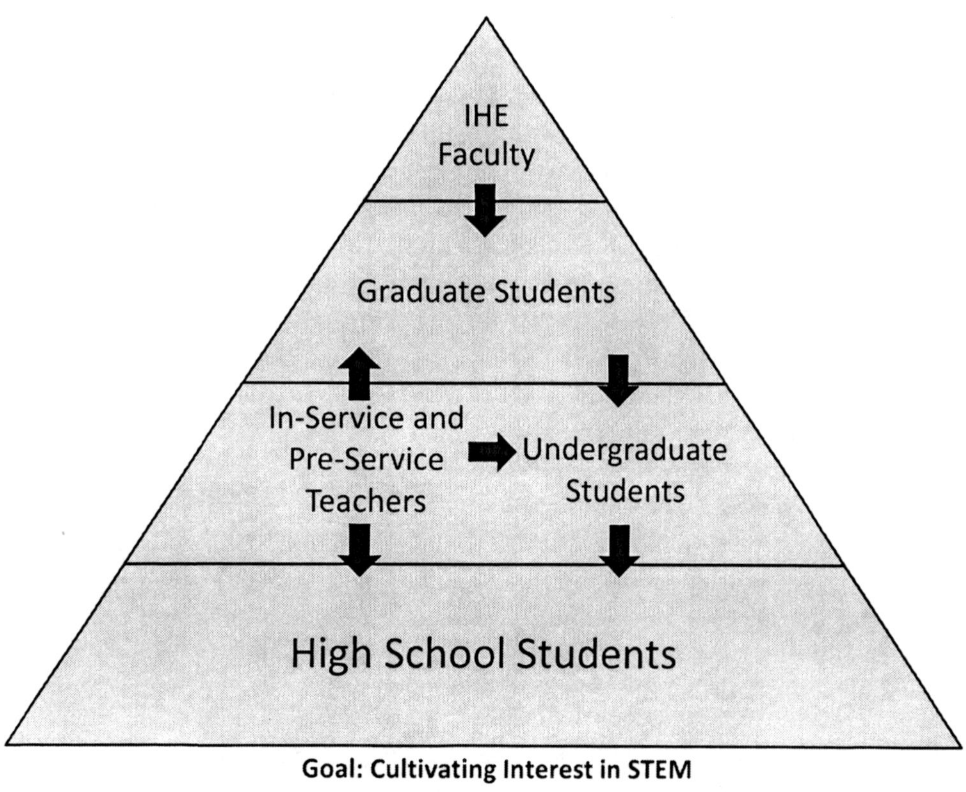

students on the design, development, and implementation of research goals, content, and materials throughout the academic year. In turn, the results of these studies were built into ongoing knowledge and experiential practices for other participants that centered around the summer experience. As part of the process, faculty-led research and instructional teams of graduate students, undergraduate students, and secondary STEM educators met to plan and conduct research experiences that would promote active learning through inclusion of mobile technologies. Faculty reviewed research proposals and served as expert advisors to the team during the design and full-scale experimentation processes. Faculty also reviewed the curriculum and lesson plans based on this research that was developed by the secondary STEM educators to ensure fidelity of content and research practices.

Graduate Students

Graduate students co-led the research and instructional teams, and interacted directly with peers, undergraduate students, in-service secondary STEM educators, and as needed, higher education faculty. The graduate students led the team in planning, research, experiential practices, finding solutions for specific tasks and problem-solving. They also were responsible for accessing and providing beneficial innovations that reflected the needs of industry partners. The graduate students in turn worked closely with undergraduate students, advising them during specific assigned stages of the experimentation process. Graduate students also mentored in-service secondary STEM educators, helping them to develop their own research projects and assisting in finding ways to transfer the experiment and content to lesson plans with a focus on increasing understanding of STEM research and its connection to classroom learning.

Undergraduates

In addition to academic year experiences, undergraduate students participated in a 10-week long program over the summer. This program was designed to give undergraduates an in-depth, authentic research experience, during which they conducted and carried out small projects under the supervision of a graduate student and faculty advisor. Each was assigned to a project that was related to their future STEM career selection. Corollary activities included trainings and workshops on graduate schools, how-to research skills, and leadership development. In addition, these undergraduates served as direct mentors to the high school students during the one-week workshop on campus and received mentoring in that process from both graduate students and secondary STEM educators.

Secondary STEM Educators

The secondary STEM educators were involved in a six-week summer program that was established to allow in-service and pre-service teachers the opportunity to participate in the development and implementation of research within their selected field of STEM. Throughout this process, these educators also created or enhanced curriculum for K-12 students that met the current learning standards and new content. The secondary STEM educators primarily worked with the graduate students to learn how to do research and supplement content knowledge. Active engagement activities focused on providing secondary STEM educators with exposure to mobile technologies and authentic practices. For part of the program, the secondary educators attended the one-week high school workshop and worked on collaborative activities with high school students, and worked with the undergraduates using mobile technology to support active and authentic experiences. Prior to the one-week workshop, secondary STEM educators' worked with the undergraduate students to develop activities for the high school students; this included designing, developing, and transferring information and activities that could be integrated into their own classroom instruction.

High School Students

High school students participated in a week-long program designed for students to examine the use of innovative mobile technologies and understand how these technologies are applicable to the real world. The purpose of the workshop was to interest students in STEM related careers through introduction to research, practice, cutting edge content, and interaction with professionals engaged in the field. Faculty and graduate students shared their knowledge and appreciation of STEM topics through presentations and participation in hands-on activities throughout the week; undergraduate students directly interacted with the secondary students on project work, sharing knowledge of possible career choices and introducing them to the higher education environment; secondary STEM educators advised and helped to supervise the hands-on constructivist based activity for these adolescents, assisting in adapting the language and goals as needed. All of these activities centered around participation in active, authentic experiences conducted by teams working in a collaborative manner.

Specific Roles within the One-Week High School Workshop

The one-week, fully integrated summer workshop served as the culminating activity of the program. This event, located on the higher education campus, involved all levels of participants and relied heavily on hierarchical mentoring and turnkey knowledge transfer. For the one-week high school summer workshop, the IHE faulty and graduate students (expert professionals) provided guidance and knowledge to the novice professionals (i.e., undergraduates) by working with them in the lab, giving guidance on where to find information, reading research notes, etc. During the workshop faculty also presented brief lectures on topics related to STEM research to the high school students.

Similarly, the faculty and graduate students (experts in content) worked one-on-one with secondary STEM educators (experts in pedagogy, moderate knowledge of content, novices in research), facilitating their ability to conduct research at the university over a six-week period. During this experience secondary educators actively participated in the graduate students' research projects and also developed instructional modules for their classes that integrated engineering research design components and other STEM related topics.

The graduate students, with the mentorship from faculty and secondary educators, served as the lead teachers for the summer workshop, providing the high school students with introductory knowledge prior to the expert faculty presentation. Graduate participants also led the activities after the presentations.

The undergraduates, as beginning professionals, facilitated and assisted the graduate students by engaging the high school students in active learning processes. They assisted with the hands-on activities, using scaffolding questions to guide secondary students to a potential solution. The undergraduates' constant presence provided an age referent positive role model for the high school students and served as a source of information about STEM and college life. The undergraduate students also facilitated increased interest and motivation in STEM by taking the high school students on a tour of the labs, describing the research they were conducting, and their interactions with peers and faculty. In addition, the undergraduates gave presentations to the secondary STEM educators on a specific STEM topic related to their research.

Three replications of the summer workshop occurred over the course of the project, hereafter referred to as Pilot, Replication I, and Replication II (pilot program, *n=20* ; Replication I, *n=25;* and Replication II, *n*=27). The majority of the high school students taking part in the program during each replication were reflective of a typical undergraduate STEM population, Caucasian or Asian males, who exhibited prior interest in engineering as a career choice. Students attending this program were already well advanced into their STEM education coursework; they were enrolled in a special engineering or science class for the upcoming school year and/or registered for an advanced mathematics course such as pre-calculus or calculus.

During the Pilot, Replication I, and Replication II, interviews (*n*=22) were conducted with the faculty, graduate, undergraduates and secondary STEM educators to verify the implementation of the active learning environment and provide feedback on perceptions of skill development. Observations (*n*=8) were performed to verify the implementation of the mobile technology devices with a mentored active learning environment throughout the duration of the week-long program. To document changes in undergraduates' expectations, previous research experience, and career plans, the undergraduate students filled out pre- and post-surveys related to the impact of the program on their career plans and their perceptions of benefits and limitations of the program. Pre, post, and follow-up surveys were used to collect data directly from the high school students to gain demographic information, insight into their attitudes toward STEM, perceptions of the program, their future goals, interest in engineering, and assessment of gains in understanding of what engineers do. Pre- and post-surveys were also used to document the secondary STEM educators' knowledge of STEM content; post-surveys also addressed perceived relevancy of the program, the presentations, and hands-on activities.

RESULTS

Results of assessed measures, observations, and interviews indicate that the use of active learning promoted by integration of mobile technology was successful. Transfer of cutting edge results of new research, and support for transfer of knowledge was noted. In addition, use of a collaborative, hierarchical mentoring model, embedded within the system, actively assisted in broadening and strengthening the outcomes. Presented below are evidence-based outcomes noted for each level of participation.

Faculty Outcomes

Faculty indicated the experience was valuable in relation to outreach, professional growth for graduate and undergraduate mentees, and also provided a positive learning experience for themselves. In terms of their own professional growth, faculty noted the need for an increased awareness of K-12 educational standards and their importance as they assist secondary STEM educators and incoming undergraduate students. Faculty expanded on this, noting the necessity of learning about the secondary STEM education standards if they were to adequately mentor a hierarchical team, for example:

"Get information about high school students curriculum beforehand" and "the high school [has] a very strict curriculum so it is hard for us to decide what is appropriate for the curriculum"

Faculty also expressed interest in learning more about pedagogical approaches and strategies as a means of ensuring that their work with graduates, undergraduates, and secondary STEM educators meet the diverse needs present in the classroom, for example:

I think it's important ... to ensure ... [lessons and information] can be told to a wide and diverse audience ... need to learn to do that because strategies could help them with students with learning and cognitive disabilities.

Faculty reported that the one-week summer workshop strengthened their perception of the need to integrate knowledge and practice across all levels to promote and ensure that cognitive and affective variables are developed that will support the ultimate goal of preparing and engaging students in STEM domains. Specifically, faculty noted that working with the graduate and undergraduate students, as well as the secondary education teachers, as a collaborative, cohesive group, helped promote a better understanding of what was needed at each level and how to better build on the experiences of each member in the group. In addition, faculty indicated the workshops and hierarchical mentoring process resulted in knowledge gain at all levels on how to better implement active learning techniques using mobile technologies to promote greater student understanding and lifelong skills.

Graduate Student Outcomes

Graduate students acknowledged growth in content and research skills, and expressly reported satisfaction with the hierarchical mentoring approach. Graduate students made special note that, through this process, they received both formal and informal training and enhanced professional experiences in their meetings with faculty. The integrated involvement of all participants in the design and use of hands-on lab experiences using mobile technologies was reported as very valuable part of this process. Involvement in group meetings was reported to be beneficial in developing not only their research skills, but also in gaining in-depth knowledge in STEM content and practice and communication, for example:

"In the group meeting we sometimes face[d] novel ways of solving [problems] together [providing us] with new ways of looking at problems." and "[The institute] provides a dynamic and flexible research environment..."

More specifically, graduate students indicated that this direct involvement with faculty members provided them with extra guidance and advice, valuable active learning experiences, and assistance with overcoming their academic weaknesses, for example:

"I am getting [better] guidance from my faculty advisor..." and "[They] guide me in my academic and research development, point out my strengths and weaknesses, help me overcome my weaknesses, [and] provide academic guidance."

Graduate students also reported that their involvement as a mentor was a valuable learning/teaching experience as well as an opportunity to disseminate their academic knowledge beyond the program, for example:

"...we enjoyed working together and we learned from each other..." and "As a mentor myself, I learned to provide both theoretical and experimental experiences."

Undergraduate Student Outcomes

Responses from undergraduate student participants indicated that overall, the program also was a positive experience. Undergraduates indicated that involvement was important to them because it allowed them to be part of higher level research projects and because the structure of the program focused on the future; the experience not only helped them with their current degree requirements, but also fostered active learning that would give them a head start in their career or further education, such as:

"I plan on designing videogames after college and this project will help me with the programming and understanding of control theory."; "This [research position] was by far the best. There were clear goals and organization. The overall goal, a futuristic vision, really intrinsically motivated me"; and "held great interest because it is one of the cutting edge research hubs ... [exposed to] a constant stream of innovative ideas."

When queried as to what were the most helpful and relevant activities, undergraduate students predominantly referred to the development of professional skills. More specifically, the leadership seminar, trainings on laboratory equipment, and meetings with advisors, peers, and graduate students were cited by undergraduates as unique opportunities to develop interpersonal skills and collaborative techniques, for example:

"The [program] worked to really stress what 'good leadership' is and how to resolve interpersonal conflicts in a professional, productive manner" and "...teaches us to work with others, they try to challenge you...help you, but you have to think and try to solve the problem yourself and in the real world that will happen...teaches responsibility."

When queried further, undergraduates provided examples of the skills they gained through participation; these included: research skills, collaboration and communication, innovation, career preparation, and knowledge application. Many students indicated the experience was very effective in developing research skills. One student noted this was a first time experience in an academic research setting, allowing "in-depth exposure to higher level math and simulation software" and providing opportunities to work "alongside a high-level academic professor who is developing cutting edge technologies." In general, the undergraduates reported the experiences were effective in preparing them to communicate and collaborate with other individuals

"The workshops facilitated collaboration between student co-workers, [it] was fantastic" and "helped [from] a business standpoint because we had lessons on how to present, write a paper professionally, and get your point across so people will understand it."

This aspect of community and collaboration was noted as particularly helpful by students who were not STEM majors.

Another cited benefit of the program for undergraduates was its ability to foster creativity that led to innovation, for example:

"it is more than likely the things done now will change. They teach you to look at the problem, solve it, and try to modernize it." and "I was able [to] present a creative idea for a different type of architecture which had various beneficial properties," and "already I have been working towards developing an adaptive display device to bring customizable dynamic business signage to small businesses affordably."

When asked for examples of how the program assisted them in preparing for the future and their careers, students noted that the program taught them time management techniques (including prioritization), how to interact in a variety of social situations, and how to work as a team. In addition, students noted that the program helped in gaining a better understanding of and new insight into research, professional industry settings, and broadened their career plans, for example:

"It's another window into the world of research"; "[The] program [gave] me a chance to experience research; I realized that I like working on research and exploring new methods [to] solving

problems"; "If I choose to go into research after industry experience, I will know what to expect."; ."..this program helped me decide to pursue a Ph.D. degree in the future, because this summer I knew what Ph.D. students [do] on a daily basis" and "At the beginning of the summer I was unsure about going to grad school, but after my experience here, I found that I quite enjoy the atmosphere and will be looking for a position similar to this in a few years."

Secondary STEM Educator Outcomes

Secondary STEM educators noted multiple positive outcomes from participation in the program. First and foremost was its relevance to their teaching population and practices; they indicated a wide range of their students would have the ability to pursue STEM careers (up to 90%) if provided the opportunity and that many could pursue a career that involved designing, building, and/or maintaining information systems in STEM domains. Secondary STEM educators noted that participation reinforced their perception that their students had a high need for skills related to innovation, critical, thinking, problem-solving, research, creativity, and resourcefulness, as well as knowledge in science, math, and technology content. These needs were reflected in the teachers' requests for assistance in the design of teaching materials and methods that would increase student interest and excitement while covering fundamental concepts. More specifically, the secondary STEM educators wanted instructional approaches that would integrate 21st Century learning skills into STEM topics and reinforce why this knowledge was important to their students future lives. Secondary STEM educators noted that information presented through the program should include even more development and sharing of sample instructional techniques and materials, specifically requesting additional demonstrations, modules, experiments, pictures, presentations, worksheets, and group work activities.

When queried on how the program provided them with experiences connected to their classroom curriculum, all the secondary educators reported that the real world hands-on applications were very useful in helping bring STEM concepts into their classroom and in having their students connect with real world use, for example:

"I [was] looking for more connections with the engineering community [prior to the program]"; With this program I can reference to [the institution] when working with my students and discuss with them about actual, real world research and application that is taking place"; "[Bringing the real world applications to my students] mak[es] what they are learning more relevant and tangible, while informing and encouraging the students to pursue a career in engineering"; "I am constantly finding new ways to connect this information to the common core [required curriculum]"; "There are lots of variations [for] activities and the field[s in STEM are] constantly growing" and "Taking research labs [in] engineering and [learning how to] translate that into hands-on experiences to be used in the classroom."

Secondary education participants reported that the knowledge they gained through their own hands-on research experience and the supporting activities with graduate students was highly beneficial, for example:

"The time spent working with the graduate student on the research experiments is the most helpful and relevant in that I get to actually participate in the project"; "I like what I have been doing at [the institute] since day one"; "I never expected to do so [many] useful [activities]" and "...what I do makes education come alive in the classroom for [my] students."

When queried as to how participation in the summer program helped prepare the secondary STEM educators to use their STEM knowledge in their profession, all of them mentioned that the invaluable first-hand experience and knowledge application strategies they learned demonstrated how to implement content into their classroom curriculum, for example:

"The first-hand experience is what I have really gained"; "Technology has helped a lot with the program activities [and learning how] to utilize them with the common core" and "I can definitely see how to use technology in the classroom, connect it to the standards, and make learning fun."

High School Student Outcomes

High school students, key participants in the summer workshop, indicated very high perceptions of its value, and noted, via end of workshop and six month follow-up feedback, that it had made a difference in their education careers. Participants indicated an increased understanding of the depth of all that encompasses a profession in engineering. Notable growth was seen in areas related to engineers as problem-solvers. Prior to participation, only 73% of the students perceived problem-solving as a key professional task; at the end of the program almost all (94%) of the students perceived problem-solving as a major component of being an engineering professional. When queried more specifically, key areas of problem-solving noted by the students included a growing understanding of the social and global relevance of the field, for example:

"Engineers could make a difference by developing new technologies that take us beyond the reach of space and time," and "If we can find a way to create an alternative energy or better yet create something that runs perpetually, we would drastically reduce our dependence on oil and eliminate the need for us to use fossil fuels."

Students also reported an increased understanding of the depth of skills encompassing a career in engineering, noting that it is a multifaceted career reflecting skills that are valuable for many different jobs, that they will work extensively with people, not just tools, and that they will have many choices about what they can do and impact within their jobs.

Self-reported changes in student attitude toward learning and studying STEM related content also were noted (See Table 1). Students' general affect toward science and math class remained high or increased after program attendance. Similar positive self-concepts were noted when students were queried about solving math problems. After participation, students also desired that their high school science and technology classes would focus more time on learning through hands-on projects and activities (i.e., modeled after the active learning environment that the program created). After participating in the summer program these students also revealed more interest in joining a science or technology club during the upcoming school year. Students' overall self-confidence and self-efficacy in STEM learning and problem-solving activities increased after participation in the hands-on, constructivist program activities, especially those that involved designing solutions in lab activities, leading a team in designing and building something, knowing where to get information to solve a problem, and communicating math or science to others. Approximately, one-third of the students reported that participation in the summer program had an impact on their career aspirations; noting that the high school summer program encouraged them to think about what they will do after graduating from high school and helped to provide motivation for applying themselves more in school.

These responses were supported by comments from the graduate and undergraduate students who directly worked with these high school students, with a special note made on the importance of hands-on experimentation and activities, for example:

Table 1. High school student's STEM self-efficacy

Item	% Agree*			
	Pilot		Replication	
	Pre	Follow-Up	Pre	Post
I like learning how things work	100	100	100	100
I look forward to science class in school	97	94	96	96
I look forward to math class in school	95	100	93	96
More time should be spent on hands-on projects in science or technology activities at school	92	94	100	96
I would rather solve a problem by doing an experiment than be told the answer	95	94	93	87
I would like to (or already do) belong to a science or technology activities club	77	94	70	78

*Numbers represent percent of students who responded "Strongly Agree" and "Somewhat Agree."

"Students learned a lot by doing"; "By creating something like the circuits and testing them, the students owned their knowledge" and "Students were given immediate feedback through hands-on experience. They had to interpret signals and test out hypotheses by manipulating and transferring the signals of the waves and voltage."

The graduate students perceived that the hands-on activities provided to the high school students had the biggest impact on change in secondary students' attitudes, noting that students were often so intrigued and engaged by the activities, they inquired as to whether they could purchase their own materials for further study. The goal for the graduate students and faculty to "excite them [the high school students] enough to want to learn more and…feel that they can learn in this domain" was met and the experience helped students become "more comfortable working with electrical schematics … building a circuit [and] achieving success through building circuits." The undergraduate students who worked directly with the secondary students also verified an increased interest and engagement as a result of the hands-on activities, for example:

[The activities] enable students to think outside the box. I saw the high school students go beyond what the activities called for by their own choice.

CONCLUSION AND IMPLICATIONS

The unique program discussed in this study integrated active learning, mobile technology, and hierarchical mentoring models to serve multiple goals and multiple stakeholders. The program's central focus was the transfer of STEM content and experience via an active learning environment facilitated by mobile technology and hierarchical mentoring. External evaluation documented that the program facilitated the process of sharing information and knowledge, and increased perceived relevance of the knowledge. By the end of the program, high school students indicated continued and increased interest in STEM learning, STEM careers, and pursing college work in STEM. Both undergraduate and high school students evidenced increased self-confidence and self-efficacy necessary for STEM activities, especially in areas related to problem-solving, working with others, and communicating ideas. Undergraduate and graduate students noted reinforced career aspirations, more interest in educational pedagogy and a broader perspective of global needs. Faculty at the higher education level had an increased awareness of the role of education standards and the needs of teachers and students, across all levels, to gain research and problem-solving skills. Current and future secondary educators gained first-hand knowledge in high stakes research and received assistance in transferring new knowledge immediately into the classroom.

All participants had positive perceptions of the program and expressed added value to the skill development offered to them. Most notable for undergraduates and secondary STEM educators were research skills and increased knowledge of and self-efficacy in innovation and implementation using new technologies. In addition, the undergraduate students indicated their experience promoted and helped to develop skills necessary for professional success in STEM in the twenty-first century; this included collaboration and communication skills, as well as content necessary for their future careers. The hierarchical model of mentoring that fostered interactions across a series of professional levels enhanced understanding of these differing needs and a greater appreciation of the outcomes. The collaborative approach sponsored through this hierarchy allowed participants to share their specific needs and to understand the needs of those they mentored and those for whom they were mentees. This communication was further enhanced through the active learning and sharing of mobile technologies. As each level of the hierarchy grew familiar with tools, outcomes and needs, this information was shared, debated, and used for problem-solving. This active engagement fostered greater career knowledge, content skills, and interest.

The findings and example of this pilot program has implications for both the discussion about and practice of active learning. Although previous literature shows support of the use of active learning, especially in authentic settings, we need more information about the importance of the roles and strategies used by participants. We also need more research which studies the impact of autonomous or collaborative learning, the need for direct vs. guided vs. constructed pedagogy, and the degree to which the instructor becomes a learner as well as a teacher. Further study and discussion also is needed on the role of mentoring as part of learning in an active setting; for example: What skills are learned and reinforced as we teach and advocate for others in active settings? What variables are related to successful mentors? How can we better teach mentoring? Finally, further studies should include the role of technology, especially 21st Century mobile technology in supporting these changes. The future of education and the future of our workforce are embedded in active participation, sharing of information, and the use of technology. We must continue to design, develop, and implement methods that support their integration.

REFERENCES

Barak, M., Harward, J., Kocur, G., & Lerman, S. (2007). Transforming an introductory programming course: From lectures to active learning via wireless laptops. *Journal of Science Education and Technology*, *416*(4), 325–336. doi:10.1007/s10956-007-9055-5

Brophy, S., Klein, S., Portsmore, M., & Rogers, C. (2008). Advancing engineering education in P-12 classrooms. *Journal of Engineering Education*, *97*(3), 369–387. doi:10.1002/j.2168-9830.2008.tb00985.x

Brown, C. J., Hansen-Brown, L. J., & Conte, R. (2011). Engaging millennial college-age science and engineering students through experiential learning communities. *Journal of Applied Global Research*, *4*(10), 41–58.

Dori, Y. J., Hult, E., Breslow, L., & Belcher, J. W. (2007). How much have they retained? Making unseen concepts seen in a freshman electromagnetism course at MIT. *Journal of Science Education and Technology*, *16*(4), 41–58. doi:10.1007/s10956-007-9051-9

Eagan, M. K., Sharkness, J., Hurtado, S., Mosqueda, C. M., & Chang, M. J. (2011). Engaging undergraduates in science research: Not just about faculty willingness. *Research in Higher Education*, *52*(2), 151–177. doi:10.1007/s11162-010-9189-9 PMID:22557706

Feldman, A., Divoll, K. A., & Rogan-Klyve, A. L. L. Y. S. O. N. (2013). Becoming researchers: The participation of undergraduate and graduate students in scientific research groups. *Science Education*, *97*(2), 218–243. doi:10.1002/sce.21051

Fisher, M., & Baird, D. E. (2006). Making mlearning work: Utilizing mobile technology for active exploration, collaboration, assessment, and reflection in higher education. *Journal of Educational Technology Systems*, *35*(1), 3–30. doi:10.2190/4T10-RX04-113N-8858

Gauci, S. A., Dantas, A. M., Williams, D. A., & Kemm, R. E. (2009). Promoting student-centered active learning in lectures with a personal response system. *Advances in Physiology Education*, *33*(1), 60–71. doi:10.1152/advan.00109.2007 PMID:19261762

George, Y. S., & Neale, D. (2006). *Report from Study Group Meetings to Develop a Research and Action Agenda on STEM Career and Workforce Mentoring*. American Association for the Advancement of Science Directorate for Education and Human Resources Programs.

Hucks, D., & Ragan, M. (2013). Technology expanding horizons in teacher education: Transformative learning experiences. In J. Keengwe (Ed.), Research perspectives and best practices in educational technology integration (pp. 61-79). Hershey, PA: IGI Global.

Kardash, C. A. M. (2000). Evaluation of undergraduate research experience: Perceptions of undergraduate interns and their faculty mentors. *Journal of Educational Psychology*, *92*(1), 191–201. doi:10.1037/0022-0663.92.1.191

Kiersma, M. E., Hagemeier, N., Chen, A. M., Melton, B., Noureldin, M., & Plake, K. S. (2012). A graduate student mentoring program to develop interest in research. *American Journal of Pharmaceutical Education*, *76*(6), 1–9. doi:10.5688/ajpe766104 PMID:22919080

Kim, S. H., Mims, C., & Holmes, K. P. (2006). An introduction to current trends and benefits of mobile wireless technology use in higher education. *AACE Journal*, *14*(1), 77–100.

Looi, C. K., Seow, P., Zhang, B., So, H. J., Chen, W., & Wong, L. H. (2010). Leveraging mobile technology for sustainable seamless learning: A research agenda. *British Journal of Educational Technology*, *41*(2), 154–169. doi:10.1111/j.1467-8535.2008.00912.x

McGrath, D. (2007, September 24). Tool 'engages' EE students. *EE Times*. Retrieved from http://www.eetimes.com/document.asp?doc_id=1167108

Michael, J. (2006). Where's the evidence that active learning works? *Advances in Physiology Education*, *30*(4), 159–167. doi:10.1152/advan.00053.2006 PMID:17108243

National Science Board. (2010). *Science and engineering indicators (NSB 10-01)*. Arlington, VA: U.S. National Science Foundation.

Newman, D., Clure, G., Deyoe, M. M., & Connor, K. (2013). Using technology in a studio approach to learning: Results of a five year study of an innovative mobile teaching tool. In J. Keengwe (Ed.), Pedagogical applications and social effects of mobile technology integration (pp. 114-132). Hershey, PA: IGI Global.

Newman, D., Coyle, B., & McKenna, L. (2013). Changing the face of ELA classrooms: A case study of TPACK professional development. In J. Keengwe (Ed.), Research perspectives and best practices in educational technology integration (pp. 270-287). Hershey, PA: IGI Global.

Newman, D., & Gullie, K. (2009). *Using constructivist methods in technology-supported learning: Evidence of student impact.* Paper presented at the Annual Meeting of the American Educational Research Association. San Diego, CA.

Newman, D., Morris Deyoe, M., Connor, K., & Lamendola, J. (2014). Flipping STEM learning: Impact on students' process of learning and faculty instructional activities. In S. Keengwe & G. Onchwari (Eds.), *Promoting active learning through a flipped classroom model.* Hershey, PA: IGI Global. doi:10.4018/978-1-4666-4987-3.ch006

Niemi, H. (2002). Active learning–A cultural change needed in teacher education and schools. *Teaching and Teacher Education, 18*(7), 763–780. doi:10.1016/S0742-051X(02)00042-2

Noonoo, S. (2012). Half of high schoolers own a smartphone or tablet, according to the report. *T.H.E. Journal.* Retrieved from http:/thejournal.com/2012/10/24/half-of-high-schoolers-own-a-smartphone-or-tablet

Olson, S., & Riordan, D. G. (2012). *Engage to excel: Producing one million additional college graduates with degrees in science, technology, engineering, and mathematics. Report to the President.* Executive Office of the President.

Otterstetter, R., Buser, S., Kappler, R., Herold, K., Mullet, N., & Mullins, N. (2011). A model for collaborative laboratory experiences between higher education and a high school technical preparation program: Promoting higher education, mentoring, and research. *Advances in Physiology Education, 35*(4), 460–463. doi:10.1152/advan.00103.2010 PMID:22139787

Prince, M. (2004). Does active learning work? A review of the research. *Journal of Engineering Education, 93*(3), 223–231. doi:10.1002/j.2168-9830.2004.tb00809.x

Rodd, J., & Newman, D. (2009). *The impact of multi-media on learning specific to user characteristics.* Paper presented at the Annual Conference of the American Evaluation Association. Orlando, FL.

Samson, P. J. (2010). Deliberate engagement of laptops in large lecture classes to improve attentiveness and engagement. *Computers & Education, 2*(1), 1–19.

Sirinterlikci, A., Zane, L., & Sirinterlikci, A. L. (2009). Active learning through toy design and development. *Journal of Technology Studies, 35*(2), 14–22.

Taylor, S. S. (2009). Effects of studio space on teaching and learning: Preliminary findings from two case studies. *Innovations in Higher Education, 33*(4), 217–228. doi:10.1007/s10755-008-9079-7

United States Department of Education. (2010). *National education technology plan 2010.* Washington, DC: Office of Educational Technology.

Universities and Colleges Admissions Service (UCAS) Media. (2013). *Eight out of ten freshers have smartphones.* United Kingdom.

Wangler, T. G., & Ziliak, E. M. (2013). Increasing student engagement and extending the walls of the classroom with emerging technologies. In J. Keengwe (Ed.), Research Perspectives and Best Practices in Educational Technology Integration (pp. 44-60). Hershey, PA: IGI Global.

Wiggins, G., & McTighe, J. (2005). *Understanding by design* (2nd ed.). Alexandria, VA: Association for Supervision and Curriculum Development.

Wilson, Z., Holmes, L., deGravelles, K., Sylvain, M., Batiste, L., & Johnson, M. et al. (2012). Hierarchical mentoring: A transformative strategy for improving diversity and retention in undergraduate STEM disciplines. *Journal of Science Education and Technology, 21*(1), 148–156. doi:10.1007/s10956-011-9292-5

Won Hur, J., & Anderson, A. (2013). iPad integration in an elementary classroom: Lesson ideas, successes, and challenges. In J. Keengwe (Ed.), Pedagogical Applications and Social Effects of Mobile Technology Integration (pp. 42-54). Hershey, PA: IGI Global.

ADDITIONAL READING

Bernier, A., Larose, S., & Soucy, N. (2005). Academic mentoring in college: The interactive role of student's and mentor's interpersonal dispositions. *Research in Higher Education*, *46*(1), 29–51. doi:10.1007/s11162-004-6288-5

Campbell, T. A., & Campbell, D. E. (1997). Faculty/student mentor program: Effects on academic performance and retention. *Research in Higher Education*, *38*(6), 727–742. doi:10.1023/A:1024911904627

Earle, R. S. (2002). The integration of instructional technology in public education: Promises and challenges. *Educational Technology*, *42*(1), 5–13.

Ertmer, P. A. (1999). Addressing first- and second-order barriers to change: Strategies for technology integration. *Educational Technology Research and Development*, *47*(4), 47–61. doi:10.1007/BF02299597

Gee, J. P. (2009). Digital media and learning as an emerging field, part I: How we got here. *International Journal of Learning and Media*, *1*(2), 13–23. doi:10.1162/ijlm.2009.0011

Hegstad, C. D. (1999). Formal mentoring as a strategy for human resource development: A review of research. *Human Resource Development Quarterly*, *10*(4), 383–390. doi:10.1002/hrdq.3920100408

Holm, G., & Dynak, J. (1994). A window of opportunity: University students mentoring high school students. *People and Education*, *2*(4), 431–441.

McCarthy, J. P., & Anderson, L. (1999). Active learning techniques versus traditional teaching styles: Two experiments from history and political science. *Innovative Higher Education*, *24*(4), 279–294. doi:10.1023/B:IHIE.0000047415.48495.05

Metcalf, D. S., & DeMarco, J. M. (2006). mLearning: Mobile learning and performance in the palm of your hand. Amherst, MA: HRD Press, Inc.

Mouza, C. (2008). Learning with Laptops. *Journal of Research on Technology in Education*, *40*(4), 447–472. doi:10.1080/15391523.2008.10782516

Naismith, L., Sharples, M., Vavoula, G., & Lonsdale, P. (2004). *Literature review in mobile technologies and learning*. Academic Press.

Newman, D., & Gullie, K. (2009). *Using constructivist methods in technology-supported learning: Evidence of student impact.* Paper presented at the Annual meeting of the American Educational research Association. San Diego, CA.

Newman, D., Reinhard, D. E., & Clure, G. (2007). *Using constructivist methods in technology supported learning: Evidence of student impact.* Paper presented at the American Educational Research Association. Chicago, IL.

Norton, P., & Wiburg, K. (1998). *Teaching with technology*. Fort Worth, TX: Harcourt Brace College Publishers.

Roschelle, J. (2003). Keynote paper: Unlocking the learning value of wireless mobile devices. *Journal of Computer Assisted Learning*, *19*(3), 260–272. doi:10.1046/j.0266-4909.2003.00028.x

Zydney, A. L., Bennett, J. S., Shahid, A., & Bauer, K. W. (2002). Impact of undergraduate research experience in engineering. *Journal of Engineering Education*, *91*(2), 151–157. doi:10.1002/j.2168-9830.2002.tb00687.x

KEY TERMS AND DEFINITIONS

Active Learning: A process that engages students in learning by gathering information, thinking, problem-solving, reflection, and self-assessing their personal degree of understanding.

Collaborative Learning: A method of teaching in which students work together to solve a problem, create a project, or explore a topic together.

Common Core Learning Standards: A set of standards developed for students in all of the United States that define the expected skills and knowledge students should have at the end of each school grade level.

Constructionism: The idea that effective learning occurs when students actively build concrete artifacts that applies to the content; stems from experiential learning.

Constructivist Learning: Learning theory that refers to the idea that students construct knowledge through their own experiences.

Cooperative Learning: A form of collaborative learning where students work together in a small group on an activity or assignment that is structured. Individuals are usually held accountable for their own work.

Group/Team Learning: A form of learning where students work as a group or a team on an activity or assignment where individuals and the group as a whole are assessed on their work.

Hierarchical Mentoring: The provision of advisement, training, and support from someone with more experience to someone with less experience. This mentoring occurs for various levels of learners who support one another in each successive rank (i.e., Graduate students mentor undergraduate students who mentor high school students who mentor middle school students who mentor elementary students).

Mentoring: A mutual partnership in which, one member has more experience than the other and it is their responsibility to organize and guide the less experienced member.

Studio Approach to Learning: A learning environment in which students actively participate in their own learning through technology-based experiments and collaborative group work; typically an integrated lecture/laboratory format.

ENDNOTES

1 This material is based upon work supported by the National Science Foundation under Grants EEC-0812056 & DUE-0717832

This work was previously published in "Promoting Active Learning through the Integration of Mobile and Ubiquitous Technologies" edited by Jared Keengwe, pages 116-134, copyright 2015 by Information Science Reference (an imprint of IGI Global).

Chapter 42
Using Project-Based Learning to Teach Sustainability Issues to Elementary Students

Ingrid Weiland
University of Louisville, USA

Elisa Pokral
Monroe County Indiana Waste Management District, USA

Kristin Cook
Bellarmine University, USA

ABSTRACT

This chapter describes a project-based learning unit on sustainability that was implemented in a fourth grade classroom by an informal educator (the second author) employed by a local waste management district. Previous instruction by the informal educator consisted of one-hour lessons that were separate units lacking the project-based learning format and transition links. This chapter describes a mixed-method case study in which pre and posttest tests as well as focus group interviews assessed students' learning as a result of participating in a cohesive project-based learning unit designed by the authors. There was a statistically significant change ($p<0.05$) in students' pre to posttest scores, and focus group interviews indicated that students could elaborate much more deeply on their ideas about sustainability after the program. The authors conclude that project-based learning can support students' understandings of sustainability while providing an engaging and enriching format to informal educator programs.

INTRODUCTION

The history of sustainability education is complex, and a variety of terms (e.g., conservation education, outdoor education, and environmental education) have been used to describe similar approaches. For the purposes of this chapter, we will use the term sustainability education unless we are quoting another author. The term "sustainability" has been defined as "development that meets the needs of the present without compromising the ability of future generations to meet their own needs" (Brundtland, 1987, p. 41). Thomas and Muga (2012) note, "while there are

DOI: 10.4018/978-1-4666-7363-2.ch042

varying definitions of sustainability from different sectors of industry, what is important is that it strives for protection of the environment, prudent use of natural resources, equitable social progress, and maintenance of economic well-being without compromising the environment and society" (p. 264). Few would argue that future generations deserve a healthy and fruitful environment in which to live, however the current rate of development in many Western countries, coupled with global population growth, causes a great strain on our environmental and natural resources. While some believe that new technologies allow for the current rate of development to continue while causing less strain on the environment, the reality is that we must concurrently learn to minimize our impact. Education for sustainability can occur in a variety of venues, including in the home, in the classroom, or in informal education settings.

Sustainability education can play a versatile role in curriculum as it lends itself to interdisciplinary instruction and can facilitate connections between informal and formal classroom instruction. Informal science education is supported as a complement to formal education (or education in classroom settings) in a variety of contexts. Sustainability education is sometimes considered informal science education as it often occurs outside of the classroom or by an educator that is not employed by the school district. Formal classroom teachers could infuse sustainability into their curriculum and standards, but often do not because the connections within the standards to sustainability are rarely explicit. Research suggests that only teachers who have a great interest in nature and the environment take the time to infuse sustainability into their curriculum (McKeowen-Ice, 2000) - a concern shared by our own informal educator as an impetus to this study. It is important to note that the *Next Generation Science Standards* (NRC, 2013) include more focus on sustainability than many of the previous state science standards.

To explore how sustainability best be approached in the elementary classroom, we posit that an informal educator could successfully teach sustainability education in a classroom most effectively through a project-based learning (PBL) framework. Given its emphasis on interdisciplinary reasoning and its flexible design structure, we assert PBL can offer a meeting point for informal and formal educators. The PBL framework allows students to resolve complex socioscientific issues while exploring multiple perspectives and a variety of resources. Ramsey (1993) states "science education should produce students who can and do participate in the investigation and resolution of science-related social issues" (p. 414). In that vein, science educators can teach social responsibility through sustainability education, which provides a complementary opportunity to integrate formal and informal education. Tal (2004) notes that the very nature of sustainability education, often focusing on local, socio-scientific issues, allows for partnerships between schools and communities.

Project-Based Learning

Research and practice calls for a constructivist approach to sustainability education to "design, develop, and test specific teaching strategies for developing environmental conceptions that address the cognitive, affective, and behavioral dimension" (Ballantyne & Packer, 1996, p. 30). Sustainability education offers the opportunity for students to engage in problem-solving and action-based activities, whereas traditional formal educational settings are often structured with linear curricula that are teacher-centered. Thus, strategies to support both aims and structures are needed. Recent science educational reform efforts for adolescents have attempted to engage students in science by utilizing project-based instruction (Rivet & Krajcik, 2008). In the PBL approach, teachers act as facilitators to student learning as students have access to technology in the classroom that allows them to cooperatively explore and guide their learning, organize their work, and manage their time. This sort of learning environment, coupled with substantial teacher support, has the potential of greatly enhancing student participation and engagement with sustainability issues, which require consideration of multiple perspectives. The Buck Institute (2013) defines project-based learning (PBL) as "an extended process of inquiry in response to a complex question, problem, or challenge. While allowing for some degree of student 'voice and choice,' rigorous projects are carefully planned, managed, and assessed to help students learn key academic content, practice 21st Century Skills (such as collaboration, communication & critical thinking), and create high-quality, authentic products & presentations" (paragraph 1). Through a meta-analysis of PBL research, Edutopia (2013) outlines the essential features as, "students learning knowledge to tackle *realistic problems* as they would be solved in the real world, increased *student control* over his or her learning, teachers serving as *coaches and facilitators* of inquiry and reflection, students (usually, but not always) working in *pairs or groups*" (emphasis in original, paragraph 1).

The features of PBL are consistent with the learning needs of many types of students and hold promise as an effective tool for teaching complex topics such as sustainability issues. Research suggests the benefits of using a PBL approach include deeper understanding and retention of subject matter, increased self-direction and motivation, and improved problem-solving, collaboration skills and attitudes toward learning (Blumenfeld, Soloway, Marx, Krajcik, Guuzdial, & Palincsar, 1991; Gallagher, Stepien, Rosenthal, 1992; Strobel & van Barneveld, 2009; Walker & Leary, 2009). Furthermore, Geier, Blumenfeld, Marx, Krajcik, Fishman, Soloway, & Clay-Chambers (2008) found that seventh and eighth graders engaged in PBL scored higher of standardized state science exams peers receiving traditional science instruction, and Schneider, Krajcik, Marx, & Solloway (2002) found that tenth and eleventh graders who participated in PBL science instruction outscored their national peers on 44% of test items on the National Assessment of Education Progress.

Based on the results of previous studies on the effectiveness of PBL in the formal classroom, we hypothesized that the implementation of a PBL unit by an informal educator would increase students' content knowledge and improve attitudes toward sustainability. The PBL format provides a consistent theme and purpose to informal programming that occurs over a period of time. Moreover, the features of PBL allow for students to explore sustainability issues in an active, student-centered environment that encourages critical thinking and action through the creation of a culminating project. Many informal educators offer programs to classroom teachers, but little is known about the effectiveness of PBL in teaching sustainability education. Our research was therefore guided by the following research question: *What can elementary students learn from a PBL unit on sustainability implemented by an informal educator?* In this chapter, we will describe the context in which the research question was investigated, as well as share the details of the PBL unit we developed. Additionally, we will describe the research methods and results with regard to student outcomes before and after having participated in the PBL unit.

ORGANIZATION BACKGROUND

This research was conducted as a partnership between two science educators and an informal educator from the local solid waste management district. In 1990, Waste Management Districts were formed by HB 1240 as a separate local governmental entity to manage solid waste in Indiana. The mission of the "District," the Monroe County Waste Management District is to secure a healthier environment in Monroe County, Indiana by eliminating waste going to final disposal through reduction of source materials, reuse of reusable materials, recovery of recyclable materials, and by offering educational resources and programs. As well, the District aims to promote sustainable materials practices throughout our communities. The District achieves this mission by working with all Monroe County residents, businesses, and community organizations. The sustainability education offered by the District serves to divert waste from the landfill, which makes the community a cleaner and healthier place in which to live in many ways (i.e. reducing waste, reusing as much as possible, recycling, and minimizing the impact on humans and wildlife). As part of the "three R's," the District's education programs seek to teach people to take care of water, soil, and air quality by avoiding the unnecessary use of pesticides, herbicides, and household chemicals. Reducing energy usage and disposing of trash properly, including hazardous materials such as paints, cleaners, electronics, and pharmaceuticals, serve to keep the earth healthy and all living creatures healthy. Connecting people to the environment helps people understand how and why they should become more environmentally friendly and inspires them to teach others simple ways to take care of the earth every day in order to make an impact in the community and in the world.

This research was conducted within the context of a pre-established relationship between Ms. Elisa and Ms. Simon (pseudonym). Ms. Elisa regularly visited schools throughout the area to provide one-hour programs related to sustainability and waste management at no cost to teachers. Ms. Simon worked as a fourth grade teacher at a semi-rural parochial school in the Midwest. As a part of its mission, the school promotes respect for and stewardship of the Earth, as well as strong content knowledge about the world around us. However, Ms. Simon stated that she did not feel comfortable in her own science content knowledge, and that she co-taught science and social studies with a third grade colleague, exchanging students during that block. Ms. Simon stated that she supplemented this science instruction with Ms. Elisa's programs, which were interdisciplinary in nature and supported some of the science content. At Ms. Simon's school, Ms. Elisa visited primarily second and third grade on average eight times per school year and kindergarten classes four times per year. Ms. Elisa, an informal educator, and the authors, both science teacher educators, designed a project-based curriculum that was composed of six instructional units wherein students studied a variety of topic areas related to ecology and sustainability education. The six-session PBL unit, designed to be implemented over a twelve-week period, was created by the researchers in collaboration with Ms. Elisa with the goal of having students learn state science content standards that relate to sustainability education, as well as maintaining Ms. Elisa's behavioral objectives as the waste management education specialist. The two researchers had personal interest and experience working in sustainability education, and were excited to examine the research-based practice of PBL within this context. They provided knowledge of PBL as well as state standards while Ms. Elisa adapted her existing programs to fit the PBL model. Ms. Simon, the formal educator, was not involved in the planning of the PBL unit, but was provided a list of standards the sessions covered and agreed to allow class time for completion of the final culminating project. She was not involved due to her own time constraints; however she did state that she hoped for an interdisciplinary program.

SETTING THE STAGE

The three authors met three times for two hours each to plan the PBL unit, which was based on a revision of individual programs already offered by Ms. Elisa. The authors met to create a cohesive unit from the individualized programs and to imbed cross-curricular content standards into the program along with a culminating project to showcase student learning. The cohesive unit begins by examining sustainability issues from a macro level and narrows focus each week to end on the individual (or micro) level. The six lessons are described below. Table 1 highlights the main goals, roles, and activities for each session.

Session 1: Envirocircles

The first session connects students to the larger scope of the environment by focusing on the concepts of interconnectedness, balance, and ecosystems. The session begins by showing the students a hula-hoop, and asking "Where do humans belong in this circle of life? In the center? On the inside? Outside?" Students begin thinking about and providing a rationale for where we stand in relation to all other beings. The instructor then initiates a discussion about circles and cycles and where we find them in the world. Students may come up with many ideas, and the instructor can provide additional examples (i.e., Chinese Yin Yang, water cycle, spiral shells, recycling symbol). Students consider by way of a group discussion how these symbols are a metaphor for humans' relationship with the Earth.

The instructor then presents the information on the different levels of ecological interactions: populations, communities, and ecosystems from the 4th grade science standards and explains the

Table 1. Sessions begin with large scale (environment, ecosystem level) to smaller scale (personal level)

Lesson/Topic	Goals	Activities
1 - Envirocircles	Students learn about interconnectedness, balance, and ecosystems	• Discussion of cycles (Yin-Yang, water cycle, recycling symbol) • Humpty Dumpty activity • Food Chain discussion • Human Knot activity • Read "Tinker's Clock" story
2 – Home and Community	Students gain sense of place and learn about healthy communities Students understand life is all around us, life needs to be protected, healthy communities need a variety of natural habitats	• Life Box • "Milky Way" poem • Discuss development of land in their community • Student draw their own backyard or "place" • Students make suggestions for making their "place" healthier • Read "Window" book
3 – Household Hazardous Wastes (HHW)	Students identify source of everyday hazardous materials Students make their own non-toxic house cleaners	• Students identify hazardous materials in their homes and backyards (review drawing from last week) • Discussion of HHW • Toxic trash quiz • Students create own non-toxic natural cleaning solutions (all-purpose cleaner and furniture polish)
4 – Perspectives on Trees	Students are exposed to multiple perspectives of an issue Students understand the complexity of environmental issues Students explore "rewilding" their yard	• Guided meditation • "Mystery Bag" activity • Students connect items to where they come from • Read "Growing Wild" • "Tree Perspective" debate (three perspectives of development: none, partial, and full) • Students rewild their yards using drawings from session 2. • Students brainstorm ideas for culminating projects
5 – Power of One	Students recognize the effect one person can have on protecting the planet Students learn about recycling	• Song, "Who Made This Mess" • Discussion about challenges to waste management • Smart Shopper game (students choose environmentally friendly purchases) • Kinesthetic exercise to learn how to wash out recyclable materials
6- Culminating Projects*	Students share their content knowledge with others	Parents were invited to culminating project presentations

*One session prior to session 6 was devoted to students practicing projects

Ecosystem Puzzle, a homemade game made of various natural scenes cut into puzzle pieces. For example, pieces include photos of leaves, a cricket, a robin, and a hawk. Students discuss amongst themselves how the different puzzle pictures are related to one another (i.e., they are all a part of a food Web which consists of producers, consumers and sometimes decomposers). The whole class discusses what can break the circle of relations (disrupt the ecosystem). The instructor can ask, "What could have happened to break the interconnectedness of these living beings?" and "How does this affect us?"

Students then participate in the Human Knot activity (New Games Foundation, 1976). This activity teaches students not only about the complexity of nature (i.e., food Webs), but also that it is easier to prevent difficult situations (being tangled in a human knot) than to solve a challenging problem. Session one concludes with a story, "The Tinker's Clock" (Strauss, 2006) to demonstrate that we are all responsible for maintaining healthy circles in life, and that the first rule of ecology is to work individually and together to responsibly protect all the parts of an ecosystem.

Session 2: Home and Community

The second session connects students to their community by focusing on the concepts of sense of place (Sobel, 2004) and healthy communities. The session begins by showing the students a "life box" (Project Wet, 2011) and asking them what would be in a life box (students are provided with film canisters that contain soil, leaves, water, "air," etc.). Students are asked to make decisions about the needs of living organisms and discuss what resources they observe organisms using in their local environments.

The instructor then has students outline where they are on the map using the song "Milky Way" (Harley, 2001a). The concept of *habitat* is introduced and students are asked what was on the land before their house or apartment was built. They discover that some sources like space and arrangement of food, water, and shelter are crucial for animals. At this point, the basic elements needed for animals and humans are shared kinesthetically by making motions to represent food, water, shelter and space. The instructor explains that each student will draw their own backyard to explore their sense of place. Afterwards, the instructor initiates a discussion with students about what is in their place that organisms there depend on (healthy areas for shelter, space, food and water). Students then brainstorm suggestions to make their place healthier.

To conclude the session, the instructor reads *Window* (Baker, 1991), a book without words, to illustrate change over time. The book initiates a discussion about students' homes, their sense of place, and healthy places. The instructor asks, "do you see your home as contributing to a healthy habitat? In what ways could your "place" be a healthier environment for plants and animals?" This session concludes with the following concepts:

- Life is all around us,
- Life should to be valued protected, and
- Healthy communities need a variety of sustainable natural habitats.

Session 3: Household Hazardous Waste (HHW)

The third session connects students to their home by focusing on hazardous waste and threats to healthy homes. The session begins by having students review their drawings from the previous session and asking students to identify potentially hazardous materials in their homes (e.g., oil, electronics, medicines, medical aids, paint, cleaners, phones, personal healthcare products, batteries). Students gain content knowledge on common substances that can be harmful to human health and the environment. Storm water pollution is also explained along with prevention measures for in-home HHW.

Students are then asked three questions from the Public Broadcasting System "Your Toxic Trash Quiz" (KERA-TV, Personal communication, 1993) and identify information required by law to be on labels, such as proper safety, storage and disposal, certain ingredients, first aid, and emergency contact information. Please note that although the quiz was published twenty years ago, the information is still current and needs no revisions. In groups, students are assigned a room of a house and asked to brainstorm HHW items that could be in that room. Discussion ensues about how the use of common HHW can affect our health and pollute our environment. Students then design and create their own cleaning solutions from organic materials by using measuring and observation skills. This activity demonstrates the connection of the health of humans and the ecosystem to the health of the home with common household products. The session concludes with the student development of rules for home and environmental health measures such as using alternatives to HHW, using smaller amounts or non-aerosol products, and making organic cleaning agents. This session provides a context for the classroom teacher to pursue a guided inquiry with their students over the coming weeks in which students investigate the effectiveness of the various organic cleaning solutions. Self-evaluations with a student-developed rubric enable the classroom to decide which cleaner best meet the aforementioned health measures while still doing a good job cleaning of variety of spills. Furthermore, an opportunity exists to involve school maintenance staff in the evaluation of the different cleaning solutions and share the students' knowledge about environmental health.

Session 4: Perspectives on Trees

The fourth session is a look at multiple perspectives of the complexity of the sustainability issue regarding the natural environment with a focus on trees and *rewilding* yards. "Rewilding" refers to the practice of returning or renewing areas of land to a natural state to increase the health of the ecosystem or to return to the original or similar habitat to the natural one. The session begins with having students experience with music the physical and emotional influence of the natural environment. Through a guided meditation, students observe the physical and emotional influence of being close to trees. Students learn about dependence on trees by humans, animals, and the ecosystem by pulling items out of a mystery bag such as: pencils, rubber balloons, vanilla extract, toothpaste, shoe polish, puzzle pieces, chocolate, apples, nuts, gum, syrup, fabric, books, and paper currency. All of these products come from trees–students learn there are more than 5,000 products made from trees. The instructor also shares the percentage of forests in the world, who owns them, what a forester does, and the length of time city trees survive. The instructor then initiates a discussion in which students hold up signs that they read to the class explaining benefits of trees.

Next, the book *Growing Wild* (Perenyi, 1991) is shared with students during which students are engaged in an interactive demonstration of the book's story about rewilding. The rewilding term connects to the previous sessions' concepts of health, connectivity, complexity, ecosystem, community, and home. During the story, students are asked questions, "Why is a mowed area nice to have? Why is a wild area important to have for both humans and other life? Which area is more fun to play in? What are alternatives to pesticides, herbicides, and fences?" Discussion of student answers aims to encourage reflection on what human's perceive as beautiful often conflicts with the needs of a healthy ecosystem.

After the Growing Wild discussion, the instructor introduces the Tree Perspective exercise based on an activity titled "Forest Consequences," (Project Learning Tree, 2000a). Three students read three different development perspectives: keeping the area wild, partial development, and full development. In groups they brainstorm questions they would need to ask in order for their type of development to be healthy and sustainable. The session concludes with the instructor asking students to review their home drawing again to write and/or draw actions they could take to rewild their own yard. Both students and the instructor share some suggestions.

Session 5: Power of One

The fifth session connects students to the power of what one person can do to protect the earth and how the health of the individual relates to the health of the natural world. The session begins with the instructor asking "Are you Earth-friendly?" The students then do kinesthetic motions to the song "Who Made This Mess" (Harley, 2001b). Afterwards, the instructor asks, "What if everything we threw out stuck to us? What would we find on you?" Three ideas are shared during session:

1. Kids can make a difference (power of one),
2. Trash is treasure, and
3. The power of one becomes the power of many people.

At this point, students learn about major challenges we face in our world related to waste. Concepts are related in individual terms (i.e., relating amount of individual lifetime waste production to that of a gymnasium filled with trash). The instructor then explains the four principles of a smart shopper in an exercise where students are asked what they would like to buy based on certain characteristics (e.g., functionality, reasonable price, durability, and earth friendly in production, consumption, and disposal). The term "smart" shopper stands for not only someone who is intelligent but someone who saves money and reduces trash. The following activity teaches students what is and is not recyclable in their county. Students are asked to pretend they are in a store and they are helping their parents make Earth-friendly choices. Children understand the science behind being earth-friendly by learning about disposables, reuse, nutrition, and how trash can become hazardous waste. The instructor reveals the second idea: that trash is treasure.

Students then participate in a kinesthetic exercise to demonstrate that certain recyclables must be rinsed before placing in a recycle bin (glass, plastic, metal cans) using motions to simulate proper recycling procedures. The process of recycling helps students to understand that humans create products out of natural resources–some renewable and some non-renewable- and that all resources are ultimately non-renewable if not protected. Throughout the sessions and as noted here, scientific vocabulary is used as needed in the context, rather than decontextualizing vocabulary at the beginning or end of the unit. In a follow up game, students match up natural resources to the materials people create from them (i.e., petroleum→ plastic, sand→glass). The session concludes with a review of the benefits of reducing, reusing, and recycling. The session ends with the reciprocal concept that the natural environment depends upon the individual to keep it healthy and vice versa: Everyone is responsible to make a difference and has the power to make that difference.

Session 6: Culminating Projects

As the unit concludes, students begin to consider their final culminating projects. Students are placed in groups of 2-4 in order to work collaboratively on a project that demonstrates their learning as a result of participating in the unit. Those with similar interests can work together and proposals for projects are listed on the board to ensure no repetition of projects. Throughout project planning, the instructor visits each team individually to discuss which topics needed further explanation. Students are offered a variety of materials and choose their own presentation media to demonstrate what they learned (see Table 2). Each student chooses topics related to environmental stewardship that interests them and investigates further information via the Internet. Students can produce PowerPoint™ presentations, brochures from Publisher™, hand-written posters, murals, role-plays, dioramas, and creative displays of information gained (again, see Table 2 for a list of sample ideas and actual projects the fourth graders completed).

Students then present their work to parents (each group was given 5-10 minutes), community members, and peers at a final assembly. Teachers could also invite other classes, teachers, parents, or administrators to the final presentations. In our case, the local newspaper came to observe the presentations and published a story about the entire unit. The resulting projects unit incorporated three of the four key features of PBL: production of tangible artifacts (i.e., project), collaboration with peers and teachers (during the creation and demonstration of their projects), and use of cognitive tools such as the Internet.

Table 2. Culminating Project Ideas

General Ideas	Fourth Graders' Ideas
Poster	Wetlands tri-fold poster, Pollution poster
Newspaper article	Recycling "Are You Smarter Than a Fourth Grader?" Game show
Diorama	Recycling and land use diorama
Play/skit	Habitat skit, recycling skit, Littering skit
Poem	"Going Green" poetry
Short story	
Advertisement	
Clay	
Activity Book	
PowerPoint™ presentation	
Brochure (Publisher™)	

CASE DESCRIPTION

This case study focuses primarily on the experiences of two educators: Ms. Elisa (a female Caucasian education director for a local waste management district) and Ms. Simon (a female Caucasian fourth-grade instructor with more than 20 years of teaching experience at the elementary grade level). Ms. Elisa is a well-established informal educator with a total of seventeen years in education–seven years teaching environmental education to grades K-8, four years as a curriculum enrichment specialist and lead teacher for elementary school age programs, and six years teaching at the university level as a cross-cultural program coordinator and adjunct faculty. As stated previously, we worked with a class of 21 fourth-grade students at a small K-12 Christian academy located in a Midwestern town of approximately 81,500 people. In Ms. Simon's fourth grade class were Asian-American (2), African-American (1), and Caucasian (18) students. The majority of these students came from a medium to high socioeconomic background and lived either in the rural surroundings of the school or in the periphery of a large neighboring university. Ms. Elisa elected to work with Ms. Simon and her fourth grade class for the purposes of this case study research because she had worked with her previously and she already had six sessions scheduled with her for the school year. Ms. Elisa was also seeking ways to market her programs to higher grade levels since she had been predominately working with K-3 teachers. She felt that piloting the PBL framework with fourth graders might give her ideas for modifying her programs for middle school students. Furthermore, research has shown that PBL is successful with secondary students (Geier, 2008; Schneider, 2002) yet we were interested in the applicability of PBL to teaching sustainability issues to elementary students. Students in the classroom agreed to participate in our study to offer us a glimpse into the learning that was fostered by the sustainability-based PBL.

Data Collection

To explore student learning as a result of the PBL unit, students were given a pre and posttest to determine their understanding of content underscored by the sessions. Questions were drawn from the released test items from the Trends in International Mathematics and Science Study (TIMSS, 2009) fourth grade standardized assessment and included such questions as:

Some things were buried in wet ground. Several years later they were dug up. Which thing is MOST likely to have stayed the same? (Please see Appendix A for entire assessment).

- An egg shell.
- A plastic cup.
- A paper plate.
- An orange peel.

Eighteen students completed both the pre and posttest. The TIMMS data (n. d.) were used to triangulate the data obtained from the focus group interviews with the students. All students were asked to participate in semi-structured interviews prior and subsequent to their participation in the PBL unit (See Appendix B). Rather than assessing students using only a standardized exam, we felt it was important to explore the cognitive effects of the PBL through a qualitative approach that took into account students' unique perspectives, learning experiences, and personal reactions to the instructors' teaching strategies. These interviews were open-ended and followed a flexible protocol; that is, the interviewer was free to modify the sequence and wording of questions, omit or add questions, and determine the amount of time and attention given to each question or topic (Bernard, 2002; Robson, 2002). Twenty-one students participated in both the pre and post interviews in groups of 3 or 4. During the interviews, which lasted between 10 and 15 minutes, students were asked to explain their conceptions of key concepts taught throughout the lesson. Upon completion of data collection, full transcripts were produced from the audio recordings and each student was assigned a pseudonym to protect their confidentiality. Finally, to further triangulate the data, the first and third authors employed the use of a reflective journal in which they took detailed field notes during observations of the classroom. The journals and observation notes served to support or challenge interpretations throughout the study and to note items for further discussion with the educators.

Data Analysis

We used a mixed method approach to both quantitatively and qualitatively assess the learning outcomes of the PBL unit on elementary students. Students' pre and posttests were analyzed using a pairwise t-test comparing pre and post scores. Additionally, student interviews and reflective journals were examined using a qualitative analytical approach that incorporated elements of grounded theory (Glaser & Strauss, 1967). This approach called for the iterative and combined use of interpretative and flexible methods of analysis such as close reading of interviews and debriefing sessions, open coding, and memo-ing (Bernard, 2002; Emerson, Fretz & Shaw, 1995). Peer debriefing sessions among all involved educators were frequently held to triangulate emerging interpretations of the data. In these sessions, discursive records of particular episodes were examined collectively, individual analyses shared, and interpretations discussed extensively. The emergent account was gradually adjusted to include any variation that surfaced from this reflective group interpretation of the data. These debriefing sessions were very important during the interpretative analysis of episodes as they contributed significantly to guarding against individual researcher biases (Robson, 2002).

Student Outcomes from a Project-Based Approach

The gain in student content knowledge was statistically significant. A paired-samples t-test (n=18) was conducted to compare results from the pre-test to the post-test, which elicited a statistically significant ($t(17)= 4.486$, $p <.005$) gain on the 14-item test (9 multiple choice–1 point each and 5 short answer–2 points each), worth 19 possible points. The mean difference from pretest (mean=12.89) to posttest (mean=15.89) was 3.17 points with a standard deviation of 3.00 points. These results were corroborated with the focus group interviews of students. Students demonstrated gains in learning throughout the PBL both in terms of an understanding of the complexity of sustainability topics and of content vocabulary and standards.

With regard to understanding complexity in sustainability topics, students cited information that they had learned from Ms. Elisa. For example, in her pre-interview when asked about her understanding of recycling, Rachel said, "They have big machines to crush it [metal] and then when you need it, you can put it in fire with these really big globes." However, in her post-interview, Angie described in more detail when asked to talk about recycling, "You might not be able to recycle this device [audio recorder] because it may have mixed materials that can't be separated. I learned that from Ms. Elisa." Both by stating that she had learned more about recycling from Ms. Elisa and indicating that she understood that some materials could be recycled while others could not, she displayed her developing grasp on the complexity of sustainability issues.

Another example of an increased understanding of the complexity of these topics came from Nathan when asked about the process of science. In his pre-interview, he claimed that the reasons scientists might have different results from the same experiment was because they "might add things at a different time or in a different order. Because no one person can do the same things as each other," which indicated that he perceived differences to be due to procedural variations. In his post-interview, he responded to the same question with "Nobody thinks exactly alike–sometimes, but not always. Like a mechanical engineer, if they are designing something, some person might have one idea of what might work and another person would think differently." This change in response was something Ms. Elisa had emphasized in the PBL when discussing multiple perspectives and subjectivities within the process of science.

The other dimension in which students showed positive gains was in their understanding of content vocabulary and standards. Content standards such as food Webs and types of organisms in the ecosystem were emphasized repeatedly in the PBL unit. When asked about food Webs in her pre-interview, Annika responded, "I don't know what that is, is that like a spider Web?" However, in her post-interview she articulated, "A food Web is overlapping food chains. If a hyena would either eat a duck or a chicken, well the duck has one food Web and the chicken has one food Web, so it'll kind of be overlapping." Likewise, Liam claimed in his pre-interview that, "Some animals only eat dead animals, like vultures just wait there until something dies. Other animals depend on living animals;" whereas later in his post-interview, he detailed that "There are three different types of organisms–there's consumers, producers, and decomposers. Decomposers rely on dead bodies to feed them. Consumers rely on the producers to help feed them. A decomposer is like an ant or a beetle, they'll start eating your body until there's only bones left." In both of these examples, students showcased their deeper understanding of the fourth grade content standards embedded in the PBL unit.

In another example of deeper understanding of content, students were asked about interactions in the ecosystem and interrelatedness of species. Lily, in her pre-interview, stated simply, "Most animals have to kill, but some animals don't." Later, in her post-interview, she had much more to say when asked again about these ecosystem interactions:

If a cow dies and mushrooms decompose, it goes into soil so the flowers can grow there and so we can have air, it takes the carbon dioxide and makes it air. So, it's kind of like the carbon cycle. If there were no decomposers, there would be no carbon cycle. Then, we would not have that much oxygen. The cow dies and the decomposers come and the carbon dioxide releases into the air. Then the plants can use the carbon dioxide to produce oxygen, so that keeps everything going. Like fungi will grow and eat on something that dies in the woods. An animal when it dies, it decays and it eventually turns into soil and the plants use it grow.

It was exciting to note these gains in student understanding of the complexity of ecological issues and content related to sustainability and interdependency issues.

Surprisingly, although we expected to see these learning gains evidenced in the final student projects, the culminating projects did not give any indication whether or not students learned content. The students chose often unrelated topics which contained little demonstration of what they learned. Students were given the freedom to select the mode with which they would demonstrate their learning from the unit. The end of session 5 (HHW) included a short brainstorm of possible projects students could complete in order to summarize their understandings. Students then selected their project, and Ms. Simon agreed to spend some time within the next two weeks of class on the development of those projects. Ms. Simon later explained that she only allowed students one hour to work on their projects, and did not give them a great deal of guidance. The culminating projects would have likely been more developed if we had introduced the idea to the students at the beginning of the unit and scaffolded the creation of the project throughout the unit. Although the students were able to demonstrate creativity, their projects contained limited information related to the unit, however if they had selected a project to work on early in the unit and worked on the project throughout the twelve-weeks, they may have produced more developed projects. Furthermore, students would likely have benefited from advanced discussion of how the projects were going to be assessed to ensure expectations were clear.

CURRENT CHALLENGES FACING THE ORGANIZATION

Ms. Elisa and Ms. Simon seemed to feel that there were some obstacles to a continual implementation of a project-based approach. Similar to other studies that investigated informal science learning (Kisiel, 2010; Kim & Fortner, 2006), both educators felt that the PBL approach took a great deal of time. Ms. Elisa stated that it was worth the time because she felt there was greater impact and visible parent involvement due to the final project. She stated that sustainability education at the elementary level is an initiation to become a lifelong learner about the environment.

I see sustainability education as a step towards continual learning of the importance of the natural environment, our connections to it, and our dependence and interdependent relationship with it. It should be an understanding that grows as we age, not something that fades away as it seems to in our education system. A variety of national polls from 2000 on (NEETF/Roper poll, 2000) indicate that a majority of the public still does not know the leading causes of such issues as water pollution, air pollution, and solid waste yet many (one 2005 national poll indicating 50%) of Americans consider themselves to be an environmentalist. The No Child Left Inside (NCLI) international movement begun by author Richard Louv states that students today can tell you more about the Amazon rain forest than the natural environment they live in and how to protect it! Disconnection to nature is linked to many illnesses in society including the impact of NOT reducing, reusing, and recycling. By the time children reach high school, they know what "pollution" is but not common pollution prevention practices.

Ms. Elisa also noted that PBL learning seemed to intensify and sometimes create a relationship with sustainability issues directly that students and teachers may not have previously had. Project-based learning can engage students and stimulate a sense of wonder that is vital to earth care. This then can perhaps provide a stronger basis for an ongoing relationship with environment. Ms. Elisa noted that once she leaves the classroom, she hopes students will continue the conversation amongst or within themselves, regardless of whether or not the teacher discusses these issues in the classroom.

However, Ms. Simon stated that she was unsure if the time spent on PBL was worth the outcome. Although she worked for a private school, she still felt pressure to teach state content standards, and she noted that the time spent on the PBL could have been more structured and efficient.

But I think that there does need to be a little more direction, a little more accountability. And more support, and in a way the evaluation can be a support for the students because at the end of the six sessions we asked them to do a project. Maybe we didn't frame the projects as content based, we did tell them that it was their chance to show what they learned, but we didn't stress that.

Ms. Simon noted that the culminating project should have been emphasized throughout the unit, and that students needed more support and resources in order to showcase their learning. Ms. Elisa also felt that more support from the classroom teacher (e.g., reinforcing concepts throughout the week) would have helped students see the unit as an integral part of their school learning. In addition to these organizational issues, Ms. Elisa mentioned that money was also an obstacle.

The obstacle that I really see is money. For instance with the [Household Hazardous Waste (HHW)] session, we had to have the students purchase the [spray] bottles. Some teachers won't want to do that because it involves time in contacting parents and collecting supplies and it does cost money, although a minimal amount. I couldn't do the HHW session that way for the other classes because [the waste management district] promotes using reusable supplies and props in education programs. I can ask the teacher to provide the supplies through school funds but again, it goes back to not really what I want to do, but what the teacher wants to do.

While time and money are common barriers to implementing project-based instruction, it is interesting to note that although students learning gains were significant, they did not develop strong culminating projects that showcased how much they actually learned. We were disappointed in the lack of content evident in the culminating projects, as students did not seem to focus on what they had learned, but more on a specific area of interest (e.g., one student created a poster all about frogs, which had not been covered in Ms. Elisa's programs). In fact, the local newspaper and parents had been invited to observe the final projects, yet we felt that that the impact of the program was not well-communicated through the projects. We believe that this issue can be addressed by more collaboration between the informal educator and the classroom teacher, as discussed in the next section. Collaboration did not occur in this project because Ms. Simon was very busy and it was difficult to schedule a time when we could all meet. We now realize that this was detrimental to the programs.

The two researchers noted that during the first two sessions, Ms. Simon was disengaged from

Ms. Elisa's instruction. She spent the one-hour period grading papers, working on her computer, and even left the classroom a couple of times. By the third session, Ms. Simon began to pay attention and support Ms. Elisa with classroom management. When asked why she became more engaged, Ms. Simon responded that she knew she had to prepare the students for the culminating projects, and therefore needed to know what they were learning. The two researchers were pleased to realize that the PBL format may have contributed to Ms. Simon's engagement and accountability with regard to the programs. We realize that a discussion about engagement and accountability would have been beneficial before the programs began, and that perhaps if Ms. Simon had been given a more-defined role, she would have been engaged from the start.

SOLUTIONS AND RECOMMENDATIONS

The obstacles of money and time to the implementation of a PBL format to informal sustainability education instruction identified by both educators are not surprising. Kim and Fortner (2006) found that content knowledge, time, standards, teachers' attitude, pedagogical knowledge were all barriers to environmental education among science teachers. However, we believe that the obstacles of money and time can be overcome. Money concerns could be addressed by applying for small grants, planning inquiries that use non-consumables, or asking students or teachers to pay a very minimal fee (recall that Ms. Elisa's programs are currently offered without a fee). Our PBL unit required some consumable materials during the HHW session (as seen in Table 1), however they were relatively cheap to acquire (i.e., water bottles, borax, vinegar). Ms. Elisa noted that since the completion of the current study, she has successfully made these adaptations. She noted,

An example of a compromise though in a session like the HHW session is to request that the teacher purchase one spray bottle and vinegar for their individual classroom or they share the vinegar among the classes. This spray bottle filled with vinegar and water is used in the classroom for regular cleaning needs and serves as a demonstration of PBL learning and responsible earth care. Students are responsible for using it and maintaining the mixture. The District may be able to fund this supply if the teacher doesn't want to because the cost per classroom is minimal. This compromise has worked with teachers since the PBL study.

Time constraints could be overcome if content was explicitly addressed through structured inquiry and more deliberate integration of cross-curricular content. In her post-interview, Ms. Simon suggested:

I think I may have demanded a little more out of the students for their final projects. Maybe, instead giving them so much freedom and open-endedness...support them in the presentations, reinforce what they were doing.

Ms. Simon recognized the need to emphasize the standards more directly throughout the unit, as well as the importance of providing a purpose to the unit from its initiation. Therefore, it is imperative to explicitly discuss the premise of project-based instruction throughout the unit so that students value the learning taking place, have ownership of their learning, and are prepared to demonstrate their learning through their culminating projects.

Both educators felt that more structured collaboration would have increased the benefits of the unit as well as provided students with the tools they needed to showcase their learning. It is therefore recommended that classroom teachers seek out informal educators with whom they can communicate regularly and build a relationship. Waste management districts, soil & water conservation departments, naturalists in parks and recreation departments and DNR naturalists, university sources, land trust agencies, and science museums educators are all possible resources. The challenge is to find informal educators who can present in an interactive, age appropriate, and academic manner. Ideally, informal educators provide expertise and resources to enhance instruction, and are able to be flexible to meet the pacing and objectives of the classroom. Furthermore, informal educators should identify classroom teachers that would be willing to take the time to plan and collaborate in advance, as well as participate and support instruction while working with the students. We also suggest that the informal educator and the classroom teacher meet in advance of the PBL unit to discuss shared goals and objectives, resources needed to implement the unit, and roles one another should take in order to support the students throughout the unit. The classroom teacher could be reinforcing content throughout the week when the informal educator is not with them, and can encourage student thought about the culminating project. Finally, the educators could meet throughout the unit to ensure that goals and objectives are being met, as well as meet once the unit has been completed to debrief the process and make suggestions for future collaborations.

FUTURE RESEARCH DIRECTIONS

Further research should examine the effect an informal educator's instruction using a PBL format versus a more traditional format of distinct individualized sessions. We did not collect control group data on the learning that occurs from Ms. Elisa's traditional sessions and therefore could not compare it to the six-session PBL format. We therefore cannot determine differential content knowledge gain due to the PBL format; however we did find that students learned about sustainability issues as a result of participating in the PBL program.

Furthermore, future research could explore the impact of explicit collaboration that emphasizes direct attention to the alignment of goals, establishment of a community of practice (Wenger, 1998), and clear expectations and duties. Both educators must be made aware of the commitment involved in an explicit collaboration, including materials needed, expected levels of communication, individual roles, and time needed for effective outcomes. This kind of collaboration could promote the infusion of the PBL model from the beginning of the unit if both educators worked to support students' conceptual understanding and their ideas for a culminating project throughout the duration of the program. We posit that more purposeful collaborative infusion of the PBL model (e.g., informal educator discusses a topic while the classroom teacher follows up after the session to brainstorm project ideas) would promote more content-rich projects.

CONCLUSION

The results of this study suggest that a PBL framework is effective for teaching sustainability concepts. A paired samples comparison of student pre and posttests indicated that the fourth graders learned science content knowledge throughout the six- lesson program. Both in terms of gains in content understanding and grasp of complexity of sustainability topics, students consistently exhibited positive trends pre to post PBL as evidenced in their focus group interviews. We believe that learning occurred within Ms. Elisa's programs due to the continuity, frequency, and intensity of instruction. Moreover, the format of PBL offers a cohesive structure to otherwise disconnected lessons–we planned the unit to transition learners from a macro level (community) to micro level (individual) through the use of a PBL format. The unit incorporated the four key features of PBL: students learning about realistic problems, student control over their learning, teachers serving as coaches and facilitators of learning, and students working in pairs or groups. Programs were planned to connect to one another and students were receiving instruction every two weeks for twelve weeks, in contrast to six instructional periods throughout the entire school year. Students were engaged and excited throughout the PBL unit, and were very enthusiastic to share their projects with their parents and the local media, however their culminating projects lacked the showcasing of the knowledge they had gained. While several constraints to implementation exist, we believe that these obstacles can be overcome through strategic collaboration among educators, and that students and educators alike can enjoy the multitude of benefits of the PBL format in teachings students about sustainability.

REFERENCES

Baker, J. (1991). *Window*. London, UK: Walker Books, Ltd.

Ballantyne, R., & Packer, J. (1996). Teaching and learning in environmental education: Developing environmental conceptions. *The Journal of Environmental Education*, *27*(2), 25–33. doi:10.1080/00958964.1996.9941455

Barak, M., & Dori, Y. (2005). Enhancing undergraduate students' chemistry understanding through project-based learning in an IT environment. *Science Education*, *89*(1), 117–139. doi:10.1002/sce.20027

Bernard, H. R. (2002). *Research methods in anthropology: Qualitative and quantitative approaches* (5th ed.). Walnut Creek, CA: Alta Mira Press.

Blumenfeld, P., Soloway, E., Marx, R., Krajcik, J., Guzdial, M., & Palincsar, A. (1991). Motivating project-based learning: Sustaining the doing, supporting the learning. *Educational Psychologist*, *26*(3&4), 369–398. doi:10.1080/00461520.1991.9653139

Culen, G. R. (2005). The status of environmental education with respect to the goal of responsible citizenship behavior. In H. Hungerford, W. Bluhm, T. Volk, & J. Ramsey (Eds.), *Essential readings in environmental education* (pp. 37–45). Champaign, IL: Stipes Publishing L.L.C.

Edutopia. (2013). *Project-based learning research review*. Retrieved August 18, 2013 from http://www.edutopia.org/pbl-research-learning-outcomes#outcomes

Emerson, R. M., Fretz, R. I., & Shaw, L. L. (1995). *Writing ethnographic fieldnotes*. Chicago: University of Chicago Press. doi:10.7208/chicago/9780226206851.001.0001

Gallagher, S. A., Stepien, W. J., & Rosenthal, H. (1992). The effects of problem-based learning on problem solving. *Gifted Child Quarterly*, *36*, 195–200. doi:10.1177/001698629203600405

Geier, R., Blumenfeld, P. C., Marx, R. W., Krajcik, J. S., Fishman, B., Soloway, E., & Clay-Chambers, J. (2008). Standardized test outcomes for students engaged in inquiry-based science curricula in the context of urban reform. *Journal of Research in Science Teaching*, *45*(8), 922–939. doi:10.1002/tea.20248

Glaser, B. G., & Strauss, A. L. (1967). *The discovery of grounded theory: Strategies for qualitative research*. Chicago: Aldine.

Harley, B. (2001a). *Milky way. On Down in the Backpack* [CD]. Santa Monica, CA: A&M Records.

Harley, B. (2001b). *Who made this mess. On Big big world* [CD]. Santa Monica, CA: A&M Records.

Kim, C., & Fortner, R. (2006). Issue-specific barriers to addressing environmental issues in the classroom: An exploratory study. *The Journal of Environmental Education*, *37*(3), 15–22. doi:10.3200/JOEE.37.3.15-22

Kisiel, J. (2010). Exploring a school-aquarium collaboration: An intersection of communities of practice. *Science Education*, *94*, 95–121.

McKeown-Ice, R. (2000). Environmental education in the United States: A survey of preservice teacher education programs. *The Journal of Environmental Education*, *32*(1), 4–11. doi:10.1080/00958960009598666

National Research Council (NRC). (2013). *The next generation science standards*. Washington, DC: The National Academies Press. Retrieved June 1, 2013 from http://www.nextgenscience.org/

New Games Foundation. (1976). *Knots*. Jackson, TN: Main Street Books.

Perenyi, C. (1991). *Growing wild: Inviting wildlife into your yard*. Hillsboro, OR: Beyond Words Publishing Company.

Project, W. E. T. (2011). The life box. In *Project WET curriculum and activity guide* (p. 69). Bozeman, MT: Project WET International Foundation.

Project Learning Tree. (2000). Forest consequences. In *PreK-8 Environmental education activity guide* (p. 138). Washington, DC: American Forest Foundation.

Ramsey, J. (1993). The science education reform movement: Implications for social responsibility. *Science Education*, *77*(2), 235–258. doi:10.1002/sce.3730770210

Report, B. (1997). *Report of the World Commission on Environment and Development: Our common future*. Retrieved June 7, 2013 from http://www.un-documents.net/our-common-future.pdf

Rivet, A., & Krajcik, J. (2008). Contextualizing instruction: Leveraging students' prior knowledge and experiences to foster understanding of middle school science. *Journal of Research in Science Teaching*, *45*(1), 79–100. doi:10.1002/tea.20203

Robson, C. (2002). *Real world research* (2nd ed.). Oxford, UK: Blackwell Publishing.

Schneider, R., Krajcik, J., Marx, R., & Solloway, E. (2002). Performance of students in project-based science classrooms on a national measure of science achievement. *Journal of Research in Science Teaching*, *39*(5), 410–422. doi:10.1002/tea.10029

Sobel, D. (2004). *Place-based education*. Barrington, MA: The Orion Society.

Strauss, K. (2006). The tinker's clock. In *Tales with tails: Storytelling the wonders of the natural world* (pp. 193–194). Santa Barbara, CA: Libraries Unlimited Publishing.

Strobel, J., & van Barneveld, A. (2008). When is PBL more effective? A meta-synthesis of meta-analyses comparing PBL to conventional classrooms. *Interdisciplinary Journal of Problem-based Learning*, *3*(1), 44–58.

Tal, T. (2004). Community-based environmental education: A case study of parent-teacher collaboration. *Environmental Education Research*, *10*(4), 523–543. doi:10.1080/1350462042000291047

The Buck Institute. (2013). *Project-based learning for the 21st century: What is PBL?* Retrieved August 15, 2013 from http://www.bie.org

Thomas, K., & Muga, H. (2012). Sustainability: The new 21st century general education requirement for engineers. In M. Rasul (Ed.), *Developments in engineering education standards: Advanced curriculum innovations* (pp. 263–285). Hershey, PA: IGI Global. doi:10.4018/978-1-4666-0951-8.ch015

Trends in International Mathematics and Science Study (TIMMS). (n. d.). Retrieved March 14, 2009 from http://nces.ed.gov/TIMSS/pdf/TIMSS4_Science_Items.pdf

Walker, A., & Leary, H. (2008). A problem based learning meta analysis: Differences across problem types, implementation types, disciplines, and assessment levels. *Interdisciplinary Journal of Problem-Based Learning*, *3*(1), 12–43.

Wenger, E. (1998). *Communities of practice: Learning, meaning and identity*. New York: Cambridge University Press. doi:10.1017/CBO9780511803932

This work was previously published in a "Handbook of Research on Pedagogical Innovations for Sustainable Development" edited by Ken D. Thomas and Helen E. Muga, pages 190-209, copyright 2014 by Information Science Reference (an imprint of IGI Global).

APPENDIX A: CONTENT ASSESSMENT

Name:__________________
Date:__________________

1. Which one of these refers only to living things?
   a. clouds, fire, rivers.
   b. fire, rivers, trees.
   c. rivers, birds, trees.
   d. birds, trees, worms.
   e. trees, worms, clouds.
2. Some things were buried in wet ground. Several years later they were dug up. Which thing is MOST likely to have stayed the same?
   a. an egg shell.
   b. a plastic cup.
   c. a paper plate.
   d. an orange peel.
3. Here is a picture of the inside of a box. If you put a worm on the X on the bottom of the box, to which corner would you expect it to go?

Light Side	**Dark Side**
Damp	Damp
Dry	Dry

   a. damp and light.
   b. dry and light.
   c. damp and dark.
   d. dry and dark.
4. Write down one reason why animals could NOT live in a world without plants.
5. If you throw each one of these things away, which will decay fastest?
   a. a glass bottle.
   b. a metal can.
   c. a plastic bottle.
   d. an apple core.
6. Minerals are used for making many things including jewelry, chalk, and concrete. Where do we get the minerals needed to make these things?
   a. from air.
   b. from wood.
   c. from rocks.
   d. from crops.
7. In a pond ecosystem, a heron, a fish, a tadpole, and a duckweed plant live together. The tadpole most likely gets its food from which of the following?
   a. sun.
   b. fish.
   c. duckweed plant.
   d. heron.

8. What covers most of the Earth's surface?
   a. water.
   b. bare rock.
   c. farm land.
   d. cities and towns.
9. What gas in the air do we need to breathe in order to live?
   a. nitrogen.
   b. oxygen.
   c. carbon dioxide.
   d. hydrogen.
   e. water vapor.
10. Are plants living or nonliving? (Check one box.)
    ◦ Living
    ◦ Nonliving
    Explain your answer.
11. Of the following, where is the best location to grow crops?
    a. Ocean shores.
    b. Field near a river.
    c. Mountains.
    d. Desert.
12. Write down two different things that people can do to help reduce air pollution.
13. Write as completely as possible why large oil spills in rivers and seas are harmful to the environment.
14. One way for animals to protect themselves is by escaping (running, flying, or swimming away). What are two other ways they protect themselves?

APPENDIX B: STUDENT FOCUS GROUP INTERVIEW QUESTIONS

1. Scientific investigations are seldom exactly the same. If differences occur, give reasons for why these differences exist.
2. How do measuring instruments, such as microscopes, telescopes, and cameras, gather accurate information for making scientific comparisons?
3. Describe why some materials are easier to recycle than others.
4. What might happen if part of a food chain is missing and why?
5. Describe a food Web. How do species depend on each other?
6. In what ways do some organisms depend on dead and decaying organisms?
7. Identify better reasons for believing something than "Everybody knows that..." or "I just know," and discount such reasons when given by others.

Chapter 43
Integration of the Computer Games into Early Childhood Education Pre-Service Teachers' Mathematics Teaching

Hatice Sancar Tokmak
Mersin University, Turkey

Lutfi Incikabi
Kastamonu University, Turkey

ABSTRACT

Mathematics experiences in the early years of education can help children be more prepared for the future. Mathematics is known to be one of the most difficult learning areas. Computer games in today's technological world seem to offer a way to educate young children in mathematics in a fun and engaging way. However, success does not rely solely on the technology itself, but also how it is used during instruction. For that reason, early childhood teachers have a responsibility to integrate technology into their instruction, and to understand such technologies with regard to content and audience. Therefore, this chapter provides an example of how to incorporate educational computer games into upper-level education courses. The instructor designed the course described in this chapter to teach early childhood teachers how to integrate educational computer games for the purpose of teaching mathematics to young children.

INTRODUCTION

Educating young children has been one of the main concerns for societies in recent years because the early years of education can help members of societies be more prepared for the future. The importance of early childhood education gives birth to the debates on what an early childhood education curriculum should include, and which methods should be used to meet curriculum goals. Many education experts advocate that one of the fields to be taught from early years is mathematics. According to Smith (2006), mathematics should be taught in early years to prepare children for the future. Moreover, Fuson (2004) states that learn-

DOI: 10.4018/978-1-4666-7363-2.ch043

ing mathematics in early years contributes to later learning and success in mathematics.

One of the main critiques of early childhood education literature also provides a clear solution on which methods should be used to meet curriculum goals. This critique is that young children's academic achievements are frequently emphasized in early childhood education. Beauchat, Blamey and Walpole (2010) criticized this emphasis by saying that it causes a reduction in the amount of playtime that a young child can engage in. Ginsburg (2007) advocates that play has a potential for providing both children's learning and enjoyment, saying that "[a]s we strive to create the optimal developmental milieu for children, it remains imperative that play be included along with academic- and social-enrichment opportunities and that safe environments be made available to all children" (p. 188). In other words, children can learn many things during the playing, and even if the aim is the children's enjoyment, play can also be used as a method for enriching education (Pramling Samuelsson & Johansson, 2006).

Making children search, find and construct their own answers, and teaching them to use technologies to effectively do these activities, is important. According to Pramling Samuelsson and Johansson (2006), creativity can be fostered if young children are given an opportunity to reflect their ideas freely, and play is a way to provide this opportunity. Pramling Samuelsson and Johansson (2006) state that dimensions such as joy, creativity, creation of meaning and children's possibilities to control and form goals are important both for play and learning and for that reason, difference of these two concepts are less well defined.

The advantages of using play for educational purposes are counted by Johnson, Cristie, and Yawkey (1999) as the following:

- **Positive Effect:** Play is fun and enjoyable,
- **Nonliterality:** Play helps children learn about the meanings of objects and actions during play,
- **Means Over Ends Orientation:** Activity gains importance, rather than goals or outcomes,
- **Learning Different Skills:** Different skills, such as writing, speaking, recognition and comprehension, can be learned during play.
- **A Broad Spectrum of Learning Opportunities:** Play can provide many learning opportunities, such as observation, exploration, experimentation, peer collaboration, teacher scaffolding, practice and teachers' instructional links.

Moreover, Pound (2008) believes that play is also an effective and fun way to teach mathematics to young children. There are two approaches advised in the literature on how play should be used (Johnson et al., 1999). One of these approaches is play-generated curriculum, and the other is curriculum-generated play. In the first one, children are observed during play and curriculum content is then extracted (Johnson et al., 1999). Sandberg and Pramling Samuelsson (2003) give the name "free play" name to this approach. In the second one, play is devised according to curriculum content, which is then taught through play (Johnson et al., 1999).

Computer games with growing audiences (Brand, Knight, & Majewski, 2003), are among the popular indoor play activities of young children (R. Clements, 2004; Cherney & London, 2006; Dwyer, 2007) and take the place of outdoor play time (e.g. R. Clements, 2004; Cherney & London, 2006; Dwyer, 2007; Li & Atkins, 2004). In today's technological world, computer games seem to provide a promise of educating children while providing them with fun times (Paraskeva, Mysirlaki, & Papagianni, 2010), and simultaneously giving young children technological skills (Sandberg & Pramling Samuelsson, 2003). However, the computer games only make this promise real if they are aligned with curriculum goals, designed according to their audience, and used effectively by instructors.

The characteristics of computers games as "rules," "goals," "objectives," "outcome and feedback," "conflict," "competition, challenge, and opposition" and "interaction" contributed to their popularity (Prensky, 2001). According to Hogle (1996), the games including these characteristics are activities in which the participants must use physical and mental skills, and follow specific rules, in order to reach a goal.

As Niederhauser and Stoddart (2001) advocated: "Computer technology, in and of itself, does not embody a single pedagogical orientation" (p. 15). For that reason, teachers should be careful when selecting educational software (Yanpar-Yelken, 2011). There are many checklists, evaluation forms, and surveys that can be used to help teachers select appropriate software for instructional purposes (e.g. Cennamo, Ross, & Ertmer, 2010; De Villiers, 2004; Heinich, Molenda, Russell, & Smaldino, 2002; Niederhauser & Stoddart, 2001; Squires & Preece, 1996). But a teacher should know how to use these tools to select the best software for their curriculum.

As stated before, knowing how to use these tools is important for effective instruction. Also, as Dembo and Gibson (1985) state, teacher performance affects students' learning during instruction. For that reason, there is a consensus about developing early childhood instructor competencies for selecting and integrating computer games into instruction effectively. Smith (2006) suggests that education instructors should provide early childhood pre-service teachers with information about the necessary pedagogy and content for an early childhood curriculum as emphasized by prior research. Mishra and Kohler (2006) added technology to content and pedagogy, and developed the TPACK (technology-pedagogy and content knowledge) framework. Shin et al. (2009) describes how TPACK "connects technology to curriculum content and specific pedagogical approaches and describes how teachers' understandings of these three knowledge bases can interact with one another to produce effective discipline-based teaching with educational technologies" (p. 4152). According to Özgün-Koca, Meagher, and Edwards (2010), developing teachers' awareness of the intersection of technology, pedagogy, and content is necessary to provide effective teaching.

Developing an understanding of pedagogy, content and technology is important, teachers' self-efficacy on teaching should be strength also (Dembo & Gibson, 1985) since teachers' performance affect the students' academic success (Finney & Schraw 2003; Pajares, 1996; Riggs & Enochs, 1990;). Likewise, a teacher's self-efficacy affects his or her performance when it comes to teaching (Dembo &Gibson, 1985). The results of the study conducted by Gibson and Dembo (1984) showed that student achievement was influenced by low or high efficacy in their teachers.

Bandura (1997) first described perceived self-efficacy in the Social Learning Theory, and described it as "beliefs in one's capabilities to organize and execute the courses of action required to produce given attainments" (p. 3). Gibson and Dembo (1984) assert that a teacher's self-efficacy consists of two components, "outcomes expectancy belief" and "self-efficacy beliefs." Moreover, they found that teachers who have these two components "....persist longer, provide a greater academic focus in the classroom, and exhibit different types of feedback than teachers who have lower expectations concerning their ability to influence student learning" (Gibson & Dembo, 1984, p. 570). Similarly, a study by Zengin (2003) showed that teachers who had high self-efficacy developed better lesson plans and focused more on student success and development, when compared to teachers who have low self-efficacy.

Moreover, Bandura (1997) states that an individual's self-efficacy is affected by the situation (or context) and the results of the actions to be performed. Hoy and Spero (2005) conducted a longitudinal study about changes in pre-service teachers' self-efficacy beliefs and found that pre-service teachers' self-efficacy increased significantly during student teaching, but decreased significantly during the first year of teaching. The results of the study into how pre-service teachers' teaching self-efficacy can be enhanced showed that the key is teaching experiences (e.g. Plourde, 2002). In addition to teaching experiences, Bursal and Paznokas (2006) found that anxiety also negatively affected a teacher's self-efficacy. According to their study results, there was a negative relationship between mathematics anxiety and confidence scores with regard to teaching elementary mathematics ($r = -.638$). Isler and Cakiroglu (2009) investigated different teachers' self-efficacy who were teaching according to new curriculum guidelines, and found that the primary teachers had significantly higher self-efficacy beliefs about the new curriculum than mathematics teachers.

Ertmer, Conklin, and Lewandowski, (2001) state that education teachers should provide opportunities for pre-service teachers that help them to increase their competence and confidence in technology integration. Bhattacherjee and Premkumar (2004) suggest that first-hand experiences may affect the way that pre-service teachers use information technology. Similarly, Ertmer et al. (2001) says that a teacher's confidence in the integration of technology is related to knowing the technology well, and for that reason, teaching them how to use the technology is important. Mishra and Koehler (2006) advocate using their framework, TPACK, to affect the confidence of pre-service or teachers with regard to the integration of technology. Moreover, pre-service teachers should understand these three main aspects (content, pedagogy, and technology) well enough to incorporate all three.

The instruction described in this chapter is consistent with the above literature in that it attempts to enhance early childhood pre-service teachers' knowledge about computer games (technology), pedagogy and content. Curriculum-generated play instruction (pedagogy) is the basis of the course design. Moreover, with teaching experience, the pre-service teachers are given a chance to practice theoretical knowledge gained during the course, and so their teaching efficacy can be enhanced.

MATHEMATICS EDUCATION IN EARLY CHILDHOOD

Much attention is paid to teaching mathematics in early childhood education courses. Smith (2006) asserts that mathematics should be taught before children are two to three years old. This is because mathematics prepares children in their early years for their futures (Smith, 2006). Similarly, Fuson (2004) and Pound (2008) state that according to research, exposing children to mathematical thinking in early years contributes to their ability to calculate, communicate and make sense of a world related to science and technology. Moreover, Fuson (2004) advocates that teaching math in a child's early years also forms a base for greater mathematics ability and keeps him or her from falling behind in mathematics.

Teaching mathematics is advocated by the literature, but the methods and subject matter are debated (e.g. Fuson, 2004; Pound, 2008; Smith, 2006). The curriculum mathematics program and the methods suggested were shown to be crucial for providing mathematics experiences to young children. The National Association for the Education of Young Children [NAEYC] (2008) isolated numbers, basic operations, geometry, measurement, patterns and algebraic thinking, and collected, analyzed and presented data for young children under the scope of mathematics teaching. Ball (1994) points out the importance of early childhood curriculum by stating that high quality early childhood education leads to young children have high-quality provision, and one of the major prerequisites for high-quality provision is an appropriate early learning curriculum.

The literature suggests that a teacher will understand the content and pedagogy in the curriculum best by emphasizing the importance of knowledge about content and pedagogy (e.g. Borko & Putnam, 1995; Shulman, 1986). Borko and Putnam (1995) state that to teach mathematics, content and pedagogy are not enough: .".. they must acquire richer knowledge of subject matter, pedagogy, and subject-specific pedagogy; and they must come to hold new beliefs in these domains" (p. 60). Similarly, Shulman (1986) proposes a framework, Pedagogical Content Knowledge (PCK) which is an understanding of both pedagogy and content. Moreover, one of the most popular topics in the current literature about pre-service teachers' education, TPACK, is an extension of Shulman's PCK. TPACK also has technology component (Mishra & Koehler, 2006; Angeli & Valanides, 2008). Niess et al. (2009) describes TPACK as "a body of knowledge teachers needed for teaching with and about technology in their assigned subject areas and grade levels" (p. 7).

As seen from the above literature, an early childhood teacher should be very familiar with the mathematics program (framework) that includes mathematics content and teaching strategies. The early childhood curriculum in Turkey includes mathematical experiences for young children between 32 and 72 months old (Milli Egitim Bakanligi [MEB], 2006). The curriculum in Turkey includes five main parts with respect to developmental areas:

1. Psychomotor development;
2. Social-Emotional development;
3. Language development;
4. Cognitive development; and
5. Self-care abilities (MEB, 2006).

The mathematical thinking experience is placed under the cognitive development, and includes such topics as numbers, basic operations and algebraic thinking, measurement, and geometry (MEB, 2006). The described behavioral objectives in the curriculum related with each topic are as follows:

1. **Numbers:**
   a. Children will count to 20.
   b. Children will count backwards from 10.
   c. Children will show how many objects have been named.
   d. Children will count correctly how many objects have been shown.
   e. Children will group objects from least to most by counting their quantity.
   f. Children will say the number of objects in a group with less than 10 in number.
   g. Children will read the numbers from 0 to 10.
   h. The children will write the numbers in a given set of objects at most 10.
2. **Basic operations and algebraic thinking:**
   a. Children will add a specified number of objects to a group of objects.
   b. Children will extract a specified number of objects from a group of objects.
   c. Children will add using objects.
   d. Children will subtract using objects.
   e. Children will solve problems which require them to add, not exceeding 10.
   f. Children will solve problems which require them to subtract, not exceeding 5.
3. **Measurement:**
   a. Children will guess the results of measurement.
   b. Children will measure with non-standard units (e.g. pen, eraser, book, their hands, and their footsteps).
   c. Children will compare the results which they guess with the actual measurements.
4. **Geometry:**
   a. Children will name the shape of each object.
   b. Children will show objects which are similar to circle, triangle, square, and rectangle.

c. Children will create different models by using circle, triangle, square, and rectangle (e.g. s/he will create a square by combining two triangles).

Incikabi and Tuna (in press) analyzed the Turkish Early Childhood Education Curriculum in terms of mathematics experiences provided within the cognitive domain in their study, and concluded that the Turkish curriculum, being different from the New York State Learning Standards, is based on constructivisim and multiple intelligence theory. Moreover, they found that early childhood curriculum in Turkey integrate playing and learning (Incikabi & Tuna, in press).

Although the literature emphasizes that technology (especially computer games) can contribute to effective mathematics learning for young children (Sedighian & Sedighian, 1996; Klawe, 1999), the curriculum in Turkey, even the primary school curriculum, does not include activities with technology (Incikabi, 2011), even though using technology in early childhood education is suggested (MEB, 2006). According to the curriculum and literature, technology can be integrated into early childhood education if pre-service teachers are trained to use the technology.

The instructional design described in this chapter aims to help pre-service teachers learn the behavioral objectives of the mathematics program and teaching mathematics with regard to technology and pedagogy. In other words, as in the curriculum, the activities provide experiences which include three areas of knowledge, namely play, technology, and content.

EDUCATIONAL COMPUTER GAMES: WHY AND HOW TO USE THEM

Computers as an educational tool for young children have been the center of many debates (Plowman, McPake, & Stephen, 2010). However, the debate over whether technology can be used to educate young children turns into how early childhood education teachers can best integrate technology into their instruction (Verenikina, Harris, & Lysaght, 2003; Plowman et al., 2010). Also, increasing their ability to select such technologies is one concern of the educators because the current generation has been born into a world dominated by technology (Verenikina et al., 2003 Plowman et al., 2010).

Rideout, Vandewater, and Wartella (2003) state that the increase in Information Communication Technology (ICT) use at home causes young children to come to school with computer skills. Similarly, D. H. Clements (2002) states that "research has substantiated that computers can help young children learn mathematics" (p.162). Mioduser, Tur-Kaspa, and Leitner (2000) studied the contribution of computer-based instruction to early reading skills acquisition. They compared it with the contribution of teacher instruction with textbooks on the same (based on three dependent variables, like children's phonological awareness, word recognition and letter recognition) of forty-six pre-school children (aged 5 - 6), who were at high risk for learning disabilities (Mioduser et al., 2000). They found that the children who were given computer-based instruction significantly improved in phonological awareness, word recognition, and letter naming skills compared with their peers who received instruction with textbooks (Mioduser et al., 2000).

Computer games are one of the most popular indoor activities of young children (R. Clements, 2004; Cherney & London, 2006; Dwyer, 2007) although their social consequences are debated (Amory, Naicker, Vincent, & Adams, 1999). Li and Atkins' (2004) study results showed that "[m]ost of the children played on a computer at least once a week if not more often. Of children who did not have a computer at home, more than half had access to a computer somewhere else" (p.1721). Li and Atkins (2004) also state that most children used the computer mostly to play games, followed by other activities such as educational software, creativity software, and other activities. Similarly, R. Clements (2004) found in her survey study, conducted online and using a total of 830 mothers, showed that 81% of children played computer games.

It is not new to use games as an educational tool. Incidents of using games as educational tools can be traced back to the 18th Century (Bopp, 2008) and now computer games are popular among children, adolescents, and even among a growing number of adults (Bopp, 2008). The popularity of computer games has generated an important body of research focused on their use as an educational tool. Most of the research supports using computer games for educational purposes (Paraskeva, Mysirlaki, & Papagianni, 2010). The reason may be that computer games are very close to the nature of early childhood education, in that learning with playfulness is important. Beauchat et al. (2010) advocate joyful playtime while teaching a young child. Similarly, Ginsburg (2007) states that play can provide both learning and enjoyment for children. According to Saracho and Spodek (2002), "In true instructional games, the concepts to be learned are intrinsic to the structure and content of game" (p. 177). Today, a technology-dominated world causes computer games to take the place of outdoor play time (e.g. R. Clements, 2004; Cherney & London, 2006; Dwyer, 2007; Li & Atkins, 2004). O'Neil, Wainess, and Baker (2005) state the benefits of computer games used for learning with the following words:

Computer games were hypothesized to be potentially useful for instructional purposes and were also hypothesized to provide multiple benefits: (a) complex and diverse approaches to learning processes and outcomes; (b) interactivity; (c) ability to address cognitive as well as affective learning issues; and, perhaps most importantly, (d) motivation for learning. (p. 455).

Kraus (1981) found in his study that second graders who play computer games one hour a day over a two week period answered correctly to twice as many questions on speed tests when compared to students who were in the control group. Sedighian and Sedighian (1996) conducted a three-step study which investigated the psychology of children as they were learning mathematics in the context of computer-based mathematical game environments. They found that computer games motivated children to learn mathematics (Sedighian & Sedighian, 1996). Similarly, Klawe (1999) states that research under the E-GEMS project examined the design and use of computer games and activities for mathematics education in grades 4 to 8, and found that games can increase both motivation and achievement in mathematics learning.

However, the literature advocates that the potential of the computer depends on some issues. Van Eck (2006) states the danger of thinking all games are effective for learning with these words: "If we continue to preach only that games can be effective, we run the risk of creating the impression that *all* games are good for *all* learners and for *all* learning outcomes, which is categorically *not* the case" (p. 17). Haugland and Wright (1997) emphasize that children's creativity can be stunted if the software is not developmentally appropriate (eg, drill-and-practice). Similarly, D. H. Clements (2002) states that computer games may be effective if they are chosen properly. Amory et al. (1999) advocates that the most suitable game types for teaching were the ones that students found interesting or useful. Moreover, according to Verenikina et al. (2003), the software should be designed to be more attractive and accessible to a young audience. Gunder (2003) says games can be effective if they can affect, form suspense, curiosity, surprise and keep a user's interest. Sedighian and Sedighian (1996) list eight elements that are vital for a game to have a synergistic role in children's learning, even in mathematics, which has been described as boring and difficult by many children: meaningful learning, goal, success, challenge, cognitive artifact, association through pleasure, attraction, and sensory stimuli.

Li and Atkins (2004) add adult supervision and software appropriateness as critical elements of the effectiveness of computers to early childhood learning. Moreover, they advocate that this is more important than the content of software: "However, previous research suggested that the value of the computer for children at this particularly young age group (eg, 3–5 years) is in its process (eg, open-ended use) rather than its content (eg, making a specific product)" (Davidson & Wright, 1994 as cited in Li and Atkins, 2004, p. 1721). Parallel to Li and Atkins (2004), Klawe (1999) points out the teacher's use and the integration of computer games are also of critical importance, as much as elements of game design. Moreover, according to Dorn (1989), a teacher's attitude towards a game and their knowledge and skill in using the game affects the outcome of instruction (as cited in Egenfeldt-Nielsen, 2004)

Sarama and Clements (2004) emphasize that education teachers should focus on how they can help early childhood teachers achieve the benefits of technology with regard to early childhood mathematics. But the question is how? Van Eck (2006) states that literature on the approaches educators adopt to integrate computer games to their instructions separates into three categories: "have students build games from scratch; have educators and/or developers build educational games from scratch to teach students; and integrate commercial off-the-shelf (COTS) games into the classroom" (p.21). Kafai (2006) delineates the approaches as Instructionist and Constructionist. She defines using COTS games to make learning fun as the Instructionist approach, and giving opportunities for students to construct their own computer games as the Constructionist approach (Kafai, 2006).

Having Learners Design Computer Games

In this approach, learners have the role of game designer, and build the games. For example, Sweedyk, deLaet, Slattery, and Kuffner (2005) share their experiences in designing a broad range of courses, including integration computers games offered to computer science (CS) students in higher education. The authors state that they wanted CS students to build games in groups. The results show that the students learned the programming language, solved game design issues, and worked in groups (Sweedyk, et al., 2005). Similarly, Leutenegger and Edgington (2005) designed a course comprising of three parts to teach programming by having CS students design computer games. In first two quarters of the course, they taught programs/programming, including Flash, ActionScript g C++ and openGL (2D graphics only); in the third quarter, the CS students designed computer games with their group members (Leutenegger & Edgington, 2005). They found that the approach used in the course improved the students' understanding of all seven basic topics (Leutenegger & Edgington, 2005). Similarly, Barbour, Rieber, Thomas, and Rauscher (2009) had children design educational computer games by using PowerPoint templates.

Designing Games or Integrating Commercial Off-the-Shelf (COTS) Games to Teach Learners

In the second approach, educators design the computer games for their instruction, and in the third approach, educators select an existing COTS game to use in the classroom. In this approach, the important issue is to be able to select the most suitable computer game for target students, content and teaching method as suggested by Sancar-Tokmak, Incikabi and Yanpar-Yelken (in press) for educational software.

Egenfeldt-Nielsen (2004) conducted an empirical study in which a historical strategy games was integrated into a history course in conjunction with traditional teaching and student group work. In her study, two teachers gave the students some basic information about history in conjunction with game (Egenfeldt-Nielsen, 2004). She concluded that students could explore the game universe freely and learn the content through the game (Egenfeldt-Nielsen, 2004).

Sancar-Tokmak (in review) also designed a course in which she wanted pre-service teachers' prepare lesson plans including integration of COST games to teach a topic selected from early childhood curriculum to young children. In her design, the pre-service teachers firstly selected a topic from early childhood curriculum; secondly they searched educational games and selected the best appropriate to their content and pedagogy; and lastly they prepared a lesson plan. The results showed that the pre-service teachers felt more confidence in integrating computer games into their instruction (Sancar-Tokmak, in review). Also, they stated their knowledge about computer games, teaching methods, lesson plan preparation, and most importantly how to integrate a technology while teaching a specific content were improved (Sancar-Tokmak, in review).

Presently, according to the literature, computer games have a potential for providing effective learning. But, the question is how we can extract their greater potential. There are different approaches presented in the literature to integrate computer games for learning. The course design described in this chapter is based on the integrating commercial off-the-shelf (COTS) games to teach learners approach, but in a constructivist way. Early childhood pre-service teachers create a curriculum that includes computer games, and teach according to their plan. During this course design, the pre-service teachers are guided, not instructed step by step, and *they* decide content, then select a game and plan how to use it during instruction.

THE COURSE DESIGN

The course design presented in this chapter is an example that explains how educational computer games can be integrated into a course offered to pre-service teachers. The course, Play in Early Childhood, is offered to early childhood pre-service teachers, and our goal is to teach them how to integrate educational games while teaching mathematics to young children. As you will see, the course goal is achieved by integrating educational games into the course design.

Curriculum-Generated Play Instruction

The instructional designers (also, the authors of this book chapter) have worked to find the best way to provide teaching experiences involving computer games to early childhood education pre-service teachers enrolled in a *Play in Early Childhood Education* course. The theoretical background of the course's design is based on curriculum-generated play, which consists of play experiences that enable children to learn concepts and skills from curriculum areas, such as literacy, mathematics and science (Johnson et al., 1999). There are two functions of curriculum-generated play: initial learning and practice/consolidation. In the first function, initial learning, play activities are arranged and then an assessment is undertaken to discover whether the children gained the target skills. In the second function, practice/consolidation, direct instruction is offered and, then, play activities are arranged in order to enable the children to practice the target skills (Johnson et al., 1999).

As you will see, the instruction is based on curriculum content, behavioral goals, and target audience. The teachers select the content, and then decide on the play (in this case, a computer game) and other activities to reach the behavioral goals of the curriculum. In the current course design, the instructors do not aim to teach a specific content by using computer games, but aim to show educators how to teach mathematics to young children by using computer games. In this design, the pre-service teachers are encouraged to fully understand the early childhood mathematics program. Then, they are encouraged to find different computer games to teach specific content.

The Goals of the Design

- **Content Knowledge:** The pre-service teachers are to understand the mathematics program in the early childhood curriculum. In the current course design, they will learn not only the topics but also the behavioral objectives, target audiences' characteristics, and activities advised in the curriculum.
- **Pedagogical Knowledge:** The pre-service teachers will learn how to use play to teach specific topics in the early childhood curriculum. In this course design, curriculum-generated play instruction is meant to be used as a base for teaching mathematics. However, instead of free-play, pre-service teachers are taught how to teach mathematics using computer game play with the curriculum-generated play instruction.
- **Technological Knowledge:** The pre-service teachers will learn about educational computer games. In this course design, the goal is to teach pre-service teachers the computer game properties, and how to select and use them for teaching purposes.
- **Incorporating Content, Pedagogical, and Technological Knowledge:** The pre-service teachers will teach specific content by using curriculum-generated play instruction with the computer games. In this course design, the pre-service teachers are prepared to use three tools to teach mathematics. Figure 1 shows the activities in TPACK framework.
- **Teachers' Self-Efficacy:** Enhancing the pre-service teachers' self-efficacy as they teach mathematics to young children is one of the aims in this course design. To do this, the pre-service teachers are expected to prepare lesson plans and follow those plans in class.

The Learning Philosophy behind the Course Design

The learning philosophy behind the described course design is social constructivism. All the activities are aimed at making pre-service teachers active and building their knowledge. The pre-service teachers are directed to search, select, decide, design, perform, and evaluate during the course. They share the knowledge they found and take other pre-services' opinions on the issues during the activities.

The framework used in the design is TPACK and the literature advocates that "TPCK development efforts need to invest on socio-cognitive constructivist ideas" (Angeli & Valanides, 2008, p. 16). According to them, the TPACK framework targets learners' conceptual ecology consisting of their knowledge base by transforming a content domain through technology (Angeli & Valanides, 2008). Moreover, Nelson, Christopher, and Mims (2009) state that social media, such as Web 2.0 tools, serve learners construct knowledge together and teachers who have developed TPACK may use these tools to design instruction.

The Phases and Activities in the Course Design

The pre-service teachers are expected to follow the curriculum-generated play instruction for their teaching practices for the course design. There are four phases in the instruction:

1. Learning the mathematics program in the national early childhood curriculum

Figure 1. Schematic view of activities in TPACK framework

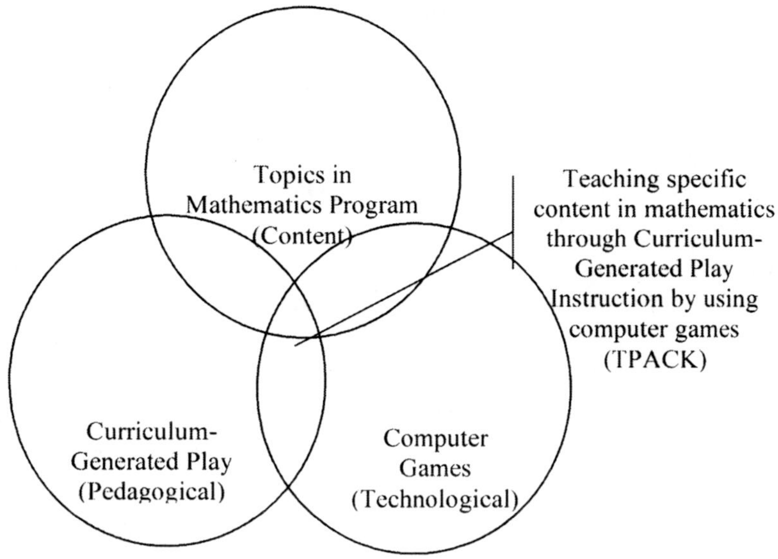

2. Computer games selection (technology selection)
3. Preparing lesson plans
4. Teaching mathematics to young children with computer games

Learning the Mathematics Program in the National Early Childhood Curriculum: The pre-service teachers are directed to examine the mathematics program in the national early childhood education curriculum. All the program topics are expected to be defined in terms of goals, behaviors, and learners' level. The pre-service teachers prepare a list which includes the topics in the mathematics program. During in-class activities, the pre-service teachers are separated into groups. They discuss the topics and choose one of them.

The pre-service teachers are expected to prepare a report on the selected topic in terms of goals, behaviors, and learners' level. They also find examples of activities used to teach the selected topics in the curriculum. In-class, they come together with their groups and share the found information. Then, they are expected to prepare a presentation, and introduce the topics in terms of goals, behaviors, learners' level, and activity examples taken from the mathematics program in the curriculum.

They are expected to demonstrate the learning philosophy adopted by the mathematics program of the national early childhood curriculum with topics they selected. They use the goals, behaviors, and activity examples given in the program to prove the learning philosophy behind it. They prepare a report on the issue and discussed it in class with classmates.

Computer Games Selection: The pre-service teachers are expected to find a computer game in order to teach the topics selected. They also prepare a presentation to introduce the selected computer game. In-class, they work with their groups to decide which computer games they would use to teach a selected topic to young children. Then, they present the computer games to the class.

Preparing the Lesson Plans: The pre-service teachers are expected to prepare a lesson plan describing how to teach topics with the computer games selected. In the plan, they are expected to explain the topics, young children's level, goals, behaviors, activities they prepare, and how to use computer games during the instruction, and lastly how to assess young children's learning. In class, the pre-service teachers come together with their groups. They discuss their plans, and prepare a new one, or choose one of the group members' plans if they want for the instruction.

Teaching Mathematics to Young Children with Computer Games: In class, they teach topics with group members according to the lesson plan they prepared with the groups. Before the class, they work with their group members, and design materials they will use during their instruction. In class, they teach, and also evaluate both themselves and the other groups' instructions. Figure 2 shows the phases of course design.

As seen from the figure, the core of the course design is learning the program in the national curriculum in terms of topics, goals, behaviors, and young children's levels. Instead of teaching the program directly, the pre-service teachers are directed to examine the program in the national curriculum. Then, they select a topic and investigate it in detail by focusing on how to teach it. However, not only the topics/goals/behaviors/young children's level in the program, but also the learning philosophy behind the curriculum is important, since the national ministry of education expects teachers to teach in accordance with this philosophy. Constructivism is the philosophy behind the curriculum in this case. For that reason, the pre-service teachers are directed to study the learning philosophy with this activity, making them demonstrate curriculum philosophy with the goals/behaviors/young children's level of the topics selected from the curriculum. Moreover, during the course this philosophy is followed by instructing pre-service teachers to search, select, decide, design, and evaluate. In this course design, the main focus is to show the pre-service teachers

Figure 2. The phases of course design

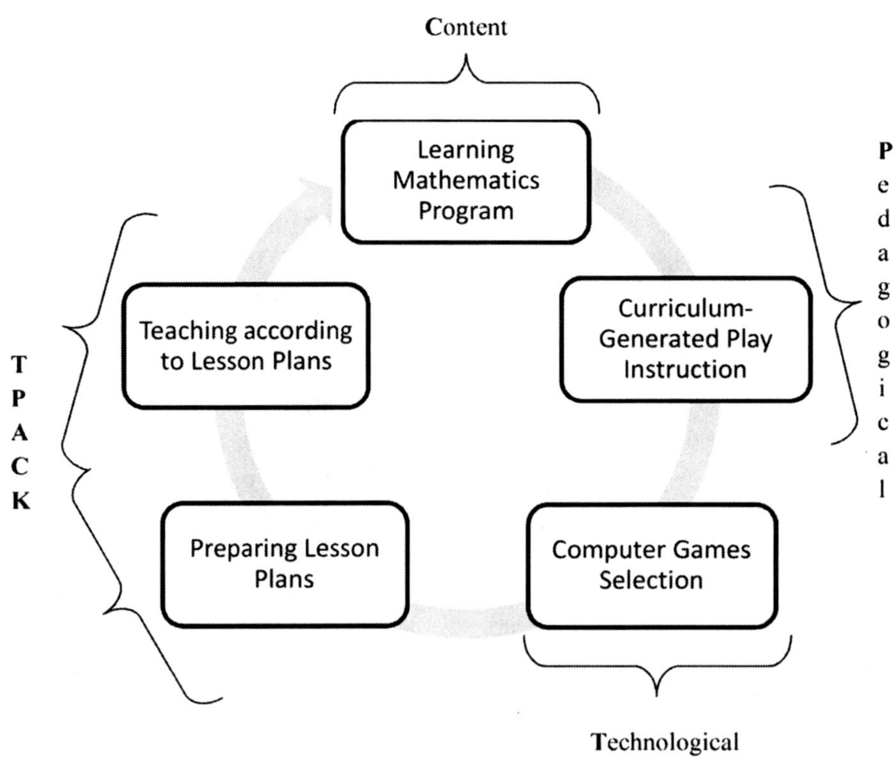

how to teach mathematics with the help of computer games by integrating computer games into instruction in accordance with constructivism.

As a last step of the course, the education teachers evaluate the course design effectiveness by conducting focus group interviews with each group. This allows them to modify the course design according to the pre-service teachers' feedback.

FUTURE TRENDS

We will discuss the future of this technology in early childhood education over the next ten years. Using this technology in early childhood education provides a range of possibilities in the development of children's technological abilities, mathematics and literacy learning by making learning entertaining.

The future of computer games can be guessed by looking at its situation today. Young children spend most of their time indoors, with activities such as TV, computer games, or other software (Clements, 2004; Lia & Atkins, 2004; Cherney & London, 2006; Dwyer, 2007). The literature calls the new generation "digital natives," since they can adapt and easily use technologies thanks to the technology – dominated world. Bennett, Maton, and Kervin (2008) summarize the conventions about the existence of the digital natives as following:

1. *Young people of the digital native generation possess sophisticated knowledge of and skills with information technologies.*
2. *As a result of their upbringing and experiences with technology, digital natives have particular learning preferences or styles that differ from earlier generations of students.* (p. 777).

Moreover, Van Eck (2006) states that one of the reasons for public interest in computer games as learning tools is the existence of "digital natives." The characteristics of "digital natives" (e.g. Bennett, Maton, & Kervin, 2008) encourage the use of technology within instruction (Van Eck, 2006). Using computer games while teaching young children provides an intersection of learning, entertainment and technology (Saracho & Spodek, 2002). When the popularity of computer games and the reasons of this popularity are examined, it may be said that this technology will continue to be the center of interest to researchers, children and parents. For that reason, the current situation, where the number of educational computer games having high quality is very limited today will change. In other words, the quality of educational computer games in terms of design, message, and educational side will be enhanced. In our view, the most explicit changes in educational games will be in their design. The avatars/characters, as well as their appearance in the educational computer games for young children, will be changed with the effects of TV cartoons and magazines. We believe that the computer game play-method will also change with the advance of the technology. Most games will be directed via voice and actions, as in Wii games. So, the interaction in the computer games interface will be enhanced with the help of advancement in simulation/augmented reality technology.

Moreover, today, finding computer games that are suitable to the behavioral objectives of curriculum are difficult for an early childhood teacher who wants to use it during instruction (Verenikina, et al., 2003). Again, future computer games will be designed more curriculum-centered in terms of both curriculum philosophy and behavioral objectives.

A number of elective courses on the design and integration of educational computer games into instruction will be offered in Teacher Education Programs. We believe that these courses will be offered under specific content areas, such as mathematics, language teaching or literacy. Today, the literature advocate, pre-service teachers, should know how to teach a specific content with pedagogy, and how they can use technology assist it (Mishra & Kohler, 2006; Angeli & Valanides, 2008; Shin et al., 2009; Özgün-Koca, et al., 2010). We believe that this trend will continue and cause educational games to become more curriculum oriented.

Today, educational games characteristics, from design to content, from play-method to interaction properties, are debated. In addition, whether the computer games are beneficial or harmful for young children is debated (Plowman, et al., 2010). We believe that the debates over computer games will continue without affecting the popularity of this technology. We are only at the beginning stages of educating pre-service teachers how to use computers games while teaching young children. These attempts reflect changes in points of view about technology, and that many are realizing that the benefits of technology can be extracted with the effective use as Verenikina, et al. (2003) state.

CONCLUSION

The use of computer games for the education of young children has been debated for years. Although there are different perspectives on whether the computer games should be used or not, the popularity of these technologies in education has been growing day by day (Prensky, 2001). However, it is the way the computer games are used that is the most important issue. Finger, Jamieson-Proctor and Albion (2010) state the importance of providing pre-service teachers with the necessary skills to teach with technology: "Pre-service teacher education programs have the responsibility for preparing future teachers who are likely to be teaching their students in a world characterized by ongoing technological changes"(p.114). As Mishra and Koehler (2006) state, the theoretical background enhances the effectiveness of an instructional design that includes technology integration. This framework shows how a pre-service teacher's understanding of the content, pedagogy, and technological knowledge can interact to produce effective instructions with educational technologies (Shin et al., 2009). For that reason, our course design follows this framework to enhance pre-service teachers' TPACK, with the integration of computer games (technology) and mathematics (content) within curriculum-generated instruction (pedagogy).

REFERENCES

Amory, A., Naicker, K., Vincent, J., & Adams, C. (1999). The use of computer games as an educational tool: Identification of appropriate game types and game elements. *British Journal of Educational Technology*, *30*(4), 311–321. doi:10.1111/1467-8535.00121

Angeli, C., & Valanides, N. (2008, March 24-28). *TPACK in pre-service teacher education: Preparing primary education students to teach with technology*. Paper presented at the 2008 Conference of the American Educational Research Association, New York City, NY.

Ball, C. (1994). *Start right: The importance of early learning*. London, UK: Royal Society of Arts.

Bandura, A. (1997). *Self-efficacy: The exercise of control*. New York, NY: Freeman.

Barbour, M., Rieber, L. P., Thomas, G., & Rauscher, D. (2009). Homemade PowerPoint games: A constructionist alternative to Web-Quests. *TechTrends*, *53*(5), 54–59. doi:10.1007/s11528-009-0326-2

Beauchat, K. A., Blamey, K. L., & Walpole, S. (2010). *The building blocks of preschool success*. New York, NY: The Guilford Press.

Bennett, S., Maton, K., & Kervin, L. (2008). The 'digital natives' debate: A critical review of the evidence. *British Journal of Educational Technology*, *39*(5), 775–786. doi:10.1111/j.1467-8535.2007.00793.x

Bhattacherjee, A., & Premkumar, G. (2004). Understanding changes in belief and attitude toward information technology usage: A theoretical model and longitudinal test. *Management Information Systems Quarterly*, *28*(2), 229–254.

Bopp, M. M. (2008). *Storytelling and motivation in serious games.* Final Consolidated Research Report of the Enhanced Learning Experience and Knowledge Transfer – Project (ELEKTRA), (European Commission contract number 027986). Retrieved from http://www.elektra-project.org/

Borko, H., & Putnam, R. T. (1995). Expanding a teachers' knowledge base: A cognitive psychological perspective on professional development. In T. R. Guskey & M. Huberman (Eds.), *Professional development in education: New paradigms and practices* (pp. 35–65). New York, NY: Teachers College Press.

Brand, J. E., Knight, S. J., & Majewski, J. (2003, November 4-6). *The diverse worlds of computer games: A content analysis of spaces, populations, styles and narratives.* Paper presented at the first Level Up Digital Games Research Conference, University of Utrecht, The Netherlands.

Bursal, M., & Paznokas, L. (2006). Mathematics anxiety and preservice elementary teachers' confidence to teach mathematics and science. *School Science and Mathematics*, *106*(4), 173–180. doi:10.1111/j.1949-8594.2006.tb18073.x

Cennamo, K., Ross, J., & Ertmer, P. (2010). *Technology integration for meaningful classroom use: A standards-based approach.* Belmont, CA: Wadsworth, Cengage Learning.

Cherney, I. D., & London, K. (2006). Gender-linked differences in the toys, television shows, computer games, and outdoor activities of 5- to 13-year-old children. *Sex Roles*, *54*(9-10), 717–726. doi:10.1007/s11199-006-9037-8

Clements, D. H. (2002). Computers in early childhood mathematics. *Contemporary Issues in Early Childhood*, *3*(2), 160–181. doi:10.2304/ciec.2002.3.2.2

Clements, R. (2004). An investigation of the status of outdoor play. *Contemporary Issues in Early Childhood*, *5*(1), 160–181. doi:10.2304/ciec.2004.5.1.10

De Villiers, R. (2004). Usability evaluation of an e-learning tutorial: Criteria, questions and case study. In G. Marsden, P. Kotze, & A. Adesina (Eds.), *Proceedings of the 2004 Annual Research Conference of the South African Institute of Computer Scientists and Information Technologists on IT Research in Developing Countries* (pp. 284-291).

Dembo, M. H., & Gibson, S. (1985). Teachers' sense of efficacy: An important factor in school improvement. *The Elementary School Journal*, *86*(2), 173–184. doi:10.1086/461441

Dwyer, J. (2007). Computer-based learning in a primary school: Differences between the early and later years of primary schooling. *Asia-Pacific Journal of Teacher Education*, *35*(1), 89–103. doi:10.1080/13598660601111307

Egenfeldt-Nielsen, S. (2004). Practical barriers in using educational computer games. *On The Horizon - The Strategic Planning Resource for Education Professionals, 12*(1), 18-21.

Ertmer, P. A., Conklin, D., & Lewandowski, J. (2001, November 8-12). *Increasing preservice teachers' capacity for technology integration through use of electronic models.* Presented at the Annual Meeting of National Convention of the Association for Educational Communications and Technology, Atlanta, GA.

Finger, G., Jamieson-Proctor, R., & Albion, P. (2010). Beyond pedagogical content knowledge: The importance of TPACK for informing preservice teacher education in Australia. *IFIP Advances in Information and Communication Technology, 324,* 114–125. doi:10.1007/978-3-642-15378-5_11

Finney, S. J., & Schraw, G. (2003). Self efficacy beliefs in college statistics course. *Contemporary Educational Psychology, 28,* 161–186. doi:10.1016/S0361-476X(02)00015-2

Fuson, K. C. (2004). Pre-K to grade 2 goals and standards: Achieving 21st-century mastery for all. In D. H. Clements & J. Sarama (Eds.), *Engaging young children in mathematics: Standards for early childhood mathematics education* (pp. 105–148). Mahwah, NJ: Lawrence Erlbaum Associates.

Gibson, S., & Dembo, M. H. (1984). Teacher efficacy: A construct validation. *Journal of Educational Psychology, 76*(4), 569–582. doi:10.1037/0022-0663.76.4.569

Ginsburg, K. R. (2007). The importance of play in promoting healthy child development and maintaining strong parent-child bonds. *Pediatrics, 119*(1), 182–191. doi:10.1542/peds.2006-2697 PMID:17200287

Gunder, A. (2003, November 4-6). *As if by magic: On Harry Potter as a novel and computer game.* Paper Presented at Digital Games Research Conference, University of Utrecht, The Netherlands.

Haugland, S. W., & Wright, J. L. (1997). *Young children and technology: A world of discovery.* Boston, MA: Allyn and Bacon.

Heinich, R., Molenda, M., Russell, J. D., & Smaldino, S. E. (2002). *Instructional media and technologies for learning* (7th ed.). Upper Saddle River, NJ: Prentice-Hall.

Hogle, J. G. (1996). Considering games as a cognitive tools: In search of effective edutainment. *ERIC ED, 425,* 737.

Hoy, A. W., & Spero, R. B. (2005). Changes in teacher efficacy during the early years of teaching: A comparison of four measures. *Teaching and Teacher Education, 21,* 343–356. doi:10.1016/j.tate.2005.01.007

Incikabi, L. (2011). The coherence of the curriculum, textbooks and placement examinations in geometry education: How reform in Turkey brings balance to the classroom. *Education as Change, 15*(2), 239–255. doi:10.1080/16823206.2011.619144

Incikabi, L., & Tuna, A. (in press). Türkiye ve Amerika eğitim sistemlerinin 60-72 aylıklar için geliştirilen okul öncesi matematik eğitimi programı açısından karşılaştırılması. *Mersin Üniversitesi Eğitim Fakültesi Dergisi, 8*(2).

Isler, I., & Cakiroglu, E. (2009). Teachers' efficacy beliefs and perceptions regarding the implementation of new primary mathematics curriculum. *Proceedings of CERME 6*, Lyon, France, 28 January – 1 February 2009, (pp. 1704-1713).

Johnson, J. E., Christie, J. F., & Yawkey, T. D. (1999). *Play and early childhood development* (2nd ed.). New York, NY: Longman.

Kafai, Y. B. (2006). Playing and making games for learning: Instructionist and constructionist perspectives for game studies. *Games and Culture, 1*(1), 36–40. doi:10.1177/1555412005281767

Klawe, M. (1999). Computer games, education and interfaces: The E-GEMS project. In the *Proceedings of the Graphics Interface Conference* (pp. 36–39)

Kraus, W. H. (1981). Using a computer game to reinforce skills in addition basic facts in second grade. *Journal for Research in Mathematics Education*, *12*, 152–155. doi:10.2307/748709

Leutenegger, S., & Edgington, J. (2007, March 7-10). A games first approach to teaching introductory programming. In *Proceedings of the 38th SIGCSE Technical Symposium on Computer Science Education.* New York, NY: ACM Press.

Li, X., & Atkins, M. S. (2004). Early childhood computer experience and cognitive and motor development. *Pediatrics*, *113*(6), 1715–1722. doi:10.1542/peds.113.6.1715 PMID:15173496

Milli Egitim Bakanligi (MEB). (2006). Okul Öncesi Eğitim Programı (36–72 Aylık Çocuklar İçin) (Ed. T. Gürkan & G Haktanır). Ankara, Turkey: Author.

Mioduser, D., Tur-Kaspa, H., & Leitner, I. (2000). The learning value of computer-based instruction of early reading skills. *Journal of Computer Assisted Learning*, *16*, 54–63. doi:10.1046/j.1365-2729.2000.00115.x

Mishra, P., & Koehler, M. J. (2006). Technological pedagogical content knowledge: A framework for teacher knowledge. *Teachers College Record*, *108*(6), 1017–1054. doi:10.1111/j.1467-9620.2006.00684.x

National Association for the Education of Young Children (NAEYC). (2008). *Early childhood mathematics: Promoting good beginnings. A joint position statement of the National Association for the Education of Young Children (NAEYC) and the National Council for Teachers of Mathematics (NCTM).* Washington, DC: NAEYC.

Nelson, J., Christopher, A., & Mims, C. (2009). TPACK and Web 2.0: Transformation of teaching and learning. *TechTrends*, *53*(5), 80–87. doi:10.1007/s11528-009-0329-z

Niederhauser, D. S., & Stoddart, T. (2001). Teachers' instructional perspectives and use of educational software. *Teaching and Teacher Education*, *17*, 15–31. doi:10.1016/S0742-051X(00)00036-6

Niess, M. L., Ronau, R. N., Shafer, K. G., Driskell, S. O., Harper, S. R., & Johnston, C.,.... Kersaint, G. (2009). Mathematics teacher TPACK standards and development model. *Contemporary Issues in Technology & Teacher Education*, *9*(1), 4–24.

O'Neil, H. F., Wainess, R., & Baker, E. L. (2005). Classification of learning outcomes: Evidence from the computer games literature. *Curriculum Journal*, *16*(4), 455–474. doi:10.1080/09585170500384529

Özgün-Koca, A., Meagher, M., & Edwards, M. T. (2010). Preservice teachers' emerging TPACK in a technology-rich methods class. *The Mathematics Educator*, *19*(2), 10–20.

Pajares, F. M. (1996). Self-efficacy beliefs and mathematical problem-solving of gifted students. *Contemporary Educational Psychology*, *21*, 325–344. doi:10.1006/ceps.1996.0025 PMID:8979868

Paraskeva, F., Mysirlaki, S., & Papagianni, A. (2010). Multiplayer online games as educational tools: Facing new challenges in learning. *Computers & Education*, *54*, 498–505. doi:10.1016/j.compedu.2009.09.001

Plourde, L. A. (2002). The influence of student teaching on preservice elementary teachers' science self-efficacy and outcome expectancy beliefs. *Journal of Instructional Psychology*, *29*, 245–253.

Plowman, L., McPake, J., & Stephen, C. (2010). The technologisation of childhood? Young children and technology in the home. *Children & Society*, *24*(1), 63–74. doi:10.1111/j.1099-0860.2008.00180.x

Pound, L. (2008). *Thinking and learning about mathematics in the early years*. Abingdon, UK: Routledge.

Pramling Samuelsson, I., & Johansson, E. (2006). Play and learning–Inseparable dimensions in preschool practice. *Early Child Development and Care*, *176*(1), 47–65. doi:10.1080/0300443042000302654

Prensky, M. (2001). *Digital game based learning*. New York, NY: McGraw-Hill Press.

Rideout, V., Vandewater, E., & Wartella, E. (2003). *Zero to six. Electronic media in the lives of infants, toddlers and pre-schoolers*. Menlo Park, CA: Kaiser Family Foundation.

Riggs, M. I., & Enochs, L. G. (1990). Toward the development of elementary teacher's science teaching efficacy belief instrument. *Science Education*, *74*(6), 625–637. doi:10.1002/sce.3730740605

Sancar-Tokmak, H. (in review). A case study on early childhood pre-service teachers' integration of computer games to instruction. *Adiyaman University Journal of Social Sciences*.

Sancar-Tokmak, H., Incikabi, L., & Yanpar-Yelken, T. (in press). Differences in the educational software evaluation process for experts and novice students. *Australasian Journal of Educational Technology*, *28*(8).

Sandberg, A., & Pramling Samuelsson, I. (2003). Preschool teachers' play experiences then and now. *Early Childhood Research & Practice*, *5*(1), Retrieved May 12, 2012, from http://ecrp.uiuc.edu/v5n1/sandberg.html

Saracho, O. N., & Spodek, B. (2002). *Contemporary perspectives on early childhood curriculum* (Vol. 1). Greenwich, CT: Information Age Publishing.

Sarama, J., & Clements, D. (2004). Building blocks of early childhood mathematics. *Teaching Children Mathematics*, *9*(8), 480–484.

Sedighian, K., & Sedighian, A. S. (1996). *Can educational computer games help educators learn about the psychology of learning mathematics in children?* Paper presented at 18th Annual Meeting of the International Group for the Psychology of Mathematics Education, Florida, USA.

Shin, T. S., Koehler, M. J., Mishra, P., Schmidt, D. A., Baran, E., & Thompson, A. D. (2009). Changing technological pedagogical content knowledge (TPACK) through course experiences. In I. Gibson, R. Weber, K. McFerrin, R. Carlsen, & D. A. Willis (Eds.), Society for information technology and teacher education international conference book (pp. 4152-4156). Chesapeake, VA: Association for the Advancement of Computing in Education (AACE).

Shulman, L. S. (1986). Those who understand: Knowledge growth in teaching. *Educational Researcher*, *15*, 4–14.

Smith, S. S. (2006). *Early childhood mathematics* (3rd ed.). Boston, MA: Allyn & Bacon.

Squires, D., & Preece, J. (1996). Usability and learning: Evaluating the potential of educational software. *Computers & Education*, *27*(1), 15–22. doi:10.1016/0360-1315(96)00010-3

Sweedyk, E., deLaet, M., Slattery, M. C., & Kuffner, J. (2005, February 23-27). *Computer games and CS education: Why and how*. Paper presented at SIGCSE'05, New York, USA.

Van Eck, R. (2006). Digital game-based learning: It's not just the digital natives who are restless.... *EDUCAUSE Review*, *41*(2), 16–30.

Verenikina, I., Harris, P., & Lysaght, P. (2003). Child's play: Computer games, theories of play and children's development. In *Proceedings of the Young Children and Learning Technologies*. Selected papers from the International Federation for Information Processing Working Group.

Yanpar-Yelken, T. (2011). *Ogretim teknolojileri ve material tasarımı* (2nd ed.). Ankara, Turkey: Ani Yayincilik.

Zengin, K. U. (2003). *İlköğretim öğretmenlerinin öz-yeterlik algıları ve sınıf-içi iletişim örüntüleri*. Unpublished Master's thesis, Dokuz Eylül University, Izmir, Turkey.

KEY TERMS AND DEFINITIONS

Commercial Off-the-Shelf Games (COTS Games): The games are available for public use in market place and produced under government contract.

Milli Egitim Bakanligi (Ministry of National Education) – MEB: The name of the National Ministry of Education in Turkey.

National Association for the Education of Young Children (NAEYC): NAEYC is a nonprofit association in the United States for the purpose of serving to improve young children well being especially in terms of education and development.

Pedagogical Content Knowledge (PCK): A framework proposed by Shulman (1986) which is an understanding of both pedagogy and content.

Self-Efficacy: Beliefs in one's capabilities to organize and execute the courses of action required to produce given attainments (Bandura, 1997, p.3).

Technological Pedagogical Content Knowledge (TPACK): A framework proposed by Mishra and Kohler (2006) as an extension of Sulman (1986)'s PCK framework. It is an understanding of three knowledge (content, pedagogy, and technology) bases can interact with one another to produce effective discipline-based teaching with educational technologies both pedagogy, and content (Mishra & Kohler, 2006).

This work was previously published in "Research Perspectives and Best Practices in Educational Technology Integration" edited by Jared Keengwe, pages 178-196, copyright 2013 by Information Science Reference (an imprint of IGI Global).

Chapter 44
Linking Education to Creating a Knowledge Society:
Qatar's Investment in the Education Sector

Alan S. Weber
Weill Cornell Medical College, Qatar

ABSTRACT

Due to the continued high price of oil and gas, the oil-rich State of Qatar has used its large budget surpluses in the last decade to finance human capacity development, including research, higher education, and the reshaping of its K-12 educational system. This chapter argues that the recent substantial educational reforms in the State of Qatar are closely intertwined with planned future economic transformation (diversification). Although Qatar possesses the world's third largest reserves of natural gas, this resource is ultimately finite and over-reliance on one major economic driver (hydrocarbons) for the bulk of GDP creates boom and bust cycles that have shaped Gulf politics and social development since the 1970s. This chapter examines Qatar's educational efforts to build a knowledge economy to transition away from a resource-rich export-based hydrocarbon economy towards economic activities linked to patents, research, trademarked technologies, skills, and knowledge products.

INTRODUCTION

Due to the continued high price of oil and natural gas, the State of Qatar has experienced large budget surpluses in the last decade that have been utilized in building infrastructure, financing human capacity development, establishing research capabilities, and funding an innovative educational experiment called Education City (now transitioning to a consolidated institution called Hamad bin Khalifa University). In 2013, 437 students received bachelor degrees from Education City universities in the science, engineering, medical, and political science fields. The national university Qatar University also offers a full range of Science, Technology, Engineering, and Mathematics (STEM) field bachelor's degrees, but only a small handful of advanced degrees (Masters and PhD) in science and technology; for example, a doctorate in pharmaceutical science. Thus education is viewed as a key sector in creating a knowledge producing regime in Qatar, but further provision for advanced education and science and ICT infrastructure is required to achieve this goal.

DOI: 10.4018/978-1-4666-7363-2.ch044

Production in Qatar's North Field unassociated gas field lying in Gulf waters (one of the world's largest gas deposits and jointly owned by Iran) has reached maturity with 14 Liquid Natural Gas (LNG) trains online; much of this gas is shipped to Europe, Japan, and Korea, and Qatar ranks as the largest exporter of natural gas internationally. Thus Qatar's sovereign wealth funds generated by export revenues which are available for investment and development rose to 112 billion Qatari Riyals in 2010 (QIA, 2012). Sovereign wealth funds are financial instruments frequently employed by oil rich nations to provide investment capital and sequester revenue to prevent rapid inflation and overheating of the economy during boom cycles. Qatar's economy grew extremely rapidly in the last decade and the official 2012 unemployment rate was only.5% (QSA, 2012, p. 2). Although the government has officially signaled its desire to diversify the economy into education and knowledge economy activities, oil and gas are still the primary generators of state income. An important factor to consider is that the majority of Qatari citizens work for the police, military or in government offices as clerks and managers. In 2011, Qataris made up only.5% of the private sector workforce (QSA, 2011, p. 11). The revenue to pay government employees derives almost exclusively from oil and gas revenues generated by State Owned Enterprises, such as Qatar Petroleum, as well as subsidiary industries based on inexpensive hydrocarbon fuel and feedstock – aluminum smelting, plastics, cement, fertilizer and petrochemicals. Thus price shocks in oil and gas markets directly affect government revenues and the state's ability to pay out salaries and benefits to citizens. Current benefits to those holding Qatari passports include free water, electricity, and health care both in public hospitals and abroad, free education from K-12 and higher education (including study abroad) and loans and subsidies for marriage, land and housing. These benefits form part of what political scientists call the "rentier state bargain" in which citizens obtain financial benefits from the state in return for political support of the status quo.

Thus, creating more industries and small businesses in areas not subject to the price fluctuations of finite natural resources (gas and oil) makes economic sense both in the short and long term. Often overlooked, however, in the discussions of knowledge economy development in the Gulf is the additional factor of national pride and Islamic identity: Arabic-speaking countries are well aware of the golden age of Islamic science in which mathematics, astronomy, and medicine were highly cultivated in Damascus, Cairo, Baghdad and later in Al Andalus at centers of learning such as Toledo and Cordoba. With the resurgence of wealth in the oil-rich Gulf, initiatives in Saudi Arabia, Qatar and UAE are partly aimed at recapturing the scientific culture and learning of a past age. A special issue of the science journal *Nature* in 2006 looked at this movement in depth, and emphasized several important points; for example, "injunctions in the Koran that it is the bounden duty of every Muslim, man or woman, to acquire knowledge" (p. 36). Conservative religious authorities have sometimes been hostile to modern scientific endeavor, believing that it promotes a secular and western view of the world. However, history provides abundant evidence that advanced scientific knowledge and Islam have always been compatible.

The previous Emir of Qatar HH Sheikh Hamad bin Khalifa Al Thani, who voluntarily abdicated in 2013 in favor of his son the Heir Apparent Tamim, along with the former Emir's wife HH Sheikha Moza bint Nasser, engineered a revolutionary change in Qatar's educational system in parallel with Qatar's rapid economic growth. The leadership of Qatar realized that a modern wealthy state whose income depends to a large extent on advanced LNG technology (primarily developed in the U.S.) requires a highly-trained technical workforce. Also, Qatar's oil wells may have already peaked in production, and similar to

Oman, further production increases will be tied to high-tech Enhanced Oil Recovery (EOR) solutions which originate in the West. Some of these technologies include steam, water, and polymer injection and hydraulic fracturing (fracking). However, many of these techniques use vast quantities of water, which present technical challenges in water-scarce Qatar, in addition to the pollution created. Thus knowledge and research will need to play a key role in the proper management of advanced systems in order to adapt them efficiently to Qatar's specific technological needs and fragile terrestrial and marine ecologies.

In order to innovate in an increasingly globalized economy and to maintain Qatar's leading edge in LNG exports, and to solve a number of serious environmental issues, Qatar will need to produce more STEM field graduates with the critical and analytical skills to solve large, complex problems regarding land use, desalination, waste management, EOR, marine resources management, chemical engineering, petroleum production, and alternative energy. Otherwise, Qatar will be forced to rely on foreign expertise for necessary technologies and expatriate labor to implement and manage them. Qatar's previous under-development in the areas of science, technology and education have proven to be an advantage in one sense, since the country has been able to start from a blank slate and employ international best practice models from the outset, for example lessons learned from the 'Singapore Model'. Qatar Foundation for Education, Science and Community Development has since its inception in 1995 sponsored six highly ranked American branch campuses and three international universities within Qatar to provide high quality educational programs to Qataris and qualified expatriates. These programs have been given unprecedented freedom in Qatar in establishing their programs to closely mirror their main campuses' curricula and institutional culture.

One explicit objective of Qatar Foundation's higher education policies is to diversify Qatar's oil and gas-based economy into knowledge production. Although other Gulf countries have also invested substantially in knowledge production and higher education–such as Saudi Arabia's graduate programs and research at KAUST and the expansion of its private higher education institutions, and Dubai's Internet City and Knowledge Village–Qatar presents a particularly interesting case since its educational policies and developmental vision are particularly well integrated with business incubation (QSTP) and a new national research fund (QNRF). Qatar Science and Technology Park (QSTP) provides assistance for spin-off research companies originating in ideas from faculty research in Education City. QSTP also sponsors government and private industry partnerships in the areas of energy and petroleum engineering. Qatar National Research Fund (QNRF) is the national funding agency of Qatar, modeled after the U.S. National Science Foundation; this program funds a wide variety of competitive research programs from humanities and social sciences to pure and applied sciences, medicine, ICT, and software engineering. All of these institutions above operate essentially under the same management and oversight of Qatar Foundation.

Some known educational challenges in the State of Qatar include the existence of large expatriate workforces, underdevelopment of primary education with low recorded PISA and PIRLS scores, student motivation in an entitlement culture, and the small percentage of advanced STEM field degrees awarded per year. Current educational policy and its linkages to science and technology development has been specifically designed (with assistance from the U.S. RAND Corporation) to meet these well-recognized challenges.

RAND Corporation formed the RAND-Qatar Policy Institute (RQPI) in 2003 to research and implement their *Education for a New Era K-12* reform of Qatar's schools. In 2010-2011, the Ministry of Education was decommissioned and all responsibility for education in Qatar was transferred to the Supreme Education Council (SEC) as part of these changes. However, RAND policies have been met with some resistance and criticisms from local educational stakeholders. In 2013, Qatar Foundation did not renew RAND's 10-year contract, amid criticisms that the 10-year reform program Education for a New Era did not significantly improve test scores, or school conditions and the professional environment for teachers and Headmasters. It is also equally possible that with successful educational reforms well underway, that RQPI was no longer necessary. The ambiguity in RQPI's departure and the lack of meaningful official public statements by both Qatar Foundation and RQPI underscores the fact that education policy is rarely subject to public debate in Qatar.

Qatar's educational policies and science policy innovations fit the definition of "Big Science," a term coined in the early 1960s to describe large scale science projects such as the Manhattan Project, Human Genome Project and CERN's Large Hadron Collider. Projects including Qatar Foundation and Qatar National Research Fund are funded on the scale of billions of U.S. dollars: for example, the operating budget of the new Sidra Hospital for women and children will be drawn from a $7.9 billion USD permanent endowment. The religious institution of *waqf*, the setting aside of charitable trusts for the future, has been successfully adapted to these projects. The Sidra Hospital, due to its charitable nature, and goal of improving women's and children's health outcomes (which have been neglected in the past, as males have traditionally been given priority in health care during resource scarcity because they are responsible for the economic health of the family), closely fits the original spirit and purpose of *waqf*. In this system, individual ownership of assets ceases and God becomes the implied owner in perpetuity.

Due to the similarities to Big Science, Qatar's educational reforms could logically be called "Big Education." The highly centralized nature of educational policy making (all levels of education are essentially now under the aegis of the Supreme Education Council, and higher education development is shared between Qatar University and Qatar Foundation) may provide the administrative structure in which Big Education and Big Science could work together and provide efficient, high-quality and targeted education that is closely aligned with Qatar's knowledge economy labor market needs. Large central bureaucracies, on the other hand, could inhibit individual teacher and school innovation and force cookie-cutter solutions onto creative educators who are working to adapt international best practices to the unique learner populations of the Gulf. By the next decade, enough empirical data should be available to assess fully Qatar's educational and knowledge economy experiments. Fortunately, institutions such as the Supreme Education Council, Qatar Foundation and Qatar Statistics Authority have improved their statistics gathering capabilities immensely, recognizing that progress can only be measured by the careful collection and assessment of empirical data and comparing performance scores to standard international benchmarks. In fact, the closing of the former Ministry of Education and the emergence of the Supreme Education Council may have in part been due to that institution's failure to capture a broad and accurate range of statistics on schools in Qatar, which made it difficult for education policy makers to understand the current state of Qatar's educational systems as well as plan for the future.

BACKGROUND: CONCEPTUAL FRAMEWORK OF THE KNOWLEDGE ECONOMY

In the 1990s, perhaps stimulated by the widespread dissemination of the Internet, the concept of the Knowledge Economy gained prominence in international development reports, especially by the World Bank, IMF, and United Nations

Development Programme. Examples of influential reports include: *Arab Human Development Report 2002* (UNDP, 2002), *Knowledge Economies in the Middle East and North Africa: Toward New Development Strategies* (World Bank, 2003), and *The Road Not Traveled: Education Reform in the Middle East and North Africa* (World Bank, 2008).

Scientist Vannevar Bush and management specialist Peter Drucker, however, had advanced similar ideas decades earlier during the rise of analog computers and after World War II. It was widely believed that advances in several knowledge sciences–communications, cryptography, and early warning systems (RADAR)–had played a decisive role in the Allied victory. Both Drucker and Bush envisioned happy and productive societies that were organized around knowledge and its production, storage and retrieval to solve real-world problems in the social and economic realms. Since the rise of the first flint weapons, mankind has recognized that the know-how to mass-produce superior technologies for making war provides a considerable economic advantage over less knowledgeable societies. Thus knowledge may be rooted in fundamental aspects of human survival, now manifest in the highly funded military research programs of advanced nations.

Globalization can be roughly defined as the integration of cross-border economic, social, and political networks such that all nation states meld into an increasingly interdependent international marketplace. Some analysts link the concept of knowledge economy to globalization because of the modern fluidity of capital, the ease of international transport of labor and goods, and the spread of American free market policies and educational paradigms even in areas which are not receptive to them. Thus knowledge economy rhetoric may be a proxy for the opening of the markets of developing countries to those with more developed knowledge regimes; for example, well established American e-commerce firms such as E-bay and Amazon.com benefit from the establishment of reliable broadband networks and the lowering of local business barriers and restrictions. Financial institutions controlled by wealthier nations such as the World Bank and IMF promote ICT development in poorer countries in terms of 'fostering innovation' and 'delivering information.' These conditions also help more established, more highly capitalized, and efficient businesses in wealthy nations to deliver commercial goods and services via the Internet, competing with and possibly destroying local businesses in less developed countries.

In order to quantify, assess, and compare knowledge economy activities across countries, the World Bank has developed an empirical framework called the Knowledge Assessment Methodology, with two indices: the Knowledge Index (KI) and Knowledge Economy Index (KEI) consisting of four pillars, themselves an aggregate of a larger number of specific variables. The Knowledge Index measures broadly a country's knowledge production as averaged scores of the three pillars of *Education, Information and Communication Infrastructure (ICT)* and *Innovation*. The Knowledge Economy Index factors in the fourth pillar, *Economic Incentive and Institutional Regime*: in other words, do the country's policies, laws and markets encourage and promote the use of knowledge? In 2012, Qatar scored a respectable 5.84 on the KEI and 5.50 on the KI, higher than the world averages and much higher than the MENA region average scores of 4.74 and 4.51 respectively (World Bank, 2013a). In 2007, Qatar specifically requested an individualized knowledge economy assessment report from the World Bank, which resulted in the document sponsored by the Planning Council and Qatar Foundation entitled *Turning Qatar into a Competitive-Based Economy: Knowledge Economy Assessment of Qatar*. In the education pillar, the low gross tertiary education enrollment was cited as a specific challenge. Also, concerns were voiced in the report suggesting that "the quality of the Qatari workforce and education system as well as the business sophistication and innovation capacity of its public and private companies will need to improve significantly in order to be able to compete in global markets outside the hydrocarbon sector" (Planning Council, 2007, p. 13).

Qatar's interest in knowledge development must be placed in the context of broader interest in knowledge economy projects in the Arabian Gulf – Saudi Arabia, the UAE, Oman and Bahrain are actively diversifying their economies with tourism, heavy industry, and financial institutions, with varying degrees of success. Yemen, although not an official Gulf Cooperation Council member, has already suffered political disruption and economic hardship from lack of development and resource depletion. Yemen is not a major oil producer, but 70% of its export revenues derive from oil, and production peaked in 2001 (EIA, 2013b). For Oman, economic diversification is a pressing necessity, as their hydrocarbon reserves may be exhausted within 15 years. Although Oman is one of the largest non-OPEC oil and gas producers in the Middle East, more and more of its hydrocarbon production relies on Enhanced Oil Recovery (EOR), such as steam injection and polymers, because of difficult extraction problems. Gas and oil revenues provide over 40% of Oman's GDP, and the government announced in 2012 that oil must remain at $104 USD per barrel in order to produce enough revenue to carry out government functions, known as the "break-even oil price" (EIA, 2013a). Thus all countries of the Gulf including Qatar are cognizant of, and some are actively preparing for, the post-oil era.

In addition to a number of smaller projects, the emirate of Dubai (a prominent member of the United Arab Emirates) has introduced four large knowledge economy projects resembling Qatar Foundation's Education City: Dubai Knowledge Village and Dubai International Academic City (both owned by TECOM Investments, a subsidiary of Dubai Holding controlled by Sheikh Al Maktoum), and Dubai Media City and Dubai Internet City. These are free trade economic zones offering tax incentives and often 100% foreign ownership to create knowledge hubs in education, media and the Internet in Dubai. The education zones have attracted Russian, Australian, Indian and Iranian institutions such as Birla Institute, Islamic Azad University, Saint Petersburg State University, and The University of Wollongong. The King Abdullah University of Science and Technology (KAUST), one of the most heavily endowed universities in the world and run by the national oil company Saudi Aramco, represents one of the key knowledge economy projects in Saudi Arabia. Researchers there have forged scientific partnerships with the American University in Cairo, Stanford University, and Woods Hole Oceanographic Institute among others.

Education has been highlighted as one of the important pillars of knowledge economy development in the Qatar National Development Strategy 2011-16, which is being used as a road map for future growth. This detailed document, which outlines development goals for both the private and public sector including health, education, and industry, is the first five-year plan for the nation based on a broader and more theoretical framework presented in the Qatar National Vision 2030. All ministries and councils are expected to follow the principles in these two documents and to coordinate their own growth and development in line with both the broader goals and specific numerical targets of the 2011-16 Development Strategy.

The national goals for education parallel similar philosophical, social and economic justifications that are common in OECD countries: preparing citizens to participate in the rapid economic growth in Qatar, providing "a solid grounding in Qatari religious, moral and ethical values, national identity, traditions and cultural heritage," and aiding in the development of a more cohesive and participatory society (GSDP, 2011, p. 122). Fundamentally, training and education will help Qataris "make better decisions about health, marriage, parenting and social responsibility" as well as supporting "innovation in science, medicine and industry" (p. 122).

While the future educational goals of Qatar's emerging knowledge economy are ambitious, the various international analyses of knowledge economy initiatives in MENA and the GCC have on the whole suggested a lack of progress towards stated goals and targets. Aubert for example (one of the authors of an early World Bank knowledge economy report) argues concerning MENA knowledge economy projects in the last decade that "efforts to build more knowledge and innovation-oriented economies have, with the exception of the UAE, been insufficient to cope with demand for jobs, particularly from the youth bulge. The reasons are two-fold. First, there has been a rather timid shift toward KE, with a consequently limited impact on the economy. Second, people's skills have not adapted to the emerging knowledge-intensive sectors, even among university diploma holders" (Aubert, Karlsson, & Utz, 2013, p. 361). However, given the wide scope of the changes envisioned in this kind of economic transformation (which may involve such fundamental reorientations as a shift in world-view), it is doubtful that a wholesale economic revolution could occur within the one decade since GCC nations began seriously inaugurating knowledge economy projects.

In addition, the report entitled *Transforming Arab Economies: Traveling the Knowledge and Innovation Road*, sponsored by the Center for Mediterranean Integration (CMI) in collaboration with the World Bank, the European Investment Bank (EIB), and the Islamic Educational, Scientific, and Cultural Organization (ISESCO), examined progress in the MENA region in diversifying economies into knowledge production. The report argued that increased expenditure on education has not resulted in better educational outcomes: "although the countries of the Arab world have invested a respectable share of their gross domestic product (GDP) in education, those efforts are not yielding the expected results: more schooling has not been synonymous with more learning in most parts of the Arab world. International achievement tests reveal that students in the Arab countries still trail those of many other countries at comparable levels of development in the acquisition of basic knowledge" (World Bank, 2013, p. xvi).

However, these analyses can be criticized from a number of perspectives: for example, the first stage in large scale educational reform will necessarily involve substantial expenditures on basic infrastructure (new schools, new evaluation and assessment organizations, laboratories, textbooks, perhaps even overhauled or new education ministries) as well as extensive funding for needs assessment, planning, and implementation, which involves many man-hours of data gathering, meetings, reports and committees. Results such as improved international test scores may not become evident until reforms actually reach the class room level, such as the implementation of national standards and curriculum restructuring, and the maturation of teacher training, certification and professional development programs. Qatar, for example, began all of these kinds of educational reforms mentioned above ten years ago, but many of them, such as school assessment programs and professional development strategies, have only become fully functional in the last 2-3 years.

EDUCATION AND THE KNOWLEDGE ECONOMY IN QATAR

In the pre-oil era (pre-1940s – the first oil shipment left Qatar in 1949), there was little necessity for education due to the Bedouin lifestyle which was based on pastoral nomadism and various subsistence activities. Settled coastal merchants, however, did need to know mathematics for keeping accounts and letter writing to facilitate business relationships. Also, all Muslims were expected to know the Quran and important Hadiths, to be able to recite and understand important passages related to religious practice and personal conduct. This knowledge could be taught informally by literate sheikhs or through the small number of *kuttab* religious schools in the Gulf. A small number of American Mission schools began operation in the Gulf in the late 19th century, notably Al Raja School in Bahrain, but rulers were wary of potential Christian proselytizing, which is still banned in Gulf countries.

Qatar never experienced an industrialization stage of development following an agricultural stage in the sense that once oil was discovered by western companies, already mature technologies and technical workers from Britain and America were rapidly imported to develop the new oil wells. The first workers who provided manual labor for the oil companies were Qatari, but labor demands soon outstripped supply, necessitating the importation of Arab regional labor, and then Asian workers from the 1980s onwards. The Middle East region is the greatest importer of labor for the gas and oil industries, with an 86.4% foreign labor force and only 13.6% local labor force (Hays, 2013, p. 19).

In addition to labor, the requisite petroleum technologies, such as basic well equipment, and more advanced enhanced recovery techniques, gas to liquid processes, and processing and refining technologies have all originated in more developed countries from the energy giants ExxonMobil, Total, Schlumberger, and Shell, etc. rather than arising from homegrown Gulf solutions. This is an important fact, since this developmental path did not instill in Qatari society a full appreciation or understanding of how technology is predicated upon basic science research. Thus, until recently the Gulf countries did not develop advanced training and education programs in geology, petroleum and chemical engineering, or any of the other related sciences, until the late 20th century–with such exceptions as the highly regarded King Fahd University of Petroleum and Minerals (est. 1963)–since students could be sent abroad to established science and engineering programs in OECD countries once oil wealth became available for public higher education.

Education for a New Era

Beginning in 2001, the State of Qatar has undergone a radical transformation of its basic educational system. The impetus for this change can be directly attributed to a dynamic and widely revered leader in Qatar, the previous Emir Sheikh Hamad bin Khalifa Al Thani who inherited an under-developed country when he overthrew his father in a bloodless coup in 1995. In the same year, he founded Qatar Foundation for Education, Science and Community Development, the umbrella organization (still chaired by his wife Sheikha Moza bint Nasser) which would launch a range of educational and scientific organizations including the branch campuses of Education City. Since that time, spending on education has been well above world averages: whereas OECD nations spend an average of 3.7% of GDP on education, Qatar dedicated 4.8% of GDP to education in 2008 (GSDP, 2011). In 2012, the former Emir pledged to increase spending on education by 35% over the next 3 years.

The U.S. RAND Corporation developed the reform framework called Education for a New Era in 2002-2003, a comprehensive overhaul of Qatar's educational system. The four main principles were school Autonomy, Accountability, Variety and Choice. Under this schema, three kinds of primary and secondary schools now exist in Qatar: Independent, Private Arabic, and International. The Independent schools were new institutions set up under guidelines of the four main RAND Corporation principles. Preliminary assessment data from 2007 indicated that Independent school students were soon outperforming their peers studying in the traditional Ministry of Education (government) schools (Zellman et al., 2009, p. xix). In 2004, the SEC began converting all government schools to the Independent model and this transformation was completed in 2010. International Schools in Qatar are often subsidiaries or branches of well-known schools in France, Britain and elsewhere. There are also a number of highly specialized schools, such as Qatar Leadership Academy and Aspire Academy for Sports Excellence, and the DeBakey School. The Michael E. DeBakey High School for Health Professions at Qatar, whose main campus is located in Houston, Texas, is particularly relevant for Qatar's educational reforms since it follows the U.S. Advanced Placement curricula and was designed to train future professionals for Qatar's growing health sciences sector. Many of the recent graduates have matriculated into the Weill Cornell Medical College in Qatar.

Qatar Foundation and Education City

Founded in 1995 by the former Emir Sheikh Hamad bin Khalifa Al Thani, and managed by his wife Sheikha Moza bint Nasser, Qatar Foundation asked Virginia Commonwealth University in 1998 to set up a Design Program on a 14-kilometer square plot of land in Al Rayyan area near Doha known as Education City. The vision was to offer high quality bachelor degrees in Qatar through individual branch campuses of recognized American and international four-year undergraduate institutions. Other campuses in Education City include: Weill Cornell Medical College in Qatar (founded 2001), Texas A&M University at Qatar (founded 2003), Carnegie Mellon University in Qatar (founded 2004), Georgetown School of Foreign Service in Qatar (founded 2005), and Northwestern University in Qatar (founded 2008). In addition, Qatar Faculty of Islamic Studies (founded 2007) is headquartered in Education City. HEC Paris operates a business program (founded 2011) as well. Harvard Law School's Institute for Global Law and Policy (IGLP) will soon open a graduate law program in Qatar. Also, Qatar Academy in Education City, which provides K-12 programs, often serves as a feeder school for Education City campuses. University College London in Qatar has inaugurated a variety of programs aimed at preserving local culture, another facet of knowledge economies (intangible heritage) which is often overlooked due to the focus in development reports on the economic dimensions of knowledge. UCL Qatar offers degrees in Archaeology of the Arab and Islamic World (MA), Conservation Studies (MSc), Library and Information Studies (MA) and Museum and Gallery Practice (MA).

There have been some notable failures in the Gulf region with the overseas branch campus model fostered by Qatar Foundation, such as the 2009 closure of George Mason University in Ras al-Khaimah (UAE), or the shutting down of Michigan State University's branch campus in Dubai International Academic City (DIAC). MSU still maintains, however, two small Master's degree programs in Dubai. But Qatar's well-funded Education City campuses, despite the normal expected growing pains, are thriving and steadily increasing their enrolments and numbers of graduates each year.

Other Qatar Foundation Initiatives

Qatar National Research Fund (QNRF) was launched in 2006 with two programs: the Undergraduate Research Experience Program (UREP) and the National Priorities Research Program (NPRP), whose research priority areas are set by the government. Lead Principal Investigators are required to be Qatar residents, but international collaboration is common and encouraged. By the end of 2011, QNRF had disbursed over $345 million USD in research grants (Cecchine et al., 2012, p. iii). The UREP program introduces Qatari undergraduate students to real-world research problems, and many results have ultimately been published in peer reviewed scientific journals. The entire process provides hands-on experience for Qataris in the production of new knowledge.

Al Saadi's detailed survey in 2010 of Qatar's research capabilities, using data from questionnaires from 23 government, industry and educational institutions, indicated low research capacity in Qatar. However, his data was gathered circa 2007-2009, when QNRF was less than 3 years old. But Al Saadi's report (a PhD thesis project) did clearly reveal several interesting patterns. Only 45% of the 953 researchers in Qatar are Qatari nationals, and represent "500 for every million citizens (a mere one tenth of the world's benchmark)" (Al Saadi, 2010, p. 4.46). Also, internationally the average ratio of research assistants and technicians to researchers is three to one, while in Qatar it is only one to one, indicating a weakness in the scientific support system. Al Saadi also revealed the lower qualifications of researchers in Qatar than international norms: "Researchers' qualifications are distributed as follows: PhDs: 541 (38.3% are Qataris); Masters: 116 (29.3% Qatari); Bachelors: 296 (64.2% Qatari). This is in contrast to other countries where most researchers hold PhDs or masters degrees" (p. 4.46). Al Saadi's analyses also uncovered low levels of research in the areas of environmental engineering, medical engineering, industrial biotechnology, nanotechnology, and animal science.

Qatar Science and Technology Park (QSTP, launched 2009) is Qatar Foundation's technology business incubator and partnered with companies such as Shell, iHorizons, ExxonMobil, and GE. Many of the research projects involve energy, energy efficiency, alternative and clean energies (specifically solar power) and digital technologies. An innovative knowledge project was the joint venture among QSTP, the national airline Qatar Airways, and the CAS Department of Biological and Environmental Sciences at Qatar University to develop airline biofuels derived from algae feedstock. Airbus, Shell and several other QSTP local partners demonstrated in a similar experimental flight of Qatar Airways, that a 50-50 blend of synthetic Gas-to-Liquids (GTL) kerosene, of which Qatar is a significant producer, and conventional kerosene is a viable aviation fuel (Weber, 2013a, p. 94). Qatar has also established four research facilities in computers (Qatar Computing Research Institute), cardiology (Qatar Cardiovascular Research Center), biomedical sciences (Qatar Biomedical Research Institute), and the environment (Qatar Environment and Energy Research Institute).

Qatar University Reforms

Qatar University (QU) with the assistance of RAND Corporation began a comprehensive reform program in 2003, chronicled in a biannual bulletin entitled *Tawasol: Qatar University Education Reform Journal*. These reforms were designed in part to graduate more and higher quality knowledge economy graduates with research experience to shift the university's focus away from arts and humanities degree programs. Being a state institution with free tuition for Qataris subsidized by oil wealth has been challenging as many students treat admission to the university as a citizenship right rather than a privilege and reward for previous academic achievement. As the President of Qatar University Sheikha Al Misnad has commented: "the community believes that any student enrolled in the university must eventually get a degree, without consideration to the potential he possesses, or to the student's self-readiness to acquire academic achievements" (2008, p. 12). These comments, and another recent speech by Al Misnad posted on Youtube, sparked angry social media campaigns against the President by Qatari students, some of whom called for her removal.

In 2009, the University launched the Qatar University Strategic Plan 2009-2012. The plan is divided into four Key Performance Areas (KPAs): academic program enhancement, expansion of research capacity, development of community outreach and services, and facilities management and internal communications (Henry, 2009, pp. 15-16). In alignment with knowledge economy goals and the development of human capacity in science and engineering outlined in the Qatar National Vision 2030, Qatar University has invested substantially in research, with assistance from the national grant-funding agency Qatar National Research Fund (QNRF). In 2010-2011, investment in research totaled $150 million USD and rose to $200 million USD in academic year 2011-2012 (QU, 2012, p. 12). Qatar University is one of the largest recipients of QNRF funding, and was awarded over $54 million USD worth of grants in the fifth cycle of the National Research Priority Program, representing 40% of all awarded proposals in Qatar (p. 12). The Qatar University reform project has established quantitative targets and performance indicators for research; for example, the number of individual research collaborations (individual grants), number of signed and effective MoUs, the degree of primary stakeholders' satisfaction with research outcomes, and the number of publications in indexed journals (QU, 2010, p. 14).

To disburse and manage these funds, several academic research centers have been created or substantially overhauled at QU, for example: the newly formed Social and Economic Survey Research Institute (SESRI) which has conducted the first systematic country-wide social science surveys of both Qataris and expatriates, and the QU Mobility Innovations Center (QMIC), formerly the QU Wireless Innovations Center (QUWIC). QMIC engages in research to develop services and software for intelligent transport systems, logistics management, and road safety. Since Qatar has some of the highest rates of motor vehicle accident mortality and morbidity in the world, road safety is a priority research area for Qatar. Other research centres at Qatar University, which collaborate with the gas and oil sectors, include the Gas Processing Center (GPC), Center of Advanced Materials (CAM) (formerly Materials Technology Unit (MTU)), Environmental Studies Center (ESC), and Central Laboratories Unit (CLU). Although these research centers focus on solving practical problems that arise in Qatar's gas and oil sectors (see the case study below on pulse chlorination), the hope is that patentable processes will result from this R & D that can be sold to other oil-producing nations, thus bringing direct royalty revenue to Qatar through the sale and lease of proprietary knowledge.

A large question looms of whether these new research monies can instill a research culture and scientific view of the world in Qatari society. The country has been struggling with providing even the most basic scientific infrastructure – importing reagents, attracting laboratory technicians (mostly expatriates), building basic specialized facilities including vivariums for animal trials and testing, and developing international partnerships, etc. The Biomedical Training Program for Nationals, which the author helped to develop at his own institution Weill Cornell Medical College in Qatar, hopes to address the lack of Qatari nationals in science management positions and to encourage science careers by training science graduates in managerial techniques and communications skills.

Unfortunately, almost ten years after the establishment of Education City, a palpable divide between this institution and Qatar University still exists, partially because of the large physical distance between the two institutions and the difficulty of travelling between the two locations which discourages intellectual exchange. In a relatively bold opinion piece in the Arabic daily *Al Sharq*, a Qatari woman voiced the complaints of some Qatar University students who feel the sharp divide between their national university and the 'elitist' American universities of Education City (HBKU). She accused the country of treating Qatar University graduates as "second class" students (*daraja thania*) compared to the opportunities and privileges of HBKU students (Al Khater, 2012).

ICT Infrastructure and Online Ecosystems

ictQATAR is the telecommunications regulator and developer for the State of Qatar. ictQATAR has an ambitious digital agenda: by 2015, QR 6.2 billion will be spent on various media and digital infrastructure initiatives, including: a new optical fibre network to expand universal high-speed broadband Internet throughout Qatar; Qatar's first communications satellite, which will be built and launched in Qatar; two additional international submarine cables to provide extra security and service (the region has suffered two recent serious Internet outages from severed marine cables); and provision of free Wi-Fi throughout the entire country (ictQATAR, 2013, p. 1). As with other ministries in Qatar, ictQATAR is essentially implementing the science, technology, education and development goals enunciated in the QNV 2030. Between 2008 and 2012, the Internet penetration rate for individuals doubled and computer and mobile device usage rates are among the highest in the MENA region. The average Qatari home contains two computers, three mobile phones, and one smart phone (p. 4). However, although there was an impressive 80% increase in households with broadband Internet since 2010 (85% of total households), over 50% of households surveyed reported low speeds between 256 Kbps and 1 Mbps (p. 5). Thus development goals and aspirations sometimes collide with development realities in Qatar, as is common elsewhere in the world.

Also ictQATAR as detailed in the *Qatar National ICT Plan 2015* is strengthening the regulatory structure of the Internet in the country with new laws as well as a framework to integrate e-learning and e-content into Qatar's classrooms, including tools to assess the effectiveness of these initiatives. In addition to a national digitized e-library which will include scanned rare books on Arabic history and language from Qatar's Heritage Library, advanced customized learning management systems will be deployed throughout schools to help parents, students, teachers, and school officials communicate and collaborate online in the learning process. Information gathering and communication between schools is currently limited due to the isolated proximities of many of Doha's schools which are hard to locate physically since many are situated in residential and commercial zones.

ictQATAR is also unveiling over a five year phased period an e-maturity and self-assessment tool which "measures the extent to which ICT is integrated and adopted in each school. Schools can evaluate their current e-maturity level, compare themselves to other schools, and develop targeted action plans to update and improve their technology" (ictQATAR, 2012, p. 40). Since national curriculum standards have now been established, and course materials developed, ictQATAR plans to facilitate the digital delivery of approved content. This scheme illustrates the benefits (increased efficiency, standardization, shared resources) of one central Internet service provider and regulator and one educational management council (SEC), a regime that the author has dubbed "Big Education." However, provision of e-content and learning management systems must simultaneously be coupled with teacher and parent training since both technophobia and lack of training in using computers among teachers have been identified as serious barriers in the educational system. Also a potentially serious technical issue is the routing of all Internet traffic through one ISP (Ooredoo) since cyber-attacks or server failure could be disastrous. Qatar's RasGas company, for example, suffered a serious disruption in computer operations in 2012 from the malicious Shamoon computer virus possibly originating in Iran or Syria. However, Qatar is addressing this issue through the installation of more fiber optic trunk lines, the launching of a satellite to provide backup services, and the gradual opening of Qatar's telecommunications market, most recently to Vodaphone, as competition to the single national carrier Ooredoo.

Arab digital content and Arabic language machine translation has been the focus of several Qatar technology stakeholders, such as the digital companies headquartered in QSTP, Carnegie Mellon University in Qatar, and Qatar Computing Research Institute (QCRI). Also ictQatar and Yahoo! are partnering to develop an Arabic online knowledge ecosystem. QCRI maintains an Arabic language technologies division to extend and preserve the Arabic language in virtual space and increase knowledge production in Arabic-speaking countries. Some of their language projects include: Arabic speech recognition in formal Arabic and other Arabic dialects, machine translation, Arabic key-word and semantic content indexing, and language education materials (QCRI, 2013). In addition, Qatar Foundation and the British Library are collaborating to jointly release one half million pages of digital documents on Arabic culture and history both in Arabic and English.

Potential Barriers to Educational Reform and Knowledge Economy Development

The Rentier economy – an economy which derives value primarily from rents on natural resources – has in and of itself been accused of breeding inefficiency, and hindering the development of other economic sectors. Qatar is generally recognized as possessing a rentier economy and political regime, although this is a complex and debated concept (Crystal, 1995; Herb, 1999; Gray, 2013). According to international educationalist Andreas Schleicher, "more generally, countries with greater total rents from natural resources tend to be economically and socially less developed, as exports of national resources tend to appreciate the currency, making imports cheap and the development of an industrial base more difficult" (Schleicher, 2012). Schleicher's arguments suggest another serious issue about education in rentier economy countries: "there is also a significant negative relationship between the money countries extract from national resources and the knowledge and skills of their school population …. Israel is not alone in outperforming its oil-rich neighbors by a large margin when it comes to learning outcomes at school, this is a global pattern that generally [holds] across 65 countries that took part in the latest PISA assessment" (Schleicher, 2012).

Education in Qatar is predominantly not co-educational, and school segregation is the norm throughout the Gulf region: Qatar University classes are segregated and the university maintains a separate male and female campus as well as separate libraries; and at the secondary level, only international schools are co-ed. According to a 2011 survey by SESRI, only 17% of Qatari parents would send a daughter to an International School (SESRI, 2011, p. 11). Acceptance for co-education, however, increases for both the primary school and university school levels, but on average only 30.3% of Qatari parents would send a daughter to a co-educational university (p. 11). A more pressing issue, however, is the acceptance of the interaction of non-related men and women in the workplace – a condition which will be required if more women enter the workforce. The best place to reduce tension and develop acceptable patterns of public male-female relations would be in schools from an early age. However, gender segregation is a deeply rooted custom in Qatar and elsewhere in the Gulf.

The lack of female employment options is clearly due in part to cultural restrictions and attitudinal barriers. Nursing for example, is objectionable due to the possibility that women may potentially work with males alone at odd hours. Many of these restrictions are claimed by conservative religious authorities to be religious in nature, yet as the bans against women driving in Saudi Arabia demonstrate, many restrictions placed on women are based on culture and customs, not Shariah law. Nursing also holds low status since the profession is mostly staffed by Philippina nurses in Qatar. According to the Supreme Health Council, the total number of nurses in Qatar in 2011 was 10,262, but only 445 were Qatari nationals (SCH, 2013). The country's only graduate nursing program, a branch campus of Calgary University in Alberta, produces about 20 Qatari nurses per year, so the shortage of Qatari nurses will continue in the near future.

At this stage, the Qatari culture is highly divided on the issue of gender segregation – for example, a female judge was appointed for the first time in Qatar in 2010. Yet in 2013, the SEC ordered that female Qatari teachers, administrators, accountants and other staff at Independent boys' schools were required to be transferred to girls' schools. Female expatriates working at boys' schools were ordered to have their contracts terminated. In addition, a proposed draft law will only allow women to work in nurseries in Qatar; thus the concept of gender-designated jobs is still prevalent. Placing gender restrictions on job categories can unfortunately lead to labour inefficiencies and gender inequality as well as shortages in specific areas.

Private and state owned enterprises in Qatar who rely on Qatar's education system to supply needed skills have expressed some concerns about the lack of education-workplace alignment. Due to the lack of adequate government training programs, Qatar Petroleum, for example, relies on educational programs abroad or one of the two technical colleges (College of the North Atlantic in Qatar, and a Qatar branch of Houston Community College) for worker training. But as Al Saadi reported in his study of technology transfer in Qatar, "according to Qatar Petroleum (QP), the education system does not adequately prepare workers. QP management pointed out that the company has to reteach basic academic subjects like math and English to secondary school graduates for at least 26 weeks before they can be enrolled in special trade training. This results in huge production loss" (2010, p. 5.35).

The introduction of the American models of education carried out in the RAND reforms, containing certain embedded cultural values, would inevitably cause clashes with a conservative Muslim culture. Along with the increasing presence of western expatriate advisors and managers at high levels of business and government, American-style education has been both quietly and publically criticized as negatively influencing public behaviors and altering Qatari and Islamic values and culture. For example, the "One of Us" campaign was started by Qatari women to encourage modest dress in Qatar (no tank tops, bare shoulders or bare legs); the penal code of Qatar specifically prohibits immodest dress in public. In addition, in 2012 Qatar University announced its new dress code, which prohibited 'visible tattoos,' 'clothes with inappropriate text or images,' and 'fad hair styles like dreadlocks and the unnatural coloring of hair.' These restrictions were clearly targeted at the kinds of styles found among western college students.

Diglossia and the English Language in Instruction

As judged by the number of academic books and articles published each year internationally, English has become the world's dominant language of science and engineering. However, Arabic is the official language of all the GCC nations and residents there speak a dialectical variant called *Khaleeji* or Gulf Arabic which is significantly different in pronunciation and vocabulary from Egyptian, Levantine, and Moroccan dialects. Most media and academic exchanges take place in a stylized form of classical Arabic called Modern Standard Arabic (*fusha*) which must be formally learned. This situation gives rise to the situation of diglossia in the Arabic language in which educated Arabs must become bilingual, and if working in an international field, trilingual due to the widespread adoption of English. Many educators in Qatar have recognized the problem of the lack of professional competence among Arabic speakers in mastering Modern Standard Arabic, for example in Northwestern University-Qatar's journalism and communications programs, where many students will proceed to a career in Arabic media. Modern Standard Arabic is an artificial language with a complex grammar system for Gulf Arabs and must be learned through study. However, MSA is the standard for Al Jazeera network programming and most other radio and television news shows across the Arabic-speaking world. NU-Q is considering options to improve student proficiency in formal Arabic such as providing more courses in Modern Standard Arabic, hiring an Arabic writing specialist and requiring passing a course called Arabic for Media. Gerd Nonneman, Dean of Georgetown SFS in Qatar, estimates that although two-thirds of Georgetown students come from an Arab background, only half of them achieve professional competence in Arabic despite the language courses offered at the university (quoted in Guttenplan, 2013). An interesting unpublished study at QU by Shaheen, Al Mansoury, & Saleem, indicated that Bedouin mothers in Qatar were abandoning their native dialects. The small survey "showed 92 percent of Qataris already speak civilised Arabic than Bedouin because this is the dominating language and the language of education" (Abedelmoaty, 2011). The distinction 'civilized Arabic' (MSA) versus 'Bedouin' Arabic used in the study is particularly interesting and troubling since it reveals evidence of high and low status language registers, and these attitudes threaten the existence of the remaining Bedouin dialects.

Lack of a Reading Culture in the Gulf

Literacy is a recent phenomenon in Qatar due to the widespread availability of public schooling only in the late 1950s and 1960s and the oral nature of Bedouin culture itself. Remarkably, Qatar has achieved almost full adult literacy in the last two decades and many younger Qataris are proficient English speakers as well. A number of reading researchers have commented on the lack of reading culture in the Gulf, obviously linked to social institutions such as the *majlis* and *diwan* (daily or weekly gatherings to share news, poetry, and to socialize) and the rise of the popularity of the cell phone for exchanging information (Weber, 2013b, pp. 11-26; O'Sullivan, 2009; Kiranmayi, 2012). The *majlis*, *diwan*, cell phone and social media groups effectively perform many of the core functions of reading. However, these communicative modes are not adequate for transmitting and retaining expert knowledge which is found in advanced textbooks and peer reviewed specialist journals only available to the general public through purchase, subscription, or access to an academic library. Educators in Qatar have therefore been active in promoting the skills and pleasures of reading to move students beyond the free knowledge of variable quality available on the Internet towards more reliable specialist materials, since reading, information assessment, and analytical/critical skills are key competencies for a future advanced knowledge worker.

Both local educators and government agencies in Qatar have been addressing low interest in reading with book fairs, inviting a floating bookshop to dock in Doha harbor, and institution-wide reading projects. The Qatar government gives free vouchers to Qatari citizens to purchase children's books each year at a large annual book exhibition. The Georgetown SFS library, with approximately 60-80,000 volumes, has opened its collections to membership from the community. Otherwise, the only other major public library in the country outside of the Qatar University Library is the relatively inaccessible Qatar National Library, or Dar Al Kutub, in a crumbling historic building downtown. A new central library, however, is being constructed in Education City which will offer full public access to both books and extensive academic research databases.

Pulse Chlorination: Successful Case Study of Collaborative Knowledge Economy Implementation

The successful scientific partnership among Qatargas, Imperial College London, and two Netherlands companies–KEMA Technical & Operational Services and Aquator BV–to develop pulse chlorination technology for sea water intake pipes provides a clear representative case study of a knowledge economy project in Qatar in action. Seawater is used throughout the Gulf for both seawater desalination and industrial cooling. However, to prevent fouling of intake pipes and equipment, biocides – primarily sodium hypochlorite ('chlorine bleach') – must be added to the water to control marine growth. However, Gulf governments have been concerned about the release of chlorine into the Gulf and subsequent toxicity to already declining shrimp and fish stocks in addition to widespread coral mortality. Thus instead of continuous chlorination, smaller doses can be pulsed to prevent macrofouling. Ironically, one of the most significant and widespread pipe fouling organisms is the pearl oyster (*Pinctada radiata*), which was the previous basis of the Gulf's wealth and trade. By studying the organism's response to chemical stress, scientists were able to kill off adult organisms with small pulses of chlorine which did not kill the bivalves directly, but triggered a switch between their aerobic and anaerobic modes causing physiological exhaustion and death (Macdonald et al., 2012, p. 292). Thus less chlorine was introduced into the environment. This development would not have been possible without research into the specific Gulf marine ecology and the life cycle of pearl oysters in order to adapt pulse chlorination to the needs of Qatar's gas and desalination industries. Although pulse chlorination has been successfully used in the Netherlands, it was introduced in Qatar into a very different marine ecosystem.

SOLUTIONS AND RECOMMENDATIONS

One potential solution to increasing knowledge economy production in Qatar would be to draw on the underutilized female labor force for knowledge workers since between 60-70% of the entrants to Qatar's universities are now women. Yet Qatari females make up only 2% of the entire labor force in Qatar (QSA, 2011). Female graduates in the Gulf normally enter fields such as education and healthcare, and their job prospects in other areas can be limited by traditional gender roles. Attitudes towards women in the workplace, however, may be changing: in the SESRI 2012 omnibus survey of 2,854 Qataris and expatriates, the most frequent answer for the question "how does education help women?" was "to secure better careers" (97% of respondents), rather than "to become a better mother or wife," "find a better husband" or "secure more freedom" (p. 20). Also, Qatar University has been successful in increasing the number of female students and graduates in engineering, a typically male-dominated field in many countries. From 1998 to 2009, female enrolment in the QU engineering program increased 181% from 177 to 497 students, and awarded bachelor's degrees rose during the same time period from 19 to 86 (a 353% increase) (Sulaiman & Almuftah, 2010, p. 4).

Male students and future male knowledge workers face very different but substantial barriers and challenges in comparison with Gulf females. First, until recently, science and technical work in Qatar has been done almost exclusively by expatriates; thus there are few local role models in science and mathematics careers. In addition, due to generous state benefits including high salary, retirement packages, and perks such as free water and electricity, Qatari males have been leaving the workforce in their 40s and 50s. From ages 25-40, approximately 85-90% of Qatari males are employed, which is close to international norms for workforce participation. However, at age 40 a sharp decline occurs such that by age 55 only 45% are employed and only 25% at age 60 (GSDP, 2011, p. 155). Thus a Qatari male who has accumulated some wealth in his early years has several financial alternatives to working: living off savings, renting property to expatriates, investments, or serving as a silent partner in a business. Highly educated and successful Qataris in technical fields may unfortunately be removing their expertise from society at a young age. This phenomenon has only been studied in depth in a small number of research articles and government reports in Qatar (GSDP, 2011; Weber, 2013c), but has been documented in depth in the UAE in a series of articles by Forstenlechner and Rutledge (2010; 2011). The main government strategy up to this point to encourage knowledge economy careers for males has been the provision of educational opportunity, i.e. new higher education programs in medicine, computer science, design, and political science through Qatar Foundation branch campuses and science research institutes.

Changing the public perception of STEM field careers could be accomplished through simple means such as increasing or subsidizing salaries–although this is not a long-term sustainable solution as it represents a resource drain–and media and social media campaigns. The greatest palliative and virtue, of course, will be *time* to allow these very recent knowledge economy projects, the vast majority of them only launched between 5-10 years ago, to mature and develop.

FUTURE RESEARCH DIRECTIONS

Due to the novelty of Qatar's ambitious educational reforms, only a small handful of books and articles on education in Qatar exist at the present time, many of them generated by RAND Corporation. Thus all forms of quantitative and qualitative research, both longitudinal and cross-sectional, should be undertaken and publically disseminated. Fortunately, knowledge economy activities have spawned the necessity of collecting fine-grained economic and social indicators which are now gathered by most government bodies in Qatar. However, more of this data needs to be made publically available to education researchers, preferably via the Internet.

CONCLUSION

Although gas and oil and their related industries will continue to dominate Qatar's short term economic future, a planned and phased transition to a knowledge economy has been set out in two recent comprehensive planning documents, the *Qatar National Vision for 2030* and the *Qatar National Development Strategy 2011-2016*. Education and research are both vital to these efforts. The K-12 system has been dramatically reformed according to a framework set out by the U.S. RAND Corporation. The implementation of this vision is also in full swing in higher education, with Education City campuses graduating future knowledge workers in the fields of medicine, computer science, design, museums management, archaeology, and business management. In addition, millions of dollars of research funds are now available in several competitive grant programs administered by the Qatar National Research Fund. Qatar University has also undergone rapid change beginning in 2003, although some backlashes to these innovations have occurred, initiated by both local students and parents, who have forced the reintroduction of Islamic dress codes, the reduction of English language foundational courses, and the return of the language of instruction of some programs back to their original Arabic.

The small size of the Qatari population, estimated at between 250,000 – 300,000, is a serious obstacle to building knowledge capacity on a large scale. Presently, a key question is how much technology and skills transfer is actually occurring from expatriate scientists and engineers to the local workforce and at what point will Qatari STEM field workers be able to replace fully the expatriates in high skills positions. Given the enormous size of Qatar's oil and gas industries (Qatar is the world's number one exporter of Liquid Natural Gas), this changeover may never be practical and some expatriate labor will probably always be required for a variety of needs ranging from manual labor to high tech positions. Also, employment in the military, police and government offices is more attractive at this point to the average Qatari job-seeker, and the status of science and engineering careers needs to be actively promoted through public information campaigns. Media efforts for recruiting Qatari nurses have met with some success at the University of Calgary in Qatar nursing school. The lack of tuition fees may possibly be causing an undervaluing of the growing higher education opportunities for Qatari youth, who can enter and leave the educational system with relative ease, or study abroad, without incurring penalties. All of these challenges can be solved if government leaders choose to forge ahead and maintain the current pathway of development: the financial resources are certainly available to attempt this economic transition, but the complex nature of Qatar's revolutionary path towards knowledge production makes possible future scenarios difficult to predict.

REFERENCES

Abedelmoaty, A. (2011, July 16). Bedouins fast diminishing: Study. *The Peninsula*. Retrieved from http://www.thepeninsulaqatar.com/qatar/159171-bedouins-fast-diminishing-study.html

Al Khater, M. (2012, May 13). Ara' wa maqalat. *Al Sharq*. Retrieved from http://www.al-sharq.com/Detail/articles/81

Al Misnad, S. (2008). Qatar University president highlights QU's ongoing progress. *Tawasol: Qatar University Educational Reform Journal, 9*, 11–14.

Al-Saadi, R. A. (2010). *Effectiveness of technology transfer in the search for sustainable development: The case of Qatar*. (Unpublished Doctoral Dissertation). Cranfield University, Cranfield, UK.

Aubert, J.-E., Karlsson, M., & Utz, A. (2013). Building knowledge and innovation-driven economies in Arab Countries: How to do it. In T. Andersson, & A. Djeflat (Eds.), The real issues of the Middle East and the Arab spring: Innovation, technology, and knowledge management (pp. 359-69). New York: Springer Science+Business Media.

Aubert, J.-E., & Reiffers, J.-L. (2003). *Knowledge economies in the Middle East and North Africa: Toward new development strategies*. Washington, DC: The World Bank.

Cecchine, G., Darilek, R. E., Harrell, M. C., Mattock, M. G., Culbertson, S., & Ortiz, D. S. … Nardulli, B.R. (2012). Sustaining the Qatar National Research Fund. Santa Monica, CA: RAND Corporation.

Crystal, J. (1995). *Oil and politics in the Gulf: Rulers and merchants in Kuwait and Qatar*. Cambridge: Cambridge UP.

EIA (U.S. Energy Information Administration). (2013a). *Oman*. Retrieved from http://www.eia.gov/countries/analysisbriefs/Oman/oman.pdf

EIA (U.S. Energy Information Administration). (2013b). *Yemen*. Retrieved from http://www.eia.gov/countries/analysisbriefs/Yemen/yemen.pdf

Forstenlechner, I., & Rutledge, E. (2010). Unemployment in the Gulf: Time to update the "social contract." *Middle East Policy, 17*(2), 38–51. doi:10.1111/j.1475-4967.2010.00437.x

Forstenlechner, I., & Rutledge, E. (2011). The GCC's "Demographic imbalance": Perceptions, realities and policy options. *Middle East Policy, 18*(4), 25–43. doi:10.1111/j.1475-4967.2011.00508.x

General Secretariat for Development Planning (GSDP). (2008). *QNV 2030 (Qatar national vision 2030)*. Doha: GSDP.

General Secretariat for Development Planning (GSDP). (2011). *Qatar national development strategy 2011-2016 (QNDS)*. Doha: GSDP.

Gray, M. (2013). *Qatar: Politics and the challenges of development*. Boulder: Lynne Rienner.

Guttenplan, D. D. (2012, June 11). Battling to preserve Arabic from English's onslaught. *New York Times*. Retrieved from http://www.nytimes.com/2012/06/11/world/middleeast/11iht-educlede11.html?_r=0

Hays Specialist Recruitment (Australia) Pty Ltd. (2013). *Oil & gas global salary guide 2013*. Retrieved from http://www.hays.com.au/oil-gas-salary-guide/

Henry, J. B. (2009). Qatar University strategic plan 2009-2012. *Tawasol: Qatar University Educational Reform Journal, 11*, 15–16.

Herb, M. (1999). *All in the family: Absolutism, revolution, and democracy in the Middle Eastern monarchies*. Albany: State University of New York Press.

ictQATAR. (2012). *Qatar's national ICT plan 2015*. Doha: Supreme Council of Information & Communication Technology.

ictQATAR. (2013). *Qatar's ICT landscape 2013*. Doha: Supreme Council of Information & Communication Technology.

Kiranmayi, N. C. (2012). Rethinking the methodology used in developing reading skill in Omani students of a design college. *International Review of Social Sciences and Humanities, 4*(1), 24–32.

Macdonald, I. A., Polman, H. J., Jenner, H. A., & Quyam, S. Q. B. M. (2012). Pulse-Chlorination®: Anti-fouling optimization in seawater cooling systems. In S. Rajagopal, H.A. Jenner, & V.P. Venugopalan (Eds.), Operational and environmental consequences of large industrial cooling water systems (pp. 287-301). New York: Springer Science+Business Media.

Nature (Ed.). (2006). Where are the new patrons of science? *Nature, 444*(2), 35-36.

O'Sullivan, A. (2009). Reading in the Gulf: Is it time for a re-focus? In Cultivating Real Readers (pp. 43-51). Al Ain, UAE: HCT Ma'rifa.

Planning Council, State of Qatar. (2007). Turning Qatar into a competitive knowledge-based economy: Knowledge economy assessment of Qatar. Doha: State of Qatar.

Qatar Computing Research Institute (QCRI). (2013). Arabic language technologies. Retrieved from http://www.qcri.com/our-research/arabic-language-technologies/arabic-language-technologies

Qatar Investment Authority (QIA). (2012). Qatar Central Bank. SWF Institute. Retrieved from http://www.swfinstitute.org/swfs/qatar-investment-authority/

Qatar Statistics Authority (QSA). (2011). *Labor Force Sample Survey*. Doha: Qatar Statistics Authority.

Qatar Statistics Authority (QSA). (2012). *QSA labor force sample survey 2012*. Doha: Qatar Statistics Authority.

Qatar University (QU). (2010). *Qatar university strategic plan 2010-2013*. Doha: Qatar University.

Qatar University (QU). (2012). *Qatar university annual report 2011/2012*. Doha: Qatar University.

Schleicher, A. (2012). Knowledge and skills are infinite – oil is not. *OECD educationtoday*. Retrieved from http://oecdeducationtoday.blogspot.com/2012/03/knowledge-and-skills-are-infinite-oil.html

Silatech/Gallup. (2012). *Qatar's rising entrepreneurial spirit: February 2012 brief*. Doha: Silatech.

Social & Economic Survey Research Institute (SESRI). (2011). *Annual omnibus survey: A survey of life in Qatar 2011*. Doha: SESRI.

Social & Economic Survey Research Institute (SESRI). (2012). *Annual omnibus survey: A survey of life in Qatar 2012*. Doha: SESRI.

Sulaiman, N. F., & Almuftah, H. (2010). Women in the engineering pipeline: What do preliminary studies say about female engineering students at QU? *Tawasol: Qatar University Educational Reform Journal*, *12*, 3–9.

Supreme Council of Health (SCH). (2013). *Healthcare statistics*. Retrieved from www.sch.gov.qa

United Nations Development Programme (UNDP). (2002). *Arab human development report 2002*. New York and Oxford, U.K.: Oxford University Press.

Weber, A. S. (2013a). Review of sustainable and renewable energy activities in the State of Qatar. In *Renewable and Sustainable Energy Conference (IRSEC), 2013* (pp. 91- 95). New York: IEEE.

Weber, A. S. (2013b). Literacy development in an Arab Gulf country. In N. Bakić-Mirić & D. E. Gaipov (Eds.), *Building cultural bridges in education* (pp. 11–26). Cambridge: Cambridge Scholars Press.

Weber, A. S. (2013c). Youth unemployment and sustainable development: Case study of Qatar. *Revista de Asistenţă Socială*, *12*(1), 1–11.

World Bank. (2003). *Knowledge economies in the Middle East and North America: Toward new development strategies*. Washington, DC: World Bank.

World Bank. (2008). *The Road not traveled: Education reform in the Middle East and North Africa*. Washington, DC: World Bank.

World Bank. (2013a). *KEI and KI Indexes (KAM 2012)*. Retrieved from http://info.worldbank.org/etools/kam2/KAM_page5.asp

World Bank. (2013b). *Transforming Arab economies: Traveling the knowledge and innovation road*. Washington, DC: International Bank for Reconstruction and Development/The World Bank.

Zellman, G. L., Ryan, G. W., Karam, R., Constant, L., Salem, H., & Gonzalez, G. … Al-Obaidli, K. (2009). Implementation of the K-12 education reform in Qatar's schools. Santa Monica, CA: RAND Corporation.

ADDITIONAL READING

Al-Jaber, H., & Dutta, S. (2008). Qatar: Leveraging technology to create a knowledge based economy in the Middle East. In *Global information technology report 2007-2008* (pp. 133–143). Geneva: World Economic Forum.

Al Misnad, S. (1985). *The Development of modern education in the Gulf*. London: Ithaca Press.

Al-Shelek, A. Z., Aquil, M., & Al-Abdulla, Y. (2009). *Political development of Qatar: From the creation of the emirate to the independence of the state*. Doha: Renoda Modern Press.

Bahgat, G. (1999). Education in the gulf monarchies: Retrospect and prospect. *International Review of Education*, *45*(2), 127–136. doi:10.1023/A:1003610723356

Brewer, D. J., Augustine, C. H., Zellman, G. L., Ryan, G., Goldman, C. A., Stasz, C., & Constant, L. (2007). *Education for a new era: Design and implementation of K-12 education reform in Qatar*. Santa Monica, CA: RAND Corporation.

Brock, C., & Levers, L. Z. (Eds.). (2007). *Aspects of education in the Middle East and North Africa*. Oxford: Symposium Books.

Bunglawala, Z. (2011). *Young, educated and dependent on the public sector: Meeting graduates' aspirations and diversifying employment in Qatar and the UAE*. Doha: Brookings Doha Center.

Donn, G., & Al Manthri, Y. (2010). *Globalisation and higher education in the Arab Gulf states*. Oxford: Symposium Books.

Foley, S. (2010). *The Arab Gulf states: Beyond oil and Islam*. Boulder: Lynne Reinner.

Gallup, Inc. & Silatech. (2010). *Silatech index: Voice of young Arabs*. Washington, DC: Gallup, Inc.

Ghabra, S., & Arnold, M. (2007). *Studying the American way: An assessment of American-style higher education in Arab countries*. Washington, DC: Washington Institute for Near East Policy.

Koren, M., & Tenreyo, S. (2012). Volatility, diversification and development in the Gulf Cooperation Council countries. In D. Held & K. Ulrichsen (Eds.), *The Transformation of the Gulf: Politics, economics and the global order* (pp. 189–217). London: Routledge.

Peterson, J. E. (2009). Life after oil: Economic alternatives for the Arab Gulf states. *Mediterranean Quarterly, 20*(3), 1–18. doi:10.1215/10474552-2009-011

Smith, S. (2004). *Britain's revival and fall in the Gulf: Kuwait, Bahrain, Qatar, and the trucial states 1950–71*. London: Routledge Curzon.

Supreme Education Council (SEC). (2008). *The TIMSS 2007 study in Qatar: Summary of key findings and options for policy and further study*. Doha: Supreme Education Council.

Weber, A. S. (2009). E-learning in the Gulf Cooperation Council countries (GCC): Problems and prospects. In V. Uskov (Ed.), *Web Based Education (WBE 2009)*. Calgary: IASTED.

Weber, A. S. (2010). *The Development and current status of web-based learning in Qatar and the GCC states*. Occasional Papers 5. Doha, Qatar: CIRS.

Weber, A. S. (2011a). What is a knowledge economy? Oil-rich nations post-oil. *The International Journal of Science in Society*, *2*, 1–9.

Weber, A. S. (2011b). Comparative analysis of national higher education policies in the Gulf Cooperation Council (Saudi Arabia, Bahrain, Oman, Qatar, UAE, Kuwait). [Honolulu, HI: HICE.]. *Proceedings of HICE, 2012*, 3153–3174.

Weber, A. S. (2011c). The Role of education in knowledge economies in developing countries. In G. Akçamete (Ed.), *Proceedings of the Third World Conference on Educational Sciences*. Istanbul: Bahçeşehir University Press.

Zahlan, R. S. (1998). *The Making of the modern gulf states: Kuwait, Bahrain, Qatar, The United Arab Emirates and Oman*. Reading: Ithaca Press.

Zellman, G. L., Constant, L., & Goldman, C. A. (2011). *K-12 education reform in Qatar*. Santa Monica, CA: RAND Corporation.

KEY TERMS AND DEFINITIONS

Big Education: Analogous to Big Science; centrally planned education initiatives characterized by big staffs, big budgets, and national or international scope.

Big Science: Large post-WW II science projects characterized by big staffs, budgets, and instruments, i.e. CERN's Large Hadron Collider.

Education City: A consortium of nine U.S., British and French branch campuses administered by Qatar Foundation and located in a 14 square-kilometer area west of Doha.

Enhanced Oil Recovery (EOR): Various techniques to extract more oil from wells which are declining in productivity.

Gulf Cooperation Council (GCC): A regional political organization formed in 1981 comprising the countries of Bahrain, Kuwait, Oman, Qatar, Saudi Arabia, and UAE. Loosely synonymous with 'Arabian Gulf' or 'Gulf.'

ictQATAR: The Supreme Council of Information and Communication Technology (est. 2004) creates Internet policy and oversees telecommunications infrastructure development.

Knowledge Assessment Methodology (KAM): A World Bank set of software tools to help countries assess their knowledge capacity, consisting of indicators which comprise the Knowledge Index (KI) and Knowledge Economy Index (KEI).

Knowledge Economy: An economic mode of production centered on information, skills, know-how, and training and education as opposed to capital or labor.

Rand Qatar Policy Institute (RQPI): Operational 2003-2013; a collaboration between U.S. Rand Corporation and Qatar Foundation to research issues in education, energy, security, and policy.

Rentier State: In rentier state theory, the state derives most of its revenue from renting resources (such as oil) to foreign powers.

Soverign Wealth Fund: A state-controlled investment fund designed to utilize surplus national revenue for investing and economic stabilization.

STEM: The fields of Science, Technology, Engineering and Mathematics, generally viewed as critical to knowledge economy development.

Qatar Foundation: Qatar Foundation for Education, Science and Community Development (founded 1995 by the former Emir of Qatar and his wife) oversees educational development and research in Qatar, including QNRF, QSTP and Education City.

QNRF: Qatar National Research Fund, the national research funding body in Qatar launched in 2006.

QSTP: Qatar Science and Technology Park (launched 2009); the technology business incubator for Qatar Foundation partnered with companies such as Shell, iHorizons, ExxonMobil, and GE.

Silatech: An Arab youth employment and entrepreneurship organization founded in Qatar by Sheikha Moza bint Nasser.

Supreme Education Council (SEC): (est. 2002) Manages all aspects of education in Qatar and replaces the former Ministry of Education.

This work was previously published in a "Handbook of Research on Higher Education in the MENA Region" edited by Neeta Baporikar, pages 52-73, copyright 2014 by Information Science Reference (an imprint of IGI Global).

Section 4
Cases and Applications

This section discusses a variety of applications and opportunities available that can be considered by practitioners in developing viable and effective STEM Education programs and processes. This section includes 19 chapters that review topics from case studies in STEM Education. Contributions included in this section provide excellent coverage of today's education and IT communities and how research into STEM Education is impacting the social fabric of our present-day global village.

Chapter 45
3D Multi-User Virtual Environments in Science Education: Potential and Challenges

Yufeng Qian
Northeastern University, USA

ABSTRACT

The purpose of this chapter is to identify the potential and challenges in science education in the use of 3D MUVE science programs. These programs offer a number of instructional benefits in motivating and engaging students and in improving their science learning and scientific inquiry. 3D MUVE is a promising media in narrowing gender and racial achievement gaps and enabling an authentic and valid assessment of science education. Like all new instructional technologies, however, the wide use and implementation of 3D MUVE technology in mainstream science classroom is still facing a number of challenges, which are mainly related to technological complexity and cost, and design difficulty in incorporating some elements critical to inquiry-based learning into the 3D MUVE environment. To overcome these identified challenges and make optimal use of the opportunities, suggestions for integrating 3D MUVE into science curriculum and classroom are made and discussed, along with future research directions.

INTRODUCTION

U.S. students are lagging behind in science literacy. In the most recent Program for International Student Assessment (PISA) in science, the ranking of U.S. students fell from 19th in 2009 to 22nd in 2012. While the average science literacy score of 15 year-old U.S. students caught up with that of the OECD (Organization for Economic Cooperation and Development) countries, the ranking of U.S. students is still trailing their peers in 15 of the 34 other OECD countries and 7 of the 31 non-OECD countries (National Center for Education Statistics, 2013). The 2011 National Assessment of Educational Progress revealed that less than a third of the nation's 8th graders were proficient in science and only 2% achieved the advanced level (National Center for Education Statistics, 2011). The science competency of U.S. students has caused a great concern. Mr. Duncan, U.S.

DOI: 10.4018/978-1-4666-7363-2.ch045

Secretary of Education, called U.S. performance on the 2012 PISA "educational stagnation," which "must serve as a wake-up call against educational complacency and low expectations" (U.S Department of Education, 2013). In response to U.S. students' consistent unsatisfactory science performance, President Obama has indicated that over the next decade, American students' achievement in science education must move to the top in international assessments, including PISA (EdWeek, 2012).

While a number of factors may be responsible for U.S. students' current middling scores on high-profiles international tests in science and a relatively high percentage of students less proficient in science on the nationwide assessment, the dominant science instruction pedagogy appears to be the most problematic (Roth & Garnier, 2007; Wallance & Louden, 2002). Heavily influenced by high-stake tests and aggravated by the relatively less time allotted to science teaching than to reading and math, which by law are the nation's educational priority, one-way delivery of curriculum content and outcomes of science inquiry process in the classroom appear to be the only choice left for time-constrained science teachers. High interest activities and real life projects, which usually take more instructional time and demand more efforts that require higher-order thinking skills, have usually been introduced as a sidebar for innovation and engagement purposes, rather than as the mainstream pedagogy for science learning (Roth & Garnier, 2007). In response to the decry against the lecture-based science education, hands-on activities have been introduced into the science curriculum in the past decade (President's Council of Advisors on Science and Technology, 2010). While seemingly more effective than one-way delivery of science content, the use of hands-on instruction is no better than cookbook receipts, instructing students on "what to do next" (Ibarra, 2006). Students are not challenged to think and act independently and thereby acquire the scientific inquiry skill that is at the heart of science and science education (National Research Council, 1996).

There has been a call for inquiry-based instruction in science education. The National Science Education Standards (NSES) include inquiry as a content to be learned and a way to learn science (National Research Council, 1996). The National Science Teachers Association recommends that all K–16 teachers use scientific inquiry as the centerpiece of the science classroom and ensure that students develop a deep understanding of science and scientific inquiry (NSTA, 2004). Despite the decade-long call for inquiry-based approach in science teaching and learning, its implementation in K-12 classroom has been problematic. Teachers' unfamiliarity with and inability to effectively design and use the inquiry-based method is one of the most important factors (Kazempur, 2009). Many teachers, unclear how to design and implement inquiry in classroom, simply substitute real scientific inquiry with traditional "cookbook" experiments (Wallence & Louden, 2002).

To address this unfortunate situation in science education, scientists, science education researchers, and school teachers have started joint efforts to explore how to optimize the use of technologies to improve science teaching and learning, including the use of emerging three dimensional multi-user virtual environment (3D MUVE) technology. As an innovative instructional tool and approach, 3D MUVE holds a great deal of potential in engaging and challenging students in a socially situated and distributed learning environment that simulates real-life, science-related issues/projects. However, like any other emerging learning tool, the design, development and implementation of 3D MUVE for science education are still facing significant challenges before its potential can be fully unleashed and realized. The purpose of this chapter is therefore to identify the potential and challenges in the use of this emerging technology in science education through reviewing and analyzing existing 3D MUVE science programs.

BACKGROUND

A 3D MUVE learning environment is a simulated, three-dimensional online space where learners, represented by avatars, can interact with each other, computer generated agents, digital artifacts, and the virtual environment in real time that is similar to real-life face-to-face interactions. Different from current mainstream online learning environments (e.g., Blackboard, MOOC), 3D MUVE has a number of significantly distinctive features that have the potential to provide a new learning platform that enables richer and deeper learning experience.

In early 2000s, immersion, engagement and presence were often recognized as the critical features of 3D virtual learning environments (McLellan, 2004; McMaham, 2003). Immersion refers to the sense of users that the world they are in is real and complete; engagement is a state of mind in which users invest substantial emotion, time, energy and effort into a task; and presence refers to users' perception of "being there" (McMaham, 2003). In looking more closely from the perceptual/psychological perspective, researchers have argued that immersion, engagement, and presence are subjective senses that are not unique only to 3D MUVE, not totally dependent on the technical capabilities, but rather on a range of contextual factors, including the user's state of mind (Lomard & Ditton, 2006; Slater, 2004). Presence, for example, is the result of a combination of a number of factors, including quality of social interaction, realism in the environment, effect of "transportation," degree of immersiveness generated by the interface, user's ability to accomplish significant actions and its social impact, and user's response to the computer as an intelligent and social agent. A user in a highly immersed environment with teleporting capabilities may not have a strong sense of presence of "being there" or "togetherness" due to the quality of social interaction (McMaham, 2003).

Concurring with Lomard and Ditton (2006), Dalgarno and Lee (2010) have argued that the dependency of these subjective senses should not be deemed unique characteristics of 3D virtual environments; instead, they have stressed the "representational fidelity" and "learner interaction" as the distinguishing characteristics of 3D virtual learning environments, arguing that it is essentially the fidelity and the interactive capability that will result in the perceptual and psychological sense of immersion, engagement, or presence.

It should be noted that while representational fidelity is a unique property of 3D MUVEs and has been essential to many high-end commercial 3D MUVE programs, such as PlayStation's Battlefield, the ultimate aim of a 3D MUVE for a learning purpose goes beyond merely providing a virtual space that is as lifelike as possible. Kjems (2010) suggested that what is essential to a virtual environment for a learning purpose is the representation of the most important features and accurate facts relevant to an object, an issue, a phenomenon, or the environment which helps decision making.

Closely related to "representational fidelity," another unique technical capability that distinguishes a 3D MUVE learning environment from any other interactive online learning environments is its ability to allow users to visualize and manipulate both representations of the concrete, tangible realities and abstract, intangible concepts in the same space. It has been widely recognized that 3D MUVE displays life-like environment in three dimensions where learners experience smooth temporal changes and interactivity, which provide a very realistic experience to learners (Dalgarno & Lee, 2010). In addition, a virtual world "bridges the gap between the concrete world of nature and the abstract world of concepts and models" (Mintz, Litvak, & Yair, 2001, p. 294), in which users can observe and manipulate normally inaccessible objects, variables, and processes. The concretization of abstract and intangible objects, concepts, ideas, or processes and the capability of allowing users to manipulate and observe the results provide a superior learning experience that enable richer and deeper understanding and application of knowledge.

Given the unique technical capabilities of 3D MUVE, there has been a surge of research interest, since the early 2000s, in exploring the potential this technology may afford to science education. Joining the efforts of science education researchers and science teachers in public school systems, several signature 3D MUVE programs have burst into the limelight of science education research, designed to improve K-12 science teaching and learning.

SciCentr, founded in 1998 by Cornell University and probably among the earliest efforts in enhancing science education using virtual worlds, hosts and maintains an online 3D MUVE science museum in the format of exhibits, games, and interactive laboratories that aims to improve the interest of young people, especially those in underserved and minority communities in science, and reach the goal of building basic scientific literacy. The design of the SciCentr is based on discovery learning and community building that promotes children's self-exploration of scientific topics of their own choice and provides an online community where children can share their passion and outcomes of a particular topic with the science community. Since 2001, SciFair, the flagship virtual world program of SciCentr, has involved more than 1,000 middle school students and teachers annually (Corbit, Bernstein, Kolodziej, & McIntyre, 2006).

Alien Rescue, developed by researchers at the University of Texas at Austin, is an online, problem-based, 3D, virtual learning environment for sixth grade space science. Role playing scientists in *Alien Rescue*, students are presented with a problem–to rescue six different species of aliens who have been displaced from their home planets by finding them suitable relocation sites within the solar system. In this program students are not guided on how to begin problem solving. They must probe available resources and figure out how to proceed on their own. To solve this problem, students need to think and act like scientists in a team, explore a variety of tools, databases, and resources, and engage in the problem-solving process and scientific inquiry-based activities, during which they must identify their own learning needs (self-directed), discriminate among several viable options and justify their selection with documenting evidence (decision-making and justification), and collaborate and negotiate effectively with team members (collaboration). *Alien Rescue* has been used as part of the science curriculum by 16 middle schools in Central Texas with a diverse ethnic base, as well as schools in twenty other states and three countries: Australia, Canada, and China (Liu, Horton, Kang, Kimmons, & Lee, 2013).

Different from the open and unguided design of *SciCentr* and *Alien Rescue*, three widely cited 3D MUVE science programs, *River City, EcoMUVE and Quest Atlantis,* have focused more on the guided inquiry process in science education. *River City*, funded by the National Science Foundation and developed by the Graduate School of Education at Harvard, is a scientific inquiry-based 3D MUVE that targets middle school students. Adopting a storyline –a familiar game design scheme among mainstream games, *River City* is set in a 19th-century city with a river running through it and its citizens facing a chronic illness. The students' task is to find out why the residents of *River City* are getting sick and what can be done to help them. The problems are interdisciplinary and integrate contents from science, history, and social studies, allowing students to experience real world inquiry skills required in disentangling multi-causal problems in a complex environment. Centered on the scientific inquiry skills and on contents in biology and ecology, *River City* guides students through making observations, posing questions, developing hypothesis, investigating, explaining, predicting, proposing answers, and communicating the results in the form of a letter to the Mayor of River City. Since 2001, *River City* has been implemented successfully in twelve states, involving hundreds of teachers and thousands of students (Nelson, Ketelhut, Clarke, Bowman, & Dede, 2005).

Building upon and continuing the research with *River City* and funded by the Institute of Education Sciences in the U.S. Department of Education, *EcoMUVE* has focused on the application of the unique visualization and teleporting affordances of 3D MUVEs to address a challenging topic of understanding complex causality ecosystems in 6th grade science classroom. In *EcoMUVE*, students can zoom in to the microscopic level, travel to different points in time, and see effects emerge across time and distance, which is difficult to achieve in the classroom learning environment. The *EcoMUVE* program has been adopted district wide at the sixth-grade level in Cambridge Public Schools in Massachusetts (Metcalf, Kamarinan, Tutwiler, Grotzer, & Dede, 2011).

Quest Atlantis, funded by the National Science Foundation and MacArthur Foundation and developed by the Center for Research on Learning & Technology at Indiana University, is another widely cited innovative science learning program. Similar to *River City* and *EcoMUVE*, *Quest Atlantis* is a 3D multi-user online learning community intended to engage children ages 9–12 in science learning. Its legend is that the people of "Atlantis" face an impending disaster as their world is slowly being destroyed through environmental, moral, and social decay. The task of the project is to save Atlantis. Leveraging 3D technologies and game-based methodologies, the problems are presented in an interactive narrative in which the ''reader'' has agency in co-determining how the story unfolds (Barab, Sadler, Heiselt, Hickey, & Zuiker, 2007). To echo the national call for inquiry-based math and science learning, *Quest Atlantis* was designed to involve students in refining questions, gathering data, evaluating information, developing plausible interpretations, and reflecting on their findings. Similar to other multi-user virtual worlds, *Quest Atlantis* is a globally distributed community with more than 20,000 participants from four continents (Barab, Arici, & Jackson, 2005).

Continuing the research effort in using 3D MUVE to enhance U.S. middle school students' performance in science and scientific inquiry skills, *SAVE Science* (Situated Assessments Using Virtual Environments) aims to improve U.S. science education by developing science assessments situated in authentic inquiry in 3D MUVE. As a joint research project, researchers from the University of Maryland at College Park, Temple University, and Arizona State University have created and are implementing game-based assessment modules for evaluating science content and inquiry skills in grades 7-8 in the Mid-Atlantic region. The conventional paper-and-pencil based, multiple-choice science assessments are unable to assess students' scientific inquiry skills, which is at the heart of science education. As a promising alterative, *SAVE Science*, using a database of student interactions in the program, is designed to assess not only students' evolving patterns of understanding of the science content but also to provide detailed information about students' problem-solving process and inquiry activities (Ketelhut, Nelson, Schifter, & Kim, 2010).

Parallel to the *SAVE Science* project, the *VPA* project (Virtual Performance Assessment Project) is another recent research venture in science assessment using 3D MUVE. Funded in 2008 by Institute of Education Sciences of the U.S. Department of Education, as well as the Gates Foundation, researchers from the Harvard Graduate School of Education are developing virtual performance assessments to assess middle school students' science inquiry skills. Its goals are to assess the feasibility of using 3D MUVE as an assessment format and provide states with valid and reliable assessments of students' performance in both science content and scientific inquiry. At present, the *VPA Project* is planning large-scale implementation targeting about 7,000 middle school students (Code, Clarke-Midura, Zap, & Dede, 2012).

POTENTIAL OF 3D MUVE IN SCIENCE EDUCATION

As evidenced in these pioneering 3D MUVE programs for science education, this emerging technology holds a great number of instructional benefits for improving science learning and is potentially a viable solution to the pressing issues facing science education in schools. Four major instructional benefits of 3D MUVE to science education have surfaced in the aforementioned exemplar programs, which will be discussed below.

1. Intrinsically Motivated, Cognitively Engaged, and Significantly Improved Learning

Schools are faced with the challenge of engaging this generation of students in formal learning in the classroom. Studies over a span of two decades reveal a consistently low level of engagement in the classroom, which has resulted in widely reported classroom boredom and escalating drop-out rate of high school students. One reason for this may be related to the widening gap between tech-savvy students and print-centric schools. Children today are growing up in a rapidly evolving digital media environment where using cutting-edge gadgets has become an integral and important part of their growing and learning experience. Despite children's massive use of digital technologies outside of the classroom, schools still continue to operate within a print-based cultural logic.

3D MUVE resembles most closely the digital game environment that consumes most of U.S. teenagers' leisure time. The fidelity of representation, a critical technological capability of 3D MUVE, could enable a user's high level of immersion and consequently, a strong sense of presence (Dalgano & Lee, 2010). Immersion and presence alone cannot automatically lead to motivation and engagement in learning a subject and completing a learning task. When learning environments are more authentic and inquiry-based, which requires and enables collaboration, they are more likely to motivate students. Motivation is the initial critical element that could boom or doom learning and will set the stage for engagement. When students are highly motivated, they are more likely to engage in higher levels of learning and thinking by investing considerable time and mental effort in the learning process (Blumenfeld, Kempler, Krajcik, & Blumenfeld, 2006). Leveraged by the high degree of resemblance of real world, a 3D MUVE program is able to create authentic, real-life situations, issues, and problems to motivate and engage learners, which is evident in the existing science programs reviewed in this chapter.

Results from the implementations of *River City* with approximately 2000 students in public school classrooms from geographical diverse urban areas indicate that both male and female students are highly motivated by the 3D MUVE program, with students reporting that they "felt like a scientist for the first time" (Clarke & Dede, 2005). At the same time a statistically significant positive impact ($p < .05$) on student science achievement has been found when compared to a paper-based curriculum (Ketelhut, Dede, Clarke, & Nelson, 2006; Nelson, 2007). Further analysis reveals that, on average, an increase of only eight inquiry activities in the *River City* resulted in a 5% increase on science content score ($p < .002$) (Ketelhut & Dede, 2007). Similarly, students in *SciCentr* rate their learning experience in 3D MUVE significantly higher than in a traditional science teaching environment (p <.05), stating that they had more fun and had learned more. Moreover, *SciCentr* appears to have the greatest impact on students who begin with neutral or negative attitudes toward science

(Corbit, Bernstein, Kolodziej, & McIntyre, 2006). This echoes the results of *River City,* which show a greater impact on learning for low achieving students in inner-city schools (Dede, Brown-L'Bahy, Ketelhut, & Whitehouse, 2003) and the results of *Alien Rescue*, in which girls had more positive attitudes toward the simulation than their male counterparts, along with a significantly higher gain score than boys (33 percentage points vs. 30 percentage points) (Liu, Horton, Kang, Kimmons, & Lee, 2013). Some of these programs have also taken a further look into the relationship between motivation/engagement and learning achievement and as a result, a significant relationship between these two factors has been found. For example, in *Alien Rescue,* student attitude towards the simulation program has contributed significantly to their posttest achievement ($B = .05$, $t(384) = 5.8$, $p < .01$.) (Kimmons, Liu, Kang, & Santata, 2012). Obviously, if well designed and wisely used, the 3D MUVE would be a viable platform to motivate and engage students in science learning, which leads to significant improvement in science learning.

2. Authentic Context for Scientific Inquiry

The 3D MUVE provides a viable platform in supporting the authentic scientific inquiry process and helping learners acquire inquiry skills defined by the National Science Education Standards (1996). As in *River City* and *Quest Atlantis*, the scientific inquiry process and skills are seamlessly embedded in the immersive probing environments. In such environments students are first exposed to a complex authentic problem, such as finding solutions to save Atlantis that is facing similar problems as Earth. In order to disentangle the complex multi-causal problems, students need to go through the process of making observations, refining questions, gathering data, evaluating information, developing plausible interpretations, and reflecting on their findings, a set of inquiry activities that are critical to science and science education. In addition to inquiry skills, the science content and skills specified in the curriculum are embedded in the inquiry activities, accomplishing a task that demonstrates students' mastery of these contents and skills.

The unique visualization capability of both concrete and abstract representations of 3D MUVE makes it a viable platform for conducting scientific inquiry. In 3D MUVEs students have opportunities to experience, observe, manipulate, and experiment a broad range of objects, variables, processes, phenomena, which are normally inaccessible, impossible or impractical in reality. In *EcoMUVE,* for example, the visualization and concretization of the causality of ecosystem makes the learning of a difficult topic more intuitive and straightforward. In addition, students can also test and observe instantly the results of an alternative reality in the 3D MUVE environment, which is impossible to achieve in a classroom learning environment. Moreover, in most 3D MUVEs, students can teleport instantly from one place to another, "physically" (via avatar) visiting a place thousands of miles away or even on the other side of the globe and meeting and chatting with people and content experts from around the world. These capabilities can create a profound sense of motivation and engagement conducive to a rich and deep inquiry experience.

The past ten years of research in 3D MUVE science programs has provided strong evidence supporting that 3D MUVE can be an effective platform to provide authentic contexts that enable inquiry-based science learning (Barab, Sadler, Heiselt, Hickey, & Zuiker, 2007; Clarke, Dede, Ketelhut, & Nelson, 2006; Corbit, Bernstein, Kolodziej, & McIntyre, 2006; Liu, Horton, Kang, Kimmons, & Lee, 2013). "Inquiry-based learning can be made accessible to all students, through the experiences and tools that are possible in virtual worlds" (Metcalf, Kamarainen, Tutwiler, Grotzer, & Dede, 2011, p. 90).

3. Authentic Assessment of Science Content and Scientific Inquiry

There has been a decry against the conventional paper-and-pencil based, multiple-choice selection tests in science education. The current standardized science tests typically present a series of isolated fact-based questions, poorly representing the complexity of how real-world science is constructed, thus cannot effectively test understanding of scientific concepts or the nature of scientific inquiry (SAVE Science, n.d.). In order to address the test format issue and include assessment of students' scientific inquiry skills, some standardized science tests have tried to include open-ended questions in an effort to contextualize test questions in detailed narration (National Assessment Governing Board, 2012). However, the lengthy text-based, open-ended questions involve a student's reading ability, and the accuracy of an answer relies on the student's literacy in both reading and science (Ketelhut, Nelson, Shcifter, & Kim, 2013). The National Science Education Standards (1996), viewing assessment as an integral part of the learning process and a primary feedback mechanism in the science education system, calls for a shift to authentic assessment that "requires students to apply scientific knowledge and reasoning to situations similar to those they will encounter in the world outside the classroom, as well as to situations that approximate how scientists do their work" (p. 78). However, the paper-and-pencil based test makes it difficult to contextualize a test question; in addition, the scientific inquiry as defined by the NSES that involves higher order skills is not easily measured by multiple-choice tests (Southerland, Smith, Sowell, & Kittleson, 2007).

SAVE Science is a pioneering effort in rethinking and innovating assessment in science education in the U.S. Taking advantage of the immersive and realistic environment enabled by the visualization capability of both concrete world and abstract concepts of 3D MUVE, *SAVE Science* is able to contextualize the assessment by placing a problem-based test in a realistic context and students have access to information related to the problem in multiple ways, just as scientists would in the real world. The *SAVE Science* has shown that integration of assessment of content with scientific inquiry is possible in 3D MUVE, where "students must use their scientific inquiry skills to solve the content-based problem, while the tools they choose and how they interpret data gathered through use of those tools relies on their understanding of the content" (Ketelhut, Nelson, Schifter, & Kim, 2013, p.179). In addition to the advantages of contextualization of the test problem and integration of science content and inquiry process, 3D MUVE-based assessment can "record all interactions that take place within the environment, producing data from students that can be analyzed to infer evolving levels of competency around science inquiry and concepts (Ketelhut, Nelson, Schifter, & Kim, 2013, p. 177)," as well as their problem-solving strategies. In addition, the qualitative content analysis from the pilot study with 43 seventh grade students in a near-urban district does reveal that students demonstrate a high level of interest and enjoyment of the program, as well as a tentative indication that "contextualizing questions does improve students' ability to answer, if students take an active part in the contextualized assessment tasks" (Ketelhut, Nelson, Schifter, & Kim, 2013, p. 187).

Similarly, taking advantage of the unique technological capabilities of 3D MUVE, Harvard Graduate School of Education's *VPA project* has adopted the evidence-centered design approach to collect data on aspects of knowledge, skills, and abilities throughout the scientific inquiry process, including students' theorizing, questioning and hypothesizing, investigating, analyzing and synthesizing, which are aligned with current national standards (Clarke-Midura, Mayrath, & Dede, 2013). The results from research studies over a 5-year span (2008 - 2013), involving over thousands of middle school students, have indicated "this type of assessment is practical and affordable at scale, as well as more valid for sophisticated performances like scientific inquiry than paper-and-pencil, item-based assessments" (Dede, 2013, p. 6). Since sophisticated technologies like 3D MUVE can capture students' interactions in an invisible and non-obtrusive way, teachers and schools can use this information for diagnostic, formative, and summative assessment purposes, which could be a great potential complement to standardized testing format. In addition, students have reported very positive experience with the assessment format in VPA. "Tests I take in school are stressful and more fact based. This was more like learning" (Code, Clarke-Midura, Zap, & Dede, 2012, p. 243).

4. Narrowed Achievement Gaps

The gender- and race-based achievement gaps in K-12 science have been a persistent issue that has caused a grave educational and social concern across the nation. The most recent 2011 Nation's Report Cards has shown a 5-point gap for 8th grade boys over girls, which represent a fairly consistent pattern of girls trailing boys in science (NAEP, 2013). Moreover, the 2009 PISA results in science reveal that the biggest gender gap in favor of boys was observed in the United States and Denmark. Regarding the racial achievement disparity in science, while it is encouraging to see the average score gap between White and Black students and White and Hispanic students narrowed from 2009 to 2011 (NAEP, 2011), numerous studies have shown that a racial-based, science achievement disparity has been substantial and persistent throughout primary and secondary education for decades in this nation (Bacharach, Baumeister, & Furr, 2003; Bankston & Caldas, 1997; National Center for Education Statistics, 1991-1999).

Making optimal use of emerging technology and media to address the national science achievement gap has been part of the effort in the aforementioned exemplary 3D MUVE science programs. *SciFair*, for example, has targeted especially middle school students from underserved, rural, and minority communities (Corbit, Bernstein, Kolodziej, & McIntyre, 2006). Similarly, *River City* has aimed to improve the educational outcomes of low-achieving, middle school students in major urban areas with high proportions of ESL and free-and-reduced-lunch population, who are disengaged from schooling and typically are difficult to motivate even by good teachers using conventional inquiry-based pedagogy (Dede, Ketelhut, & Ruess, 2002). In the design of *River City*, researchers intentionally created a lead female figure to model science success for girls. To emotionally evoke and contextually engage students, the design of the key region in *River City* is similar to the tenement area with which most immigrants, minority groups, and low SES students are familiar. The results from the implementation of *River City* over thousands of inner-city, middle school students have shown that this program was successful with both girls and minorities in affective (self-efficacy, motivation and enjoyment) and cognitive areas (science content and inquiry). More interestingly, the program has a significant effect on the lowest achieving group of students (bottom third), who as a whole improved their content knowledge by 63% (Dede, Ketelhut, & Nelson, 2004).

In a similar effort, *Alien Rescue*, considering the gender-based differences in preferences for and attitudes towards game-based learning, has used "saving aliens" as the theme, which may emotionally engage more girls than boys. The results show that while boys outperformed girls on both the pretest and the posttest, the latter had a significantly higher gain score than boys and more positive attitudes toward the simulation than their male counterparts. This finding is encouraging and promising, suggesting that the preexisting gender-based disparity in science achievement could be minimized in a technology-based learning environment and the preexisting gaps "may be mediated by altering curricular decisions to support more egalitarian learning" (Kimmons, Liu, Kang, & Santana, 2012, p. 366).

CHALLENGES OF USING 3D MUVE IN SCIENCE EDUCATION

As with all new instructional technology, the wide use and implementation of 3D MUVE technology in mainstream science classroom at present is still facing a number of challenges. They are mainly related to technological complexity and cost, required computing power that is not well supported by current technology infrastructure at school, and design difficulty in incorporating some elements critical to inquiry-based learning into the 3D MUVE environment.

1. Technological Challenge of Complexity and Cost

Educators will face enormous technological challenges in developing 3D MUVE programs efficiently and effortlessly in order to use them as a mainstream method in the science classroom. Design and development of 3D MUVE programs is a specialized, time-consuming task. As discussed previously, fidelity of representation and visualization of both concrete and abstract representations are two unique technological features of 3D MUVE that distinguish it from any other types of online learning environments. However, creating a high degree of representational fidelity and visualization for a learning environment requires high end sophisticated programming skills that need to involve software engineers or programmers in game design. The complexity and cost associated with the development of 3D MUVE may have partially explained the current existence of a limited number of full-fledged 3D MUVE programs for science education as reviewed in this chapter, which were almost all federally funded. Apparently, the complexity and associated cost of design and development of a full-fledged 3D MUVE for science learning has become an enormous hurdle that calls for the funding, support, and collaboration from every aspect of society (i.e., government, technology industry, and education).

However, there has been no lack of effort from the public domain in developing open source based 3D MUVE platforms for educational purposes, including Second Life, Open Cobalt, and OpenSimulator. While current and existing, open-source 3D MUVE platforms are quite viable in providing a virtual life-like classroom environment that enables immersive, presence, co-presence experience and collaboration opportunities, they are however limited in the technological capability in visualization of abstract concepts, data, phenomenon, which is central to science learning and scientific inquiry process. In addition, other features that are essential to 3D MUVE for science learning are basically not supported by the open source programs, including data collection mechanism for assessment purpose. Thus, current open source 3D MUVE platforms can help provide an immersive and collaboration environment, but they are presently not able to fully enable scientific inquiry processes and activities.

Initiating the effort in supporting science research, Intel, the leading company in 3D product development (including 3D Internet) in the technology industry, has been working on developing ScienceSim, a 3D virtual environment platform for science learning and research since 2009 (Intel Lab, 2009). Its goal is to enable researchers from virtually anywhere to participate in experiments in real time that involves visualization and analysis of large, complex data sets from different sources. Another objective of ScienceSim is to provide a "turnkey kit" free to researchers, with which they can use to develop innovative applications involving collaborative visualization (Intel Lab, 2009). Free high-end 3D MUVE programs like ScienceSim would significantly reduce the technological complexity and cost of developing 3D MUVE programs and leverage the use of 3D MUVE in science classroom in the near future.

Reminiscent of many other online technologies, the present challenge to 3D MUVE development due to its technological complexity and cost is deemed temporary. When the Web was initially used circa 1995, only Web developers who had the knowledge of HTML and later JavaScript language were able to build Web sites. Nowadays, with the "readymade" Web solutions, almost everyone can create a highly customized interactive and media-rich Web site and instantly publish it. To realize its full potential, 3D MUVE platform development must be similarly transformed to simplify the design and development process for non-expert users by providing template/modular solutions with an extremely user-friendly and highly customizable content creation system. Intel's pioneering effort in developing ScienceSim signals the start of transformation of current sophisticated 3D MUVE technology to daily classroom use.

2. Instructional Design Challenge of Scaffolding

Design and implementation of some instructional strategies that are essential to constructivist inquiry-based learning has been another major challenge in 3D MUVE science programs. Given the authenticity, complexity and thus higher demand on learners' affective, cognitive, and social ability in an inquiry-based science learning environment, scaffolding is deemed paramount in supporting students' success in this learning environment. Scaffolding is a modeling and coaching process in which an expert such as a teacher provides guidance and support to help a novice (e.g., a learner) progress systematically to the next potential level of development (the zone of proximal development). The support is gradually reduced and eventually removed/faded until learners are able to function independently, autonomously and competently (Lepper, Drake, & O'Donnell-Johnson, 1997). Many research attempts have been undertaken to develop theoretical frameworks and specific instructional strategies to scaffold students' learning in a technology-enhanced, constructivist learning environment (Brush & Saye, 2001; Dabbagh, 2002; Lee & Dalgarno, 2011; McLoughlin, 2002; Sharma & Hannafin, 2007; Winnips, 2001). According to Kim and Hannafin (2011), for example, a technology enhanced, problem-based learning environment needs to provide four types of scaffoldings:

1. *Procedural scaffolds* that guide students in "how to do something," such as the scientific inquiry steps and processes in science learning;
2. *Conceptual scaffolds* that provide support to students' understanding of the content;
3. *Metacognitive scaffolds* that assist students in reflecting on and self-monitoring of their thinking and learning process; and
4. *Strategic scaffolds* that help students to view an issue from multiple perspectives and consider alternative approaches to addressing the same issue.

Encompassing all types of scaffolding to support the learning of science content and inquiry in 3D MUVE has encountered a number of implementation issues. Finding the right balance is probably the most challenging. Scaffolding is all about the right type of help that comes at the right time, in the right place, and in the exact amount. Providing too much scaffolding could deviate from the nature and ultimate goal of authentic learning, in which learners are encouraged to acquire scientific inquiry skills through active self-directed exploration and discovery, reflection, and collaboration. Conversely, too little scaffolding could cognitively overwhelm and therefore demotivate and even defeat learners, which could result in their loss of motivation and inability to complete the learning task. As reported in a study on the *River City* program, some students reported not knowing where to focus their attention in the environment and difficulty in keeping track of the many sources of information encountered (Nelson, Ketelhut, & Schifter, 2010). Similarly, some students using the *Quest Atlantis* program demonstrated the "frustrated engagement," due to "the lack of navigational and operational competence to complete the quests" (Lim, Nonis, & Hedberg, 2006, p. 222), which results in their inability to complete the task. There were also numerous occasions where some students were taken off track and lost their focus on the quests by moving around aimlessly in the 3D world or distracted by some objects/elements while working on a quest. The availability of procedural scaffolding could have greatly kept students' mental resources on the task, but the directions on "click here for the next step" would be apparently instructive to learners that deviates from the open-ended design of an authentic learning environment (Lim, Nonis, & Hedberg, 2006).

Parallel to the delicate balancing act of scaffolding and authenticity of a learning environment, when to "fade" scaffolds is another pressing issue in designing 3D MUVE programs. Fading, which involves gradually reducing and eventually removing scaffolds as the learner's capabilities increase, is a defining feature of scaffolding that distinguishes itself from other types of learning support (Collins, Brown, & Newman, 1989; Guzdial, 1994; Lepper, Drake, & O'Donnell-Johnson, 1997; Stone, 1998). In a face-to-face learning environment, a teacher can intuitively and spontaneously decide the type and amount of scaffolds he/she will need to provide to assist a student by observing, monitoring, and assessing a student's behavior, performance, and interaction and determine when to fade support. In online environments, however, the difficulty is apparent in achieving this level of interaction dynamic and calibrated support based on ongoing diagnosis of learners' behavior and performance. Current 3D MUVE science programs have been able to design anticipated learning needs and embedded scaffoldings to support procedural, conceptual, metacognitive, and strategic needs in the inquiry process to assist learners' understanding of the content and the inquiry skills. However, the design and implementation of "fading" appears to be missing, which has been scarcely discussed in the existing 3D MUVE programs.

Attempting to address the design issue of fading, Dalgarno and Lee (2010) proposed using configurable parameters as an alternative to the current pre-determined scaffoldings based on anticipated learning needs. With all the system-based scaffoldings embedded in the learning environment, configurable parameters give learners options whether and to what degree a type of support is needed, such as "make all possible support available," "suggest support only when I am clearly on the wrong track," or "never suggest support." However, a significant issue identified in the research into the provision of scaffolding in a 3D MUVE program is that inexperienced/low achieving learners in this learning environment were not able to make appropriate procedural or metacognitive decisions by seeking guidance and making use of the scaffolds embedded in the learning environment. Thus, this group of learners, who do not necessarily make appropriate self-regulatory decisions, may still face the challenge of choosing the right option.

RECOMMENDATIONS FOR IMPLEMENTATION

It is evident that 3D MUVE has manifested significant instructional benefits in motivating and engaging students and improving science learning and scientific inquiry; it is also a promising media in narrowing the gender and racial achievement gaps if intentionally designed to address the issue. However, like many other emerging technology and media in the history of instructional technology, the current design complexity and costs associated with the development of 3D MUVE applications, as well as the dominant science instructional practice, have made them largely prohibitive in mainstream classrooms. These challenges cannot be fully resolved, at least in the near term. How to minimize the identified challenges so as to make optimal use of the opportunities? The following section makes suggestions regarding the implementation of 3D MUVE in K-12 science curriculum, along with future research directions.

1. Making Optimal Use of 3D MUVEs to Motivate and Engage All Learners and Narrow Achievement Gaps

A promising theme emerging from these pioneering 3D MUVE science programs is that such learning environments have shown a great deal of potential in reengaging girls and low achieving students in science learning and in narrowing the gender- and race-based achievement gaps (Dede, Ketelhut & Nelson, 2004; Kimmons, Liu, Kang, & Santana, 2012). A greater equality in learning outcomes and participation rates among girls and low social economic status minority students has been found with some programs, including *Alien Rescue* and *River City,* which have been intentionally designed to motivate and engage these two groups of leaners. *Alien Rescue*, for example, considering the gender-based differences in game-based learning, has used "saving aliens" as the theme, which may emotionally engage more girls than boys. A report by the American Association of University Women (AAUW) (2000) reveals that girls prefer role-play games that closely simulate real life, and suggests a number of design considerations for girl-friendly games, including customizable, personalizable female characters, intricate games with rich narrative, opportunities for interaction, communication, and positive social action. Future design of 3D MUVE science programs could take full advantage of its technological capabilities to implement all these design features in support of girl-friendly science learning. While current research on 3D MUVE programs has shown convincing qualitative evidence and promising quantitative data on narrowing gender- and race-based gaps in attitude towards and achievement in science, more rigorously designed experimental studies with a large sample size are needed to explore further the effect of 3D MUVE on narrowing these gaps.

2. Blending Virtual and Real Learning Environments to Augment Science Learning

To take full advantage of the demonstrated power of 3D MUVE for science education and bypass the current hurdles in development complexity and cost, it seems more realistic at present to incorporate those full-fledged, 3D MUVE science programs into appropriate grade levels in K-12 science classrooms. For example, *EcoMUVE* can be used as a 4-week unit on the topic of ecosystem, with *Alien Rescue* as a 3-week unit on the topic of space science in 6th grade science curriculum. Similarly, the fully developed assessment units in *SAVE Science* and *VPA* can be incorporated into middle school science curriculum to assess students' science inquiry knowledge, skills, and abilities. The "Sheep Trouble" in *SAVE Science,* for example, can be used to assess 7th grade students' understanding of concepts of adaptation and structure/function that underlie beginning speciation in the life science curriculum; and "Save the Kelp" in *VPA* can be used to assess 8th grade students' scientific inquiry skills. As federally funded and supported multi-year research projects, the researchers of these 3D MUVE science programs are looking for middle schools interested in participating in their research projects (see for example, *Quest Atlantis* at http://atlantisremixed.org/, and *SAVE Science* at http://www.savescience.net/). The research results from cooperation between science education researchers and educators will greatly benefit the wider implementation and further development of 3D MUVE in science learning nationwide.

In the same vein, to overcome the challenge of implementing scaffolding and its fading element, which is critical to an inquiry-based learning in a fully online 3D MUVE program, it seems more effective if such a program is integrated into the science classroom where a human expert/tutor (e.g., a science teacher or a more abled peer) is present, monitoring students' progress, diagnosing and assessing their challenges, and providing timely and calibrated support. It is believed that a combination of both virtual and real learning environments may significantly enhance students' science learning experience in ways that neither modalities could accomplish on its own. Future research is needed to explore the appropriate framework and best practices in integrating 3D MUVE into science curriculum and examine its effect on students' outcomes in affective, cognitive, and social domains.

3. Bridging Formal and Informal Science Learning via 3D MUVE

In addition, the 3D MUVE appears to be an ideal supplemental tool that connects formal science learning in class and learning in an informal setting, such as after-school programs, or leisure time playing at home. A consistent theme among the existing 3D MUVE science programs is that they are being implemented with great success in the informal setting, as teaching aids or supplemental activities in K–12 science classes. The *Quest Atlantis* program (now part of the international Atlantis Remixed Project), for example, has involved 50,000 children on six continents, with 50,000 Quests submitted and over 100,000 Missions completed, many of which were chosen by students to complete in after school time (Atlantis Remixed, n.d.). Its newly developed Family Quest program is designed to support parental involvement in their children's science learning by providing a free game-like 3D platform where parents and children spend time together solving real-life complex science problems. Instead of grousing and competing the amounts of time available in school with reading and math, which by law are the nation's educational priority, science educators should make the most use of the 3D MUVE's abundant features and popularity, and connect it with after-school activities. By building a continuum between classroom instruction and after-school or at-home activities in 3D MUVE, the passion and informal learning that occur in the 3D game-playing environment will transfer into the classroom and significantly increase student engagement in the formal learning setting. By educating parents about the passion and energy children have with this media and the significant instructional benefits of 3D MUVE for science learning, schools can develop a powerful support system from parents that has the potential to help take the science performance of students in this nation to a new level. Future research effort is needed to explore the appropriate instructional design and integration frameworks for connecting formal classroom and informal afterschool and at home science learning using 3D MUVE programs.

FUTURE RESEARCH DIRECTIONS

Current research studies on 3D MUVE have focused on examining students' affective and cognitive outcomes in this learning environment (Barab, Sadler, Heiselt, Hickey, & Zuiker, 2007; Clarke, & Dede, 2005; Code, Clarke-Midura, Zap, & Dede, 2012; Corbit, Bernstein, Kolodziej, & McIntyre, 2006; Ketelhut, Nelson, Schifter, & Kim, 2010; Liu, Horton, Kang, Kimmons, & Lee, 2013); the social domain of learning outcomes, such as the collaboration and communication skills and their impact on other domains of learning outcomes (i.e., affective and cognitive), has scarcely been discussed in the existing 3D MUVE science programs. Collaboration and communication are deemed vital to scientific inquiry. The United States' national science standards have characterized scientific inquiry as "the diverse ways in which scientists study the natural world" and identified "communicating the results" as an essential inquiry activity scientific inquiry (National Research Council, 1996, p. 23). Clark, Weinberger, Jucks, Spitulnik, and Wallace (2003) suggested that the quality of interaction affects motivational as well as cognitive processes that are vital to inquiry-based learning. A successful collaborative learning is featured with individual's active reflection, peers' constructive criticism, constant guidance of a more competent peer, effective negotiating, and solutions/evidence encompassing multiple perspectives. For the future development and research of 3D MUVE for science learning, attention should be paid to the design of collaboration activities, individual learner's collaboration skills, and the impact of the quality of collaboration on affective and cognitive domains of learning.

In addition, there appears to be a need to leverage the use of the 3D MUVE in all subject areas in K-12. The unique technological capabilities of 3D MUVE, such as representational fidelity, visualization of both concrete and abstract representations, make this media a promising alternative to the conventional study of science (and math) in text illustration and two dimensional representation. It is not surprising that science and math, thus far, are the most frequent subjects found in the research literature. The 2009 research study on 3D virtual learning environment in K-12, conducted by National Center on Accessing the General Curriculum, reported that over two-thirds (22 out of 31) of the studies found in the literature are applications in science, less than one-fourth (6 out of 31) in math, with only one study in the humanities (Strangman, Hall, & Meyer, 2009). While 3D MUVE seems to be intuitively most beneficial to the study of science (and math), it is important to recognize that the powerful learning experience, such as motivation, immersion, engagement, presence, co-presence, and communication and collaboration locally and globally, only enabled in this learning environment, can offer equally significant instructional benefits across the entire curriculum. For example, the ability to situate students in a virtual environment where they can visualize historical or fictional events could be beneficial to the learning of topics in history and social studies. Similarly, this media is powerful to foreign language education, in which a student has the opportunity to interact with individuals from the target language and culture in real time (Schwienhorst, 2002). 3D MUVE, in general, provides a powerful and dynamic learning environment that enables immersive, authentic, experiential, collaborative learning opportunities in all subject areas and its implementation in K-12 and the associated research effort should scale up to all curricular.

CONCLUSION

It is urgent for the U.S. students to improve their participation and performance in science and catch up with their international peers so as to compete globally. The existing exemplary 3D MUVE science programs have demonstrated significant instructional benefits in motivating and engaging students and improving their science learning and scientific inquiry. Clearly it is a promising media in narrowing the gender and racial achievement gaps. If designed and used appropriately, 3D MUVE has the potential to transform the U.S. science assessment format by enabling an authentic measurement of students' science content and inquiry. As with all emerging instructional technology, all the challenges, mainly related to technological complexity and cost, and the design challenge of scaffolding, may fade as time goes by. As 3D MUVE technology continues to evolve, the relevant research base grows, and the use and implementation of this media spreads across the curricular, 3D MUVE will become a powerful lever for advancing science teaching and learning in K-12 education.

REFERENCES

American Association of University Women. (2000). Tech-savvy: Educating girls in the new computer age. Washington, DC: Author.

Atlantis Remixed. (n.d.). *Get started*. Retrieved from http://atlantisremixed.org/

Bacharach, V. R., Baumeister, A. A., & Furr, R. M. (2003). Racial and gender science achievement gaps in secondary education. *The Journal of Genetic Psychology*, *164*(1), 115–126. doi:10.1080/00221320309597507 PMID:12693748

Bankston, S. J., & Caldas, S. J. (1997). The American school dilemma: Race and scholastic performance. *The Sociological Quarterly*, *38*(3), 423–429. doi:10.1111/j.1533-8525.1997.tb00485.x

Barab, S., Arici, A., & Jackson, C. (2005). Eat your vegetables and do your homework: A design-based investigation of enjoyment and meaning in learning. *Educational Technology*, *45*, 15–21.

Barab, S., Sadler, T., Heiselt, C., Hickey, D., & Zuiker, S. (2007). Relating narrative, inquiry, and inscriptions: Supporting consequential play. *Journal of Science Education and Technology*, *16*(1), 59–82. doi:10.1007/s10956-006-9033-3

Blumenfeld, P. C., Kempler, T. M., Krajcik, J. S., & Blumenfeld, P. (2006). Motivation and cognitive engagement in learning environments. In Sawyer, R. K. (Ed.), The Cambridge handbook of the learning sciences (pp. 475-488). New York, NY: Cambridge.

Brush, T. A., & Saye, J. W. (2001). The use of embedded scaffolds with a hypermedia-supported student-centered learning activity. *Journal of Educational Multimedia and Hypermedia*, *10*(4), 333–356.

Clark, D., Weinberger, A., Jucks, I., Spitulnik, M., & Wallace, R. (2003). Designing effective science inquiry in text-based computer supported collaborative learning environments. *International Journal of Educational Policy, Research, and Practice*, *4*(1), 55–82.

Clarke, J., & Dede, C. (2005). Making learning meaningful: An exploratory study of using multi-user environments (MUVEs) in middle school science. Paper presented at the annual meeting of the American Educational Research Association, Montreal, Canada.

Clarke, J., Dede, C., Ketelhut, D. J., & Nelson, B. (2006). A design-based research strategy to promote scalability for educational innovations. *Educational Technology*, *46*(3), 27–36.

Code, J., Clarke-Midura, J., Zap, N., & Dede, C. (2012). Virtual performance assessment for serious games and virtual worlds. In H. Wang (Ed.), *Interactivity in e-Learning: Cases and frameworks*. New York, NY: IGI Publishing.

Collins, A., Brown, J. S., & Newman, S. E. (1989). Cognitive apprenticeship: Teaching the crafts of reading, writing, and mathematics. In L. B. Resnick (Ed.), *Knowing, learning, and instruction: Essays in honor of Robert Glaser* (pp. 453–494). Hillsdale, NJ: Lawrence Erlbaum Associates.

Corbit, M., Bernstein, R., Kolodziej, S., & McIntyre, C. (2006). *Student project virtual worlds as windows on scientific cultures in CTC SciFair*. Retrieved from http://www.scicentr.org/cgibin/download.aspx?get=Resources/CorbitEtAlPCST2006.pdf

Dabbagh, N. (2003). Scaffolding: An important teacher competency in online learning. *TechTreds*, *47*(2), 39–44. doi:10.1007/BF02763424

Dalgarno, B., & Lee, M. (2010). What are the learning affordances of 3-D virtual environments? *British Journal of Educational Technology*, *41*(1), 10–32. doi:10.1111/j.1467-8535.2009.01038.x

Dede, C. (2013). Opportunities and challenges in embedding diagnostic assessments into immersive interfaces. *Educational Designer*, *2*(6), 1–22.

Dede, C., Brown-L'Bahy, T., Ketelhut, D. J., & Whitehouse, P. (2003). Distance learning (Virtual learning). In H. Bidgoli (Ed.), *The internet encyclopedia* (pp. 549–560). New York, NY: John Wiley & Sons.

Dede, C., Ketelhut, D. J., & Nelson, B. (2004, April). Design-based research on gender, class, race, and ethnicity in a Multi-User Virtual Environment. Paper presented at the American Education Research Association, San Diego.

Dede, C., Ketelhut, D. J., & Ruess, K. (2002). Motivation, usability, and learning outcomes in a prototype museum-based multi-user virtual environment. In P. Bell, R. Stevens, & T. Satwicz (Eds.), *Keeping learning complex: The proceedings of the fifth international conference of the learning sciences*. Mahwah, NJ: Erlbaum.

EdWeek. (2012). *U.S. education pressured by international comparisons*. Available online at http://www.edweek.org/ew/articles/2012/01/12/16overview.h31.html

Guzdial, M. (1994). Software-realized scaffolding to facilitate programming for science learning. *Interactive Learning Environments*, *4*(1), 1–44. doi:10.1080/1049482940040101

Ibarra, H. (2006). Teaching them to fish. In R. E. Yager (Ed.), Exemplary science in grades 5–8: Standards-based success stories (pp. 69-84). Arlington, VA: National Science Teachers Association Press.

Intel Lab. (2009). *ScienceSim: A virtual environment for collaborative visualization and experimentation*. Retrieved from http://www.intel.com/content/www/us/en/research/intel-labs-sciencesim-paper.html

Kazempour, M. (2009). Impact of inquiry-based professional development on core conceptions and teaching practices: A case study. *Science Educator*, *18*(2), 56–68.

Ketelhut, D. J., & Dede, C. (2007). *Alternative assessments of students' understanding of scientific inquiry via a multi-user virtual environment*. Paper presented at the Distributed Learning and Collaboration (DLAC-II) Symposium. Singapore, Singapore.

Ketelhut, D. J., Dede, C., Clarke, J., & Nelson, B. (2006). *A multi-user virtual environment for building higher order inquiry skills in science*. Paper presented at the American Educational Research Association. San Francisco, CA.

Ketelhut, D. J., Nelson, B., Schifter, C., & Kim, Y. (2010). Using immersive virtual environments to assess science content understanding: The impact of context. In *Proceedings of the iadis international conference on cognition and exploratory learning in the digital age* (pp. 227-230). Timisoara, Romania: Academic Press.

Ketelhut, D. J., Nelson, B., Schifter, C., & Kim, Y. (2013). Improving science assessments by situating them in a virtual environment. *Education Science*, *3*(2), 172–192. doi:10.3390/educsci3020172

Kim, M. C., & Hannafin, M. J. (2011). Scaffolding problem solving in technology-enhanced learning environments (TELEs): Bridging research and theory with practice. *Computers & Education*, *56*(2), 403–417. doi:10.1016/j.compedu.2010.08.024

Kimmons, R., Liu, M., Kang, J., & Santana, L. (2012). Attitude, achievement, and gender in a middle school science-based ludic simulation for learning. In P. Resta (Ed.), *Proceedings of society for information technology & teacher education international conference 2012* (pp. 1891-1898). Chesapeake, VA: AACE.

Kjem, E. (2010). Real world 3D modeling. *Public Service Review European Union*, *20*, 300.

Lee, M., & Dalgarno, B. (2011). Scaffolding discovery learning in 3D virtual environments: Challenges and considerations for instructional design. In S. Hai-Jew (Ed.), *Virtual immersive and 3D learning spaces: Emerging technologies and trends* (pp. 138–169). Hershey, PA: IGI Global.

Lepper, M. R., Drake, M. F., & O'Donnell-Johnson, T. (1997). Scaffolding techniques of expert human tutors. In K. Hogan & M. Pressley (Eds.), Scaffolding student learning: Instructional approaches and issues (pp. 108–144). Cambridge, MA: Brookline.

Lim, C. P., Nonis, D., & Hedberg, J. (2006). Gaming in a 3D multi-user virtual environment (MUVE): Engaging students in Science lessons. *British Journal of Educational Technology*, *37*(2), 211–231. doi:10.1111/j.1467-8535.2006.00531.x

Liu, M., Horton, L., Kang, J., Kimmons, R., & Lee, J. (2013). Using a ludic simulation to make learning of middle school space science fun. *International Journal of Gaming and Computer-Mediated Simulations*, *5*(1), 66–86. doi:10.4018/jgcms.2013010105

Lomard, M., & Ditton, T. (2006). At the heart of it all: The concept of presence. *Journal of Computer-Mediated Communication, 3*(2).

McLellan, H. (2004). Virtual realities. In D. H. Jonassen (Ed.), *Handbook of research for educational communications and technology* (pp. 461–497). Mahwah, NJ: Lawrence Erlbaum.

McLoughlin, C. (2002). Learner support in distance and networked learning environments: Ten dimensions for successful design. *Distance Education*, *23*(2), 149–162. doi:10.1080/0158791022000009178

McLoughlin, C., & Oliver, R. (1999). Pedagogic roles and dynamics in elematics environments. In M. Selinger & J. Pearson (Eds.), Telematics in education: Trends and issues (pp. 32-50). Kidlington, UK: Pergamon.

McMaham, A. (2003). Immersion, engagement, and presence: A method for analyzing 3-D video games. In M. Wolf & B. Perron (Eds.), *The video game, theory reader* (pp. 77–78). New York, NY: Routledge.

Metcalf, S. J., Kamarainen, A., Tutwiler, M. S., Grotzer, T. A., & Dede, C. J. (2011). Ecosystem science learning via multi-user virtual environments. *International Journal of Gaming and Computer-Mediated Simulations*, *3*(1), 86–90. doi:10.4018/jgcms.2011010107

Mintz, R., Litvak, S., & Yair, Y. (2001). 3D-virtual reality in science education: An implication for astronomy teaching. *Journal of Computers in Mathematics and Science Teaching*, *20*(3), 293–305.

National Assessment Governing Board. (2012). *The Nation's report card releases results from innovative science assessment*. Retrieved from http://www.nagb.org/newsroom/naep-releases/science-hots-icts.html

National Center for Education Statistics. (2011). The nation's report card. *Science, 2011*. Retrieved from http://nces.ed.gov/nationsreportcard/pubs/main2011/2012465.aspx

National Center for Education Statistics. (2013). *Selected findings from PISA 2012*. Available online at http://nces.ed.gov/surveys/pisa/pisa2012/

National Center for Education Statistics. (1991-1999). Digest of education statistics. Washington DC: Author.

National Research Council. (1996). *National science education standards: Observe, interact, change, learn*. Washington, DC: National Academy Press.

National Science Teachers Association. (2004). *NSTA position statement*. Retrieved from http://www.nsta.org/about/positions/inquiry.aspx

Nelson, B., Ketelhut, D. J., Clarke, J., Bowman, C., & Dede, C. (2005). Design-based research strategies for developing a scientific inquiry curriculum in a multi-user virtual environment. *Educational Technology*, *45*(1), 21–27.

Nelson, B., Ketelhut, D. J., & Schifter, C. (2010). Exploring cognitive load in immersive educational games: The SAVE science project. *International Journal of Gaming and Computer-Mediated Simulations*, *2*(1), 31–39. doi:10.4018/jgcms.2010010103

President's Council of Advisors on Science and Technology. (2010). *Prepare and inspire: K-12 education in science, technology, engineering, and math (STEM) for America's future*. Retrieved from http://www.whitehouse.gov/sites/default/files/microsites/ostp/pcast-stemed-report.pdf

Roth, K., & Garnier, H. (2007). What science learning looks like: An international perspective. *Educational Leadership*, *64*, 16–23.

Schwienhorst, K. (2002). The state of VR: A meta-analysis of virtual reality tools in second language acquisition. *Computer Assisted Language Learning*, *15*(3), 221–239. doi:10.1076/call.15.3.221.8186

Science, S. A. V. E. (n.d.). *Theoretical and research context for SAVE Science*. Retrieved from http://www.savescience.net/project/context-for-project/

Sharma, P., & Hannafin, M. J. (2007). Scaffolding in technology-enhanced learning environments. *Interactive Learning Environments*, *15*(1), 27–46. doi:10.1080/10494820600996972

Slater, M. (2004). How colorful was your day? Why questionnaires cannot assess presence in virtual environments. *Presence (Cambridge, Mass.)*, *13*(4), 484–493. doi:10.1162/1054746041944849

Southerland, S. A., Smith, L. K., Sowell, S. P., & Kittleson, J. M. (2007). Resisting unlearning: Understanding science education's response to the United States' national accountability movement. *Review of Educational Research*, *31*(1), 45–77. doi:10.3102/0091732X07300046045

Stone, C. A. (1998). The metaphor of scaffolding: Its utility for the field of learning disabilities. *Journal of Learning Disabilities*, *31*(4), 344–364. doi:10.1177/002221949803100404 PMID:9666611

U.S. Department of Education. (2013). The threat of educational stagnation and complacency. Retrieved from http://www.ed.gov/news/speeches/threat-educational-stagnation-and-complacency

Wallance, J., & Louden, W. (2002). *Dilemmas of science teaching: Perspectives on problems of practice*. New York, NY: Routledge.

Winnips, J. C. (2001). *Scaffolding-by-Design as a model for online learner support*. Retrieved from http://rilw.emp.paed.uni-muenchen.de/2001/papers/winnips.pdf

ADDITIONAL READING

Bacharach, V. R., Baumeister, A. A., & Furr, R. M. (2003). Racial and gender science achievement gaps in secondary education. *The Journal of Genetic Psychology*, *164*(1), 115–126. doi:10.1080/00221320309597507 PMID:12693748

Bankston, S. J., & Caldas, S. J. (1997). The American school dilemma: Race and scholastic performance. *The Sociological Quarterly*, *38*(3), 423–429. doi:10.1111/j.1533-8525.1997.tb00485.x

Barab, S., Arici, A., & Jackson, C. (2005). Eat your vegetables and do your homework: A design-based investigation of enjoyment and meaning in learning. *Educational Technology*, *45*, 15–21.

Barab, S., Sadler, T., Heiselt, C., Hickey, D., & Zuiker, S. (2007). Relating narrative, inquiry, and inscriptions: Supporting consequential play. *Journal of Science Education and Technology*, *16*(1), 59–82. doi:10.1007/s10956-006-9033-3

Blumenfeld, P. C., Kempler, T. M., Krajcik, J. S., & Blumenfeld, P. (2006). Motivation and cognitive engagement in learning environments. In Sawyer, R. K. (Ed.), The Cambridge handbook of the learning sciences (pp. 475-488). New York, NY: Cambridge.

Brush, T. A., & Saye, J. W. (2001). The use of embedded scaffolds with a hypermedia-supported student-centered learning activity. *Journal of Educational Multimedia and Hypermedia*, *10*(4), 333–356.

Clark, D., Weinberger, A., Jucks, I., Spitulnik, M., & Wallace, R. (2003). Designing effective science inquiry in text-based computer supported collaborative learning environments. *International Journal of Educational Policy, Research, and Practice*, *4*(1), 55–82.

Clarke, J., Dede, C., Ketelhut, D. J., & Nelson, B. (2006). A design-based research strategy to promote scalability for educational innovations. *Educational Technology*, *46*(3), 27–36.

Code, J., Clarke-Midura, J., Zap, N., & Dede, C. (2012). Virtual performance assessment for serious games and virtual worlds. In H. Wang (Ed.), *Interactivity in e-Learning: Cases and frameworks*. New York, NY: IGI Publishing.

Collins, A., Brown, J. S., & Newman, S. E. (1989). Cognitive apprenticeship: Teaching the crafts of reading, writing, and mathematics. In L. B. Resnick (Ed.), *Knowing, learning, and instruction: Essays in honor of Robert Glaser* (pp. 453–494). Hillsdale, NJ: Lawrence Erlbaum Associates.

Dabbagh, N. (2003). Scaffolding: An important teacher competency in online learning. *TechTreds*, *47*(2), 39–44. doi:10.1007/BF02763424

Dalgarno, B., & Lee, M. (2010). What are the learning affordances of 3-D virtual environments? *British Journal of Educational Technology*, *41*(1), 10–32. doi:10.1111/j.1467-8535.2009.01038.x

Dede, C. (2013). Opportunities and challenges in embedding diagnostic assessments into immersive interfaces. *Educational Designer*, *2*(6), 1–22.

Dede, C., Ketelhut, D. J., & Nelson, B. (2004, April). Design-based research on gender, class, race, and ethnicity in a Multi-User Virtual Environment. Paper presented at the American Education Research Association, San Diego.

Dede, C., Ketelhut, D. J., & Ruess, K. (2002). Motivation, usability, and learning outcomes in a prototype museum-based multi-user virtual environment. In P. Bell, R. Stevens, & T. Satwicz (Eds.), *Keeping learning complex: The proceedings of the fifth international conference of the learning sciences*. Mahwah, NJ: Erlbaum.

Guzdial, M. (1994). Software-realized scaffolding to facilitate programming for science learning. *Interactive Learning Environments*, *4*(1), 1–44. doi:10.1080/1049482940040101

Ibarra, H. (2006). Teaching them to fish. In R. E. Yager (Ed.), Exemplary science in grades 5–8: Standards-based success stories (pp. 69-84). Arlington, VA: National Science Teachers Association Press.

Kazempour, M. (2009). Impact of inquiry-based professional development on core conceptions and teaching practices: A case study. *Science Educator*, *18*(2), 56–68.

Ketelhut, D. J., Nelson, B., Schifter, C., & Kim, Y. (2013). Improving science assessments by situating them in a virtual environment. *Education Science*, *3*(2), 172–192. doi:10.3390/educsci3020172

Kim, M. C., & Hannafin, M. J. (2011). Scaffolding problem solving in technology-enhanced learning environments (TELEs): Bridging research and theory with practice. *Computers & Education*, *56*(2), 403–417. doi:10.1016/j.compedu.2010.08.024

Kimmons, R., Liu, M., Kang, J., & Santana, L. (2012). Attitude, achievement, and gender in a middle school science-based ludic simulation for learning. In P. Resta (Ed.), *Proceedings of society for information technology & teacher education international conference 2012* (pp. 1891-1898). Chesapeake, VA: AACE.

Kjem, E. (2010). Real world 3D modeling. *Public Service Review European Union*, *20*, 300.

Lee, M., & Dalgarno, B. (2011). Scaffolding discovery learning in 3D virtual environments: Challenges and considerations for instructional design. In S. Hai-Jew (Ed.), *Virtual immersive and 3D learning spaces: Emerging technologies and trends* (pp. 138–169). Hershey, PA: IGI Global.

Lepper, M. R., Drake, M. F., & O'Donnell-Johnson, T. (1997). Scaffolding techniques of expert human tutors. In K. Hogan & M. Pressley (Eds.), Scaffolding student learning: Instructional approaches and issues (pp. 108–144). Cambridge, MA: Brookline.

Lim, C. P., Nonis, D., & Hedberg, J. (2006). Gaming in a 3D multi-user virtual environment (MUVE): Engaging students in Science lessons. *British Journal of Educational Technology*, *37*(2), 211–231. doi:10.1111/j.1467-8535.2006.00531.x

Liu, M., Horton, L., Kang, J., Kimmons, R., & Lee, J. (2013). Using a ludic simulation to make learning of middle school space science fun. *International Journal of Gaming and Computer-Mediated Simulations*, *5*(1), 66–86. doi:10.4018/jgcms.2013010105

Lomard, M., & Ditton, T. (2006). At the heart of it all: The concept of presence. *Journal of Computer-Mediated Communication, 3*(2).

McLellan, H. (2004). Virtual realities. In D. H. Jonassen (Ed.), *Handbook of research for educational communications and technology* (pp. 461–497). Mahwah, NJ: Lawrence Erlbaum.

McLoughlin, C. (2002). Learner support in distance and networked learning environments: Ten dimensions for successful design. *Distance Education*, *23*(2), 149–162. doi:10.1080/0158791022000009178

McLoughlin, C., & Oliver, R. (1999). Pedagogic roles and dynamics in elematics environments. In M. Selinger & J. Pearson (Eds.), Telematics in education: Trends and issues (pp. 32-50). Kidlington, UK: Pergamon.

McMaham, A. (2003). Immersion, engagement, and presence: A method for analyzing 3-D video games. In M. Wolf & B. Perron (Eds.), *The video game, theory reader* (pp. 77–78). New York, NY: Routledge.

Metcalf, S. J., Kamarainen, A., Tutwiler, M. S., Grotzer, T. A., & Dede, C. J. (2011). Ecosystem science learning via multi-user virtual environments. *International Journal of Gaming and Computer-Mediated Simulations*, *3*(1), 86–90. doi:10.4018/jgcms.2011010107

Mintz, R., Litvak, S., & Yair, Y. (2001). 3D-virtual reality in science education: An implication for astronomy teaching. *Journal of Computers in Mathematics and Science Teaching*, *20*(3), 293–305.

National Research Council. (1996). *National science education standards: Observe, interact, change, learn*. Washington, DC: National Academy Press.

Nelson, B., Ketelhut, D. J., Clarke, J., Bowman, C., & Dede, C. (2005). Design-based research strategies for developing a scientific inquiry curriculum in a multi-user virtual environment. *Educational Technology, 45*(1), 21–27.

Nelson, B., Ketelhut, D. J., & Schifter, C. (2010). Exploring cognitive load in immersive educational games: The SAVE science project. *International Journal of Gaming and Computer-Mediated Simulations, 2*(1), 31–39. doi:10.4018/jgcms.2010010103

Roth, K., & Garnier, H. (2007). What science learning looks like: An international perspective. *Educational Leadership, 64*, 16–23.

Schwienhorst, K. (2002). The state of VR: A meta-analysis of virtual reality tools in second language acquisition. *Computer Assisted Language Learning, 15*(3), 221–239. doi:10.1076/call.15.3.221.8186

Sharma, P., & Hannafin, M. J. (2007). Scaffolding in technology-enhanced learning environments. *Interactive Learning Environments, 15*(1), 27–46. doi:10.1080/10494820600996972

Slater, M. (2004). How colorful was your day? Why questionnaires cannot assess presence in virtual environments. *Presence (Cambridge, Mass.), 13*(4), 484–493. doi:10.1162/1054746041944849

Southerland, S. A., Smith, L. K., Sowell, S. P., & Kittleson, J. M. (2007). Resisting unlearning: Understanding science education's response to the United States' national accountability movement. *Review of Educational Research, 31*(1), 45–77. doi:10.3102/0091732X07300046045

Stone, C. A. (1998). The metaphor of scaffolding: Its utility for the field of learning disabilities. *Journal of Learning Disabilities, 31*(4), 344–364. doi:10.1177/002221949803100404 PMID:9666611

Wallance, J., & Louden, W. (2002). *Dilemmas of science teaching: Perspectives on problems of practice*. New York, NY: Routledge.

Winnips, J. C. (2001). *Scaffolding-by-Design as a model for online learner support*. Retrieved from http://rilw.emp.paed.uni-muenchen.de/2001/papers/winnips.pdf

KEY TERMS AND DEFINITIONS

3D MUVE: Abbreviation for three-dimensional multi-user virtual environment. 3D MUVE is a visually immersive 3-Dimentional online environment where individuals – represented by avatars (cartoonish-looking characters), meet, socialize and interact with each other, with computer-based agents, digital artifacts, and the environment in real time, just as they might in the real world.

3D MUVE Science Programs: K-12 science learning programs in the 3D MUVE environment. In this study, they refer specifically to the following federally funded and/or large education foundation supported multi-year programs: *SciCentr, Alien Rescue, River City, EcoMUVE, Quest Atlantis, SAVE*, and *VPA*.

Representational Fidelity: One of the unique technological features of 3D MUVE. It allows for the representation of the most important features and accurate facts relevant to an object, an issue, a phenomenon, or the environment which helps learners' decision making in a 3D MUVE learning environment.

Scientific Inquiry: Defined by the National Research Council in 1996, scientific inquiry refers to "the diverse ways in which scientists study the natural world and propose explanations based on the evidence derived from their work. Scientific inquiry also refers to the activities through which students develop knowledge and understanding of scientific ideas, as well as an understanding of how scientists study the natural world" (National Science Education Standards, 1996, p. 23).

Visualization: Refers to 3D MUVE's unique technological feature that allows users to visualize and manipulate both representations of the concrete, tangible realities and abstract, intangible concepts in the same space.

This work was previously published in "Exploring the Effectiveness of Online Education in K-12 Environments" edited by Tina L. Heafner, Richard Hartshorne, and Teresa Petty, pages 302-324, copyright 2015 by Information Science Reference (an imprint of IGI Global).

Chapter 46
Learning about Sustainability in a Non-Formal Laboratory Context for Secondary Level Students:
A Module on Climate Change, the Ozone Hole, and Summer Smog

Nicole Garner
Institute for Science Education, University of Bremen, Germany

Maria de Lourdes Lischke
Institute for Science Education, University of Bremen, Germany

Antje Siol
Institute for Environmental Research and Sustainable Technologies, University of Bremen, Germany

Ingo Eilks
Institute for Science Education, University of Bremen, Germany

ABSTRACT

This chapter discusses a project of curriculum development for the non-formal educational sector. The project aims at student learning about sustainability issues in a chemistry-related context. For this purpose, non-formal laboratory-based learning environments are developed. The learning environments center round half- or one-day visits of secondary school students in a university laboratory and are networked with the formal school syllabus in chemistry and science education respectively. All modules integrate the non-formal laboratory event about issues of sustainability with teaching materials for preparation and assessment tasks in school to fulfill part of the school curriculum in chemistry or science teaching. This chapter discusses the project of developing respective modules, the structure thereof, and initial findings from their application. The discussion is illustrated by a module on environmental problems connected to the chemistry of the atmosphere, namely climate change, the hole in the ozone layer, and the phenomenon of summer smog.

DOI: 10.4018/978-1-4666-7363-2.ch046

INTRODUCTION

The global economy of the last few decades can be characterized, at least in the Western countries, by continuous growth, and in most cases a great improvement in the quality of life. A lot of innovations from science and technology have simplified life. Today, it is hard to imagine life without modern health care, materials, or energy supply. Chemistry and chemical industry contributes greatly to this development (Bradley, 2005). However, this development had and still has its price. Mankind has to cope with growing scarcity of resources, availability of clean water, climate change, and many other problems (Mortensen, 2000). In order to deal with these issues, emissions have to be reduced and both energy and available raw materials must be used as efficiently as possible (Vinten, 1994). Again, chemistry is central to the response to these challenges. One way in which chemistry may answer these challenges is constructed around the idea of a *Green* or *Sustainable Chemistry* as a guiding framework for contemporary chemistry research, development, and industrial production (Centi & Perathoner, 2009; Höfer & Bigorra, 2007).

With the growing importance of sustainability issues in science and technology research and industrial production, it is suggested that respective topics play a more prominent role in education in general (Wheeler & Bijur, 2000), and chemistry education in particular (Burmeister, Rauch & Eilks, 2012). Student learning, about issues of sustainability and the environment, is needed to develop a balanced view towards contemporary chemistry and shaping respective attitudes (Ware, 2000). Knowledge and skills are needed to enable students to assess new chemistry-based products and technologies in their life and society and to act appropriately (Hjeresen, Schutt & Boese, 2000; Dawe, Jucker & Martin, 2005; Arbuthnott, 2009; Karpudewan, Ismail & Mohamed, 2011). Unfortunately, research has shown that students have a lack of understanding sustainable issues such as global warming, ozone and greenhouse effect (Howard, Brown, Chung, Jobson & Van-Reken, 2013). This may be caused due the fact that learning about sustainability issues is barely represented in many science curricula in general, and in secondary chemistry education in particular (Burmeister et al., 2012). The reasons for this range from teaching and learning materials not being sufficiently available, via a lack of adequate experiments and laboratory equipment in schools, towards deficits in teacher education (Burmeister, Schmidt-Jacob & Eilks, 2013).

The project discussed in this paper intends to rectify this situation by implementing innovations into chemistry teaching via non-formal settings, combined with curriculum development for secondary chemistry education, and contributions to teacher continuous professional development. The project is called "Sustainability and chemistry in non-formal student laboratories." It aims at developing and implementing innovative non-formal laboratory teaching and learning environments for secondary school students which allow for contention with sustainability issues connected to the science and chemistry curricula in schools respectively. This paper discusses the basic issues of the project, the structure of the teaching and learning environments, and illustrates the description utilizing an exemplary case focusing environmental problems connected to the chemistry of the atmosphere, namely climate change, the hole in the ozone layer, and the phenomenon of summer smog.

ORGANIZATION BACKGROUND

"Sustainability and chemistry in non-formal student laboratories" is a cooperative project between the Universities of Bremen and of the Saarland, both in Germany. The project is driven both by researchers in chemistry and environmental sciences together with domain-specific researchers and curriculum experts from the field of chemistry education. The cooperation was established to secure access to authentic and up-to-date research in the field of sustainable chemistry as well as to structure learning environments in line with modern educational theory. The project is funded with about 250.000 € by the Deutsche Bundesstiftung Umwelt (German Environmental Funds) for the years 2012-2014.

The central focus of the project is to develop half- and full-day non-formal laboratory-based learning environments on issues of sustainability in chemistry related contexts. All together a minimum of ten different learning environments, each five per partner, for the non-formal student laboratories in the universities of Bremen and the Saarland are to be developed and implemented (Table 1). The target groups are secondary science and chemistry classes of grade 5-13 (age range 10-19 years).

The development of the learning environments follows a cyclical process of design, testing, evaluation, and revision. In this means the innovation project can be understood as a project of Action Research triggered innovation of classroom practice (Eilks, 2014; Mamlok-Naaman & Eilks, 2012). The laboratory environments are designed based on available information about the science background, knowledge from educational theory and documented teaching and learning experiences, e.g. about available school-type and university-type experiments. The initial designs are pre-tested and reflected upon with small groups of secondary school students or student teachers. Beginning with the first complete design, the learning environments are cyclically tested and refined with secondary school classes visiting the university's student laboratory.

Within the cyclical process of refining, the case is constantly evaluated based on a qualitative paradigm as suggested for classroom-based curriculum innovations by Bodner, MacIsaac and White (1999). A qualitative paradigm is chosen since the development is cyclical and the project takes place in authentic teaching and learning situations. To achieve valid interpretations Eilks and Ralle (2002) suggested data collection with different tools and a coverage of feedback from all relevant groups of within the developmental process. Thus, data is collected about the expectations as well as experiences from both the teachers' and students' point of view by applying respective questionnaires. Additional data stems from short open feedback reports of the teachers and classroom observation reports from the accompanying university staff. In this way evaluation takes into consideration all the viewpoints of the teachers, students, and accompanying university staff as well it uses different types of data collection. Data evaluation is driven by an interpretative ap-

Table 1. Overview of the modules developed at the University of Bremen and their connection to chemistry-related sustainability issues

Grade	Module	Sustainability Focus
5th/6th	Fragrances from flowers und fruits	Consideration of resources and the use of renewable raw materials
7th/8th	Chemistry of the atmosphere	Human influences on the planet and its climate
9th/10th	Biodiesel produced by vegetable fats	Renewable energy sources to protect resources and limit climate change
9th - 13th	Parabens in cosmetics	Alternative ingredients in products and more sustainable routes for chemical compounds
11th - 13th	Synthesis of vanillin	Natural products, synthetic alternatives, and assessment of different pathways for industrial chemical production
	Click-Chemistry	Applications of Green Chemistry principles in modern chemistry research and industry
	Zeolites and molecular sieves -Applications in industry and daily life	

proach by qualitatively analyzing the raw data and triangulating them to construct meaning. Driving criteria of the analysis and triangulation process are to achieve plausibility, credibility, relevance and importance in the conclusions as suggested by Altheide and Johnson (1994). This is done by a negotiation process of data interpretation within a group of researchers.

SETTING THE STAGE

Non-Formal Learning Environments and Science Education

Scientists develop new knowledge through a continuous interaction of experimental observations and scientific theories; hence science without laboratory work is virtually unthinkable (Hofstein & Lunetta, 2003). That is why laboratory work is assigned a central role in any student-oriented science education (Tobin, 1990; Abrahams, 2011). However, the positive effect of practical work on learning is not self-evident (Hofstein, Kipnis & Abrahams, 2012). In the school setting, experiments often take place under difficult conditions: the laboratories are less equipped, time is limited, and the teacher-student ratios are unfavorable. Experiments are often conducted as a pure demonstration without any inquiry-character which makes them illustrations of content but does not use them to challenge students' thinking (Tobin, 1990; Hofstein & Kind, 2012).

Out-of-school settings provide good conditions for a different type of experimentation in science education. They show potential for supporting science learning. Time and equipment is less restricted; the teacher-student ratio is more comfortable (Rennie, 2007; Coll, Gilbert, Pilot & Streller, 2012). Out-of-school activities are commonly divided into *informal* and *non-formal* settings (OECD, 2012). The main difference is to whether learning has a specific structure (even if it includes open tasks) and whether it is connected to any kind of a formal syllabus or curriculum. If all activities are free of choice and rarely connected to formal learning, like unstructured leisure visits of a museum or zoo, the environment is called *informal*. If there is a specific program and maybe a connection to a certain curriculum, out-of-school activities are to be called *non-formal*. In this means, our project focuses the development of *non-formal* laboratory learning environments. The learning environments are organized for visits of regular secondary school science classes to the university laboratory. The learning focuses a certain topic that is connected to the official school curriculum. For each learning group a specific module is structured that is before and after the laboratory visit both prepared and continued in the regular school science lessons.

Werquin (2010) suggests a greater recognition of non-formal learning settings by highlighting the main benefits of out-of-school learning settings clearly. Motivation, a positive perception of learning, supporting career orientation by identifying interests, and continuous professional development of the accompanying teachers are described as the main benefits. Eshach (2007) adds that non-formal learning may offer enjoyable learning and create scientifically literate school leavers. These advantages and benefits are not easily transposable as it needs a good connection between in- and out-of-school learning. Thus, linking outer- and inner-school learning will be the challenge to benefit from the alternate learning setting (Bybee, 2001). Gallacher and Feutrie (2003), Eshach (2007), Hofstein and Rosenfeld (1996), Orion and Hofstein (1994), Rennie (2007), Rennie and McClafferty (1996), and Braund and Reiss (2006) suggest the following points in order to make optimum use of non-formal learning environments in science education:

- Flexible and individually adaptable programs ease the integration of non-formal learning activities into formal school curricula.

- A preparatory learning phase in school is necessary to raise effective learning in the non-formal setting.
- The working materials for the non-formal environment need to be adjustable to the current student's performance and knowledge level.
- The learning environment should be student-centered, inquiry-based, interactive, and provoke cooperative learning.
- After the non-formal learning activity the contents and topics should be picked up in school again.
- National Science Education Standards or governmental syllabi shall be met to bridge formal and non-formal education.

Educational Philosophy of the Learning Environments

In order to create effective non-formal university student laboratory environments dealing with issues of sustainable development this project tries to take into account all of the parameters described in the previous section. All modules developed in the project are modular in construction to ensure a flexible use and individual adaptability to the learning objectives and the students' prior knowledge. In each module a pool of experiments and instructions is offered. In negotiation with the accompanying university staff, the teachers select those experiments which fit best their objectives and the students' abilities. All modules encompass the non-formal student learning environment, but also consist of a set of teaching and learning materials for a preparatory phase in school prior to the visit as well as suggestions and materials for reflection and assessment (Figure 1). These materials focus both the science content necessary to complete the laboratory experience as well as learning about the significance of the topic for sustainable development (see Table 1). This means that all materials not only cover the learning of the science background but also a reflection of the topic in the foreground of its ecological, economic and societal implications as the three most prominent dimensions of most sustainability models (Burmeister et al., 2012). The materials for preparation and assessment enable the linkage between the non-formal laboratory visit and formal teaching in the school setting as well as the laboratory experience with education for sustainable development.

In the theory of situated cognition (Greeno, 1988) learning is suggested to be most effective if it is embedded into meaningful contexts. Such contexts might stem from the life-world of the

Figure 1. Structure of the learning modules

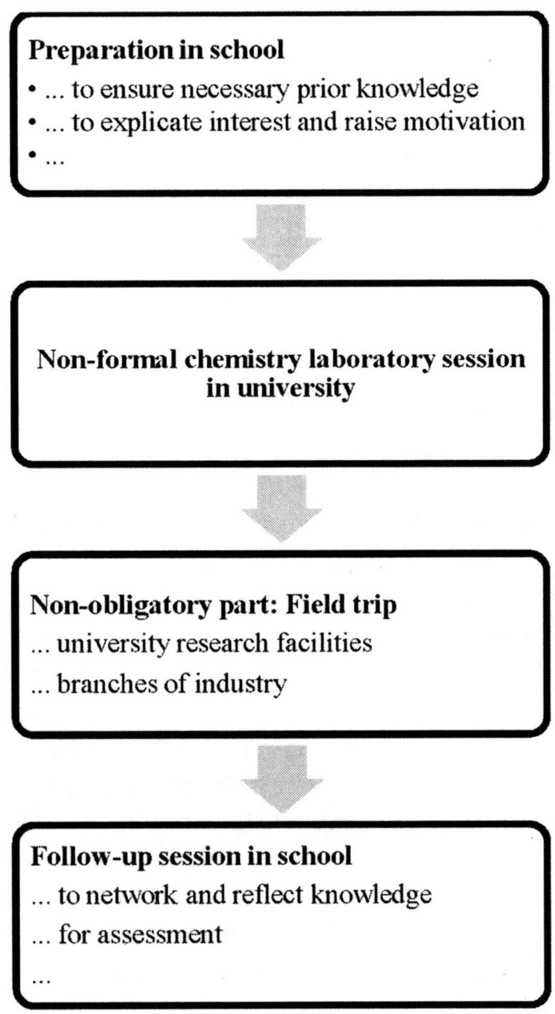

students, the society they live and operate, from technology, or a potential later workspace (Parchmann, Gräsel, Bär, Nentwig, Demuth & Ralle, 2006; Rannikmäe, Teppo & Holbrook, 2010). Contexts that are bound to chemical technology, research and industry (e.g. Hofstein & Kesner, 2006) as well as to societal relevant issues (e.g. Hofstein, Eilks & Bybee, 2011) provide the most promising frameworks to connect chemistry learning with all the different dimensions that make the learning of science relevant (Stuckey, Mamlok-Naaman, Hofstein & Eilks, 2013). Accordingly, this project operates a context-based and societal-oriented approach to science learning. The contexts are current and authentic practices of research and industrial applications to promote a more sustainable chemistry for the future. The spectrum of examples ranges from daily-life natural and industrial products (such as vanillin, plastics and fuels) to authentic and controversial societal issues (such as climate change and renewable energy supply).

Concerning the pedagogy, a student-orientated (Eilks, Prins & Lazarowitz, 2013) and inquiry-based (Hofstein, Kipnis & Abrahams, 2013) educational paradigm is applied. The tasks for learning are problem-based and provoke investigations of questions, scenarios and problems. Science education literature (Hofstein, Navon, Kipnis & Mamlok-Naaman, 2005; Berg, Bergendahl, Lundberg & Tibell, 2003) suggests that such an approach is advantageous for learning about the nature of science, achieving a deeper understanding of concepts, supports the development of problem-solving and evaluation skills, and contributes enhancing motivation. In practice, the tasks in the laboratory learning environment range from guided to open inquiry questions. Operating the inquiry tasks asks for open minded thinking and structured problem-solving. The laboratory instructions are designed in such a way to allow for differentiation. The experimental instructions offer different levels of complexity and differentiated learning aids are available. By using these learning aids student of different abilities will be able to complete the different tasks successfully on their own. Self-directed and autonomous learning becomes possible and can be adapted to the students' skills and prior knowledge. The students' tasks are given to small groups of students to enable cooperative learning processes.

All the tasks are designed to be solved by autonomous, self-determined and cooperative work of the students. Nevertheless, scaffolding by the accompanying university staff and structured aids are available if necessary. It has been suggested that failures during inquiry- based experimentations are essential to learning success (Lunetta, Hofstein & Clough, 2007). The inquiries are in most cases split into steps of 15-20 minutes. Thus, trying out potentially unsuccessful attempts will not exceed time limitations. The laboratory environment is directed by *researcher's booklets* given to the students. The booklets include all necessary worksheets, regulations and safety instructions. All experiments start with a short presentation of the problem that needs to be solved by the students. Space is available to write down hypotheses, ideas, sketches, observations and notes.

A non-mandatory part of each module is a field trip into research laboratories in the university or branches of industry that fit the thematic background of the module and that operate sustainability strategies in an authentic research or industry context. These trips are intended to make the context of learning even more authentic and allow for further career orientation. Finally, all the modules are structured in a way that contents and contexts are in line with the national German science education standards as well as the regional syllabi in question.

The teacher handbooks for each module include all this information. The handbook covers all the experiments, even those that were not selected for the laboratory visit, a theoretical introduction and a selection of work sheets for preparing the visit and to assess it. This teacher handout consists of the following chapters:

- **General information on the project:** The general structure and the central focus of the project are outlined.
- **Necessary prior knowledge:** A description is given which chemistry knowledge and skills are necessary to operate the laboratory visit in this specific module successfully.
- **Connection of the topic with the national science education standards.**
- **Specific scientific and sustainability background:** The teachers get information about the specific scientific knowledge the students will learn in their module and how the specific example is embedded into any kind of innovation aiming on contributing to sustainable development. This information is meant to update the teachers' subject matter and sustainability related knowledge, too.
- **Information about inquiry-experimentation:** The philosophy and different approaches to inquiry-based learning in the laboratory are outlined.
- **Description of the experiments:** An overview about the experiments is given. Objectives, operations, and sample solutions are briefly portrayed.
- **Embedding the laboratory visit into the school curriculum:** Materials and scenarios for the preparation phase, exemplary selections of experiments for certain sustainability foci, and materials for reflection and assessment are discussed.
- Master copies for all experiments and worksheets.

AN EXEMPLARY CASE IN PRACTICE: CHEMISTRY OF THE ATMOSPHERE

Background and Scope

One of the modules developed during this project deals with chemistry-related problems of the atmosphere. Global climate change, the hole in the ozone layer, and summer smog are among the great environmental and economic challenges in the 21th century (Mortensen, 2000). Researchers undertake major efforts to analyze and counteract these atmospheric problems (Lefohn, Shadwick, Oltmans, 2010, McEvoy, Gibbs & Longhurst, 2000; Alonso, Bond & Dumesic, 2010). Therefore, the problems are authentic, relevant, controversial, and openly discussed in the media. This offers prototypic issues, within the topic, for societal oriented science education aiming on general educational skills and participatory learning (Eilks, Nielsen & Hofstein, 2014). The reasons for the greenhouse effect, ozone hole, and summer smog are multifaceted and complex (Monks, 2007, Gallego-Alvarez, 2012; Tang, Wilson, Solomon, Shao & Madronich, 2011). Understanding these reasons may contribute more thorough environmentally conscious behavior and thinking among students.

The module "Chemistry of the atmosphere" focuses on grade 7-8 teaching (age range 13-15). In these grades topics like air, air quality, and environmental problems in the atmosphere are typically embedded in all the German regional chemistry or science syllabi respectively. In order to successfully work in the laboratory the teachers have to assure that different basic contents were learned by the students:

- The composition of air.
- States of matter, their changes, and explanations on the particulate level.
- A first model of atoms and molecules.
- Basic issues of chemical reactions, especially combustion.
- The various layers of the atmosphere.
- Basic knowledge about photosynthesis.
- Energy conversion, especially from heating radiation to light radiation and vice versa.

In the teacher guide, different scenarios are outlined to cover either the whole range of potential environmental problems in the atmosphere or to concentrate with the students on selected aspects. Objectives from applying the module encompass:

- Learning about the basics of the greenhouse effect and climate change, the role and effects of ozone both in stratosphere and troposphere, and about the role of aerosols and clouds and their effects in the atmosphere.
- Gaining experience and confidence in doing experiments in a chemical laboratory, in conducting science inquiries and problem-solving activities, and practicing the documentation process of chemical investigations.
- Becoming aware about the influence of humans on generating air pollutants, recognizing that air pollution is a problem that has a profound effect on the climate and life on earth, and to become aware of the complexity and susceptibility of atmospheric processes.
- Reflecting upon our own actions for a sustainable action in everyday life and society and their potential contribution to reducing environmental problems.
- Developing skills in cooperative working in a team of three or four.

The Learning Environment

It is the decision of the teacher what the students should perform during the laboratory visit. The whole module could be confined to one part of the thematic contents that will be performed intensively but giving an overview is also possible. One of the outlined scenarios aiming at an overview and the distinction of the different environmental problems in the atmosphere is called "The atmosphere as a patient." This scenario outlines a parallel from environmental problems in the atmosphere to diseases in the human body. The atmosphere may be treated as a patient. The patient's diseases are called greenhouse effect, ozone hole, and summer smog. On this base the three environmental problems are reflected in the foreground of symptoms of the disease, causes of the disease, and potential cure. The symptoms are climate change, the hole in the ozone layer, and bad air quality in cities. The causes are emissions by carbon dioxide, chlorofluorocarbons (CFCs), and dust particles. The cure for a more sustainable development is based in reduced emissions and alternative technologies.

Based on this scenario the students prepare for their visit in the non-formal student laboratory. In the laboratory a selection of respective experiments is offered to the students:

- **Greenhouse Effect:**
    - **E5:** Absorption of heat radiation by different gases
    - **E6:** Model experiment of the greenhouse effect
- **Ozone:**
    - **E8:** Positive effects of ozone: Ozone in the stratosphere
    - **E9:** Negative effects of ozone: effects on plants
- **Aerosols:**
    - **E11:** Formation of aerosols and their effects on temperature

- **E12:** Formation of aerosols and their effects on light intensity

In the practical part of the module the students examine the atmospheric diseases with regard to causes and symptoms. Visitors can learn about basics of the greenhouse effect through an experiment about absorptions of heating radiation by different gases. The effects of increasing carbon dioxide emissions are shown by an experimental model of the greenhouse effect. Other experiments are on the topic of ozone. Ozone may be both a pathogen and remedy. One the one hand it protects earth's surface and life against hazardous UV radiation in stratosphere. On the other hand ground level ozone acts like a pathogen and causes summer smog and dying plants. The last of the three topics is about aerosols and clouds. Clouds may generated by accumulation of particles. Experiments show how clouds emerge. The effects of clouds and dust on light intensity and temperature are also examined. The students become acquainted with the symptoms of atmospheric diseases during the non-formal chemistry laboratory in university. Identifying the pathogens is the major learning objective in post-processing after experimentation. In addition to this they have to learn something about how to fight against these diseases and what each and every student is able to do.

Exemplary Experimental Tasks

Model Experiment of the Greenhouse Effect

A simple experiment may demonstrate the greenhouse effect. It can be used to show how an increased carbon dioxide value may influence temperature on earth. Therefore the earth and the atmosphere are portrayed utilizing a model. The earth and its atmosphere are demonstrated by a poly-styrene box with a black inner floor. Clouds and water vapor in the atmosphere are modeled by using a crystallizing dish filled with water, and the sun by a lamp (Figure 2; Parchmann & Jansen, 1996). The box needs to be pierced by a thermometer near the 'ground'. It needs also to be covered by the transparent film. The temperature is measured inside the box. Afterwards the lamp is turned on. After waiting for two minutes the temperature is measured again. The experiment can be repeated with different amounts of carbon dioxide inside the box as well as with other greenhouse gases, e.g. methane. Preparing this experiment, students will learn that the higher the concentration of carbon dioxide in the air the greater temperature rise.

Negative Effects of Ozone: Effects on plants

Negative effects of ozone on the process of photosynthesis may be shown by another small

Figure 2. Model experiment of the greenhouse effect (experimental set up)

experiment. Ozone destroys the chlorophyll that causes the green color of leaves. Without a sufficient amount of chlorophyll plants metabolism is disturbed. The plants death is an inevitable result of this. Chlorophyll may be isolated out from leaves by using an extraction technique with ethanol. Therefore leaves are crushed with a pestle and mortar and overlaid with ethanol. Extracted chlorophyll dyes ethanol green immediately. The ethanol-chlorophyll mixture colors brown after ozone gas has transported through the mixture (Figure 3). This brown color demonstrates clearly that chlorophyll is destroyed by ozone.

Figure 3. Ethanol-chlorophyll mixture before (left) and after (right) exposition to ozone

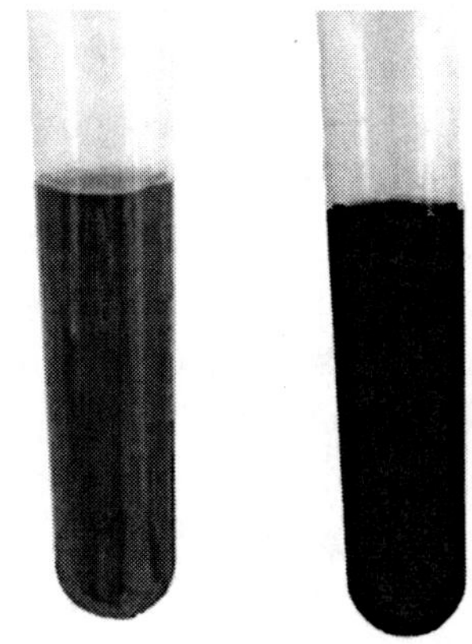

Formation of Aerosols and its Effects on Light Intensity and Heat Radiation

Clouds may reduce light intensity and heat radiation. These effects are shown clearly by a simple experiment. To this end, light intensity of a lamp is measured by a luxmeter. This value needs to be compared with the light radiation that is influenced by aerosol clouds. A closed, light transmissible vessel which is filled with aerosol cloud has to be placed between the lamp and the luxmeter (Figure 4; Ammermann, Kaminski & Pietzner, 2012). Measurements of influences of heat radiation are equal to this, except a thermometer is used instead of a luxmeter.

Getting the aerosol cloud inside a vessel can be achieved on several ways:

1. A gasoline oxygen-ozone mixture is combined with essential oils (e.g. eucalyptus oil) inside the vessel, or
2. A burning incense stick is placed into the vessel before closing the vessel.

Expectations and Experiences

The various modules of this project have been prepared since February 2012. More than 600 students and 20 teachers have participated in the project so far and carried out experiments within the non-formal chemistry laboratory in Bremen. The module "Chemistry of the atmosphere" has been offered since January 2013. Two groups with a total of 32 students so far examined the greenhouse effect, summer smog, atmospheric ozone and aerosols.

In all the visits, both teachers and students are invited to contribute to a survey prior to and after visiting the university laboratory. The pre-questionnaire for the teachers consists of 15 Likert-items and five open questions, while the student's questionnaire consists of 16 Likert-items and two open tasks. All questions focus the prior

Figure 4. Experiment on effects of aerosol clouds on light intensity

expectations of the participants towards the visit in the non-formal university student laboratory. Similar structured teacher and student questionnaires inquire into the teachers' and students' personal experiences and reflections after the visit in the non-formal university student laboratory.

In the responses from the pre-questionnaire, the teachers supported a need for more intense experimentation in science classes. The teachers indicated that it is difficult for them to conduct appropriate experiments in their schools because of insufficient equipment and facilities. There was hope that the visit in the non-formal university laboratory would contribute to and enrich the practice of laboratory work in their classes. The teachers expected the non-formal university laboratory to contribute to raising motivation in science learning. They attributed motivational potential to the societally relevant aspects of the experience, such as providing students with insights into university education as well as chemistry which is relevant to everyday life. The teachers only partially expected relevant content learning which would fulfill the school curriculum and governmental syllabus. Within their point of view, the visit to non-formal learning environments should have other benefits beyond cognitive school achievements. This offers a contrast with the students' point of view, in that they expected better marks after visiting the university laboratory. It seems that teachers also appreciate content that goes beyond the typical school topics and objectives. For the teachers it was more important to use the non-formal laboratory context to make the students aware of the relevance of chemistry, than to fulfill part of their formal curriculum. Nearly 90% of the teachers agreed with this statement.

From the student perspective, students expected to have a pleasant laboratory and research experience in the non-formal chemistry laboratory.

I am looking forward to the visit of the university laboratory because I like very much to do experiments. I believe that there is a big difference between in preparing experiments here [in the university]. I am glad to gain a new experience in the university and to feel like a real researcher during the experiments. (Translated from the open tasked questionnaire)

This quote was typical for many students. Only one percent was not excited to visit the university laboratory. If the visit meets these expectations, this finding allows positive consideration that visiting the non-formal lab has potential to affect students' attitudes towards chemistry and science learning. The students connect their positive expectations mainly with their hope to do interesting experiments; especially those that cannot be done in schools (e.g. experiments with ozone are no longer allowed in German school laboratories). Students seem to be aware that school conditions are far from being optimal for doing open experiments. The missing availability of experimental material and chemicals were criticized by many students, also the 45-minutes timing of the science lessons was criticized as they believed that it hindered open experimental work. Thus, it was exactly these aspects that the students expected from the university student laboratory. The students explicitly expressed their view that there is a gap in open and problem-based experiments in school and their hope for a different experience in the non-formal laboratory. However, the students also expected to gain a better understanding of chemical issues in school and later better grades by having visited the non-formal education environment. The majority of students did not want to see the non-formal chemistry laboratory separated from formal learning in school. They expected something more tangible, in terms of getting better marks in school, however, that is inevitable.

The teacher's and students' feedback was very positive throughout. It was quite the same among the different modules and grade level of the students. After the visit, the overwhelming majority of teachers and students enjoyed the unfamiliar, non-formal atmosphere of visiting the university laboratories. Most teachers and students gave very positive feedback. Some students mentioned that working in the non-formal university laboratory was exhausting. That is why it was suggested that laboratory sessions should not exceed 2-3 hours. The teachers gave similar feedback. In contrast to the students the teachers placed big emphasis on the quality of care, organization and the connection to the school curriculum and the official syllabus. Some selected aspects that were mentioned often in the open part of the questionnaire were:

- Interesting and motivating experiments were carried out within laboratory session.
- The students were allowed to do all the experiments autonomously.
- Lots of experiments were done during the session. The students liked to do lot of experiments. Some students additionally liked to prepare their experiments on their own which they had not experienced prior to this in the school context.
- The staff to student ratio in the laboratory was much better than the teacher to student ratio in school. Students got assistance quickly when required.
- Students liked the open work in small groups.

CONCLUSION AND RECOMMENDATIONS

The students enjoyed the unfamiliar, non-formal atmosphere of visiting the university chemistry laboratory. Orion and Hofstein (1991) suggested that the development of a more positive student attitude towards learning science could be fostered by visiting informal and non-formal learning environments. This is in line with our findings. Only a very small minority of students were not looking forward to the visit or were disappointed after it. Together with the positive feedback in the open questionnaires, there is promising potential to contribute to students developing more positive attitudes towards chemistry, science and technology. There are also indications that visiting the learning environment in this non-formal university laboratory can serve as a motivating factor in science education. Connecting science learning to authentic and innovative issues from the sustainability debate embedded into the non-formal learning experience was motivating and meaningful to the learners.

Also the teachers responded well to the program. The teachers followed their students' behavior in the laboratory with great interest. Through this they learned about new strategies of sustainable chemistry, they became familiar with new experiments, of which at least part of, can be implemented into practical work in the school science classroom, and they experienced how motivating the topics from the sustainability debate and activities of an inquiry nature can be for their students. From this perspective there is hope that the project contributes to teacher continuous professional development and through this pathway helps in implementing issues of sustainable development more thoroughly into school science education in the future.

Limitations in the initiative lie in the limited regional range of the project. It is also clear that often the effects of such visits are short-term if the visits are only singular. Some studies in this area indicated a clear cognitive gain of visits in non-formal educational environments, while others were not able to support these findings (Stronck, 1983). The same applies to the motivational effects. DeWitt and Storksdieck (2008) explained this finding was due to the short term nature of most field trips and non-formal learning events that may not be suited to create lasting cognitive and motivational effects. However, there is little research investigating whether a repeated visit in such a non-formal learning environment will have more durable effects. This project has the scope to examine this as it offers non-formal learning environments for a range of topics, all of them with connection to chemistry-related issues of sustainability and sustainable development, for all grades in the lower and upper secondary level. Nevertheless, more research is needed to show how often and how regular visits are needed to contribute to long-term effects on motivation, attitudes and cognitive gains by the non-formal learning environment as well as in contention with topics from the sustainability debate.

ACKNOWLEDGMENT

We gratefully acknowledge the funding of the Deutsche Bundesstiftung Umwelt (DBU) for supporting the developments within the described project.

REFERENCES

Abrahams, I. (2011). *Practical work in secondary science: A minds-on approach*. London, UK: Continuum.

Alonso, D. M., Bond, J. Q., & Dumesic, J. A. (2010). Catalytic conversion of biomass to biofuels. *Green Chemistry*, *12*, 1493–1513. doi:10.1039/c004654j

Altheide, D., & Johnson, J. (1994). Criteria for assessing interpretative validity in qualitative research. In N. Denzin & Y. Lincoln (Eds.), *Handbook of qualitative research* (pp. 485–499). Newbury Park, CA: Sage.

Ammermann, S., Kaminski, L., & Pietzner, V. (2012). Aerosols – Relevant for climate and medicine. *Chemistry in Action*, *98*(Winter), 7–15.

Arbuthnott, K. D. (2009). Education for sustainable development beyond attitude change. *International Journal of Sustainability in Higher Education*, *7*, 152–163. doi:10.1108/14676370910945954

Berg, C. A. R., Bergendahl, V. C. B., Lundberg, B. K. S., & Tibell, L. A. E. (2003). Benefiting from an open-ended experiment? A comparison of attitudes to, and outcomes of, an expository versus an open-inquiry version of the same experiment. *International Journal of Science Education*, *25*, 351–372. doi:10.1080/09500690210145738

Bodner, G., MacIsaac, D., & White, S. (1999). Action research: Overcoming the sports mentality approach to assessment/evaluation. *University Chemistry Education*, *3*(1), 31–36.

Bradley, J. D. (2005). Chemistry education for development. *Chemical Education International, 7*.

Braund, M., & Reiss, M. (2006). Towards a more authentic science curriculum: The contribution of out-of-school learning. *International Journal of Science Education*, *28*, 1373–1388. doi:10.1080/09500690500498419

Burmeister, M., Rauch, F., & Eilks, I. (2012). Education for sustainable development (ESD) and chemistry education. *Chemistry Education Research and Practice*, *13*, 59–68. doi:10.1039/c1rp90060a

Burmeister, M., Schmidt-Jacob, S., & Eilks, I. (2013). German chemistry teachers' understanding of sustainability and education for sustainable development - An interview case study. *Chemistry Education Research and Practice*, *14*, 169–176. doi:10.1039/c2rp20137b

Bybee, R. W. (2001). Achieving scientific literacy: Strategies for insuring that free-choice science education complements national formal science education efforts. In J. H. Falk (Ed.), *Free-choice science education: How we learn science outside of school* (pp. 44–63). New York, NY: Teachers College Press.

Centi, G., & Perathoner, S. (2009). From green to sustainable chemistry. In F. Cavani, G. Centi, S. Perathoner, & F. Trifiro (Eds.), *Sustainable industrial processes* (pp. 1–72). Weinheim, Germany: Wiley-VCH. doi:10.1002/9783527629114.ch1

Coll, R. K., Gilbert, J. K., Pilot, A., & Streller, S. (2012). How to benefit from the informal and interdisciplinary dimension of chemistry in teaching? In I. Eilks & A. Hofstein (Eds.), *Teaching chemistry – A studybook* (pp. 241–268). Rotterdam, The Netherlands: Sense.

Dawe, J., Jucker, R., & Martin, S. (2005*). Sustainable development in higher education: Current practice and future developments* (Report for The Higher Education Academy). Retrieved May 27, 2013 from http://www.heacademy.ac.uk/assets/was%20York%20%20delete%20this%20soon/documents/ourwork/sustainability/sustdevinHEfinalreport.pdf.

DeWitt, J., & Storksdieck, M. (2008). A short review of school field trips: Key findings from the past and implications for the future. *Visitor Studies, 11*(2), 181–197. doi:10.1080/10645570802355562

Eilks, I. (2014). Action research in science education: From a general justification to a specific model in practice. In T. Stern, F. Rauch, A. Schuster, & A. Townsend (Eds.), *Action research, innovation and change* (pp. 156–176). London, UK: Routledge.

Eilks, I., Nielsen, J. A., & Hofstein, A. (2014). Learning about the role of science in public debate as an essential component of scientific literacy. In C. Bruguière, A. Tiberghien, P. Clément (eds.), *Topics and trends in current science education* (pp. 85-100). Dordrecht, The Netherlands: Springer.

Eilks, I., Prins, G. T., & Lazarowitz, R. (2013). How to organise the chemistry classroom in a student-active mode. In I. Eilks & A. Hofstein (Eds.), *Teaching chemistry – A studybook* (pp. 183–212). Rotterdam, The Netherlands: Sense.

Eilks, I., & Ralle, B. (2002). Participatory action research in chemical education. In B. Ralle & I. Eilks (Eds.), *Research in chemical education - What does this mean?* (pp. 87–98). Aachen, Germany: Shaker.

Eshach, H. (2007). Bridging in-school and out-of-school learning: Formal, non-Formal, and informal education. *Journal of Science Education and Technology, 16*, 171–190. doi:10.1007/s10956-006-9027-1

Gallacher, J., & Feutrie, M. (2003). Recognising and accrediting informal and non-formal learning in higher education: An analysis of the issues emerging from a study of France and Scotland. *European Journal of Education, 38*, 71–83. doi:10.1111/1467-3435.00129

Gallego-Alvarez, I. (2012). Indicators for sustainable development: Relationship between indicators related to climate change and explanatory factors. *Sustainable Development, 20*, 276–292. doi:10.1002/sd.483

Greeno, J. G. (1988). Situations, mental models, and generative knowledge. In D. Klahr & K. Kotovsky (Eds.), *Complex information processing: The impact of H. A. Simon*. Hillsdale, NJ: Lawrence Erlbaum.

Hjeresen, D. L., Schutt, D. L., & Boese, J. M. (2000). Green chemistry and education. *Journal of Chemical Education, 77*, 1543–1547. doi:10.1021/ed077p1543

Höfer, R., & Bigorry, J. (2007). Green chemistry – A sustainable solution for industrial specialties applications. *Green Chemistry, 9*, 203–212. doi:10.1039/b606377b

Hofstein, A., Eilks, I., & Bybee, R. (2011). Societal issues and their importance for contemporary science education: A pedagogical justification and the state of the art in Israel, Germany and the USA. *International Journal of Science and Mathematics Education, 9*, 1459–1483. doi:10.1007/s10763-010-9273-9

Hofstein, A., & Kesner, M. (2006). Industrial chemistry and school chemistry: Making chemistry studies more relevant. *International Journal of Science Education, 28*, 1017–1039. doi:10.1080/09500690600702504

Hofstein, A., & Kind, P. M. (2012). Learning in and from science laboratories. In B. J. Fraser, K. G. Tobin, & C. J. McRobbie (Eds.), *Second international handbook of science education* (pp. 189–207). Heidelberg, Germany: Springer. doi:10.1007/978-1-4020-9041-7_15

Hofstein, A., Kipnis, M., & Abrahams, I. (2013). How to learn in and from the chemistry laboratory. In I. Eilks & A. Hofstein (Eds.), *Teaching chemistry – A studybook* (pp. 153–282). Rotterdam, The Netherlands: Sense.

Hofstein, A., & Lunetta, V. N. (2003). The laboratory in science education: Foundations for the twenty-first century. *Science Education*, *88*, 28–54. doi:10.1002/sce.10106

Hofstein, A., Navon, O., Kipnis, M., & Mamlok-Naaman, R. (2005). Developing students' ability to ask more and better questions resulting from inquiry-type chemistry laboratories. *Journal of Research in Science Teaching*, *42*, 971–806. doi:10.1002/tea.20072

Hofstein, A., & Rosenfeld, S. (1996). Bridging the gap between formal and informal science learning. *Studies in Science Education*, *28*, 87–112. doi:10.1080/03057269608560085

Howard, K. E., Brown, S. A., Chung, S. H., Jobson, B. T., & Van Reken, T. M. (2013). College students` understanding of atmospheric ozone formation. *Chemistry Education Research and Practice*, *51*(14), 51–61. doi:10.1039/c2rp20074k

Karpudewan, M., Ismail, Z. H., & Mohamed, N. (2011). Green chemistry: Educating prospective science teachers in education for sustainable development at school of educational studies. *Journal of the Social Sciences*, *7*, 42–50. doi:10.3844/jssp.2011.42.50

Lefohn, A. S., Shadwick, D., & Oltmans, S. J. (2010). Characterizing changes in surface ozone levels in metropolitan and rural areas in United States for 1980-2008 and 1994-2008. *Atmospheric Environment*, *44*, 5199–5210. doi:10.1016/j.atmosenv.2010.08.049

Lunetta, V. N., Hofstein, A., & Clough, M. P. (2007). Learning and teaching in the school science laboratory: An analysis of research, theory and practice. In S. K. Abell & N. G. Lederman (Eds.), *Handbook of research on science education* (pp. 393–442). Mahwah, NJ: Lawrence Erlbaum.

Mamlok-Naaman, R., & Eilks, I. (2012). Different types of action research to promote chemistry teachers' professional development - A joined theoretical reflection on two cases from Israel and Germany. *International Journal of Science and Mathematics Education*, *10*, 581–610. doi:10.1007/s10763-011-9306-z

McEvoy, D., Gibbs, D. C., & Longhurst, J. W. S. (2000). The employment implications of low-carbon economy. *Sustainable Development*, *8*, 27–38. doi:10.1002/(SICI)1099-1719(200002)8:1<27::AID-SD120>3.0.CO;2-O

Monks, P. S. (2007). Chemistry of the atmosphere. In R. M. Harrison (Ed.), *Principles of environmental chemistry* (pp. 8–79). Cambridge, UK: RSC Publishing. doi:10.1039/9781847557780-00008

Mortensen, L. L. (2000). Global change education – Educational resources for sustainability. In K. A. Wheeler & A. Perraca Bijur (Eds.), *Education for a sustainable future* (pp. 15–34). Dordrecht, The Netherlands: Kluwer. doi:10.1007/978-1-4615-4277-3_3

OECD. (2012). *Higher education and adult learning - Recognition of non-formal and informal learning*. Retrieved May 28, 2013 from http://www.oecd.org/edu/skills-beyond-school/recognitionofnon-formalandinformallearning-home.htm.

Orion, N., & Hofstein, A. (1991). The measurement of students' attitudes towards scientific field trips. *Science Education, 75*, 513–523. doi:10.1002/sce.3730750503

Orion, N., & Hofstein, A. (1994). Factors that influence learning during a scientific field trip in a natural environment. *Journal of Research in Science Teaching, 31*(10), 1097–1119. doi:10.1002/tea.3660311005

Parchmann, I., Gräsel, C., Bär, A., Nentwig, P., Demuth, R., & Ralle, B. (2006). Chemie im kontext: A symbiotic implementation of a context - based teaching and learning approach. *International Journal of Science Education, 28*, 1041–1062. doi:10.1080/09500690600702512

Parchmann, I., & Jansen, W. (1996). Der Treibhauseffekt als Folge der Waermeabsorption von Gasen. *Chemie Konkret, 3*, 6–11.

Rannikmäe, M., Teppo, M., & Holbrook, J. (2010). Popularity and relevance of science education literacy: Using a context - based approach. *Science Education International, 21*, 116–125.

Rennie, L. J. (2007). Learning science outside of school. In S. K. Abell & N. G. Lederman (Eds.), *Handbook of research and science education* (pp. 125–170). Mahwah, NJ: Lawrence Erlbaum.

Rennie, L. J., & McClafferty, T. P. (1996). Science centres and science learning. *Studies in Science Education, 27*, 53–98. doi:10.1080/03057269608560078

Stronck, D. R. (1983). The comparative effects of different museum tours on children's attitudes and learning. *Journal of Research in Science Teaching, 20*, 283–290. doi:10.1002/tea.3660200403

Stuckey, M., Mamlok-Naaman, R., Hofstein, A., & Eilks, I. (2013). The meaning of 'relevance' in science education and its implications for the chemistry curriculum. *Studies in Science Education, 49*, 1–34. doi:10.1080/03057267.2013.802463

Tang, X., Wilson, S. R., Solomon, K. R., Shao, M., & Madronich, S. (2011). Changes in air quality and tropospheric composition due to depletion of stratospheric ozone and interactions with climate. *Photochemical & Photobiological Sciences, 10*, 280–291. doi:10.1039/c0pp90039g PMID:21253665

Tobin, K. G. (1990). Research on science laboratory activities, in pursuit of better questions and answers to improve learning. *School Science and Mathematics, 90*, 403–418. doi:10.1111/j.1949-8594.1990.tb17229.x

Vinten, G. (1994). The sustainable company: The need for environmental concern. *Sustainable Development, 2*(2), 1–8. doi:10.1002/sd.3460020201

Ware, S. A. (2001). Teaching chemistry from a societal perspective. *Pure and Applied Chemistry, 7*(7), 1209–1214. doi:10.1351/pac200173071209

Werquin, P. (2010). *Recognising non-formal and informal learning. Outcomes, policies and practices*. Paris: OECD.

Wheeler, K. A., & Perrraca Bijur, A. (2000). *Education for a sustainable future*. Dordrecht, The Netherlands: Kluwer. doi:10.1007/978-1-4615-4277-3

This work was previously published in a "Handbook of Research on Pedagogical Innovations for Sustainable Development" edited by Ken D. Thomas and Helen E. Muga, pages 229-244, copyright 2014 by Information Science Reference (an imprint of IGI Global).

Chapter 47
Analysis of Discourse Practices in Elementary Science Classrooms using Argument-Based Inquiry during Whole-Class Dialogue

Matthew J. Benus
Indiana University Northwest, USA

Brian M. Hand
University of Iowa, USA

Morgan B. Yarker
University of Iowa, USA

Lori A. Norton-Meier
University of Louisville, USA

ABSTRACT

This chapter discusses an analysis of discourse practices found in eight different elementary science classrooms that have implemented the Science Writing Heuristic (SWH) approach to argument-based inquiry. The analysis for this study involved examining a segment of whole-class talk that began after a small group presented its claim and evidence and ended when the discussion moved on to a new topic, or when a different group presented. The framework for the analysis of this whole-class dialogue developed through an iterative process that was first informed by previous analysis, review and modification of other instruments, and notable anomalies of difference from this data set. Each classroom was then rated using the Reform Teaching Observation Protocol (RTOP), which provided a score for the extent to which the teacher was engaged with reform-based science teaching practices. Our analysis shows that elements of whole-class dialogue in argument-based inquiry classrooms were different across varying levels of RTOP implementation. Overall, low level RTOP implementation (little evidence of reformed-based practice) had a question and answer format during whole class talk that rarely included discourse around scientific reasoning and justification. Higher levels of RTOP implementation were more likely to be focused on student use of scientific evidence to anchor and develop a scientific understanding of "big ideas" in science. These findings are discussed in relation to teacher professional development in argument-based inquiry, science literacy, and the teacher's and students' grasp of science practice.

DOI: 10.4018/978-1-4666-7363-2.ch047

INTRODUCTION

For quite some time now, reform documents have set the benchmarks for science literacy by stressing the importance of using inquiry-based approaches in the science classroom (AAAS, 1993; NRC, 1996, 2012). Yet, what drives and supports an inquiry-based approach in the classroom is still up for discussion and currently being thoroughly examined through research. Kuhn (1991) gives one good idea of what it means to actually be engaged in scientific inquiry, stating "Scientific inquiry is fundamentally a knowledge building process in which explanations are presented to the community so they can be critiqued, debated, and revised" (1991, p. 4). Deriving Kuhn's definition of scientific inquiry leads us to understand that to engage in argument-based inquiry means to use evidence in support of one's claim, which can help to focus and drive dialogues of critique and consensus (Duschl & Osborne, 2002; Osborne, Erduran, & Simon, 2004). Dialogue "is about a shared inquiry, a way of thinking and reflecting together. It is not something you do to another person. It is something you do with people" (Isaacs, 1999, p. 9) and dialogic conversation can be a way to generate knowledge (Alexopoulou & Driver, 1997; Ford, 2008; Kelly & Green, 1998; Schein, 1993). While the research community recognizes the value of dialogic communication in inquiry-based learning approaches, research that examines communication patterns found in classrooms using argument-based inquiry is still needed. This chapter will explore the talk patterns within whole-class dialogue in fifteen elementary classrooms that utilize argument-based inquiry. The two research questions guiding this study are:

1. What are the key factors that contribute to whole-class discussion found in classrooms using argument-based inquiry, and
2. How do these key factors develop across levels of implementation of argument-based inquiry?

ARGUMENTATION IN PRACTICE

Argumentation is conversational dialogue. This can be reasoned through by exploring the meanings of the words dialogue and argumentation. The word dialogue is from two Greek roots: dia and logos. This roughly translates to be "meaning flowing through" (Isaacs, 1993, p. 25). Conversation, which begins all dialogues, means "to turn together" (Isaacs, 1993, p. 35). Argumentation then is a conversational dialogue where meanings can flow through and turn together within those engaged. Argumentation "initiates change, it transforms the significance of material, it enables reflection and action, it brings divergent voices together in interaction..." (Mork, 2005, p. 18; referencing Costello & Mitchell, 1995). It is worth noting the difference in argumentation versus argument. Argumentation by many researchers is considered a discourse process (e.g. Jimenez-Aleixandre & Erduran, 2008; Osborne, et al., 2004) while argument is centering around producing or influencing a particular outcome (Cavagnetto, 2010; Toulmin, 1958). In this study, we focus on argumentation.

Scientific argumentation is about evaluating and critiquing the construction of scientific claims, evidence, and explanation (Duschl, Schweingruber, & Shouse, 2007). Once reasoned through by being constructed and deconstructed, knowledge is claimed because it carries with it tentative, but stable, evidence that is deemed trustworthy by the community of science (Gross, 1990). In actual practice, scientists engage in the process of argumentation to understand "why" (Duschl, 1990) their ideas matter. As these scientists engage in argumentation, they modify their ideas by making or critiquing claims supported with evidence (NRC, 2012). The net effect of their engagement further strengthens their understanding within their domain of study.

SCIENTIFIC ARGUMENTATION IN CLASSROOMS

This process of argumentation can and should also be a critical component of instruction in our science classrooms (Driver, Asoko, Leach, Mortimer, & Scott, 1994; Driver, Newton, & Osborne, 2000; Duschl & Osborne, 2002; Kuhn, 1993). In particular, "why" questions can help to engage students into the core practices found in communities of scientists. The reason for this is that "why" questions set up conversations around claims, data, and evidence, all of which are core essential components of scientific argumentation (Brinker & Bell, 2008). The process of engaging in scientific argumentation in the science classroom is centered around what Lemke (1990) calls "learning to talk science." This does not mean a science classroom is dominated by the teacher talking science (unfortunately this is still widely being practiced today (see Crawford, 2005)) but rather the classroom community with the teacher engaged in an ongoing conversational dialogue of scientific argumentation.

Over the last decade there have been several studies focusing on dialogic elements of scientific argumentation in science classrooms (e.g., Cavagnetto, Hand, & Norton-Meier, 2010; Driver, Newton, & Osborne, 2000; Duschl & Osborne, 2002; Hogan, Nastasi, & Pressley, 1999; Kelly & Duschl, 2002; Martin & Hand, 2009; Martins, Mortimer, Osborne, Tsatsarelis, & Jiménez-Aleixandre, 2001; Naylor, Keogh, & Downing, 2007; Simon, Erduran, & Osborne, 2006; von Aufschnaiter, Erduran, Osborne, & Simon, 2008; Zohar & Nemet, 2002). Without question, these studies imply, suggest, and/or stress that it takes time and skill to engage students in some element of conversational dialogue in order to aid them in their construction of knowledge.

Classrooms that engage in scientific argumentation also engage in "activity, reflection, and conversation" (Fosnot, 1989, p. 29). In later writings Fosnot states, "The learners [rather than the teacher] are responsible for defending, providing, justifying, and communicating their ideas to the classroom community. Ideas are accepted as trustworthy only insofar as they make sense to the community and thus rise to the level of 'taken-as-shared,'" (Fosnot, 1996, p. 29-30). Being responsible for nurturing this "taken-as-shared" experience where meanings can flow through and turn together (Isaacs, 1993) is a significant responsibility of the teacher. McNeill (2009) states, "The teacher can be essential for helping students explain phenomena, justify the claims they are making, and debate the strength of alternative explanation" (p. 259). Scientific argumentation in full classroom practice is thought to "transforms the common monologic discourse of the school science classroom" (Osborne, 2007, p. 12) that is often wrought with "monolithic paths of logic or pre-ordained discovery" (Yerrick, 2000, p. 814), which is lead or voiced by the teacher.

SUSTAINED UNDERSTANDING AND PEDAGOGICAL PRACTICE OF SCIENTIFIC ARGUMENTATION

Without doubt, transforming schools and classrooms toward the practice of scientific argumentation is slow to happen because the science curriculum and instruction is often orchestrated around predetermined investigations that serve to verify what is already known (Duschl, 1990; Lemke, 1990). Scientific argumentation in classrooms certainly will draw on expert understandings, but at its core is the ability to engage in the meta-thinking skills and practices around the evaluation, critique, and construction of evidence as discussed in documents such as the NSES (NRC, 1996), A Framework for K-12 Science Education: Practices, Crosscutting Concepts, and Core Ideas (NRC, 2012), Taking Science to School (Duschl, Schweingruber, & Shouse, 2007), and Benchmarks for Science literacy (AAAS, 1993). Certainly, classrooms that engage in high levels of scientific argumentation would entail individuals coherently constructing ideas for others so meaning can flow in and through others as understandings turn and evolve within the community of learners. To do this, teachers need sustained professional development in the ways to approach and practice scientific argumentation. They also need ongoing support that reassures them that the strength of what one knows comes from one's ability to coherently construct and critique one's understanding with self and others.

Unfortunately, sustained professional development around what is scientific argumentation and how one learns is scarcely noted in the literature. When Newton et al. (1999) surveyed 14 experienced science teachers, the teachers reported that they needed more professional development time to confidently manage and facilitate elements of argumentation.

There are two examples of well-researched professional development programs on argumentation; one is called Evidence and Argument in Science (IDEAS) (Osborne, Erduran, & Simon, 2004) from King's College in London, England, and the second is called the Science Writing Heuristic (SWH) (Hand, Wallace, & Yang, 2004; Keys, Hand, Prain, & Collins, 1999) from the University of Iowa, USA. Both research programs have produced numerous reports and guides on implementing argumentation in science classrooms (Norton-Meier, Hand, Hockenberry, & Wise, 2008; Osborne, Erduran, & Simon, 2004b) as well as having produced peer-reviewed papers (Cavagnetto, Hand, & Norton-Meier, 2010; Hand, Wallace, & Yang, 2004; Keys, Hand, Prain, & Collins, 1999; Martin & Hand, 2009; Simon, Erduran, & Osborne, 2006; von Aufschnaiter, Erduran, Osborne, & Simon, 2008) on its effects towards enhancing student learning through engagement in elements of scientific argumentation.

Professional development work in Iowa (e.g. Martin & Hand, 2009; others in press) has shown that shifting classroom activities to include elements of science argumentation not only takes time and practice (at least 18 months) but also that the teacher needs to hold a sustainable understanding that student learning happens through ongoing engagement in scientific argumentation. The professional development work in London (Osborne, Erduran, & Simon, 2004) acknowledged the ways in which the strength of a teacher's initial understanding of argumentation "determines their development, particularly in the short term" (p. 31). In a two-year study of 8th grade science teachers, related to the initial study, Simon, Erduran, and Osborne (2006) found that their extended period of professional development helped to adapt classroom practice toward the use of argumentation.

In classroom practice, talking and reading are more likely to occur than writing. This is not to suggest that writing is not important, but rather to point out that writing and especially revising one's writing is fostered by additional conversations and readings. The literature in science argumentation regarding student learning has mostly been focused on the evaluation of the written processes/products of arguments (e.g. Bell & Linn, 2000; Kelly, et al., 2000; McNeill, 2009; McNeill, Lizotte, Krajcik, & Marx, 2006; Sampson, Grooms, & Walker, 2011; Schweizer & Kelly, 2005; von Aufschnaiter, Erduran, Osborne, & Simon, 2008: Sandoval & Reiser, 2004; Zohar & Nemet, 2002) rather than analysis of the depth and nature of conversations leading up to and following student writing.

Yore, Bisanz, and Hand (2003) point out that elements of Toulmin's Argument Pattern (TAP) (often used to study writing samples) are nearly impossible to pattern and sequence in detail because of the speed in which conversation happens. Understanding whole classroom dialogue goes well beyond the notions of two or more people talking. Approaches to dialogue are many and Burbules and Bruce (2001) reminds us that "no single approach holds the patent on dialogue" (p. 1112). However, there are certainly factors that help to characterize dialogic approaches in scientific argumentation. Effort is needed to provide insight into the extent to which students understand the depth of the question(s), evidence, ideas, and extent of interactions consistent with practicing science argumentation.

ELEMENTS OF CONVERSATIONAL DIALOGUE

The role of questioning, how the class reasons and uses evidence, and how classrooms conversationally "turn together" and integrate ideas seem to be elements strongly supported in the literature and most recently in the framework for K-12 science education (NRC, 2012). Below are five key factors that help to guide whole-class dialogue in classrooms practicing scientific argumentation.

Issues of Questioning

Questioning is essential to scientific argumentation. Boyd and Rubin (2006) talks about contingent questioning and how it takes students from where they are and launches them further. These questions are generally authentic (Nystrand, Gamoran, Kachur, & Prendergast, 1997) and stay well within reach of the student's zone of proximal development (Vygotsky, 1986). These questions are also more likely to be voiced by students to further clarify, elaborate, and/or extend ideas because the teacher in dialogically organized instruction is thought to ask fewer questions and encourage more conversational talk among the class (Boyd & Rubin, 2006, p. 144).

Depth of Exchange

Exchanging ideas takes time. As classroom talk turns together, Boyd and Rubin (2006) also suggests that students must have critical turns of talk that are linguistically extended for 10 seconds or more. These extended turns help in formulating and connecting our ideas for others as well as one's self (Scott & Ametller, 2007; Wells & Mejía Arauz, 2006). Extended exchanges discourage a back-and-forth conversation volley and tend to give more ideas and insights that sustain thinking together (Issacs, 1999). In these sorts of exchanges, ideas from others are expanded, modified, and/or reframed.

Classroom Interactions

Classrooms that turn together interact in ways that are dynamic and open in the classroom community (Burbules & Bruce, 2001). There is an ongoing openness to different points of view. The community spends time "probing and revoicing each other's thinking" (Varelas & Pineda, 1999, p. 26) as a common understanding is constructed. Viewpoints of others are not only included but also acknowledged during dialogue (Scott, Mortimer, & Aguiar, 2006).

Evidence-Based Ideas

Justification of a claim requires evidence if we expect the classroom community to be aligned to the social practices of scientists (Kuhn & Reiser, 2006). In classroom practice, Kuhn (1991) suggests that some do not feel that claims need to be justified with evidence. Some do not understand what counts as evidence (Sadler, 2004). However, the social aspect of scientific argumentation requires that claims and evidence not only be presented, but subjected to ongoing discussion and critique (Kuhn & Reiser, 2006, p. 5).

Conversational Pattern

A dialogically organized conversation around one idea is "relatively unpredictable" (Boyd & Rubin, 2006, p. 146) and negotiated by the classroom on the fly (Boyd & Rubin, 2006; Nystrand, et al., 1997). Classroom roles are "neither distinct nor stable" (Burbules & Bruce, 2001, p. 19) and all can have a voice in the conversation. Turns of talks are structurally coherent and socially engaging

(Boyd & Rubin, 2006). The conversation varies in density and quality, and has an ebb and flow as ideas build (Varelas & Pineda, 1999).

METHODOLOGY AND PARTICIPANTS

Data used for this study were chosen from an existing collection of K-5 classroom videos. This collection of videos comes from classrooms whose teachers have undergone professional development in argument-based inquiry from one to six consecutive years. The overarching focus of the professional development is to help these in-service teachers understand that learning is a process of negotiation, and that by using argument-based inquiry in their science classrooms, "big ideas" in science may build and develop through on-going classroom negotiation. Their annual Professional Development (PD) consisted of five days of summer workshops and at least five visits from PD staff in their classroom during the school year. The teachers were not given any curriculum to teach. However, they were supported through PD during the academic year to embed argument-based inquiry into their existing classroom curriculum. While engaged in the PD process the teachers used the Science Writing Heuristic (SWH) (Keys, et al., 1999) in their classrooms, which we view as one approach to establishing a classroom community that practices argument-based inquiry.

As part of the research component of the PD, teachers were asked to record a video of their classrooms. This was a lesson of their choosing. For this particular study, we choose only video segments of whole-class conversation after a small group of students presented their written claim and evidence about a concept in science. We identified fifteen classroom videos that met this criterion, and a total of fifty-six segments from these classrooms. We used no more than five segments from any one classroom video. A segment began after small group presentation of claim and evidence ended and whole-class conversation about the presentation began. The segment ended when the discussion moved on to a new topic or a new presenting group.

Initially we did a comparative analysis (Charmaz, 2002; Dey, 1999; Glaser & Strauss, 1967), within and between transcript segments in order to establish codes and patterns. During this process, we explored several published instruments and research products that could further inform our coding and analysis. In particular, we found that Section V "Discourse Factors" of Marshall et al. (2009) instrument helped to guide and shape some of our early thinking as we grounded our descriptions in our K-5 classroom.

After many rounds of revision, informed by our data set as well as the literature, we settled on three levels of criteria and five whole-class dialogue factors. The criteria ranges from level 1 (little to no dialogic interaction) to level 3 (high dialogic interaction) we settled on the five key factors to be: conversational patterns, complexity of questions, evidence-based ideas, depth of idea exchange, and classroom interaction. Table 1 represents the whole class dialogue factors used in this study.

Once the framework was established, each segment was coded with a dialogue level for each of the five key dialogue factors. The first two authors of this study coded the segments individually and then discussed the differences in ratings until a consensus was reached for each of the segments in this study.

Additionally, every video used in this study was scored by experienced scorers using a modified version (Martin & Hand, 2009) of the Reform Teaching Observation Protocol (RTOP) (Sawada & Piburn, 2000). Using this protocol gave us a sense of how well each classroom engaged in "reformed" (Sawada, et al., 2000) instruction. The quantitative RTOP score was then categorized into qualitative descriptions; i.e., high, medium, and low RTOP scores to indicate high, medium, and low implementation of reformed instruction.

Table 1. Whole class dialogue factors

Factor	Level One	Level Two	Level Three
Complexity of Question	In most cases, questions were asked to explain explicit knowledge.	In most cases, questions ask students to explain their comprehension of ideas.	In many cases, questions challenged students to explain, reason through, and/or justify.
Depth of Idea Exchange	Even if opportunities existed, ideas were rarely discussed beyond initial response.	Ideas were discussed for several turns of talk but were usually limited to comparing/checking/understanding some smaller element of the "big idea."	Ideas were discussed over many turns of talk to help understand many elements/viewpoints of the "big idea."
Classroom Interactions	Students did not ask a student to justify reasoning or evidence. Teacher may occasionally ask for justification and/or reasoning.	Occasionally student(s), and often the teacher, asked follow-up responses that required student(s) to justify reasoning or evidence.	Students often, and teacher may or may not as often, ask follow-up responses that required students to justify reasoning or evidence.
Evidence-based Ideas	There is little discussion of the claim/evidence presented.	There is some discussion of the claim/evidence presented.	There is extensive discussion of the claim/evidence presented.
Conversational Pattern	Student conversation not well connected to previous turns of talk, with very short conversations about student ideas. Generally Q&A format.	Student conversation at least occasionally is connected to previous turns of talk. Some medium-length conversation occurs about a student idea.	Student conversation was consistently integrated with previous turns of talk. Lengthy discussions occur about a student idea.

RESULTS

Analyses of the data lead us to compare the criteria-based level of dialogue (i.e., level 1, level 2, and level 3) with the classroom RTOP ranking (i.e., low, medium, and high). The findings of this analysis are reported by RTOP score. For each level, we provide an overall summary of implementation and then provide more detailed descriptions of key factors noting what makes a level different than another. We also provide an actual segment of a classroom transcript for each level to further illustrate our findings.

SUMMARY OF LOW IMPLEMENTATION

In low implementation classrooms, we did not see any instances of what we would call level 3 talk. *Conversational patterns* were mostly very structured question/answer situations and generally led by the teacher with fractured connections to previous turns of talk. The teacher's goals seem to be to check student responses to *low-complexity questions* and compare them with his or her own ideas about the concept; therefore there was little talk about *evidence-based ideas*. The teacher and students rarely asked follow-up questions, which hindered the *depth of idea exchange* and tempered whole *classroom interaction* (see Figure 1).

Questioning and Use of Evidence

Quality of the question was generally quite low with seven segments containing mostly level 1, three segments containing mostly level 2, and no segments containing mostly level 3; however, there were a few instances of higher level questions being asked that required students to explain their conception. However, regardless of the question type being asked, student responses tended to foster a recalling of some piece of knowledge that the questioner accepted without any follow-up

Figure 1. Frequency of whole-class talk categories for low SWH implementation classroom segments

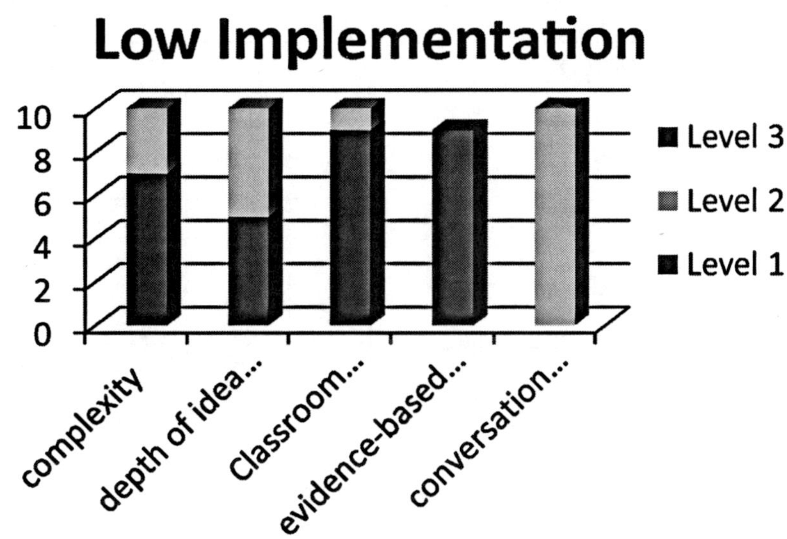

or elaboration. This lack of critique of student response suggests that *evidence-based ideas* were not well developed, because nine segments were level 1, and there were no level 2 or level 3. Questions that only required a response for a particular piece of content with no follow-up (see Tables 2 and 3) did not encourage ideas to be reasoned through by the classroom community, hindering *classroom interactions*. Additionally ideas being discussed were not grounded in the relationship between claim and evidence. This limited the amount of reasoning during classroom interaction.

Depth of Idea Exchange and Conversational Pattern

Depth of idea exchange was generally very low, with five segments at level 1, five segments at level 2, and no segments at level 3. Even when multiple turns of student talk were present, opportunities for students to reason through their ideas seemed limited. In each turn of student-student talk there were too few words and a lack of sufficient follow-up for a developed idea exchange. Overall, *conversational patterns* were generally very structured in low implementation classrooms, with all ten segments at level 2, and no segments at level 1 and level 3.

Example of Low Implementation

In Table 2, a group is presenting on the United States' third manned mission to the moon (Apollo 13) whose mission was aborted due to a malfunction. The teacher is asking questions about the spaceship malfunction and receiving responses from the students. When Kenny elaborated on an idea the teacher did not take the opportunity to discuss Kenny's idea beyond his initial response. This seems to indicate that the teacher was more interested in checking to see if the class agreed with her own idea rather than understanding the students' ideas.

In Table 3, a fourth-grade class is discussing animal adaptations. Sarah asks a question that seemed to request further explanation of the groups claim. Instead, the students responded by repeating their claim with no explanation. Neither the teacher nor students challenged the group's response; rather, the teacher allowed the conversation to move on by asking for other questions. We consider these two examples to be coded as

Table 2. Example of level one whole class dialogue

Line	Person	Words Spoken
119	Teacher	What do they need oxygen for?
120	Joyce	To breathe.
121	Teacher	So if the oxygen was leaking, is that a problem?
122	Students	Yep.
123	Teacher	Ok
124	Kenny	Cuz it was going into space.
125	Teacher	Ok.
126	Joyce	They shut down the command module… down to save battery power so they could get back to Earth. They lived in the lunar module.
127	Teacher	What's the lunar module?
Grade 4, 025-04-01-03-2009		

a *missed opportunity* that seems to establish an environment with limited opportunity for students to discuss their ideas beyond the initial response.

SUMMARY OF MEDIUM LEVEL IMPLEMENTATION

In medium ranked classrooms, teachers and students both are asking questions with some level of complexity. When a level 2 or level 3 *complex question* is asked, there is generally some level of *in-depth idea exchange* that stems directly from *evidence-based ideas*. There tends to be less idea exchange when the questions are less complex, because these questions are rarely focused on the relationship of evidence to the claim. Conversation patterns are generally structured by the teacher, or not as neatly structured if the students are leading the discussion; however, in either case, there are some instances of *classroom interaction* that involve many students sharing their ideas about the claim and evidence.

Table 3. Example of level one whole class dialogue

Line	Person	Words Spoken
	Jack	What was the best animal adaptation? Claim: the best animal adaptation is the tail for a possum. For a alligator, fish and a possum…
209	Teacher	Sara?
210	Sara	I didn't get your claim.
211	Jack	The best animal adaptation is the possum's tail.
212	Teacher	Any other comments? Sheri?
213	Sheri	I think they did pretty good as far as their claim.
Grade 4, 027-03-01-05-2008		

Questioning, Evidence, and Idea Exchange

In medium level classrooms, whole class dialogue that supported *deep idea exchange* was developing. In these classrooms, the student turns of talk tended to show *evidence of ideas* at least occasionally being responded to by other students and consistently being listened to and responded to by the teacher. At least for the teacher, their turns of talk usually reflected a response to one or more turns of student talk. *Question complexity* by both students and teacher tended to vary. Follow-up to question responses varied from no elaboration to extended dialogue. When questions asked were about understanding what something meant, there was often extensive classroom interaction among students over many turns of talk (see Tables 4 and 5). In the questions and ideas found in these turns of talk, it seems as though students at this level were starting to work towards being able to argue for their ideas. We noticed that when questions were more focused on a call or need to have evidence, talk tended to stop after a response was given (see Table 4 and Figure 2).

Table 4. Example of level two whole class dialogue

Line	Person	Words Spoken
333	Maria	Um do you know uh, if like, Iowa is in the deciduous forest?
334	Group Member	No. It has a map in there but it didn't have the United States. Lance?
Grade 6, 010-01-01-01-2007		

Table 5. Example of level two whole class dialogue

Line	Person	Words Spoken
31	Danny	Uh do you also think that larger objects can make high pitch because the jar the jar the big jar, that one also made a high pitch?
32	Alyssa	In the jar, it was like the one with no water in it made a high sound and the one almost full made a low sound. And that's what you meant with the jars.
33	Danny	How full was the jar?
34	Alyssa	It's the one over there.
35	Luke	So I think it differs with how much water you put in each jar. Cause one group, cause one group (mumble) cause one jar was overflowing.
36	Hunter	Yeah
37	Peyton	But the one with the strings had a shorter string, the higher the pitch was and the longer the string the lower it was.
38	Luke	But we are talking about the jars.
39	Peyton	I know but you said it might affect them not a lot and I'm saying cause that the jars have a little bit of water and a lot and with the strings that has a little shorter string makes a higher pitch and the longer one has a lower pitch.
40	Luke	But what our group did is that the shorter string had a really high sound and the longer ones had the same. All was the same it had a high sound.
41	Peyton	That's not what we got.
42	Luke	I'll try it.
Grade 5, 012-10-08-04-2009		

Conversational Interactions and Evidence

In each of the medium implementation classrooms, *conversation patterns* tended to be conversational and most often monitored and/or mediated by the teacher. There were a few cases where the teacher was not actively involved in the discussion, which sometimes led to conversations where students were not practicing "listen first" skills. Student conversations in these situations would have flair-ups of multiple simultaneous conversations that mainly focused on a group's claim and/or evidence. In other cases, the teacher was not actively involved in the conversation because the students were able to maintain the discussion on their own, focusing clearly on exchanging ideas (see Table 4). Overall, what we begin to see emerging is the fine balance of how to conversationally express, listen, respond, and negotiate through discussion of *evidence-based ideas*. At this level, it seems as though these elements sometimes get out of balance, thus making an expression of an idea more exuberant, listening first more challenging, and/or reacting rather than responding more likely. In this way, conversation at this level is sometimes bumpy, because of the varied use and quality of evidence and critique.

Figure 2. Frequency of whole-class talk categories for medium SWH implementation classroom segments

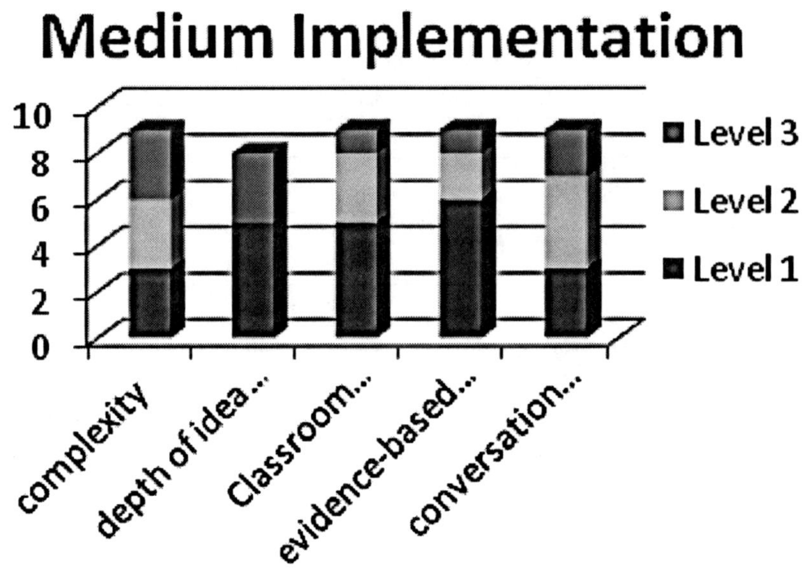

Example One of Medium Implementation

In Table 4, the presenting group is discussing the deciduous forest ecosystem. The rest of the class groups have been exploring other ecosystems. Maria asks if Iowa was in the deciduous forest. This question first suggests that Maria was listening and thinking about what was presented. Second, she asked a question that either called for a factual response or a speculation of what might be. In this case, the presenting group said no and said that their source did not have the United States in it. In terms of listener understanding, it is murky if no meant, "I don't know" or "No, it does not." The response was not grounded in evidence but lack of information. More importantly, the lack of follow-up by Maria, other students, or the teacher we see as a missed opportunity to have an evidence-based conversation, which would have explored and reasoned through the possibility of a deciduous forest in Iowa.

Example Two of Medium Implementation

In Table 5, the class was discussing evidence about the characteristics of objects that have a high or low auditory pitch. In this segment, one group claimed that a large jar generated a low pitch and a small jar generated a high pitch. Danny asked if they thought it was possible for a large jar also to generate a high pitch, because they had experienced that during their experimentation. The students discuss for several turns of talk the reasons that the two groups heard different things, which eventually lead to discussion about water levels in the jars. In this example and characteristic to this level, follow-up type questions tended to either seek clarification (see Table 5 lines 33-35) or to further challenge an idea (see Table 5 line 39). When an idea was challenged, it often brought in other students to engage in the developing negotiation. In this example, because the students spent several turns of talk sharing their ideas, they were able to develop an explanation as to the reasons that two groups came up with two different results; this in turn helped them to negotiate the concept of pitch.

In Table 5, line 37 when Peyton was making an attempt to establish a relationship of length of string to amount of water, Luke refocused the conversation back to the jars. While we see this refocus as acceptable, the detail that Peyton's outside evidence was brought into the conversation and not used then or later in the class session points out the ways in which ideas when exchanged at this level are more likely to be situated in only the group's evidence and do not take in a broader array of related evidence. Although this example clearly has a particular group arguing for their own ideas over extended and developed turns of talk, we consider this, too, a missed opportunity of reasoning through commonalities in and between groups as the class works toward understanding the properties of sound.

SUMMARY OF HIGH LEVEL IMPLEMENTATION

In high level classrooms whole class dialogue that supported student exchange of ideas was consistently evident. Questioning consistently supported a well-developed conversation that involved students reasoning through and justifying their claims and evidence (see Figure 3).

Complexity of Questions and Interactions

For classrooms at high implementation, the complexity of the question seems to be less important than the ways in which the response of the question is carried out. Questions that begin these conversations are often focused on the presenting groups' claim and evidence. The teacher and the students in these classrooms usually ask follow-up questions or make follow-up statements to any type of question that is asked. This questioning practice encourages student ideas to be reasoned through over many turns of talk.

Figure 3. Frequency of whole-class talk categories for high SWH implementation classroom segments

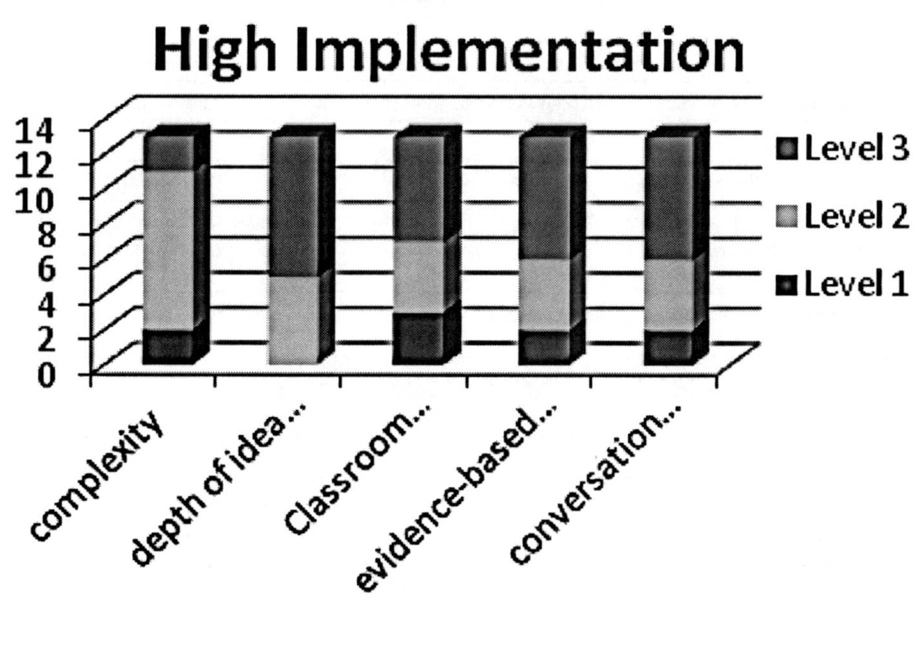

Deep Exchanges of Ideas and Evidence

Follow-ups to questions in these classrooms were important to foster the depth of ideas and evidence. Follow-up questions by students generally clarified or challenged an ongoing idea. If the teacher stepped into these exchanges, the teacher often asked a student how he/she felt about the presenting group's ideas. The explanation by the classmate often included their own personal ideas or the ideas of their group that often stimulated more questions and/or lengthier negotiations. During these deep idea exchanges, "big ideas" were driven by more than one group's claim and evidence. Not only were the presenting groups ideas circulated but so were the ideas held by other prepared groups and individuals.

Conversational Pattern

Communication in high implementation classrooms was very conversational. Conversations were about building upon student's own evidence-based ideas. Most of the time, students were able to maintain the discussion on their own, focusing the "big idea" with minimal guidance from the teacher (see Table 6). When the teacher was involved with the conversation, often it was to further challenge and/or scaffold student-generated ideas. Sometimes when the teacher was not involved in conversation over many turns of student talk, the conversation became intense because student's ideas were being challenged. On these occasions, the teacher was able to remind students to exhibit "listen first" skills, be respectful and/or thoughtful in responding. In this way, the teacher did not need to manage or control the conversation, but rather helped in extending, patching, and/or consolidating ideas in process.

Table 6. Example of level three whole class dialogue

Line	Person	Words Spoken
45	JoAnn	I disagree that you could see a seed in a seed.
46	Nellie	Why?
47	JoAnn	It just seems weird that-
48	Nellie	I saw a seed inside my seed.
49	JoAnn	It's different how you can see a seed in a seed, and I think that's make believe or something.
50	Cindy	It's not make believe, JoAnna.
51	JoAnn	I'm still talking, you know.
52	Teacher	In other words, what she's really saying to you is that what? You need to let her finish, right? Remember what we talked about in discussion, Nellie. You've got to listen first, and then she needs to have her chance, and then you can have hers.
53	JoAnn	I think it's make believe because I have never seen a seed inside a seed before, and that's kind of weird how a seed can be in a seed, and I just don't really think it's true.
54	Nellie	It's true. If you got the seed that I got, then you would be able to see a seed.
55	JoAnn	Well, our group saw a stem in it, and….
56	Nellie	And we saw a seed.
57	JoAnn	A seed in a seed. I just don't think that's true.
58	Cindy	JoAnn, the lima bean is sort of a seed on the outside, and then, JoAnn, on the inside and hanging out partly was a stem.

continued on following page

Table 6. Continued

Line	Person	Words Spoken
59	JoAnn	Well you guys didn't say anything about the stem in there.
60	Nellie	Well, we didn't want to. And there's a seed in a seed.
61	JoAnn	I just don't think that can happen.
62	Cindy	Sometimes it can.
63	Nellie	Yep. Sometimes it can't, sometimes it can.
64	JoAnn	I just don't get it.
65	Nellie	Sometimes it can. Erin.
66	Teacher	Let's go to someone else, and we can come back to you, JoAnn.
67	JoAnn	Are you sure that the little seed you saw wasn't a plant?
68	Erin	I kind of disagree with you saw a seed inside of a seed, because the plant has to be inside of the seed, and the seed is supposed to be part of the plant, and the plant is supposed to be growing from the seed, and that little thing…
69	Nellie	[interrupting] I think that worms grow out of that little seed.
70	Erin	That's what I'm trying to say. That little ball probably had the plant inside of it, so it's probably not the seed, really, it's probably just the shell to keep the seed from cracking and breaking all the way through it so what I think she's trying to say- it's a coat that's really hard that looks like a seed, but it's not really a seed, it's actually a coat. The stem from...
71	Teacher	What do you think about that? She's saying there's a coat. Nellie, listen to what she's saying. What do you think-
72	Cindy	I kind of agree with that.
73	Teacher	You need to tell her.
74	Cindy	I kind of agree with that because that outer skin of it is kind of a shield to keep it from getting…the dirt from…
75	Erin	I'm not talking about that. I'm talking about…
76	Cindy	I know. But Erin, the skin on the outside probably protects the seed from inside the lima bean from getting dirty and cracked. I think it's probably trying to keep it together so then the seed won't go out.
Grade 2, 015-08-01-01-2008		

Example of High Implementation

Table 6 is a second grade classroom that is discussing what is inside a seed. Several members of the class are challenging the presenting group's claim that there is a seed inside a seed. In line 49 JoAnn asks "How you can see a seed in a seed?" JoAnn remains persistent in denying support (line 49, 53, 57, 61, 64, and 67) that a seed is inside a seed. Over seventeen turns of talk happen because the "how" evidence appears to be lacking for JoeAnn. Then in line 68, Erin provides further support for JoeAnn's point and further explains with a related line of support.

Eventually after thirty more turns of talk (not included here), their discussion of observational data is left unresolved. The last presenting group of this session thinks that there are lungs inside the seed. A similar discussion ensues like the first. In the end, the teacher asks the students to recap their unresolved questions, reminds them that the posters they presented will stay in front of the room (this is to insure the ideas remain public), and they will be exploring, in the next class session, nonfiction books to see what they "say about whether seeds are alive or not and if there's something inside the seeds" (line 205).

In this sample excerpt we provide from Table 6 only a few questions were asked. However these questions and the ensuing discussion were about students reasoning through and justifying one group's evidence. The teacher encouraged students to listen to each other and to draw other students into the conversation. The effect of whole class conversation helped to build ideas and viewpoints over many turns of talk. In the end of this session none of the ideas were resolved, but many ideas were expressed and challenged to guide their future look into seeds and plants.

DISCUSSION

This study provides us with a snapshot of one moment in time for each of these fifteen classrooms. All of these classrooms were engaged in an authentic attempt to incorporate argument-based inquiry (Forman, Barnhart, Deafenbaugh, & Ewing, 2010). In each case, students presented their claim and evidence and some level of classroom conversation ensued. The conversations that happened within each classroom varied. However, the overall evaluation for any one classroom fell into one of the three levels fairly consistently. Dialogical patterns tend to be localized (Pratt, 1987) and this was generally the case for this data set. For example, no segment rated at Low-Level Implementation (LLI) had evidence of level 3 talk, and rarely did a High-Level Implementation (HLI) classroom have elements of level 1 talk. Medium-Level Implementation (MLI) saw no single level of talk dominate. In all cases, the teacher, by virtue of using this approach, which supports student talk, provided their own version of support toward classroom conversation (Eichinger, et al., 1991; Osborne, 2001).

From our perspective, the teacher was both the navigator and lynchpin in HLI that held together overall classroom discourse. As navigator, the teacher insured that the axis of conversation was maintained and developed around the "big idea" in science. The teacher acted as a lynchpin by insuring the five factors of whole-class dialogue did not fly off the axle and ground the conversation.

In high-level classrooms the sophistication of idea exchange, classroom interaction, and the sustained use of evidence to guide conversation was more monitored than sustained by the teacher. In these high level classrooms there was what Burbules and Bruce (2001) calls a "reciprocal relation" in which " prerogatives of questioning, answering, commenting, or offering reflective observations on the dynamic are open to all participants" (Burbules & Bruce, 2001, p. 19). The classroom community often took questions as starting points for conversation. These questions were more likely to come from students, but a teacher's question was just as likely to be acted upon in a similar fashion. The quality of questions in HLI was not critical in fostering sustained turns of student talk. In HLI the students and sometimes the teacher helped to integrate previous turns of talk. The teacher was most likely to encourage uncontested underrepresented evidence to be brought in to conversation.

In this sense, conversation in HLI classrooms had a "natural flow of conversation" (Schein, 1993, p. 30) and students were afforded opportunities for "sustained expression" (Dillon, 1981, p. 16) of their thoughts. Throughout the HLI classrooms, we saw in varying degrees flare-ups, pauses, wandering, focus and refocus, unidirectional flow, multi-voices, ignoring, and pressing (Scardamalia & Bereiter, 1991; Schein, 1993; Scott & Ametller, 2007; vanZee, Iwasyk, Kurose, Simpson, & Wild, 2001; Windschitl, 2004). These were not examples of behavioral classroom management in crisis but rather examples of what the National Science Education Standards call "reasoned arguments" (NRC, 1996, p. 50). For these reasoned arguments teachers set up expectations which required students "to explain and justify their understanding, argue from data and defend their conclusions, and critically assess and challenge the scientific explanations of one another" (NRC, 1996, p. 50). The argumentation found in these classrooms was certainly with its ebbs and flows–as are most all engaged conversations. However, this was consistent with McNeill's et al. (2006) findings in that we also saw, even in HLI classrooms students struggling to reason through how to use or what counts as evidence during argument-based inquiry.

In low-level classrooms we saw many textbook descriptions of IRE or IRF cycles (Mehan, 1979) or what Isaacs (1999) calls a "back-and-forth volley" (p. 365). Isaacs sees this as a good example of "people thinking alone, the ping-pong nature of conversation dominating - not dialogue" (Isaacs, 1999, p. 365). These LLI classrooms also had a dominance of knowledge level questions, asked by both teacher and students, which were adequately addressed but unlike HLI did not sustain further turns of talk beyond the IRF/IRE cycle.

In HLI we saw "genuine questions" (Scott, 1998, p. 66) in which the response could not follow a typical IRE or IRF pattern because it served as a "thinking device" (p. 66) for the community of learners. In these cases, genuine questions led to extensive turns of talk supported by evidence. Regardless of questioning, in LLI we saw very few sustained student-student turns of talk that included evidence. In this way, Kuhn (1991) is correct. LLI Classrooms we analyzed were poor at bringing together their claims and evidence and conversation did not naturally flow. In HLI we consistently saw elements of discourse around the claim and evidence that seemed "natural," "coordinated," "constructed," and "reasoned" (Kuhn, 1991; McNeill, et al., 2006; Osborne, Erduran, & Simon, 2004). Although LLI is not the ideal, we did see elements of Driver et al.'s (2000) notion of dialogic argumentation beginning to emerge, but differences in ideas were not given sufficient space for critique, consensus, and agreement because the conversation was truncated. In HLI, claims were supported by evidence and questions were asked that challenged ideas that were being expressed. There were attempts to persuade or convince classmates though expressions of doubt, alternate viewpoints, and the recognition that some concepts were still not well understood by the classroom community (Driver, et al., 2000; Newton, Driver, & Osborne, 1999).

The mere existence of MLI classrooms provides recognition that inquiry-based approaches are part of a dynamic process for the teacher and student, PD team, and whole learning community. MLI classrooms can be seen to be either classrooms in transition or classrooms that have settled into a form of practice that was not stable in low level or high-level practices. Duschl and Osborne (2002) points out that "normal classroom discourse is predominantly monologic" (p. 55). We are comfortable in suggesting that the MLI classroom is not usually monologic and just beginning to disengage from monologic activities. In MLI classrooms we see "change both in the structure of classroom activities and the aims that underlie them" (Duschl & Osborne, 2002, p. 55). In MLI those changes are well in place as evidenced by the increasing likelihood of student-student dialogue over more complex questioning and a greater withholding of teacher feedback, evaluation, and forced directions. The most obvious change from LLI to MLI that we see is in the form of questions. Knowledge recall questions are replaced with more complex questioning that cannot be responded to over one turn of talk. Additionally, the teacher is supportive of letting the idea develop before moving the conversation along. During MLI it seems as though the students are beginning to construct arguments for each other rather than just for the teacher or not at all (Berland & Reiser, 2009). However, as Wertsch (1979) suggests, students cannot internalize argumentative discourse until they can make counter arguments for themselves. The major difference between MLI and HLI is that in HLI we usually hear persistent evidence-based whole-class dialogue and in MLI the dialogue generally emerges and fades. While much more research is needed to properly understand the role and predominance of MLI classrooms in our study, the very existence of an MLI classroom indicates that PD can help teachers move beyond a low level implementation of argument-based inquiry; however it is unclear if all classrooms will reach a sustained high level of whole-class dialogue while engaging in argument-based inquiry. In summary, one can most easily differentiate LLI from MLI by the emergence of more "genuine questions" (Scott, 1998, p. 66) that are given sustained support through the "teacher's uptake of student ideas" (Ametller & Scott, 2009, p. 14; also see Collins, 1982). More importantly, student uptake of these questions over several turns of conversational talk begins to emerge. In HLI classrooms, what is consistently noticed is that students always seemed to have something more to say to each other (Eichinger, et al., 1991) and that classroom dialogue collaboratively built understanding as the dialogue progressed (Wells & Mejía Arauz, 2006). The whole-class dialogue in these HLI classrooms was left unsettled, yet it seemed satisfying to the classroom participants. Students seemed intrigued by others reasoned claims, evidence, and ideas. In LLI classrooms when the conversation ended or moved on, it seemed like it was "won" or "settled" and the discussion of one's idea was a moderately interactive spectator event rather than an opportunity to build knowledge with each other.

CONCLUSION

The implications of this research indicate that there are specific dialogical characteristics that occur in the science classroom that can contribute to highly effective implementation of argument-based inquiry and reformed teaching practices. In our discussion, we attempted to create a picture of what is happening across each level of implementation and how existing research also supports our evidence. The focus in the discussion was on the classroom's practice. However, we fully understand that it is ultimately the teacher that gives the authority for more than monologic instruction in any given classroom. "In learning to talk science we must buy into, and learn to work with, the conceptual tools, epistemological framing, ontological perspectives, and forms of reasoning of the scientific community" (Scott, 1998, p. 74). Without question, the first person to begin the process of engaging the classroom in the practices of an authentic scientific community must be the teacher. Maintaining a steadfast approach to helping students of elementary age to develop reasoned scientific arguments takes time, courage, and ongoing PD to fully understand and implement. This research is only scratching the surface of the ways in which classrooms engage in argument-based inquiry. Many research questions surface from this study. For example, how are students' writing samples influenced through engagement in varying levels of whole class dialogue in argument-based inquiry? What influences why some classrooms develop into HLI and others do not? How successful are students in classrooms that practice the highest level of whole class dialogue? These questions and many others are needed to understand more fully the overall impact on student learning from classrooms engaged in argument-based inquiry.

REFERENCES

Alexopoulou, E., & Driver, R. (1997). Small group discussions in physics: Peer interaction modes in pairs and fours. *Journal of Research in Science Teaching*, *33*(10), 1099–1114. doi:10.1002/(SICI)1098-2736(199612)33:10<1099::AID-TEA4>3.0.CO;2-N

American Association for the Advancement of Science. (1993). *Benchmarks for scientific literacy*. Washington, DC: AAAS.

Ametller, J., & Scott, P. (2009). *Classroom discourse, dialogic teaching and teacher professional development in science education: A selective review of the literature*. Retrieved April 16, 2010, from https://www.ntnu.no/wiki/download/attachments/8324755/Final+draft_lit_review_Leeds.doc

Bell, P., & Linn, M. C. (2000). Scientific arguments as learning artifacts: Designing for learning from the web with KIE. *International Journal of Science Education*, *22*, 797–817. doi:10.1080/095006900412284

Berland, L. K., & Reiser, B. J. (2009). Making sense of argumentation and explanation. *Science Education*, *93*(1), 26–55. doi:10.1002/sce.20286

Boyd, M., & Rubin, D. (2006). How contingent questioning promotes extended student talk: A function of display questions. *Journal of Literacy Research*, *38*(2), 141–159. doi:10.1207/s15548430jlr3802_2

Bricker, L. A., & Bell, P. (2008). Conceptualizations of argumentation from science studies and the learning sciences and their implications for the practices of science education. *Science Education*, *92*(3), 473–498. doi:10.1002/sce.20278

Burbules, N. C., & Bruce, B. C. (2001). Theory and research on teaching as dialogue. In V. Richardson (Ed.), *Handbook of Research on Teaching* (4th ed., pp. 1102–1121). Washington, DC: American Educational Research Association.

Cavagnetto, A., Hand, B. M., & Norton-Meier, L. (2010). The nature of elementary student science discourse in the context of the science writing heuristic approach. *International Journal of Science Education*, *32*(4), 427–449. doi:10.1080/09500690802627277

Cavagnetto, A. R. (2010). Argument to foster scientific literacy: A review of argument interventions in K-12 science contexts. *Review of Educational Research*, *80*(3), 336–371. doi:10.3102/0034654310376953

Charmaz, K. (2002). Qualitative interviewing and grounded theory analysis. In J. Gubrium & J. A. Holstein (Eds.), *Handbook of Interview Research: Context and Method* (pp. 675–694). Thousand Oaks, CA: Sage Publications.

Costello, P. J. M., & Mitchell, S. (1995). Introduction: Argument: Voices, text and contexts. In P. J. M. Costello & S. Mitchell (Eds.), *Competing and Consensual Voices: The Theory and Practice of Argument*. Clevedon, UK: Multilingual Matters Ltd.

Crawford, T. (2005). What counts as knowing: Constructing a communicative repertoire for student demonstration of knowledge in science. *Journal of Research in Science Teaching*, *42*(2), 139–165. doi:10.1002/tea.20047

Dey, I. (1999). *Grounding grounded theory: Guidelines for qualitative inquiry*. San Diego, CA: Academic Press.

Dillon, J. T. (1981). To question and not to question during discussion: Non-questioning techniques. *Journal of Teacher Education*, *32*(15), 15–20. doi:10.1177/002248718103200605

Driver, R., Asoko, H., Leach, J., Mortimer, E., & Scott, P. (1994). Constructing scientific knowledge in the classroom. *Educational Researcher*, *23*(7), 5–12.

Driver, R., Newton, P., & Osborne, J. (2000). Establishing the norms of scientific argumentation in classrooms. *Science Education*, *84*(3), 287–312. doi:10.1002/(SICI)1098-237X(200005)84:3<287::AID-SCE1>3.0.CO;2-A

Duschl, R. (1990). *Restructuring science education: The importance of theories and their development*. New York, NY: Teachers College Press.

Duschl, R., & Osborne, J. (2002). Supporting and promoting argumentation discourse. *Studies in Science Education*, *38*, 39–72. doi:10.1080/03057260208560187

Duschl, R. A., Schweingruber, H. A., & Shouse, A. W. (Eds.). (2007). *Taking science to school: Learning and teaching science in grades K-8*. Washington, DC: National Academies Press.

Eichinger, D. C., Anderson, C. W., Palinscar, A., & David, Y. M. (1991). *An illustration of the roles of content knowledge, scientific argument, and social norms in collaborative problem solving*. Paper presented at the American Educational Research Association. Chicago, IL.

Ford, M. (2008). Disciplinary authority and accountability in scientific practice and learning. *Science Education, 92*(3), 404–423. doi:10.1002/sce.20263

Forman, E., Barnhart, B., Deafenbaugh, L., & Ewing, M. (2010). *Fostering constructive criticism in a high school biology classroom: Understanding the social dynamics of argumentation*. Paper presented at the Annual Meeting of the National Association for Research in Science Teaching. Philadelphia, PA.

Fosnot, C. T. (1989). *Enquiring teachers, enquiring learners: A constructivist approach to teaching*. New York, NY: Teachers College Press.

Fosnot, C. T. (1996). Constructivism: A psychological theory of learning. In C. T. Fosnot (Ed.), *Constructivism: Theory, Perspectives, and Practice* (pp. 8–33). New York, NY: Teachers College Press.

Glaser, B. G., & Strauss, A. L. (1967). *The discovery of grounded theory: Strategies for qualitative research*. New York, NY: Aldine. doi:10.1097/00006199-196807000-00014

Gross, A. G. (1990). *The rhetoric of science*. Cambridge, MA: Harvard University Press.

Hand, B., Wallace, C. S., & Yang, E. M. (2004). Using the science writing heuristic to enhance learning outcomes from laboratory activities in seventh grade science: Quantitative and qualitative aspects. *International Journal of Science Education, 26*, 131–149. doi:10.1080/0950069032000070252

Hogan, K., Nastasi, B. K., & Pressley, M. (2000). Discourse patterns and collaborative scientific reasoning in peer and teacher-guided discussions. *Cognition and Instruction, 17*, 379–432. doi:10.1207/S1532690XCI1704_2

Isaacs, W. (1999). *Dialogue and the art of thinking together*. New York, NY: Currency/Doubleday.

Isaacs, W. H. (1993). Taking flight: Dialogue, collective thinking, and organizational learning. *Organizational Dynamics, 22*(2), 24–39. doi:10.1016/0090-2616(93)90051-2

Jiménez-Aleixandre, M., & Erduran, S. (2008). Argumentation in science education: An overview. In S. Erduran & M. Jiménez-Aleixandre (Eds.), *Argumentation in Science Education: Perspectives from Classroom-Based Research* (pp. 3–27). Dordrecht, The Netherlands: Springer.

Kelly, G., Chen, C., & Prothero, W. (2000). The epistemological framing of a discipline: Writing science in university oceanography. *Journal of Research in Science Teaching, 37*(7), 691–718. doi:10.1002/1098-2736(200009)37:7<691::AID-TEA5>3.0.CO;2-G

Kelly, G., & Duschl, R. (2002). *Toward a research agenda for epistemological studies in science education*. Paper presented at the Annual Meeting of the National Association for Research in Science Teaching. New Orleans, LA.

Kelly, G., & Green, J. (1998). The social nature of knowing: Toward a sociocultural perspective on conceptual change and knowledge construction. In B. Guzzetti & C. Hynd (Eds.), *Perspectives on Conceptual Change* (pp. 145–182). Mahwah, NJ: Erlbaum.

Keys, C., Hand, B., Prain, V., & Collins, S. (1999). Using the science writing heuristic as a tool for learning from laboratory investigations in secondary school. *Journal of Research in Science Teaching, 36*(10), 1065–1084. doi:10.1002/(SICI)1098-2736(199912)36:10<1065::AID-TEA2>3.0.CO;2-I

Kuhn, D. (1991). *The skills of argument*. Cambridge, UK: Cambridge University Press. doi:10.1017/CBO9780511571350

Kuhn, D. (1993). Science as argument: Implications for teaching and learning scientific thinking. *Science Education*, *77*(3), 319–337. doi:10.1002/sce.3730770306

Kuhn, L., & Reiser, B. J. (2006). *Structuring activities to foster argumentative discourse*. Paper presented at the American Educational Research Association. San Francisco, CA.

Lemke, J. (1990). *Talking science: Language, learning and values*. Norwood, NJ: Ablex.

Marshall, J. C., Horton, B., Smart, J., & Llewellyn, D. (2009). *EQUIP: Electronic quality of inquiry protocol*. Retrieved from http://iim-web.clemson.edu/wp-content/uploads/2009/02/equip-2009.pdf

Martin, A. M., & Hand, B. (2009). Factors affecting the implementation of argument in the elementary science classroom: A longitudinal case study. *Research in Science Education*, *39*(1), 17–38. doi:10.1007/s11165-007-9072-7

Martins, I., Mortimer, E., Osborne, J., Tsatsarelis, C., & Jiménez-Aleixandre, M. P. (2001). Rhetoric and science education. In H. Behrendt, H. Dahncke, R. Duit, W. Gräber, M. Komorek, A. Kross, & P. Reiska (Eds.), *Research in Science Education - Past, Present, and Future* (pp. 189–198). Dordrecht, The Netherlands: Kluwer Academic Publishers.

McNeill, K. L. (2009). Teachers' use of curriculum to support students in writing scientific arguments to explain phenomena. *Science Education*, *93*(2), 233–268. doi:10.1002/sce.20294

McNeill, K. L., Lizotte, D. J., Krajcik, J., & Marx, R. W. (2006). Supporting students' construction of scientific explanations by fading scaffolds in instructional materials. *Journal of the Learning Sciences*, *15*(2), 153–191. doi:10.1207/s15327809jls1502_1

Mehan, H. (1979). *Learning lessons: Social organization in the classroom*. Cambridge, MA: Harvard University Press.

Mork, S. M. (2005). Argumentation in science lessons: Focusing on the teacher's role. *NorDiNa*, *1*(1), 17–30.

National Research Council. (1996). *National science education standards*. Washington, DC: National Academy Press.

National Research Council. (2012). *A framework for K-12 science education: Practices, crosscutting concepts, and core ideas*. Washington, DC: The National Academies Press.

Naylor, S., Keogh, B., & Downing, B. (2007). Argumentation and primary science. *Research in Science Education*, *37*, 17–39. doi:10.1007/s11165-005-9002-5

Newton, P., Driver, R., & Osborne, J. (1999). The place of argumentation in the pedagogy of school science. *International Journal of Science Education*, *21*(5), 553–576. doi:10.1080/095006999290570

Norton-Meier, L., Hand, B., Hockenberry, L., & Wise, K. (2008). *Questions, claims and evidence: The important place of argument in children's science writing*. Portsmouth, NH: Heinemann.

Nystrand, M., Gamoran, A., Kachur, R., & Prendergast, C. (1997). *Opening dialogue: Understanding the dynamics of language and learning in the English classroom*. New York, NY: Teachers College Press.

Osborne, J. (2007). Towards a more social pedagogy in science education: the role of argumentation. *Revista Brasileira de Pesquisa em Educação em Ciências*, *7*(1). Retrieved July 12, 2010, from http://www.fae.ufmg.br/abrapec/revista/index.html

Osborne, J., Erduran, S., & Simon, S. (2004). Enhancing the quality of argumentation in school science. *Journal of Research in Science Teaching*, *41*(10), 994–1020. doi:10.1002/tea.20035

Osborne, J. F. (2001). Promoting argument in the science classroom: A rhetorical perspective. *Canadian Journal of Science. Mathematics and Technology Education*, *1*, 271–290. doi:10.1080/14926150109556470

Osborne, J. F., Erduran, S., & Simon, S. (2004b). *Ideas, evidence and argument in science: In-service training pack, resource pack and video.* London, UK: Nuffield Foundation.

Pratt, M. L. (1987). Linguistic utopias. In N. Fabb, D. Attridge, A. Durant, & C. MacCabe (Eds.), *The Linguistics of Writing: Arguments between Language and Literature* (pp. 48–66). New York, NY: Methuen.

Sampson, V., Grooms, J., & Walker, J. (2011). Argument-driven inquiry as a way to help students learn how to participate in scientific argumentation and craft written arguments: An exploratory study. *Science Education*, *95*(2), 217–257. doi:10.1002/sce.20421

Sandoval, W. A., & Reiser, B. J. (2004). Explanation-driven inquiry: Integrating conceptual and epistemic scaffolds for scientific inquiry. *Science Education*, *88*(3), 345–372. doi:10.1002/sce.10130

Sawada, D., & Piburn, M. (2000). *Reformed teaching observation protocol (RTOP)*. Technical Report No. IN00-1. Tempe, AZ: Arizona Collaborative for Excellence in the Preparation of Teachers.

Sawada, D., Piburn, M., Turley, J., Falconer, K., Benford, R., Bloom, I., & Judson, E. (2000). *Reformed teaching observation protocol (RTOP) training guide*. ACEPT Technical Report No. IN00-2. Tempe, AZ: Arizona Collaborative for Excellence in the Preparation of Teachers.

Scardamalia, M., & Bereiter, C. (1991). Higher levels of agency for children in knowledge-building: A challenge for the design of new knowledge media. *Journal of the Learning Sciences*, *1*(1), 37–68. doi:10.1207/s15327809jls0101_3

Schein, E. H. (1993). On dialogue, culture, and organizational learning. *Reflections: The SoL Journal*, *4*(4), 27–38. doi:10.1162/152417303322004184

Schweizer, D. M., & Kelly, G. J. (2005). An investigation of student engagement in a global warming debate. *Journal of Geoscience Education*, *53*(1), 75–84.

Scott, P. (1998). Teacher talk and meaning making in science classrooms: A Vygotskian analysis and review. *Studies in Science Education*, *32*, 45–80. doi:10.1080/03057269808560127

Scott, P., & Ametller, J. (2007). Teaching science in a meaningful way: Striking a balance between 'opening up' and 'closing down' classroom talk. *The School Science Review*, *88*(324), 77–84.

Scott, P. H., Mortimer, E. F., & Aguiar, O. G. (2006). The tension between authoritative and dialogic discourse: A fundamental characteristic of meaning making interactions in high school science lessons. *Science Education*, *90*, 605–631. doi:10.1002/sce.20131

Simon, S., Erduran, S., & Osborne, J. (2006). Learning to teach argumentation: Research and development in the science classroom. *International Journal of Science Education*, *28*(2-3), 235–260. doi:10.1080/09500690500336957

Toulmin, S. (1958). *The uses of argument*. Cambridge, UK: Cambridge University Press.

vanZee, E. H., Iwasyk, M., Kurose, A., Simpson, D., & Wild, J. (2001). Student and teacher questioning during conversations about science. *Journal of Research in Science Teaching*, *38*, 159–190. doi:10.1002/1098-2736(200102)38:2<159::AID-TEA1002>3.0.CO;2-J

Varelas, M., & Pineda, E. (1999). Intermingling and bumpiness: Exploring meaning making in the discourse of a science classroom. *Research in Science Education*, *29*(1), 25–49. doi:10.1007/BF02461179

von Aufschnaiter, C., Erduran, S., Osborne, J., & Simon, S. (2008). Arguing to learn and learning to argue: Case studies of how students' argumentation related to their scientific knowledge. *International Journal of Science Education*, *45*(1), 101–131.

Vygotsky, L. S. (1986). *Thought and language*. Cambridge, MA: MIT Press.

Wells, G., & Mejía Arauz, R. (2006). Dialogue in the classroom. *Journal of the Learning Sciences*, *15*, 379–428. doi:10.1207/s15327809jls1503_3

Wertsch, J. V. (1979). From social interaction to higher psychological processes: A clarification and application of Vygotsky's theory. *Human Development*, *22*(1), 1–22. doi:10.1159/000272425

Windschilt, M. (2004). *What types of knowledge do teachers use to engage learners in "doing science"? Rethinking the continuum of preparation and professional development for secondary science educators*. Paper presented at the Meeting High School Science Laboratories: Role and Vision. Washington, DC: National Academy of Sciences.

Yerrick, R. (2000). Lower track science students' argumentation and open inquiry instruction. *Journal of Research in Science Teaching*, *37*(8), 807–838. doi:10.1002/1098-2736(200010)37:8<807::AID-TEA4>3.0.CO;2-7

Yore, L. D., Bisanz, G. L., & Hand, B. M. (2003). Examining the literacy component of science literacy: 25 years of language arts and science research. *International Journal of Science Education*, *25*(6), 689–725. doi:10.1080/09500690305018

Zohar, A., & Nemet, F. (2002). Fostering students' knowledge and argumentation skills through dilemmas in human genetics. *Journal of Research in Science Teaching*, *39*(1), 35–62. doi:10.1002/tea.10008

Chapter 48
A Research of Employing Cognitive Load Theory in Science Education via Web-Pages

Yuan-Cheng Lin
Elementary School Teacher of Tainan City, Taiwan

Ming-Hsun Shen
Graduate Institute of Science Education, National Kaohsiung Normal University, Taiwan

Chia-Ju Liu
Graduate Institute of Science Education, National Kaohsiung Normal University, Taiwan

ABSTRACT

This study adopted Cognitive Load Theory (CLT) to investigate the influences of multimedia presentations on achievements of science learning and the correlations between eye-movement models under distinct multimedia combinations and learner-controlled modes. Three units from the Science Education Website set by the Ministry of Education (Tainan) to assist student learning were employed: Air and Combustion," "Heat Effects toward Substances," and "Healthy Diet." This multifunctional website offers teaching resources, interesting experiments, inquiry experiments, virtual animations, multi-assessments, and supplementary materials, which are highly interactive and simulative. Six classes of fifth graders (n=192) participated in this study. Our findings showed that the combination of multimedia elements apparently influenced students' performance; the "animation + narration" group performed evidently better than the "animation + subtitles" group. When the animated subject matters were in small segments under the Segmentation Principle, multimedia presentations still brought affections to learning achievement, suggesting that the modality effect on students' learning exists constantly. Regarding the eye-movement models, this study focused mainly on discussing the "active-control mode" and "multimedia combination forms." These eye movement data supplemented the evidences gained to identify the relevant results. In conclusion, inappropriate multimedia combinations may interfere with learning. More functions and information inputs do not guarantee better learning effects.

DOI: 10.4018/978-1-4666-7363-2.ch048

INTRODUCTION

In recent years, with constant renovation of the technology, the extent of its progress is great enough to cause a reformation to education. A number of teachers and experts are convinced that information technology can promote education reformation and infuse new ideas into the traditional teaching (Lin & Liu, 2007; Sun, Lin & Yu, 2008). Many previous studies suggested that the cognition and learning over scientific conceptions may be affected by special ability, operation in labs, and model-handling (Chuang & Liu, 2012; Lin & Atkinson, 2011; Liu, Lin, Liu, & Chang, 2012; Sun, Lin & Yu, 2008). These factors cannot be fully implemented via the current science teaching methods applied in the elementary schools. Therefore, technological back-ups are necessary for science teaching/learning. Likewise, the tendency of applying multimedia to academic areas is growing stronger (Sankey, 2003). Designers of the multimedia supplementary teaching materials integrate words, pictures, sounds, and animations to create interesting audio/video effects to attract students; the presentation styles from "nonstop playing" to "user-controlled playing" enable learners to control the learning pace according to personal interests and needs. However, are multimedia e-materials helpful to science learning? Suppose multimedia e-materials are helpful to science learning, then what is the best combination of multimedia elements for effective and efficient learning?

Cognitive Load Theory (CLT) holds that teaching designs should definitely take into consideration human cognitive structures and the possible limits which may take place when pursuing efficiency (Paas, Renkl, & Sweller, 2003). It further reveals different sources of working memory load (Jeroen, van Merrienboer & Sweller, 2005; Paas, Renkl & Sweller, 2003), including the complexity of the subject matters (intrinsic cognitive load), teaching designs (extraneous cognitive load), and the total effort (germane cognitive load) made for learning. Germane cognitive load will appear when extra or excessive information such as detailed notes or reviews are given to students. While appropriate teaching designs are inadequate, additional extraneous cognitive load and a decline in learning will be resulted. The cognitive capacity limit of learners will be exceeded if the given information is strange, excessive, or bringing intrinsic cognitive load due to the complexity of teaching materials (Jeroen, van Merrienboer & Sweller, 2005). Thus, proper material presentation can not only diminish extraneous cognitive load but can also help learners concentrate on the subtleties of the subject matter, integration, and building up schemas. The correlations among three kinds of cognitive load are shown in Figure 1.

Abundant literatures show the impact on the combination of multimedia elements, cognitive load demanded and achievement made during the learning process. Many literatures indicate that combinations of multimedia e-materials create modality effect (Jeroen, van Merrienboer & Sweller, 2005; Liu, Lin, Liu, & Chang, 2012); that is, when different information processing channels (i.e. animation + narration) are triggered to sense the stimulation from multimedia, the learning effects are better than the ones from single processing channel (i.e. animation + subtitles).

Furthermore, Mayer (2001) brought up the dual-channel assumption-words are received through the verbal channel, while pictures are received through the visual channel. Different presentations are encoded through pictures or word formats. However, words are related to either printed or narrated text and can be processed via either verbal or visual channel. In addition, Mayer (2001) defined multimedia learning as building the mental model from materials that involve verbal and pictorial presentations. That is, the verbal presentations are related to spoken text like narration or written text in books, and the pictorial presentations are related to static and dynamic visualizations such as graphs, animations, etc (van Gog & Scheiter, 2010).

Figure 1. Correlations among three kinds of cognitive load

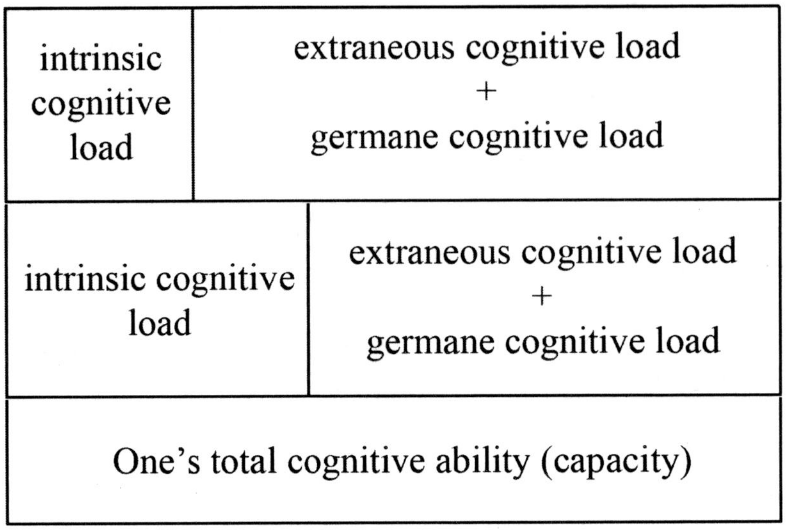

Segmentation Principle refers to properly cut multimedia material segments which allow learners to fully control the timing to get from one segment into the next. With this device, the amount of information in each segment will be lessened, giving learners sufficient time to digest, organize, and integrate the information picked up from each segment. In contrast, continuously played multimedia materials may deprive learners of advanced processing due to inadequate time. Mayer and Moreno (2003) pointed out that acquiring complicated conceptions may simplify the teaching materials; yet, the better first step is to leave learners enough time to comprehend information segment by segment, which may lower the cognitive load for them.

Furthermore, van Gog and Scheiter (2010) reminded that eye tracking could provide insights in the allocation of visual attention, and it was applied in studies on multimedia learning more and more often. Importantly, eye tracking is suitable for exploring differences of attention processes evoked by different types of multimedia materials. So, this suggests that eye tracking technology has the potential for researches about multimedia teaching/learning had taken an eye-tracking perspective in discussing the issue of multimedia teaching/learning. He also suggested that the potential approach to data analysis through eye tracking technology may advance the contribution to multimedia teaching/learning. For example, when people interact with visual environment, they make a sequence of fixations and saccade; fixations typically last about 200-500 ms and saccade is the transition from one fixation point to the other one. The use of eye movement pattern served as a memory cue while they were viewing the animation. Hence, in order to have a close study of the effects of multimedia materials on learning, this research utilized an eye tracker to record participants' eye movements during the whole process of multimedia teaching/learning on the screen. After the eye movement experiment, the data collected was analyzed to identify the tiny differences and correlations between cognitive processing and attention allocation. For instance, total fixation time showed how long the students gazed at different materials, visualization versus text, while they were learning. It's a useful measure and it indicated that how learners allocated their attention during viewing the screen. That is, we employed eye movement data to demonstrate the cognitive process during the learner viewed the multimedia materials.

Multimedia presentations provided education and entertainment with multi- sensory information (Schnotz & Lowe, 2003). Some studies showed that presenting information simply in different formats does not necessarily promote students' understanding over the content (Liu, & Shen, 2011; Mayer & Moreno, 2003). Therefore, this study focuses on the effects of different types of multimedia combinations while students were watching the web-pages of science education. Furthermore, utilizing eye tracking technology to analyze students' interactions with multimedia materials is the main feature of this study. Chuang and Liu (2012) suggested that eye movement data can provide clues about how multimodal information is processed. The cognitive process for multimedia materials should be a valuable issue, and the focus of the recent researches on the multimedia teaching/learning has also been shifted from delivery view to cognitive-process view (Liu, & Chuang, 2011; Tsai, Hou, Lai, Liu, & Yang, 2012).

Based on this eye-mind hypothesis, we infer that there are some correlations between what one is looking at and what one is thinking (Just & Carpenter, 1984). Anderson et al. (2004) hypothesized that eye tracking technology could be applied to understanding cognition, and some indicators were further analyzed to advance the understanding toward multimedia teaching/learning. In the experiment of this study, students viewed the screen and made a sequence of gaze which included fixation and saccade. Each fixation typically lasted for about 200-500 ms at the same location, and then saccade appeared-they shifted the location from one fixation point to the next one. We collected eye movement data from fixations and saccades while people interacted with visual environment and regarded them as indicators. Many studies related to multimedia materials analyzed these indicators to understand cognitive processing and allocation of attention. Eye fixations indicate the visual information to be perceived and processed by brain. According to Jacob and Karn (2003), the fixation count is related to the viewer's efficiency in searching relevant information. The greater the fixation count is, the less efficient the viewer's information searching is; the duration of fixation on particular area can be an indicator to identify the viewer's areas of interest (AOI). The longer duration may mean that the viewer is facing a more difficult task; the frequency of fixation on these AOI is related to the importance of the content shown on the computer screen. In addition, the scan path extracted from eye movement data can illustrate the location of attention on the screen over time, and it can be used to determine the AOI (Huber & Krist, 2004). Consequently, eye tracking studies show that eye movement data have the potential to reveal a viewer's cognitive process, which directs one's attention to the perception of information in a combination of visual and verbal formats (Liu, Lai, & Chuang, 2011).

Eye tracking technology supports a microscopic analysis of multimedia teaching/learning. Eye movement data can reveal important insights into the ongoing learning process. In addition, Eshach (2011) suggested that the eye-tracking technology not only enriched our perspectives in the science education research field but also benefited early science education researches, especially for young children. Moreover, Eshach suggested that eye tracking technique is a very interesting methodology and an instrument which is not employed in the science education. However, he believed that eye tracking technique could open the doors to another issue which science education community might be interested in. From this viewpoint, eye movement data such as fixation duration and fixation count may offer real-time measure of cognitive processing during the learning via multimedia (Henderson, Brockmole, Castelhano, & Mack, 2007; Liu, Lai, & Chuang, 2011).

Based on the literatures reviewed above and motives inspired, the major goal of this study is to explore and examine whether the modality effect is eliminated when the content of multimedia animation teaching materials are divided into several segments according to segmentation principle. Furthermore, after the learners are fully instructed to control (active-control mode, passive-control mode) the timing to get into the next segment, to what extend will the influence be brought on learning achievement and the time consumed through learner-controlled functions? Therefore, the purposes set for this study are:

1. To investigate the influences upon the learning achievement resulted from "learner-controlled modes" and "multimedia combination forms" after the multimedia teaching material is designed via segmentation principle.
2. To find out the influences over the eye movement models resulted from "learner-controlled modes" and "multimedia combination forms" after the multimedia teaching material is designed via segmentation principle.

MATERIAL AND METHODS

The material and methods applied to this study included research methodology, working on the Science Education and Learning Web, and experimental research.

Via quasi experiment research method, this study manipulates two independent variables-(a) learner-controlled playing modes (active and passive control mode), (b) multi-media combination forms (animation + subtitles, animation + narration, and animation + subtitles + narration)-to probe the influences on the dependent variables, learning achievement and eye movement models. The research structure is shown in Figure 2, the flow chart of the two phases of this study. Questionnaires were conducted to investigate the influences of two variables-learner-controlled modes and multimedia combination modes on the learning achievement. One hundred and ninety-six 5th graders, who were divided into three groups, participated in this study. About 40 students in each group were randomly selected to join in the eye tracking experiment. Finally, these behavioral and eye movement data were statistically analyzed, which revealed some important results to support our discussion.

Figure 2. Flowchart of the research structure

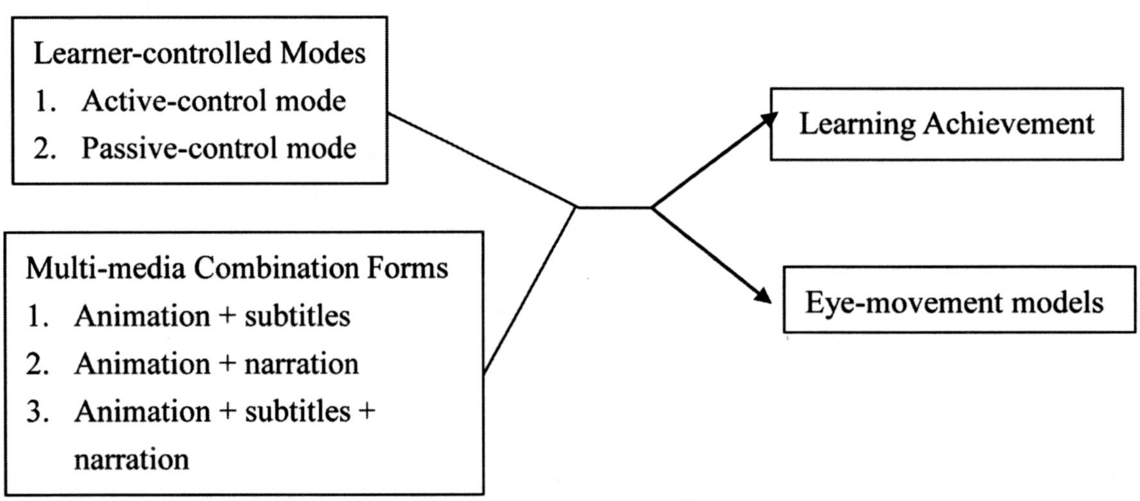

FUNCTIONS OF THE SCIENCE EDUCATION AND LEARNING WEB

This website is one of the sub-webs amid the six major science education and learning webs set up by the Ministry of Education. The prospect of the learning webs is to provide teachers and students from the 1st grade to the 6th grade with assistance and multi teaching resources. Every sub-web contains teaching resources, interesting experiments, inquiry experiments, virtual animations, multi-assessments, and supplementary materials (i.e. awarded science works in the past years and the history of science), which is expected not only to offer sufficient resources but also to effectively help students develop their science conceptions. The teaching experiment practice for this study utilized three instructional units in the said wed, which included "Air and Combustion," "Heat Effects toward Substances," and de with assistance and multi teaching resources. Every suand content-validity since they have received reliable identity from many science educators and experienced teachers in elementary schools. The designs of the web are presented as in Figure 3 and Figure 4.

Forms of Multimedia Combination

Forms of multimedia combination in this study comprised of "animation + subtitles a combination in this study comprised of content-validity since the instructional animation was divided into several segments depending on the content volume. In the form of "animation + subtitles," the subtitles appeared on the bottom of the screen to match up the animation; in the form of "animation + narration," the animation was presented with descriptive narration; in the form of "animation + subtitles + narration," the animation appeared together with narrated subtitles.

Learner-Controlled Modes

Segments of the instructional animation included "active-control mode" with functions of flipping pages forward and backward (see Figure 5) and "passive-control mode" with single function-going to the next page (see Figure 6).

Figure 3. Front page of the science education web

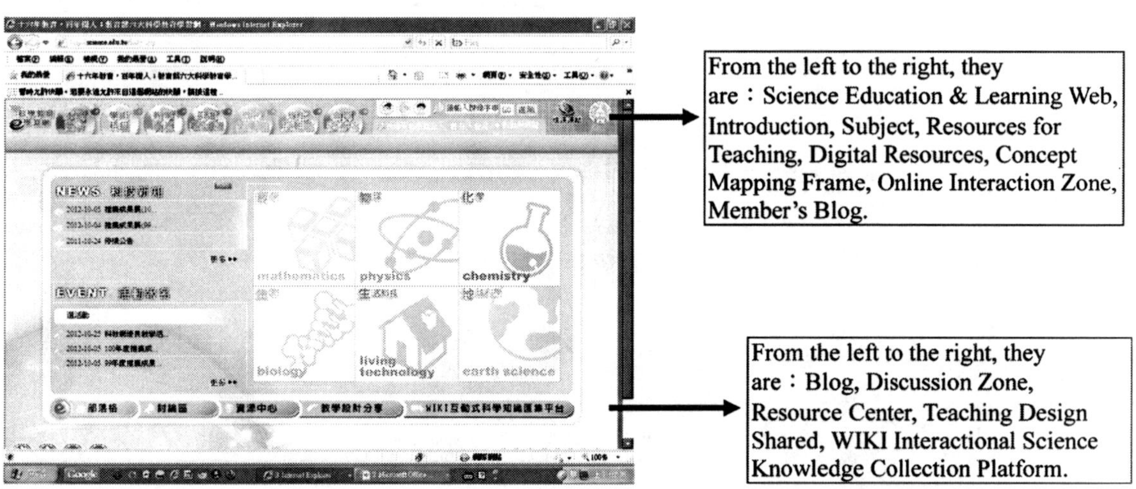

Figure 4. Instructional screen of heat effects toward substances

Figure 5. Active-control mode with functions of flipping pages

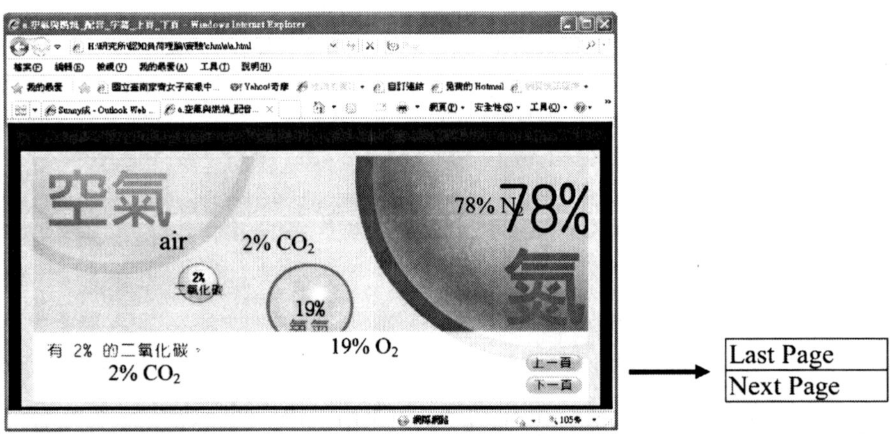

Figure 6. Passive-control mode with single function-going to the next page

RESEARCH OF THE EXPERIMENT

Experiment Design

The independent variables in this study were "learner-controlled modes" (active-control mode and passive-control mode) and "multimedia combination forms" (animation + subtitles, animation + subtitles + narration, and animation + narration). To avoid creating deviation in the experiment due to the differences among the samples themselves, repeated-measurement was practiced with the selected three units: he differences among the samples themselves, repeated-measurement was practical. Every subject in each experiment group was given instructions of the subject matters in all three multimedia combination forms. Each multimedia combination form delivered instructions only for one of the three units; therefore, the experiment design of this study is a 2×3 dual-factor-mixture type.

Subjects Selected for the Experiment

Six classes (A, B, C, D, E, and F) of fifth graders were sampled as the subjects for the experiment, which summed up to 250 students. The final sample size was 192, excluding students having extra learning aids and being absent. All six classes were comprised of students with usual academic status. All of these students were able to operate computers and surf the internet as they had received one computer lesson a week for three years. Researchers of this study randomly assigned classes A, B, and C to be the passive-control-mode team and classes D, E, and F to be the active-control-mode team, and then randomly divided each class into three small groups. The first small groups from each class (A1, B1, and C1) were then tagged as Group 1 of the passive-control-mode team. Likewise, as the formation continued, six groups came out of two teams. Thus, we had three pairs of groups; each pair comprised one group from each team. Subject matters contained in three multimedia combination forms were assigned to each pair, respectively. The groupings and topic allocation are shown in Table 1 and Table 2. In addition, we randomly selected 40 students from the three groups to record their scan paths of eye movement simultaneously through the eye tracker. The eye movement data was analyzed to understand the differences between the eye movement models.

Experimental Environment

The system of the experimental environment adopted for this study were Intel Core 2 Duo computers equipped with 1024×768 LCD screens, sound cards, earphones, and Windows XP operation systems loaded with DDR2 800 SDRAM 2048 MB. The experiment was practiced by one person operating one computer with an earphone on. The computers used were installed in advance with animated teaching materials from the Science Education Web by researchers. Additionally, this study adopted iView X RED portable eye-tracker and sample rate 60 Hz. When students received

Table 1. Grouping for the experiment

		Multi-Media Combination Form (MMCF)		
		animation + subtitles (A+S)	**animation + subtitles + narration (A+S+N)**	**animation + narration (A+N)**
Learner-controlled Mode (LCM)	passive-control (PC) N1=96	Group1 (A1, B1, C1)	Group 2 (A2, B2, C2)	Group 3 (A3, B3, C3)
	active-control (AC) N2=96	Group 4 (D1, E1, F1)	Group 5 (D2, E2, F2)	Group 6 (D3, E3, F3)

Table 2. Allocation of instruction units (I.U.) and multimedia combination forms

	I.U.	1			2			3		
	MMCF Class	A+S	A+S+N	A+N	A+S	A+S+N	A+N	A+S	A+S+N	A+N
PC Mode	A	A1	A2	A3	A3	A1	A2	A2	A3	A1
	B	B1	B2	B3	B3	B1	B2	B2	B3	B1
	C	C1	C2	C3	C3	C1	C2	C2	C3	C1
AC Mode	D	D1	D2	D3	D3	D1	D2	D2	D3	D1
	E	E1	E2	E3	E3	E1	E2	E2	E3	E1
	F	F1	F2	F3	F3	F1	F2	F2	F3	F1

the eye tracking experiment, they viewed the same screen and their eye movement data were recorded by eye-tracker additionally. The eye-tracker did not interfere with their viewing but captured the attention location of eyes naturally on the screen (Henderson, 2003).

Assessment Tools for Learning Achievement

Every unit ended up with an assessment to evaluate the learning achievement. Questions of the test were given by the researchers and the counselors of science area from elementary schools in New Tainan City. All twelve questions were in the form of multiple choices to test both memory and comprehension; six for memory test and six for comprehension test. Every question counted one point when being correctly answered.

1. **Question Planning:** Twenty questions were drafted by the researchers according to the subject matters in the three units-Air and Combustion, Heat Effects toward Substances, and Healthy Diet.
2. **Pretest:** Having planned the questions, a pretest was conducted over 68 fifth graders from two classes in Yi-zai Elementary School in the New Tainan City. The result of the pretest was then analyzed.
3. **Selection of the Questions:** After the analysis, twelve questions with discernment over.25 and difficulty degree between.40~.80 were reserved as formal questions respectively for each unit.
4. **Analysis for Reliability and Validity:** The reserved questions with the internal consistency of Kuder-richardson reliability of.826 and the average difficulty degree of.618 were identified. As to the validity, it was confirmed by experienced science teachers that the test result of each unit could be the representative one for the desired test fields.

RESULTS

The Influences Brought by "Learner-Controlled Modes" and "Multimedia Combination Forms" on Learning Achievement

Table 3 shows the mean values and standard deviations of learning achievement from the two groups of learners applying PC mode and AC mode and receiving instructions from three kinds of multimedia combinations. While Table 4 presents the summarized result of the variable-analysis over the learner-controlled modes and multimedia combination forms.

The analytic result in Table 4 indicates that the cross-effect ($F = .59$, $p = .551 > .05$) on the learning achievement gained from "learner-controlled modes" and "multimedia combination forms" did

Table 3. Mean values and standard deviations of the learning achievement from different LCMs and MMCFs

LCM	Groups		
	A+S	A+S+N	A+N
ACM	8.10 (2.76)	8.22 (2.54)	8.48 (3.26)
PCM	6.91 (2.53)	7.50 (2.14)	7.59 (2.71)

Table 4. Summarized result of the variable-analysis over the learner-controlled modes and multimedia combination forms

Variable Sources		*SS*	*df*	*MS*	*F*	*p*
LCM (A)	SSa	125.62	1	125.62	10.00	.002**
MMCF (B)	SSb	27.98	2	13.99	3.11	.046*
A×B	SSab	5.38	2	2.69	.59	.551
Deviations	SSs/a	2385.87	190	12.55		
Deviations	SSbs/a	1711.97	380	4.50		

*p<0.05 **p<0.01 ***p<0.001

not reach a significant level. Thus, a major-effect examination was executed. Table 5 demonstrates a summarized analysis, in which the major effect of "learner-controlled modes" was at a significant level (*F =3.112, p =.046 <.05)*, indicating it has evident influence on learning achievement. Comparing the results shown in Table 5 and Table 3, learners performed better with the "active-control mode" than with the "passive-control mode." Also, the major effect of "multimedia combination forms" appeared significant (*F =10, p =.002 <.05*), showing that they are influential in a learning achievement. Via comparisons with Helmert's comparing method within a group, data gained and shown in Table 6 suggest obvious differences (*p =.023 <.05*) of learning effects between A+S and A+N; the learning effects achieved by A+N is evidently better than those by A+S. Yet, no significantly different learning effects was found between A+S and A+S+N (*p =.023 <.05*) and between A+S+N and A+N (*p =.428 >.05*).

Table 5. Summarized variable-analysis of the major effects from the mixed design

Major Effects	*SS*	*df*	*MS*	*F*	*p*
Learner-control Modes	27.98	2	13.99	3.11	.046*
Deviation	1717.35	382	4.49		
Multi-Media Combination Forms					
ACM vs. PCM	41.87	1	41.87	10.00	.002*
Deviation	795.29	190	4.18		

*p<0.05

Table 6. Post-comparisons (learning effects) for the variable-analysis of the major effects from the mixed design

Comparing Method	(I-J) Mean Difference	Standard Deviation	Significance	Post Hoc
A+S vs. A+S+N	-.35	.18	.059	(A+S)<(A+N)
A+S vs. A+N	-.53(*)	.23	.023	
A+S+N vs. A+N	-.18	.23	.428	

*p<0.05

The Influence Brought by "Active-Control Model" and "Multimedia Combination Forms" and Viewed via Eye-Movement Models

In this study, we gathered 40 students and divided them into three groups, animation + subtitles (group a), animation + subtitles + narration (group b), and animation + narration (group c). Ten students were excluded due to error eye shifting. Thus, the eye movement data analyzed were collected from the rest 30 students. According to the data, we found some extremely important results shown in Table 7.

First, significant differences appeared among the three groups in the total time spent in viewing the multimedia web ($F(2,27) = 7.42$, $p < .01$, $\eta^2 = 0.35$). Clearly it showed that the mean of TT in A+S+N condition was greater than the one in A + S condition, and the TT in A + N condition was greater than the one in A+ S condition. This result suggested that the narration significantly affected the time consumed in surfing the multimedia web. The multimedia presented with narrations motivated the viewers to look at the animations for a longer time.

Second, significant differences ($F(2,27) = 5.14$, $p<.05$, $\eta^2 = 0.27$) showed up in the fixation count, and obviously the FC of group c was greater than that of group a. According to the related study, FC is related to the viewer's efficiency in searching for relevant information (Jacob & Karn, 2003). Therefore, this result showed that participants in A+ N condition had more fixation counts than those in A+S condition. We inferred that A+ N condition offered the participants verbal context and extra time to view the animation. However, the greater the fixation count is, the less efficient information searching is. So, we had to further identify another indicator of eye movement to make sure if students did search for clues from the animation efficiently.

Third, no significant differences were found in the total fixation duration (TFD: $F(2, 27) =$

Table 7. Mean, standard deviations, and one-way analyses of variance (ANOVA) for different indicators of eye movement on three dependent variables

	A+S (a) Mean(SD)	A+S+N (b) Mean(SD)	A+N (c) Mean(SD)	F	η^2	Post Hoc
TT	282.14(67.89)	364.27(43.82)	366.75(43.78)	7.42**	0.35	(a)<(b) (a)<(c)
FC	380.71(95.26)	490.09(140.51)	556.00(97.50)	5.14*	0.27	(a)<(c)
TFD	161.17(55.18	192.06(57.34)	226.53(53.35)	3.22	0.19	
FF	1.53(0.16)	1.41(0.39)	1.48(0.20)	0.42	0.03	

*p<.05 **p<.01

Note: TT means Total Time; FC means Fixation Count; TFD means Total Fixation Duration; FF means Frequency of Fixation.

3.22, p =.05, η^2 *= 0.19)* and the frequency of fixation (FF: $F(2, 27) = 0.42, p = .66, \eta^2 = 0.03$). According to the previous study, TFD is related to the difficulty of a task. Therefore, the insignificant differences of TFD among three groups indicated that the students didn't have obstacles in learning from different combination forms of subject matters; that is, the difficulty of tasks in different combination forms was almost the same. In addition, the FF is related to the key factors of the sighted area, and the result showed that the differences of FF in different combination forms were insignificant. That is, although students viewed the teaching materials in different multimedia combination forms, they paid attention to the same spot on the screen.

Finally, Figure 7 illustrates the key part students viewed frequently in different conditions. The pictures showed the key part of the material on the screen that students viewed frequently, and the result helped us find out the key elements for the multimedia materials.

DISCUSSION

The Influences of "Learner-Controlled Modes" and "Multimedia Combination Forms" on Learning Achievement

The data of the research result suggested that the influences brought by different learner-controlled modes onto the major effects of learning achievement were significant, and active-control mode was better than passive-control mode for learning. The learners controlled the learning schedules in which the instructional multimedia animations had been divided into several segments according to segmentation principles; yet, since the active-control mode was equipped with functions of flipping pages, which offered the learners more choices of access, formation, and information integration. The learning effects active-control mode created were superior to those generated by passive-control mode with the exclusive function of going to the next page.

Furthermore, the major effects of learning achievement from different multimedia combination forms were significantly different from each other. The performance of A+ N group was evidently better than that of A+S group; however, no differences in learning achievement were observed between A+ S+ N group and A+N group. This phenomenon suggested that in a plot of learning via segmented multimedia animations, the learning effects gained by dual channels (visual and audio) were better than those by single channel (visual) though each instructional segment brought less intrinsic cognitive load, indicating the modality effect was still there.

Crystal result of this study proved that the effects from learner-controlled modes and multimedia combination forms did affect the learning achievement. Moreover, the effect of active-control mode was better than passive-control mode,

Figure 7. Samples of scan path of the area frequently viewed by students. The left one was in A + S condition, the middle one was in A + S + N condition, and the right one was in A + N condition.

and the achievement of learning in A+N group was significantly better than that in A+S group. However, we extremely eager to further depict how students achieved better learning through the multimedia web of science education.

The Differences Caused by Multimedia Combination Forms from Eye Movement Models

According to the eye-movement data analysis, we found that the influences of TFD and FF among different multimedia combination forms were not significant. The results revealed that students spent equal efforts to view the multimedia web in different multimedia combination forms. The multimedia materials were understood easily by different groups. However, why did the students have different learning achievement, and which factor affected the results? Via the experiment, we obtained some evidences to answer the questions.

According to the study of Jacob and Karn (2003), the greater the fixation count is, the less efficient information searching is. From the statistical result of FC, we found that the counts of fixation in A+N group were significantly greater than those in A+S group. It suggested that the students in A+ N group viewed the screen more often than those in A+S group; yet the duration of fixation was short. This result of FC can be inferred that the students did not search for the information efficiently. However, the students in A+N group had better learning achievement. This contradictory phenomenon inspired us to assume that the greater FC was not resulted from less efficient searching but from other factors. Data collected showed that the FC from A + S+ N and A+N groups were not significantly different, which indicated that the students in A+S+N group only listened to the narration and viewed the animation simultaneously and yet ignored the subtitles either consciously or unconsciously. That is why these two groups had the same learning achievement.

The TT of the three groups spent in learning was significantly different from each other. It showed that the students in A+N group watched the animation and listened to the narration at the same time. Since there were no subtitles, they had more time to look at the animation, reinforcing their cognition over science. This evidence explained why the students in A+N and A+S+N group seemed to spend more time in searching necessary information but on the other hand gained better learning achievement.

CONCLUSION

Teaching programs in the three units, "Air and Combustion" in "Heat Effects" toward substances time in "Healthy Diet," set in the Science Education Web were adopted for the experiment in this study. The multimedia subject matters of each unit had been divided into smaller segments in accordance with Mayer's (2001) segmentation principles and learner-controlled playing functions were given for the purpose of investigating the influences of "learner-controlled modes" and "multimedia combination forms" on learning achievement and eye movement model. The conclusions are summarized as the following:

The behavioral results of the experiment suggested that the cross-effects of "learner-control modes" and "multimedia combination forms" on the learning achievement were not significant. Yet, the "learner-controlled modes" alone was significantly influential in learning, and the learning effects of "active-control mode" was explicitly superior to those of "passive-control mode." In addition, the "multimedia combination forms" had evident influences on learning achievement, since whichever learner-control mode was adopted, the outcome was that the animation + narration group's learning performances were better than the animation + subtitles group's. That is, even if the animated subject matters had been divided into small segments, the multimedia combination forms still caused influences on learning achievement; the modality effect still existed.

This study also revealed that under the situation of playing animations segment by segment, animations described with narration and subtitles simultaneously did not cause obvious influences on learning, suggesting a disappearing of redundancy effect. The researchers believe the reason was that when animations, subtitles, and narration were presented at the same time, the amount of information may be too large for visual system to process; however, the learners could choose to temporarily ignore the subtitles and focus on animations and narration only. Evidence from eye-movement data explained why the students in animation + subtitles + narration group and animation + narration group spent equal time and performed equally well. According to the models from Mayer's (2001) Multimedia Cognition Theory, the integration of animation (image model) and narration (word model) is the key for comprehension; animation and narration go to different sense channels, which can individually processes information without causing over-load and promote the effectiveness of info-integration. Therefore, learners are most likely to watch the animation and meanwhile listen to the narration without necessarily concerning with the connections between the subtitles and animations. So, even though learners may read the subtitle, it does not do much influence upon learning effects since the key integration of image and word models has been completed through the simultaneous presentation of animations and narrations. This result reasonably explains why the animation + subtitles + narration mode did not cause much affection for learning achievement. The eye-movement data also indicate the same phenomenon.

Synthesizing the results gained from learning achievement and the eye-movement models, we found that after the animations were divided into small segments according to segmentation principles, both of the "learner-controlled modes" and "multimedia combination forms" had influences on eye-movement models and learning achievement. Therefore, describing animations via subtitles was the last choice to be made; while presenting animations with matching narration was the best to promote learning effects. This shows that inappropriate audio/video combinations not only interfere with learning effects but also lead to unnecessary waste of time, suggesting that more functions and information do not always bring better learning achievement. The researchers of this study designed animated teaching materials according to Mayer's (2001) segmentation principles first, and then measured the learning effects and eye-movement models to investigate how different multimedia teaching materials affected learning. This is what makes this study different from the former ones, and the results of this study can be a reference for the practical designs of multimedia materials for a science education web. However, since the scale of experiment design was set specifically for 5th and 6th graders with certain science conceptions, extra considerations may be needed when applying this study to researches over other levels of students or other academic regions. In addition, since similar researches and studies are not many, more data supports are waited to make our findings more of pedagogical value for the coming researchers.

REFERENCES

Anderson, J. R., Bothell, D., & Douglass, S. (2004). Eye movements do not reflect retrieval processes - Limits of the eye-mind hypothesis. *Psychological Science*, *15*(4), 225–231. doi:10.1111/j.0956-7976.2004.00656.x PMID:15043638

Chuang, H.-H., & Liu, H.-C. (2012). Effects of different multimedia presentations on viewers' information-processing activities measured by eye-tracking technology. *Journal of Science Education and Technology*, *21*(2), 276–286. doi:10.1007/s10956-011-9316-1

Eshach, H. (2011). Science for young children: A new frontier for science education. *Journal of Science Education and Technology*, *20*(5), 435–443. doi:10.1007/s10956-011-9324-1

Gerjets, P., Scheiter, K., & Cierniak, G. (2009). The scientific value of cognitive load theory: A research agenda based on the structuralist view of theories. *Educational Psychology Review*, *21*, 43–54. doi:10.1007/s10648-008-9096-1

Gog, T., Paas, F., Marcus, N., Ayres, P., & Sweller, J. (2009). The mirror neuron system and observational learning: Implications for the effectiveness of dynamic visualizations. *Educational Psychology Review*, *21*, 21–30. doi:10.1007/s10648-008-9094-3

Henderson, J. M., Brockmole, J. R., Castelhano, M. S., & Mack, M. (2007). Visual saliency does not account for eye movements during search in real-world scenes. In R. van Gompel, M. Fischer, W. Murray, & R. Hill (Eds.), *Eye movements: A window on mind and brain* (pp. 537–562). Oxford, UK: Elsevier. doi:10.1016/B978-008044980-7/50027-6

Huber, S., & Krist, H. (2004). When is the ball going to hit the ground? Duration estimates, eye movements, and mental imagery of object motion. *Journal of Experimental Psychology. Human Perception and Performance*, *30*(3), 431–444. doi:10.1037/0096-1523.30.3.431 PMID:15161377

Hyönä, J. (2010). The use of eye movements in the study of multimedia learning. *Learning and Instruction*, *20*(2), 172–176. doi:10.1016/j.learninstruc.2009.02.013

Jacob, R. J. K., & Karn, K. S. (2003). Commentary on Section 4 - Eye tracking in human-computer interaction and usability research: Ready to deliver the promises. In J. Hyönä, R. Radach, R. R. H. DeubelA2 - J. Hyönä & H. Deubel (Eds.), The mind's eye (pp. 573-605). Amsterdam, The Netherlands: North-Holland.

Just, M. A., & Carpenter, P. A. (1984). Using eye fixations to study reading comprehension. In D. E. Kieras & M. A. Just (Eds.), *New methods in reading comprehension research* (pp. 151–182). Hillsdale, NJ: Erlbaum.

Kirschner, F., Paas, F., & Kirschner, P. A. (2009). A cognitive load approach to collaborative learning: United brains for complex tasks. *Educational Psychology Review*, *21*, 31–42. doi:10.1007/s10648-008-9095-2

Lin, L., & Atkinson, R. K. (2011). Using animations and visual cueing to support learning of scientific concepts and processes. *Computers & Education*, *56*, 650–658. doi:10.1016/j.compedu.2010.10.007

Lin, Y. C., & Liu, C. J. (2007). A study on learning effects among students with different learning styles using chemistry education website. In *Proceedings of the NARST 2007*, New Orleans, LA.

Liu, C. J., Lin, Y. C., Liu, B. Y., & Chang, Y. Y. (2012). A research of applying cognitive load theory to science education webpage. In *Proceedings of the 2012 IEEE International Conference on Advanced Learning Technologies (ICALT 2012)*, Rome, Italy.

Liu, C.-J., & Shen, M.-H. (2011). The influence of different representations on solving concentration problems at elementary school. *Journal of Science Education and Technology*, *20*(5), 621–629. doi:10.1007/s10956-011-9293-4

Liu, H. C., & Chuang, H. H. (2011). An examination of cognitive processing of multimedia information based on viewers' eye movements. *Interactive Learning Environments*, *19*(5), 503–517. doi:10.1080/10494820903520123

Liu, H. C., Lai, M. L., & Chuang, H. H. (2011). Using eye-tracking technology to investigate the redundant effect of multimedia web pages on viewers' cognitive processes. *Computers in Human Behavior*, *27*(6), 2410–2417. doi:10.1016/j.chb.2011.06.012

Mayer, R. E. (2001). *Multimedia learning*. Cambridge University Press. doi:10.1017/CBO9781139164603

Mayer, R. E., & Moreno, R. (2003). Nine ways to reduce cognitive load in multimedia learning. *Educational Psychologist*, *38*(1), 43–52. doi:10.1207/S15326985EP3801_6

Paas, F., Renkl, A., & Sweller, J. (2003). Cognitive load theory and instructional design: Recent developments. *Educational Psychologist*, *38*(1), 1–4. doi:10.1207/S15326985EP3801_1

Renkl, A., Hilbert, T., & Schworm, S. (2009). Example-based learning in heuristic domains: A cognitive load theory account. *Educational Psychology Review*, *21*, 67–78. doi:10.1007/s10648-008-9093-4

Sankey, M. D. (2003). *Visual and multiple representations in learning materials: An issue of literacy*. Retrieved July 25, 2009, from http://www.usq.edu.au/users/sankey/Resources/CreatEd2003.pdf

Schnotz, W., & Lowe, R. (2003). External and internal representations in multimedia learning - Introduction. *Learning and Instruction*, *13*(2), 117–123. doi:10.1016/S0959-4752(02)00015-4

Sun, K.-S., Lin, Y.-C., & Yu, C.-J. (2008). A study on learning effect among different learning styles in a web-based lab of science for elementary school students. *Computers & Education*, *50*(4), 1411–1422. doi:10.1016/j.compedu.2007.01.003

Sweller, J. (2009). Cognitive bases of human creativity. *Educational Psychology Review*, *21*, 11–19. doi:10.1007/s10648-008-9091-6

Tsai, M.-J., Hou, H.-T., Lai, M.-L., Liu, W.-Y., & Yang, F.-Y. (2012). Visual attention for solving multiple-choice science problem: An eye-tracking analysis. *Computers & Education*, *58*(1), 375–385. doi:10.1016/j.compedu.2011.07.012

van Gog, T., & Scheiter, K. (2010). Eye tracking as a tool to study and enhance multimedia learning. *Learning and Instruction*, *20*(2), 95–99. doi:10.1016/j.learninstruc.2009.02.009

van Merrienboer, J. J. G., & Sluijsmans, D. M. A. (2009). Toward a synthesis of cognitive load theory, four-component instructional design, and self-directed learning. *Educational Psychology Review*, *21*, 55–66. doi:10.1007/s10648-008-9092-5

van Merri¨enboer, J. J. G., & Sweller, J. (2005). Cognitive load theory and complex learning: Recent developments and future directions. *Educational Psychology Review*, *17*(2).

This work was previously published in the International Journal of Online Pedagogy and Course Design (IJOPCD), 4(2); edited by Chia-Wen Tsai and Pei-Di Shen, pages 19-34, copyright 2014 by IGI Publishing (an imprint of IGI Global).

Chapter 49
A Comparative Study on Undergraduate Computer Science Education between China and the United States

Eric P. Jiang
University of San Diego, USA

ABSTRACT

With the rapid growth of the Internet and telecommunication networks, computer technology has been a driving force in global economic development and in advancing many areas in science, engineering, health care, business, and finance that carry significant impacts on people and society. As a primary source for producing the workforce of software engineers, computer scientists and information technology specialists, computer science education plays a particularly important role in modern economic growth and it has been invested heavily in many countries around the world. This chapter provides a comparative study of undergraduate computer science programs between China and the United States. The study focuses on the current curricula of computer science programs. It in part is based on the author's direct observation from his recent visits to several universities in China and the conversations he had with administrators and faculty of computer science programs at the universities. It is also based on the author's over two decades experience as a computer science educator at several public and private American institutions of higher educations. The education systems in China and the United States have different features and each of the systems has its strengths and weaknesses. This is likely also true for education systems in other countries. It would be an interesting and important task for us to explore an innovative computer science education program, which perhaps blends the best features of different systems and helps better prepare graduates for the challenges working in an increasingly globalized world. We hope the study presented in this chapter provides some useful insights in this direction.

DOI: 10.4018/978-1-4666-7363-2.ch049

1. INTRODUCTION

It is of no doubt that computer technology has thoroughly integrated into our society. For many people around the world, computers or related computing technology has now become an inseparable part at the work place and in their personal lives. We use email and cell phones for communication, GPS for directions, tablets for reading, calculators for computations, and computer software tools for both business tasks and personal leisure. Computer technology has also been playing a key role in modern economic development and in advancing many areas in science, engineering, health care, business and finance that carry significant impacts on people and society.

As a primary educational discipline for producing the workforce of software engineers and computer scientists, who are responsible for the rapid growth of computer technology, computer science studies the scientific principles and practical approaches of various computing systems, from cell phones, tablets, laptops, desktops, main frames to supercomputers. Computer science education offers a foundation that permits graduates to adapt new ideas and new technologies and produce various innovative software products and engineering solutions. In order to further promote modern economic development, many countries around the world have been investing heavily in their higher education systems and in particular, in the computer science education and research programs. For instance, in the last fifteen years or so, the number of college computer science programs in China has grown very rapidly and in fact, the discipline of computer science and technology has already become the largest undergraduate major in China (ACCE, 2005).

It is perfectly reasonable that individual countries set their own education standards and program curriculum requirements. As a matter of fact, like human languages, there should not be a universal computer science curriculum model that is acceptable and works well in all countries. Each individual program has its strengths and perhaps also its weaknesses. This chapter provides a preliminary comparative study of undergraduate computer science programs between China and the United States. We believe this is a meaningful exercise because the programs in both countries share a similar base of fundamental courses while they are different in terms of the structure and scope of program requirements. The differences between the programs reflect a great deal about their culture differences. As the two largest economies in the world today, the United States and China play a significant role in promoting global economic growth. It would be very beneficial for both countries (perhaps also other countries in the world) to learn from each other about computer science curriculum development and to explore new ideas and strategies in providing computer science students with the academic training and experiences needed for their professional success and leadership in today's global environment.

The chapter is organized as follows. In Section 2, we present and discuss three representative undergraduate computer science program models in American institutions of higher education. They represent programs in large research universities, public regional universities and small private liberal arts universities, respectively. In Section 3, we present and discuss the undergraduate computer science programs from three Chinese institutions, representing different institution types in China: top tier national research universities, second tier regional comprehensive universities and third tier technology focused universities. Then, in Section 4, we compare computer programs from both countries in terms of some factors that are closely related to curriculum design and development. We provide some concluding remarks in Section 5 and future research directions in Section 6.

2. UNDERGRADUATE COMPUTER SCIENCE EDUCATION IN THE UNITED STATES

Computer science started as a distinct academic discipline in the 1950s in the United States and its first computer science degree program was formed in 1962 at Purdue University. Over the years, computer science undergraduate education in the United States has been based on computing curriculum guidelines set by the two well-known professional technology societies: the Association for Computing Machinery (ACM) and the Institute of Electrical and Electronics Engineers (IEEE). The latest complete computer science curricular volume from ACM and IEEE was released in December of 2001 (AIJTF, 2001). In order to reflect the current state of computer science and highlight future directions of the field, a joint task force from ACM and IEEE has been undertaking the development of the next new curricular volume and the volume is expected to be completed in 2013 (Sahami et al., 2011).

A typical computer science (CS) undergraduate curriculum has three components: CS core, CS electives, and general education requirement. The CS core is a fixed set of mandatory courses in computer science that all students must complete, while the CS electives are a list of other courses in the field or in other closely related fields and students can choose some from the list to complete for fulfilling program required units in electives. These two components would typically include several supporting mathematics courses such as discrete mathematics and calculus. The additional general education component is a set of broad courses in other disciplines, with some of them being electives, required for an undergraduate degree. For computer science programs, even within the United States, courses in a curriculum can vary quite significantly from institution to institution, depending on educational mission and type, program size, graduate expectations and local business and industry needs.

In the following subsections we describe three representative computer science undergraduate curricular models with different institution types and present a sample curriculum for each type. Note that this discussion does by no means cover all possible college types in the United States, and we only intend to present some fundamental and perhaps unique characteristics of American computer science programs. In the discussion, we assume that a one-unit course credit has the contact time somewhere between 50 minutes and an hour per week, over the course of a 15-week semester, which is a typical setting for an American institution with a semester academic calendar.

Computer Science Programs in Research Universities

Computer science programs in this category typically have a relatively large faculty that is capable of providing considerable depth and breadth in the field (AIJTF, 2001). As a result, these programs are able to offer a wide range of major courses, in particular the upper division elective courses on some cutting edge computer science topics.

In addition, a deep level of mathematical sophistication is one of the usual goals of research university's computer science programs. The programs typically require a sequence of calculus and one or two additional courses in mathematics.

University of California at Berkeley, or UC Berkeley, is a very well-known public research university and it offers 350 undergraduate and graduate programs in a wide range of disciplines. Currently the university has about 36,000 students. In computer science, UC Berkeley offers two programs: BA (bachelor in arts) and BS (bachelor in science). There is no difference in the required computer science course content between these two programs, but the BS program requires a few more courses in mathematics and engineering whereas the BA program requires a few more courses in humanities and social sciences. There is also a special admission policy for the BA computer science program at UC Berkeley, which students need to complete all lower division course requirements in order to be admitted as computer science majors. The current curriculum structure of its computer science program (UCB, 2012) is outlined in Table 1. Note that UC Berkeley requires a total of 120 semester units for an undergraduate degree and it implies that the courses from the curriculum model counts about 50% of the total units.

Computer Science Programs in Regional Universities

Like national universities, regional universities in the United States offer a full range of undergraduate programs and may also provide some graduate programs or courses at master's level. These universities focus more on undergraduate education and generally grant fewer than 50 percent of their degrees in liberal arts disciplines. In addition, in comparison to research universities, research expectations on faculty are somewhat lesser and student academic admission criteria are generally lower.

California State Polytechnic University at Pomona, or Cal Poly Pomona, is a regional public university with its current enrollment of about 22,000 students. It offers 64 undergraduate majors and 27 graduate programs at master's level. The university uses a shorter quarter system and in Table 2, which illustrates its computer science curriculum structure (CSPUP, 2012), we have converted all course units into their equivalent semester ones for easy comparison. As it is the same with UC Berkeley, Cal Poly Pomona requires a total of 120 semester units for graduation, but the percentage of the combined course units in its curriculum structure listed in Table 2 is slightly higher and it reaches to 59%.

Computer Science Programs in Small Liberal Arts Universities

Although a number of liberal arts universities in the United States offer graduate degree programs, these institutions emphasize undergraduate education. In addition, a liberal arts curriculum aims to expose students to the fundamental principles of a discipline and its applications in multiple domains and to promote critical reasoning and problem solving skills that are necessary for life-long learning after graduation. Students at liberal arts universities generally take a large portion of their course work outside their discipline and some of them pursue a second major or minor in another field of study. The programs at these schools can be small with a limited faculty resource and as a result, the number of computer science course offerings in these programs will be less than those at larger schools.

University of San Diego, or USD, is a private liberal arts university with about 8,300 students at its current enrollment. The university offers about 40 undergraduate degree programs and a number of master's and doctorate programs. The computer science program at USD is relatively small and

Table 1. Curriculum structure of CS (BA/BS) at UC Berkeley

Lower Division Mandatory	Upper Division Mandatory	Electives	Math Mandatory
16 units	8 units	20-22 units	16 units
Comp Programs (4)	Algorithms (4)	2 courses out of 9 (8-10)	Calculus I (4)
Data Structures (4)	Operating Systems (4)	2 UD CS courses (8)	Calculus II (4)
Machine Structures (4)		1 Tech Elective (4)	Linear Algebra & Diff Equations (4)
Digital Electronics (4)			Discrete Math & Probability (4)

Table 2. Curriculum structure of CS (BS) at Cal Poly Pomona

Lower Division Mandatory	Upper Division Mandatory	Electives	Math Mandatory
18.9 units	20.2 units	13.5 units	18.2 units
Intro to CS (2.7)	Automata (2.7)	5 courses out of 15	Discrete Structures (2.7)
Intro to Programming (2.7)	Algorithms (2.7)		Calculus I (2.7)
Comp Logic (2.7)	Comp Architecture (2.7)		Calculus II (2.7)
Data Structure & Algorithm I (2.7)	Programming Languages (2.7)		Calculus III (2.7)
Data Structure & Algorithm II (2.7)	Operating Systems (2.7)		Linear Algebra (2.7)
C++ Programming (2.7)	Software Engineering (2.7)		Calculus Several Variables (2)
Comp Organization & Assembly Program (2.7)	Seminar (1.3)		Stat Methods (2.7)
	Comp & Society (2.7)		

its curriculum structure (USD, 2012) is shown in Table 3. The university requires a total of 124 semester course units for an undergraduate degree, and as a typical liberal arts school, the computer science students at USD take only about 40% of their courses in the discipline.

The three computer science curriculum models presented in Tables 1-3 have different features that in part reflect their individual institution types, degree focus, faculty resources and student level of academic preparation. As a top research university, UC Berkeley generally admits students with strong academic credentials and high motivation levels and as a result, its curriculum offers a relatively small number of both introductory offerings and upper-division mandatory courses. For the same reason and it is also due to its fairly large faculty and program size, the curriculum requires a relatively large portion of major courses in electives and it also offers a decently large selection of electives in multiple subfields of computer science. That is quite the opposite in some aspects for the program at Cal Poly Pomona. As a regional public school, the program provides a fairly heavy load of introductory offerings and mandatory major courses to meet the needs of its students. In addition, a substantial requirement of mathematics courses in the program at Cal Poly Pomona is typical for a technology school. Different from UC Berkeley and Cal Poly Pomona, University of San Diego is

Table 3. Curriculum structure of CS (BA) at University of San Diego

Lower Division Mandatory	Upper Division Mandatory	Electives	Math Mandatory
12 units	18 units	9 units	11 units
Comp Programming I (3)	Digital Hardware (4)	3 courses out of 10	Discrete Math (3)
Comp Programming II (3)	Operating Systems (3)		Calculus I (4)
Data Structure & Algorithms (3)	Object Oriented Programming (3)		Calculus II (4)
Assembly Programming (3)	Automata (3)		
	Software Engineering (3)		
	Capstone Project (2)		

a representative liberal arts school that requires a considerably fewer courses in the discipline of study, and since it has a small computer science program, electives are offered in rotation in the curriculum.

3. UNDERGRADUATE COMPUTER SCIENCE EDUCATION IN CHINA_

The computing discipline in China was established in 1956 and its computer education programs were offered in early 1960s at only a few institutions (Li & Lunt, 2006). With the development over sixty years in the past, especially due to the rapid growth in the last decade and a half, the discipline has become the largest undergraduate major in China (ACCE, 2005). Among several computing fields of study, computer science and technology is the primary and most popular one in China. It is the equivalent computer science major in the United States.

Despite the vastly increased access to higher education in China these days, admission to undergraduate programs at China's universities remains very competitive and it is primarily based on student scores on annual national examinations. Before 1998, higher education in China had been a privilege enjoyed by just a small elite group of high school graduates. In 1985, for instance, only 0.1 percent of the population received a college education in China (Wilson et al., 1988). Started in 1998 and driven by the huge economic growth and the desire to expand educational opportunities for a larger portion of China's young talents, China has been making a transition from an elite education model to a mass education system and as the result, it has dramatically increased university enrollments (Zhang & Lo, 2010). As a matter of fact, in the last decade China doubled the number of colleges and universities to 2,409, and now it produces around eight million graduates a year from universities and community colleges (Bradsher, 2013).

The current undergraduate curricular guidelines for computer science and technology programs in China were proposed in 1996. The guidelines are based on the ACM/IEEE Computing Curricula 1991 (AIJTF, 1991) and they emphasize computer software and computer applications (Zhang & Lo, 2010). In order to meet the needs of continuing economic growth in China, in particular in the high tech sector, and also the needs of continuing growth of its higher education system, an effort from China's Ministry of Education is under way to develop the strategic directions and revised curriculum guidance for China's undergraduate computing disciplines.

As in the United States, there are some noticeable curricular variations at different institutions in China. For the rest of the section, we present the computer science and technology sample curriculum structures at three universities in China. They represent the computer science programs of different institution types of the country: top tier national research universities, second tier regional comprehensive universities and third tier technology oriented universities.

Recently we had opportunity to visit these three universities and a few additional ones, and to meet with administrators and faculty of computer science and technology departments at the universities. The universities we visited include Shanghai Jiao Tong University, Fudan University, Shanghai University, Chang An University, Xian Jiao Tong University, Xian University of Technology, and Guilin University of Technology. All of these institutions are public, which is the presently dominant institution type of higher education in China and among them, Shanghai Jiao Tong University, Xian Jiao Tong University and Fudan University are in the group of top nine universities in China, which is designated by China's Ministry of Education on the basis of past excellence in research and education. Shanghai University ranks as a second tier school and the rest of the institutions belong to groups of the third tier or lower in China.

Most universities in China use a semester system. Each course unit requires the contact time between 40 and 50 minutes per week, over the course of a semester between 15 and 18 weeks. In terms of time spent in classroom, it is comparable to a course unit in the United States.

Shanghai Jiao Tong University is one of the oldest and most prestigious and selective universities in China. The university was founded in 1896 and over the years it has grown substantially. Currently the university offers 63 undergraduate programs, 250 masters and 203 doctorate programs. It has an enrollment of about 36,000 students that include 16,800 undergraduates, 17,500 graduate students and 1,500 international students. The computer science and technology program at the university is particularly well known for its consistent superior performance at the ACM International Collegiate Programming Contest, an annual undergraduate programming competition participated by many universities around the world, including three world championships received by its student delegates at the event.

Table 4 illustrates the curriculum structure of the computer science and technology program at Shanghai Jiao Tong University (SJTU, 2011). The program core consists of 124 units, which is about 57% of the total of 219 units required for an undergraduate degree from the university.

The computer science curriculum at Shanghai Jiao Tong University shown in Table 4 could serve as an undergraduate computer science sample curriculum of top ranked Chinese universities. Shanghai University described below represents a different group of institutions of higher education in China. Although the original Shanghai University was founded in 1922 as a small liberal arts institution, the university was significantly expanded in 1994 when it merged with other three local technical colleges. Shanghai University has now become a comprehensive, regional and second tier research school. Currently it offers 71 undergraduate programs and a number of graduate programs and has a total enrollment of roughly 37,000 students that include 25,000 undergraduates, 9,000 graduate students, and 3,000 interna-

Table 4. Curriculum structure of CS (BS) at Shanghai Jiao Tong University

Lower Division Mandatory	Upper Division Mandatory	Electives	Math Mandatory
24 units	42 units	24 units	34 units
Program Design (4)	Algorithms & Complexity Analysis (3)	3 courses out of 8 (A) (9)	Calculus I (6)
C++ Program (2)	Complier Principles (3)	1 course out of 9 (B) (3)	Calculus II (4)
Data Structures (4)	Operating Systems (3)	Engineering Practice & Innovation II (2)	Math Anal I (6)
Embedded System Principles (4)	Software Engineering (2)	Engineering Practice & Innovation III (2)	Math Anal II (6)
Comp Organization & Architecture (4)	Database Principles (3)	Engineering Practice & Innovation IV (2)	Disc Math (2)
Computable Theory (3)	Comp Networks (3)	3 labs out of 4 (6)	Linear Algebra (4)
Digital Electronic Tech (3)	Engineering Practice & Innovation I (2)		Probability & Stat (3)
	CS Internship (2)		Math Methods in Physics (3)
	Comp Organization Lab (2)		
	Operating Systems Lab (2)		
	Senior Thesis (17)		

tional students. It should be noted that Shanghai University is perhaps the only school in China for adopting a much similar quarter system as used by some American universities. As we did with Cal Poly Pomona in Table 2, all course units listed in Table 5, which presents the computer science and technology curriculum structure at Shanghai University (SU, 2011), have been converted to their equivalent semester units.

The total of required course units for an undergraduate degree at Shanghai University is 193. From Table 5 we can see that the combined major course units from the program is roughly 127 and it represents 66% of the total units. It is also interesting to note that a number of non-traditional courses or practicums (which counts a total of 37 units) are integrated into the curriculum and as described in their descriptions (SU, 2011), these courses are especially designed for helping students apply what they have learned in the classroom or elsewhere to develop software solutions for some simulated or real world problems. These courses focus on the hand-on practice in developing computer software and hardware, and provide a variety of opportunities for students with teamwork environment, off-campus internships, large-scale application projects and integrated senior thesis tasks. For instance, students can choose one from the five courses called "software development practicum" offered by the program in their senior year and each of the courses focuses on a particular application domain such as embedded systems, e-commerce and information security. Depending on the interests of individual students, these courses can also help prepare them to participate in the annual ACM international programming contest, and other events in China such as the annual national information security and e-commerce design and development competitions. Furthermore, as a quite unique characteristic of the program, the practicum courses are spread out in the curriculum across the four-year program, and most of them are reasonably planned in the senior year.

Xian University of Technology, the third Chinese university we present in the section represents a different institution type from the previous two. It is a smaller and also technology focused institution of higher education. It currently ranks as a third tier university in China. The university offers

Table 5. Curriculum structure of CS (BS) at Shanghai University

Lower Division Mandatory	Upper Division Mandatory	Electives	Math Mandatory
35.9 units	49.4 units	18 units	23.3 units
Comp Tech (2)	Comp Networks (4)	6 courses out of 49	Calculus I, II, III, IV(10.7)
Programming in C (2.7)	Operating Systems I, II (5.3)		Discrete Math I, II (5.3)
Object Oriented Programming (3.3)	Software Engineering (2.7)		Linear Algebra (2)
Digital Logic (4)	Database I, II (5.3)		Probability & Stats (3.3)
Electronic Tech (2.7)	Compiler (3.3)		Numerical Computation (2)
Data Structures I, II (5.3)	System Architecture (2.7)		
Comp Organization I, II (5.3)	Internship (0.7)		
PC Programming (3.3)	Software Development Practicum (2.7)		
Intro to Practice (1.3)	Software Application Project (2.7)		
Programming Practicum (3.3)	Senior Thesis (20)		
Hardware Projects (2.7)			

around 60 undergraduate programs and multiple graduate programs and it has 16,000 undergraduates and 5,300 graduate students.

The undergraduate curriculum structure of the computer science and technology program at Xian University of Technology is presented in Table 6 (XUT, 2012). The combined unit count for the computer science major courses is about 122.5 and it represents 63% of the total of 195 units required for graduation at the university.

It can be seen from Tables 4, 5, and 6 that computer science curricula in Chinese universities are quite comparable each other in terms of including several essential computer science foundation courses, incorporating a number of practicums into programs and requiring a deep training in mathematics. They can, however, still vary in some curriculum aspects between institutions. For instance, as a top ranked university in China, Shanghai Jiao Tong University sets a little more than a quarter of the major required course units in mathematics in its computer science and technology program. In addition, the program also has a relatively small introductory sequence and requires (in the Chinese standard) fewer mandatory major courses and more units in major electives. This feature is reflected in the programs at top ranked universities in the United States such as UC Berkeley. The computer science curriculum at Xian University of Technology, on the other hand, sets a relatively large number of lecture-based mandatory major courses and in particular arranges them heavily in the third and fourth years. For the computer science and technology program at Shanghai University, it includes a significant requirement of practicum courses and offers a very sizeable selection of major elective courses for it computer science students.

4. COMPARATIVE ANALYSIS OF COMPUTER SCIENCE PROGRAMS BETWEEN CHINA AND THE UNITED STATES

Undergraduate computer science curricula in China and in the United States are similar in the sense that they all include a core of foundation courses in

Table 6. Curriculum structure of CS (BS) at Xian University of Technology

Lower Division Mandatory	Upper Division Mandatory	Electives	Math Mandatory
22 units	59. 5 units	18 units	23 units
Programming in C (4)	Operating Systems (4.5)	7 to 10 courses out 12	Calculus I (5.5)
Program Design (1)	PC Principles & Appls (5)		Calculus II (6)
Software Development Tech (2.5)	Comp Organization (4.5)		Linear Algebra (2.5)
Digital Logic (4)	Automata (3)		Probability & Stats (3)
Algorithm Analysis (3)	Software Engineering (3)		Discrete Math (4)
Data Structures (5)	Comp Networks (4.5)		Numerical Methods (2)
Object Oriented Programming (2.5)	Database (4.5)		
	Compiler (4.5)		
	Comp System Architecture (2.5)		
	Comp Graphics (2.5)		
	Practicum/Internship (4)		
	Senior Thesis (17)		

computer science. That is very reasonable as most of the programs in both countries are influenced by the ACM/IEEE Computing Curricula guidelines (AIJTF, 2001; AIJTF, 1991). However, as we have presented in Section 3, the structure and scope of computer science degree requirements still vary noticeably at different universities in China and in the United States, respectively, and they vary even more significantly between universities in China and in the U.S.. This is quite expected since every computer science program should adapt general guidelines and strategies to match the characteristics of the particular institution and to meet the needs of local high tech industry and national economic development of the particular country. Every curriculum, however, regardless of the characteristics and expectations of the institution, must provide students with a rigorous grounding in the fundamentals of computer science (AIJTF, 2001) and prepare them for lasting success in their future endeavors.

In the rest of the section, we compare the computer science programs in China and the United States, and in particular the sample curriculum structures presented in Section 2 and Section 3. The discussion is based on a number of the factors that can influence curriculum design and development (AIJTF, 2001).

Institution Type

There are a variety of types in American institutions of higher education. Some of them are public, other are private; some are research oriented and offer both BS and BA degrees in computer science (e.g., UC Berkeley), others focus on undergraduate education; some are engineering and technology schools offering only BS degree in computer science (e.g., Cal Poly Pomona), others belong to liberal arts universities offering only BA degree in computer science (e.g., University of San Diego). In addition, the admission criteria can be, based on national prestige rankings, quite different from institution to institution. All these institutional characteristics have an impact on curriculum design and degree requirement planning, which will be further discussed in details in the following subsections.

In China, public universities have been the dominant institution type. Some of the universities in China are referred to as comprehensive institutions (e.g., Shanghai University) and they offer a broad range of study programs, whereas others are considered technical schools (e.g., Xian University of Technology) with a primary focus on education in science and engineering. For the computer science and technology discipline, almost all Chinese universities offer only BS undergraduate degrees. As in the United States, Chinese universities are ranked annually by some professional organizations. But it is a bit different from the United States; admissions to universities in China are more competitive and are primarily based on student ranked scores in annual national college entrance examinations. Moreover, for education in computer science and many other disciplines in China, students are expected to concentrate on a single field of study and the program curricula contain only a small portion of general or broad education courses fall outside the field. As a result, few university students in technical fields in China pursue a second major or a distant minor of study.

Academic Qualification of Entering Students

For both countries, a strong academic preparation level of entering students in an undergraduate program usually corresponds to a high national ranking of the university or program. Hence it can vary quite substantially at different institutions. This especially is the case in China because, as we have mentioned before, its admission policies focus primarily on scores on college entrance examinations. Taking the student academic credentials and motivation levels into consideration, many computer science programs in both countries customize the introductory courses to meet the needs of their specific student groups and in addition, programs at top ranked universities often offer a small set of mandatory courses and a large set of electives in their curricula. These are reflected in the programs from both UC Berkeley (Table 1) and Shanghai Jiao Tong University (Table 4). Among all curricula discussed in Section 3, the program at Shanghai Jiao Tong University has a substantially deeper requirement in supporting mathematics.

Mandatory Course Requirements

One of the biggest differences between the education system in China and the United States is that Chinese universities usually require significantly more course units than American counterparts for undergraduate degrees. For instance, the three Chinese universities presented in Section 3 require an average of 202 semester units, while the three American universities presented in Section 2 require considerably fewer units, with an average of only 121 semester units. We believe that this fully reflects the differences about education traditions between the two countries. Schools in China generally emphasize an extensive training of content knowledge while in the United States, schools emphasize inquired based learning and are usually more flexible with simpler content (Bao et al., 2009). The different degree requirements in course units also apply to computer science curriculum. For a typical computer science program in China, it requires roughly 120 semester units in major mandatory and elective courses, including those in supporting mathematics. In comparison to the requirement of a similar program in the United States, this represents nearly twice many or more course units, depending on the institution type. Such curricula with a heavy load of required units are referred to as discipline based (AIJTF, 2001). It should be pointed out that many computer science programs in China include a significant fraction of the units for practicums, internships, and a senior thesis, which may carry somewhere between 20 and 35 units. The requirement for a senior thesis or a capstone project is also included in curricula at some American universities. For instance, a senior capstone project is integrated in the program at University of San Diego but it is with a much smaller scale in units (Table 6).

Besides cultural traditions, there are certain educational effects for having more or less required major courses in a computer science curriculum. On one hand, academic programs with a large number of required major courses aim to provide students with considerable breadth in computer science education and to help them learn more technical content knowledge and skills in a broad range of application fields. As computers and computer technology have become ubiquitous in our society and the field of computing often intersects with other disciplines, offering a wide variety of cutting-age computing principles and applications courses (e.g., in information security, apps programming and e-commerce) as mandatory ones or electives in a curriculum can be very valuable. On the other hand, a successful computer science graduate needs many skills beyond technical ones. Some of the specific skills may be obsolete in just a few years. An effective computer science curriculum should help students gain an enduring understanding of theory and practice that will allow them to maintain their currency over the long term (AIJTF, 2001). But this may not be achieved by just taking more major courses. Liberal arts curricula in the United States, for instance, emphasize breadth of a student's educational experience and intellectual discourse to a greater degree than other technical and engineering programs by requiring their students to take a significant number of courses in the disciplines outside their primary field of study.

The author had a few conversations on this topic with his colleagues at the Chinese universities he visited. Some of them were aware of the heavy load of units in their programs while others were surprised to learn the substantially fewer course units required by a typical American computer science curriculum. We all agreed that computer science students need to be inspired, motivated and capable of learning more and new skills on themselves and computer science education should provide an effective learning environment to help them acquire these skills. The Chinese colleagues also concurred that the load of required courses in their curricula might be too heavy, especially those mandatory units, and having a relatively small number of required major courses in a curriculum (as a future curriculum updating direction) would allow their students to have time to explore and engage an in-depth study of another discipline that has applications of computing such as psychology, biology, economics and business.

Supporting Mathematics Course Requirements

Computer science has a much closer relationship with mathematics than many other scientific disciplines and in fact, mathematical techniques and reasoning are integrated to most areas (if not all) of computer science. As recommended in the ACM/IEEE guidelines (AIJTF, 2001), it is critical for computer science programs to include enough training in mathematics so that students understand the theoretical underpinnings of the field.

Programs in both countries seem to follow closely with the ACM/IEEE recommendations in regard to the mathematics requirement. They all contain a course in discrete mathematics, which introduces discrete structures and applications in computer science. In addition, most of the programs include the courses in linear algebra and in probability and statistics, which are heavily used in some computer science topics such as computer graphics, computer vision, and data mining. Comparing to the American institutions, academic programs in computer science in China generally have a more rigorous requirement in mathematics, which, in our opinion, is reasonable and beneficial.

Project Based Course Requirements

Project based courses in a computer science program could include practicums or project labs where application projects are focused that may involve both design and implementation, internships where projects in an industry setting are engaged, and a senior thesis or a capstone project where either solo or team-projects are conducted and they apply a wide range of knowledge and skills in either proposing some original research work or solving a substantial problem. Computer science is a vibrant field with the enormous pace of change over time and anyone who is doing some serious work in developing computer technology needs to keep learning and updating his knowledge and skills. Such project courses, which are highly recommended by the ACM/IEEE computing curricula guidance (AIJTF, 2001), provide the platforms for students to strengthen problem solving and technical skills and acquire an understanding of effective professional practice, and the courses should also help promote life-long learning after graduation.

It seems that most programs in China have incorporated the project courses into their curricula as it is indicated in Tables 4, 5, and 6. However, there are some variations in integrating the courses in these curricula. Some of the programs require a heavy load of such courses in the junior and senior years. For example, the curriculum at Shanghai Jiao Tong University (Table 4) sets 59% of its required upper division units in project courses. Other programs arrange their project courses through all four academic years, as shown in the curriculum at Shanghai University (Table 5). In comparison, there are only a fraction of the programs in the United States require project based courses in their curricula, which include a capstone project or a senior thesis. It seems that the practice that integrates project courses into computer science programs is both important and beneficial for producing capable graduates and for general computer science programs in the United States, this project-enriched curriculum practice perhaps needs to be considered or further strengthened.

5. CONCLUDING REMARKS

We have presented a comparative study of undergraduate computer science education between China and the United States. The study focuses on the current curricula of computer science programs. It in part is based on the author's direct observation from his recent visits to several universities in China and the conversations he had with administrators and faculty of computer science programs at the universities. It is also based on the author's over two decades experience as an educator in computer science at several public and private American institutions of higher education.

The education systems in China and the United States have different features and they reflect a great deal about the differences between the traditions and cultures. As we have discussed in the chapter, the differences are also indicated in their individual computer science curricula. It appears that each of the systems has its strengths and weaknesses and it would be both important and beneficial for both countries (and perhaps other countries as well) to learn from each other about how to develop an effective computer science curriculum.

At the present time, the United States is still the world's leader in technology and innovation. In terms of computer technology workforce production, it remains competitive (though no longer dominant) in global markets. A recent KPMG survey of technology executive worldwide predicts that China will be on par with the United States as a future technology innovation leader (KPMG, 2012). The survey also indicates that China's innovation investment has fostered an environment for the development of disruptive technologies that is growing by leaps and bounds. In the broad areas of science, technology and mathematics, China is already graduating over one million college graduates a year while the United States graduates fewer than half that number (James & Tanden, 2012). The rise of China and other developing countries such as India in technology sector shows the significant challenges facing the United States to retain its position as a global leader in innovation and a global pacesetter in the education of science and technology talents.

To take the challenges, the United States must take immediate action to play catch up and make a much stronger and firmer commitment to education. It should invest significantly more in all education levels from K-12 to university. The United States should also, in particular, support and increase education and research funding to computer science and other critical-need areas of science, technology, engineering and mathematics. In addition to monetary investment, we need pay attention to how we educate the next generation of American workers to ensure our continued economic competitiveness well into the 21st century (James & Tanden, 2012).

For undergraduate computer science programs in the United States, there might be several areas that can be improved upon. In many US research universities, a significant portion of their undergraduate computer science courses are taught by graduate teaching assistants or teaching lecturers who usually do not conduct research. Effective classroom teaching and supportive student mentoring are very critical in improving student learning. The quality of computer science programs at the institutions can be further enhanced if the programs are able to have more experienced regular faculty involved in undergraduate education by developing and offering more courses that reflect faculty's research expertise and recent advances in computer science and by including more undergraduate students in faculty's research projects. This approach for improvement may work for all disciplines but it is particularly important for computer science education as computer technology evolves at a very rapid pace and computer science students need to grab every opportunity to learn as much as possible the latest development in the field and to update their knowledge and skills of computing. Another related area for curriculum improvement that should apply to most computer science programs of different institution types in the United States is to further augment their curricula with more creative cutting edge content and project enriched courses, as we have discussed in Section 4. It appears that, in comparison to a typical undergraduate computer science program in China, many programs in the United States do not have the requirement of practicum courses or a senior thesis or a capstone project for their graduates. We believe strongly that these suggested curricular changes should make the American computer science education further competitive and productive.

As a rising economic superpower, China has been investing very heavily in its higher education, particularly in the computer science technology arena (Mu, 2010). Over the last few years, China has successfully produced a significant number of university graduates in science, technology, engineering and mathematics and attracted many international technology and engineering firms for establishing research and development centers in China. Looking forward, by 2030, China will have 200 million college graduates, which is more than the entire United States workforce (James & Tanden, 2012).

In order for Chinese graduates to compete successfully in global job markets, China need to improve the quality of their academic programs and to make some substantial revisions in their educational approaches and classroom instructions. As we have discussed in this chapter, the education system in China emphasizes extensive content knowledge learning. From elementary schools to colleges, Chinese students go through rigorous problem solving instruction in all science and mathematics fields throughout their school years and become skillful at solving content-based problems. It remains unclear, however, whether such training is transferable beyond the specific content areas and problems types taught (Boa et al., 2009). To make education system as an effective incubator for technology innovator thinkers, educational programs in China should consider integrating more inquiry-based instructional methods in classrooms (Lawson, 1995) to further promote critical thinking and scientific reasoning abilities for their students.

There might be several areas for Chinese undergraduate computer science programs to be improved upon. In the last few years, research activities and publication records have become a dominant performance criterion in many Chinese universities for faculty pay raise and promotions, especially in its top ranked schools. As a result, some faculty do not pay much of their attention to classroom instruction and student learning. This situation can be changed if faculty evaluation policies in these programs are updated to also encourage and appreciate the education contributions made by individual faculty members. Another area for potential improvement is related to curriculum development. As it is indicated in our discussion in Section 4, undergraduate computer science programs in China may consider revising their academic curricula by cutting out a substantial portion of required major courses. The reduction would allow and encourage their students to explore more interdisciplinary fields of computer technology, have a more broad educational experience and to foster a better environment for them to adapt new ideas and technology.

It is of no doubt that the quality of computer science programs has a direct impact on the information technology workforce they produce. The technology jobs of the future will be competed for by the brainiest from all countries around the world. Ideally, computer science graduates need to develop both content knowledge and transferable reasoning and problem solving skills. It would be an interesting and important task for educators and researchers from both countries (and perhaps also worldwide) to explore an innovative and balanced computer science education program, which perhaps blends the best features of different systems and focuses on the knowledge, skills and competencies that help better prepare graduates for the challenges working in an increasingly globalized world. We hope the discussion and findings from this chapter provide some useful insights in this direction.

6. FUTURE WORK

There are several directions in which this study could be expanded. One direction is a further investigation on a number of representative curricula at the course level in both countries. We would examine and compare course content, instruction format and student assessment tools. Another direction is to compare faculty pedagogical practices and effectiveness in and out classroom under both education systems. That would include the faculty role in mentoring student research work, course projects, and graduation theses and managing off-campus internships. In addition, we would like to look into general job opportunities available for computer science graduates in both countries and their job readiness and performance after graduation.

REFERENCES

ACM/IEEE-CS Joint Task Force on Computing Curricula (AIJTF). (1991). ACM/IEEE Computing Curricula 1991 Final Report. Retrieved from http://dl.acm.org/citation.cfm?id=103710

ACM/IEEE-CS Joint Task Force on Computing Curricula (AIJTF). (2001). ACM/IEEE Computing Curricula 2001 Final Report. Retrieved from http://dl.acm.org/citation.cfm?id=103710

Advisory Committee of Computing Education (ACCE) of China. (2005). The report of strategic study of computing curriculum in China. *China University Teaching*, *5*, 7–10.

Bao, L. et al. (2009). Learning and Scientific Reasoning. *Science*, *323*, 586–587. doi:10.1126/science.1167740 PMID:19179514

Bradsher, K. (2013, January 16). China makes a great leap into higher education. *The New York Times*. Retrieved from http://www.wopular.com/china-makes-great-leap-higher-education

California State Polytechnic University at Pomona (CSPUP), College of Science. (2012). *Computer Science Degree Curriculum Sheet*. Retrieved from http://www.cse.calpoly.edu

James, M., & Tanden, N. (2012, August 22). U.S. Education must keep up with China's, India's Bold Programs. *US News and World Report*. Retrieved from http://www.usnews.com/opinion/articles/2012/08/22/us-education-must-keep-up-with-chinas-indias-bold-programs

KPMG Global Technology Innovation Survey. (2012). *China projected to be on par with US as a future tech innovation leader, Cloud, mobile to drive breakthroughs in coming years*. Retrieved from http://www.kpmg.com/global/en/issuesandinsights/articlespublications/press-releases/pages/china-projected-to-be-on-par-with-us.aspx

Lawson, A. E. (1995). *Science teaching and the development of thinking*. Belmont, CA: Wadsworth.

Li, X. M., & Lunt, B. M. (2006). Undergraduate Computing Education in China: A brief status and perspective. *Proceedings of the 7th ACM Conference on Information Technology Education (SIGITE)* (pp. 35-38). Retrieved from http://delivery.acm.org/10.1145/1170000/1168823/p35-xiaoming.pdf?ip=131.91.215.205&acc=ACTIVE%20SERVICE&CFID=284901811&CFTOKEN=60674018&__acm__=1361978772_1f51090239ba5ef479ebb4fb0f00df4c

Mu, R. P. (2010). *China – UNESCO Science Report 2010: The current status of science around the world* (pp. 379–399). Retrieved from http://www.uis.unesco.org/Library/Documents/UNNESCOSR10-eng.pdf

Sahami, M., et al. (2011). Setting the Stage for Computing Curricular 2013: Computer Science – Report from the ACM/IEEE-CS Joint Task Force. *Proceedings of the 42nd ACM Technical Symposium on Computer Science Education* (pp. 161-162). Retrieved from http://handbook5.com/s/setting-the-stage-for-computing-curricula-2013-computer-science-e18234-pdf.pdf

Shanghai Jiao Tong University (SJTU), School of Electronic Information and Electrical Engineering. (2011). *Computer science & technology major*. Shanghai, China: Shanghai Jiao Tong University.

Shanghai University (SU), School of Computer Science and Engineering. (2011). *Handbook of undergraduate program*. Shanghai, China: Shanghai Jiao Tong University.

University of California at Berkeley (UCB), College of Engineering. (2012). *Information for current and prospective L&S CS students*. Retrieved from http://www.eecs.berkeley.edu/csugrad/index.shtmll#cs

University of San Diego (USD), College of Arts and Sciences. (2012). *Computer science major*. Retrieved from http://www.sandiego.edu/cas/math-cs/program/cs_major/

Wilson, J. et al. (1988). Computer science education in the People's Republic of China in the late 1980's. *Communications of the ACM*, *31*, 956–964. doi:10.1145/63030.63031

Xian University of Technology (XUT), School of Computer Science and Engineering. (2012). *Undergraduate curriculum plan for computer science and technology*. Xian, China: Xian University of Technology.

Zhang, M., & Lo, V. (2010). Undergraduate computer science education in China. *Proceedings of the 41st ACM Technical Symposium on Computer Science Education* (pp. 396-400). Retrieved from http://fusion.grids.cn/career/attachments/china-SIGCSE2010.pdf

This work was previously published in "International Education and the Next-Generation Workforce" edited by Victor C. X. Wang, pages 208-223, copyright 2014 by Information Science Reference (an imprint of IGI Global).

Chapter 50
Teaching a Socially Controversial Scientific Subject: Evolution

Hasan Deniz
University of Nevada Las Vegas, USA

ABSTRACT

This chapter explores teachers' and students' acceptance and understanding of evolutionary theory by using conceptual ecology (Toulmin, 1972) as a theoretical lens. Demastes, Good, and Peebles (1995) describe the conceptual ecology for evolutionary theory. Acceptance of evolutionary theory is part of this conceptual ecology, and this conceptual ecology also contains the following five components: (1) prior conceptions related to evolution (understanding of evolutionary theory); (2) scientific orientation (degree to which the learner organizes his/her life around scientific activities); (3) view of the nature of science; (4) view of the biological world in competitive and causal terms as opposed to aesthetic terms; and (5) religious orientation. A complex web of connections among components of conceptual ecology for evolutionary theory influences one's acceptance and understanding of evolutionary theory. Therefore, studying the relationship between acceptance and understanding of evolutionary theory as a part of the conceptual ecology for evolutionary theory is more promising than studying acceptance of evolutionary theory in isolation. Moreover, studying acceptance of evolutionary theory as an integral part of the conceptual ecology may enable us to explain why some teachers and students show a high degree of acceptance and others show a low degree of acceptance.

INTRODUCTION

Evolution as a unifying theme in biology education has been supported by major science education policy documents and understanding of evolution has been considered as an important part of scientific literacy (American Association for the Advancement of Science, 1993; National Research Council, 1996). Major science education organizations such as National Association of Biology Teachers (2011) and National Science Teachers Association (2003) in the United States have also supported evolution as a unifying theme in biology. More recently evolution has been identified as one of four disciplinary core ideas in life sciences in "A Framework for K-12 Science Education: Practices,

DOI: 10.4018/978-1-4666-7363-2.ch050

Crosscutting Concepts, and Core Ideas" (NRC, 2012). According to this new framework which will serve as the basis for the next generation of science education standards biological evolution includes four components:

- Evidence of Common Ancestry and Diversity
- Natural Selection
- Adaptation
- Biodiversity and Humans

The framework takes a learning progressions approach and describes what students need to know by the end of grades 2, 5, 8, and 12 for each component. It is clear that evolution continues to be considered as a major or overarching idea in life sciences curriculum, but it is less clear to what extent evolution has been taught in actual classroom settings and how students handle learning the evolution content.

There is a difference between teaching a socially controversial scientific subject and a non-socially controversial scientific subject.

Educators need to consider these five integrated domains when teaching evolution:

- The conceptual domain
- The epistemic domain
- The socio-cultural domain
- The religious domain
- The legal domain

The first three domains are important to consider in teaching other science content as well, but the last two domains become particularly important when it comes to teaching evolution.

The Conceptual Domain

The conceptual domain includes both scientifically accepted major evolutionary ideas and students' alternative conceptions about evolution. It is well known that students hold intuitive conceptions of the natural world and these conceptions are often in conflict with the scientifically accepted conceptions (Driver, 1981; NRC, 2007). After decades of research on misconceptions we now know that students' minds are not *tabula rasa* or *empty vessels*. Students do have alternative conceptions. Students' alternative conceptions need reorganization in order to accommodate the scientifically accepted contemporary views. Many researchers emphasized that students' prior conceptions might interfere with the learning (e.g., Bransford, et al., 1999; Chinn & Brever, 1993; Pintrich, et al., 1993). Pintrich et al. (1993) suggested that prior knowledge can play two contradictory roles during the learning process. They contemplated that prior knowledge can either impede the learning process through students' alternative frameworks, or it can facilitate it by providing students a conceptual basis for evaluating the validity of newly encountered ideas.

Ausubel's frequently quoted statement captures the importance of students' intuitive conceptions during the learning process in a dramatic way. This quote appears before the preface of the book that Ausubel co-authored with Novak and Hanesian (Ausubel, Novak, & Hanesian, 1978).

If had to reduce all of educational psychology to just one principle, it would say this: The most important single factor influencing learning is what the learner already knows. Ascertain this and teach him accordingly.

Common students' alternative conceptions about evolution include the following ideas (Werth, 2012):

- All evolutionary change is adaptive.
- Evolutionary change is progressive.
- Evolutionary change is teleological (goal-directed).

Educators need to be aware that such ideas are common among students when teaching about major evolutionary concepts. Apart from these common misconceptions, many students come to the classroom with a strong tendency to believe the evolutionary theory is not true. For majority of these students, their apprehensive attitude toward evolution is related to factors such as students' religious perspectives, naïve nature of science views, and social-cultural backgrounds. Therefore, evolution instruction that embraces constructivist notions of learning should consider these factors. It is true that public school science classes are not a place to teach religion but disregarding the importance of students' religious perspective while learning about evolution is against the fundamental assumptions of constructivist pedagogy.

Science education community has long been trying to find answers to the following questions. Can students have a robust conceptual understanding of evolutionary theory without accepting its validity? Is there a relationship between understanding and acceptance of evolutionary theory? Researchers reached into different conclusions about the nature of this relationship. Some researchers found no relationship between understanding and acceptance of evolutionary theory (Sinatra, Southerland, McCounaughy, & Demastes, 2003) while others reported a positive relationship (Deniz, et al., 2008, 2011; Rutledge & Warden, 2000). Deniz et al. (2009, 2012) found that students' acceptance of evolutionary theory significantly increases after students improve their understanding of evolutionary theory.

The Epistemic Domain

Nature of science (epistemology of science) refers to values and beliefs specific to scientific knowledge and its development (Lederman, 1992). Students' nature of science views or epistemological beliefs about science determine the rules and criteria used for differentiating what counts as good scientific knowledge. Therefore, students' epistemological beliefs play a crucial role during the learning process.

It was acknowledged that there is no agreed-upon single definition of NOS among philosophers of science, historians of science, scientists, and science educators, but certain aspects of NOS are uncontroversial and relevant to K-16 education (Abd-El-Khalick & Akerson, 2004; Schwartz, Lederman, & Crawford, 2004). These NOS aspects include but are not limited to conceptions that scientific knowledge is empirically-based, tentative, subjective, inferential, socially and culturally embedded, and depends upon human creativity and imagination. Students should have appropriate nature of science understandings before they develop a robust conceptual understanding of any science subject. This is particularly true for developing a deeper scientific understanding about evolutionary theory. Dagher and BouJaoude (1997) found that allowing students to discuss their alternative conceptions about evolution and teaching them about nature of science is more likely to increase their understanding of evolution. Martin-Hansen (2008) reported that nature of instruction reduced the percentage of students feeling apprehensive about learning evolution from 38% to 12% in a first-year college course. Martin-Hansen (2008) also reported that the nature of science instruction did not help some students (Young Earth Creationists) reduce their apprehension level toward learning evolution. Verhey (2005) reported interesting findings with regard to the impact of instruction on understanding of evolutionary theory. Verhey (2005) found that strong emphasis on evolution did not significantly increase student conceptions of evolutionary theory, but comparing "intelligent design" with evolutionary theory, with a special emphasis on nature of science led to extensive change toward scientifically accepted conceptions.

Table 1 describes the nature of science views that students need to develop to be able to place evolutionary theory in its appropriate standing in science.

Staver (1999) reported that Kansas State Board of Education considered an alternative version of nature of science provided by Creation Science As-

Table 1. Description of the nature of science aspects

NOS Aspect	Description
Bounded NOS	Science is a limited way of knowing. Science cannot answer moral and ethical questions. Scientists do not invoke supernatural explanations when doing science.
Empirical NOS	Scientific knowledge is based on empirical evidence. Knowledge claims in science are made with evidence and observations of nature. However, this empirical base does not provide a secure base for science because observations are influenced by scientists' creativity, and personal and theoretical subjectivity.
Inferential NOS	There is a difference between observation and inference. Observations are descriptive statements about the nature that are available to the senses. Observes can agree upon observation statements with relative ease. Inferences are interpretations of observations. They are not immediately available to the senses.
Creative NOS	Creativity and imagination of scientists play a major role in the scientific inquiry. The role of creativity and imagination is not limited to any specific phase of the scientific inquiry. Creativity and imagination are of importance before, during, and after data collection. Creativity and imagination allow scientist to build theoretical models that inferentially explain the natural phenomena.
Subjective NOS	Scientists try to achieve objectivity, but absolute objectivity is not possible in science. Theoretical orientations of scientists make them unavoidably subjective. In addition to scientists' theoretical orientations their personal characteristics and social and cultural backgrounds contributes to subjectivity of scientists. All these factors influence scientists' choice of research questions, methods of research, observations, and interpretations of their observations.
Tentative NOS	Scientific knowledge is tentative but durable. Scientific knowledge is subject to change with the availability of new evidence and with the interpretation of the old evidence, but this change does not happen on the daily basis. Science is not concerned with finding the final truth.
"Scientific Method"	There is not a general and universal scientific method that is followed by all research scientists to solve scientifically oriented questions.
Social and cultural NOS	Science is a human activity. It is influenced by social and cultural factors. These social and cultural factors include social composition, religion, worldview, political and economic factors. Science is not only influenced by these factors but also it influences these factors.
Social NOS	Science is no longer a solitary pursuit. Scientific knowledge is constructed through social negotiation. Despite their individual differences members of a scientific community of practice share common traditions, values, and theoretical frameworks. This social dimension enhances the objectivity of scientific knowledge. The double-blind peer-review process used by scientific journals is a major component of this NOS aspect.
Theory/ Law	There is no hierarchical relationship between theories and laws. Laws are mathematical descriptions of natural phenomena. Theories do no turn into laws. Different usages of the word theory are problematic. In science, theories are extremely well-supported web of hypotheses that are constructed to explain natural phenomena. However, everyday use of the word theory refers to some sort of a wild idea which may or may not have an empirical support.

sociation for Mid-America while deciding whether or not the evolutionary theory should be included in the revised Kansas state science standards. It is interesting to note that both proponents and opponents of evolutionary theory use their own version of nature of science to justify whether or not evolutionary theory should be included in the science education standards. The way in which the nature of science is conceptualized is inextricably connected to the curricular decision making about evolution. For example, if one accepts Creation Science Association for Mid-America version of nature of science evolution does not qualify to be a scientific theory which is worthy of inclusion in the curriculum.

Nature of science items included in a proposed draft of Kansas state standards by the Kansas State Board of Education (Staver 1999).

1. Good science is science that is verifiable, falsifiable, and repeatable.
2. Historic science, which includes the study of past events such as the origin of life and the universe, is not good science because

these ideas are not testable, as the past is not verifiable, falsifiable, or repeatable.

3. Scientific law is considered to be more important than scientific theory.
4. Inductive reasoning is emphasized over and above deductive reasoning, which is downplayed.

Many religious people consider evolution as a form of atheism. They think that acceptance of evolution and belief in God are not compatible. Improving people's nature of science views can help them resolve this unnecessary conflict. Nature of instruction emphasizing the boundaries of science can help certain people come to the conclusion that their acceptance of evolution does not harm their religious beliefs. Science is a limited way of knowing. Science cannot answer questions about the supernatural. Another important nature of science understanding that can help religious people reduce their apprehensive attitude toward evolution is recognizing the difference between methodological and philosophical materialism. Scientists operate according to methodological materialism. In other words, they do not invoke supernatural explanations when doing science and explaining natural phenomena. Although scientists may or may not have personal religious beliefs they have to be silent about the existence of supernatural when doing science. Contrary to the methodological materialism, proponents of the philosophical materialism do not accept supernatural. In other words, philosophical materialists deny the existence of God and therefore they are atheists. All scientists are methodological materialists but not all scientists are philosophical materialists. In fact, some scientists accept the theistic evolution position.

The Socio-Cultural Domain

Lerner (2000) examined how different states within the United States treated evolution in their state science education standards. Lerner (2000) reported that only 10 states have very good-excellent evolution education standards, 14 states have good standards, 7 states have satisfactory standards, 6 states have unsatisfactory standards, and 13 states have standards that are worse than unsatisfactory. Evolution education is vulnerable to social influences more than any other subject in science. The poor condition of evolution education standards in many states can be interpreted in light of attitude surveys toward evolution. According to surveys administered by Gallup in 2010 and 2012, forty and forty six percent of Americans respectively endorsed the view that "God created human beings pretty much in their present form at one time within the last 10,000 years or so." Thirty-eight and thirty-two percent agreed that "human beings have developed over millions of years from less advanced forms of life, but God guided this process." Sixteen and fifteen percent agreed that "human beings have developed over millions of years from less advanced forms of life, but God had no part in this process," and six and seven percent respectively did not report any opinion in 2010 and 2012. Percentage of people supporting Young Earth Creationist view increased from 40% to 46% and percentage of people supporting theistic evolution position decreased from 38% to 32% within the last two years. Percentage of people supporting evolution without any involvement from God remained pretty much the same within the last two years. However, examination of the poll results between 1982 and 2012 indicates no dramatic change within the last thirty years.

Berkman and Plutzer (2010) conducted a comprehensive study titled "National Survey of High School Biology Teachers." They used three positions endorsed by the National Academy of Sciences (1998, 2008) to frame their study:

- There is no debate whether evolution occurred within the scientific community,
- Evolution is necessary to understand other topics in biology, and

- Evolution should be used a unifying theme in teaching biology.

Berkman and Plutzer (2010) turned these three positions into Likert-type survey questions:

1. When I teach evolution (including answering student questions) I emphasize the broad consensus that evolution is a fact, even as scientists disagree about the specific mechanisms through which evolution occurred.
2. It is possible to offer an excellent general biology course for high school students that includes no mention of Darwin or evolutionary theory.
3. Evolution serves as the unifying theme for the content of the course.

These questions were administered to a national sample of 926 biology teachers. According to the study, 74% of teachers agreed with the first question, 21% of teachers disagreed with the question, and 5% did not respond. Twelve percent of teachers agreed with the second question, 83% of teachers disagreed with the second question, and 5% did not offer any answer. As for the third question, 62% of biology teachers agreed with the third question, 34% of biology teachers disagreed, and others did not respond. As it can be understood from these numbers a significant percentage of American biology teachers remains aloof to evolution despite the endorsements from AAAS (1993), NRC (1996), NABT (2011), and NSTA (2003).

Miller, Scott, and Okamoto (2006) compared the acceptance of evolutionary theory among adults in 34 countries. Countries such as Iceland, Denmark, Sweden, France, Japan, and United Kingdom have 75 percent or more acceptance rates of evolutionary theory. Countries such as United States and Turkey have acceptance rates of 40 percent and 25 percent respectively. If a concept has little leverage within a cultural milieu, it will not be readily acceptable and it will be difficult for that concept to be included in the school curriculum. Costa (1995) stated that successful transition of students from their own world to school science depends on the compatibility of family and school cultures.

Deniz et al. (2008) found that Turkish preservice biology teachers whose parents have more education are more likely to accept evolution as a scientifically valid theory. This makes sense considering the fact Turkish education system is historically modeled based on Western education principles especially after the declaration of Republic of Turkey in 1923.

According to Aikenhead and Jegede (1999) when the culture of school science is in sync with students' social and cultural values science instruction tend to happen smoothly. However, if there is a conflict between the culture of science and students' socio-cultural values science instruction tends to damage students' socio-cultural values by forcing them to abandon their indigenous values. For this reason, Aikenhead and Jegede (1999) called for developing culturally sensitive curricula and teaching methods to be able to avoid the clash between students' cultural values and the culture of Western science. Cobern (1996) also underscored this point by stating that affective and contextual factors are evoked more strongly when a person struggles to learn a topic which does not have much leverage within that person's worldview.

The Religious Domain

Researchers reported negative relationships between religious orientation and acceptance of evolutionary theory. Grose and Simpson (1982) found that students in an introductory college biology class who perceived that their church generally influenced their thought had lower acceptance of evolutionary theory. Osif (1997) reported that that religiosity of biology teachers is positively correlated with the rejection of the evolutionary theory. Miller, Scott, and Okamoto (2006) found that religious beliefs were almost twice as much predictive of attitude toward evolution in the United States compared to European countries. Obviously, in both the United States and Europe, the correlation between religious beliefs and attitude toward evolution was negative, but this correlation was much bigger in the United States. This indicates that religious beliefs are negatively correlated with attitude toward evolution but the strength of the correlation varies in different countries. Deniz et al. (2011) found that preservice Turkish biology teachers' acceptance of evolutionary theory is negatively correlated with their religiosity. They also found that preservice teachers with lower acceptance of evolutionary theory and strong religious beliefs are less likely to teach evolution in their future classrooms.

Table 2 describes how students' religious beliefs may influence their learning about evolution and their acceptance-rejection of evolutionary theory.

Table 2. Students' learning approach and their evolution rejection-acceptance position according to their religiosity

	Learning Approach	Evolution Rejection-Acceptance Position
Extremely Religious	Students prefer to stick to their religious values. They are alienated from the culture of science and they reject to learn science.	**Young Earth creationism**: Proponents of this position hold literal interpretations of Bible and they think that Earth is less than 10,000 years old. They reject scientific findings with regard to the age of Earth and biological evolution.
	Students prefer to stick to their religious values but at the same time, they are not fully alienated from the culture of science. Prior conceptions of students in this category preclude them from accepting evolutionary theory.	**Old Earth creationism**: Proponents of this position do not interpret six days of Biblical creation as twenty-four hours but as longer time periods. They accept other modern science but they reject biological evolution. **Intelligent design creationism**: Proponents of intelligent design holds the view that the evolution of organs such as vertebrate eye is too complex to be explained naturally and this complexity requires an intelligent designer. They do accept natural selection but they do not accept that mutation and natural selection can explain evolution of species.
	Students prefer to stick to both their religious values and the culture of science. Students develop two conceptual frameworks one for school science and one for everyday experiences. At first, students in this category consciously or unconsciously keep two conceptual frameworks separate. Students will activate one conceptual framework or another depending upon the context. Later, students in this category can make a concerted effort to reconcile these two conceptual frameworks. Prior conceptions of students in this category may preclude them from learning evolution.	**Theistic evolutionism**: Proponents of this position think that God creates through the natural processes and mechanisms. They do accept that natural selection and mutation can explain evolution of species. They are methodological materialists but not philosophical materialists.
	Students prefer to assimilate into the culture of science and they are silent about the existence of God. Prior conceptions of students in this category will not preclude them from learning evolution. The extent of their learning will be determined by the quality of instruction and their personal characteristics such as motivation and metacognitive awareness.	**Agnostic evolutionism**: Proponents of this position accept evolution but they neither reject nor accept the existence of God. They are methodological materialists but they avoid accepting or rejecting philosophical materialism.
Not Religious	Students prefer to assimilate into the culture of science and they reject the existence of God. Prior conceptions of students in this category will not preclude them from learning evolution. The extent of their learning will be determined by the quality of instruction and their personal characteristics such as motivation and metacognitive awareness.	**Materialistic evolutionism**: Proponents of this position accept evolution and they reject the involvement of God with evolution. They are both methodological and philosophical materialists.

The Legal Domain

Who should determine the curriculum in public schools? Should the citizens as taxpayer decide what to teach in schools? Should educators as professionals decide what to teach? Should this decision be left to scientific community? The controversy over who should determine the curriculum in public schools was historically being tackled by the judicial system in the United States.

Table 3 shows that The United States Supreme Court has consistently ruled in favor of evolution.

Public opinion polls consistently indicated that people support teaching alternatives such as creationism and intelligent design to evolution in the United States (Plutzer & Berkman, 2008). It seems like alternatives to teaching evolution were beaten in the courtrooms, but the public desire to introduce these alternatives into the curriculum is still strong. Similarly, such desires are also present even in Britain, which enjoys about 75% evolution acceptance rate according to the international study conducted by Miller, Scott, and Okamoto (2006). According to a survey conducted by Ipsos MORI (2006) for the British Broadcasting Corporation (BBC), when participants are allowed to pick one or more choices among creationism, intelligent design, and evolution, 44% of the participants said that creationism should be taught in science lessons in British schools, 41% indicated that they would like to see intelligent design taught in schools, and 69% wanted evolution as part of science curriculum.

CONCLUSION

Students' learning about evolution can be conceived in two different ways: *learning as conceptual change* and *learning as cultural border crossing*.

According to *learning as conceptual change* approach, students are supposed to move from their alternative conceptions toward scientifically accepted conceptions. *Learning as conceptual change* model supports replacing students' alternative conceptions with the scientifically accepted conceptions.

Table 3. Sample court cases about evolution in the United States

Date	Case	Detail	Final Ruling
1925	Tennessee vs. Scopes	Scopes, a science teacher, admitted using a text dealing with human evolution. This was in violation with Tennessee law.	The Tennessee Supreme Court decided that as a public employee it was teachers' responsibility to teach whatever the state determines. The case was not taken to the United States Supreme Court.
1968	Epperson vs. Arkansas	Epperson, tenth grade biology teacher, adopted a textbook including a chapter about evolution. This was in violation of a state law that made illegal to teach human evolution in a public school or university. Arkansas Charcery Court decided that school teachers should enjoy a substantial degree of academic freedom. The Arkansas State Supreme Court overturned the ruling.	The United States Supreme Court ruled in favor of Epperson.
1987	Edwards vs. Aguillard	Aguillard challenged a Lousiana law that required the teaching of creation science along with evolution.	The United States Supreme Court found the law in violation of the First Amendment.
2005	Kitzmiller vs. Dover Area School District in Pennsylvania	The school board of Dover Area School District introduced intelligent design in the curriculum through a mandated disclaimer to be read by teachers. This was challenged by a group of science teachers and some parents.	District Court for the Middle District of Pennsylvania ruled that introducing creationism and intelligent design in the curriculum is unconstitutional because creationism and intelligent design are religion not science.

This type of learning can be represented as

$C_1 \rightarrow C_2$.

Posner et al. (1982) described the conditions required for conceptual change:

1. **Dissatisfaction:** This condition requires that the learner fails to make sense of some event with his/her existing conception.
2. **Intelligibility:** This condition necessitates that the learner has some understanding of the new conception.
3. **Plausibility:** This condition is satisfied when the learner accepts the new conception.
4. **Fruitfulness:** This condition emphasizes that the learner should be able to use the new conception to explain novel situations as well as the situations that were formerly explained by the new conception.

As it can be understood from this description, learning process is unidirectional in this model. This conceptual change model was revised in response to criticisms (e.g., Cobern, 1996; Solomon, 1987). The revised model acknowledged the importance of intuition, emotion, motivation, and social factors (Strike & Posner, 1992). However, the model still stayed as unidirectional at its core.

Aikenhead and Jegede (1999) criticized *learning as conceptual change* model because the model aims to assimilate students into the culture of Western science at the expense of their indigenous culture. Aikenhead and Jegede (1999) conceptualized the transition between a student's everyday concepts and scientific concepts as *cultural border crossing*. According to this model, students can construct both indigenous and scientific conceptions of natural phenomena side by side and they can make transitions from indigenous conceptions to scientific conceptions and vice versa depending upon the context.

This type of learning can be represented as

$C_1 \leftarrow\rightarrow C_2$.

It is reasonable to expect that biology teachers' acceptance of evolutionary theory will influence how they treat evolution in their own classes. Biology teachers who accept evolutionary theory are more likely to use evolution as an overarching theme in their teaching. In fact, a significant number of studies indicated that acceptance of evolutionary theory is a good predictor of instructional approach taken toward evolution (Aguillard, 1999; Eve & Dunn, 1990; Shankar & Skoog, 1993).

Berkman and Plutzer (2010) reported that there is a correlation between the adoption of the NAS positions and the instructional time spent on evolution. They found that biology teachers who strongly support all three positions spent 20 hours a year on evolution, whereas biology teachers who do not support these positions spent only 11 hours on evolution.

Learning is a complex process. Controversial nature of teaching evolution adds to this complexity making the learning process even more complicated. Teaching evolution requires paying attention to all five domains described in this chapter. These domains altogether shape how students make sense of evolutionary theory in biology classrooms. Conceptual, epistemic, socio-cultural, religious, and legal factors influence the learning process in both conceptualizations of learning. It is up to biology teachers to decide what conceptualization of learning can help their students learn evolution in a meaningful way.

REFERENCES

Abd-El-Khalick, F., & Akerson, V. L. (2004). Learning as conceptual change: Factors that mediate the development of preservice elementary teachers' views of nature of science. *Science Education*, *88*(5), 785–810. doi:10.1002/sce.10143

Aguillard, D. (1999). Evolution education in Louisiana public schools: A decade following Edwards v. Aguillard. *The American Biology Teacher*, *61*(3), 182–188. doi:10.2307/4450650

Aikenhead, G. J., & Jegede, O. J. (1999). Cross-cultural science education: A cognitive explanation of a cultural phenomenon. *Journal of Research in Science Teaching*, *36*(3), 269–287. doi:10.1002/(SICI)1098-2736(199903)36:3<269::AID-TEA3>3.0.CO;2-T

American Association for the Advancement of Science. (1993). *Benchmarks for science literacy: A project 2061 report*. Oxford, UK: Oxford University Press.

Ausubel, D. P., Novak, J. D., & Hanesian, H. (1978). *Educational psychology: A cognitive view* (2nd ed.). New York, NY: Holt, Rinehart, and Winston.

Berkman, M. B., & Plutzer, E. (2010). *Evolution, creationism, and the battle to control America's classrooms*. Cambridge, UK: Cambridge University Press. doi:10.1017/CBO9780511760914

Bransford, J., Brown, A., & Cocking, R. (Eds.). (1999). *How people learn: Brain, mind, experience, and school*. Washington, DC: National Academy Press.

Chinn, C. A., & Brewer, W. F. (1993). The role of anomalous data in knowledge acquisition: A theoretical framework and implications for science instruction. *Review of Educational Research*, *63*, 1–49.

Cobern, W. (1996). Worldview theory and conceptual change in science education. *Science Education*, *80*(5), 579–610. doi:10.1002/(SICI)1098-237X(199609)80:5<579::AID-SCE5>3.0.CO;2-8

Costa, V. B. (1995). When science is another world: Relationships between worlds of family, friends, school, and science. *Science Education*, *79*(3), 313–333. doi:10.1002/sce.3730790306

Dagher, Z. R., & BouJaoude, S. (1997). Scientific views and religious beliefs of college students: The case of biological evolution. *Journal of Research in Science Teaching*, *34*(5), 429–445. doi:10.1002/(SICI)1098-2736(199705)34:5<429::AID-TEA2>3.0.CO;2-S

Deniz, H., Cetin, F., & Yılmaz, I. (2011). Examining the relationships among Turkish preservice biology teachers' acceptance of evolution, religiosity and teaching preference for evolution. *Journal of Science Education and Technology, 31*(4), 2.1-2.8

Deniz, H., Donnelly, L., & Yilmaz, I. (2008). Exploring the factors related to acceptance of evolutionary theory among Turkish preservice biology teachers: Toward a more informative conceptual ecology for biological evolution. *Journal of Research in Science Teaching*, *45*(4), 420–443. doi:10.1002/tea.20223

Deniz, H., Shrader, P. G., & Keilty, J. (2012). *Impact of evolution instruction on understanding and acceptance of evolutionary theory and the nature of relationships among understanding, acceptance, and religiosity.* Paper presented at the Annual Meeting of National Association for Research in Science Teaching. Indianapolis, IN.

Driver, R. (1981). Pupils' alternative frameworks in science. *European Journal of Science Education*, *3*(1), 93–101. doi:10.1080/0140528810030109

Eve, R. A., & Dunn, D. (1990). Psychic powers, astrology & creationism in the classroom? *The American Biology Teacher*, *52*(1), 10–20. doi:10.2307/4449018

Gallup (2012). *Evolution, creationism, intelligent design*. Retrieved from http://www.gallup.com/poll/21814/evolution-creationism-intelligent-design.aspx

Grose, E. C., & Simpson, D. (1982). Attitudes of introductory college biology students toward evolution. *Journal of Research in Science Teaching*, *19*(1), 15–23. doi:10.1002/tea.3660190103

Ipsos, M. O. R. I. (2006). *BBC news*. Retrieved from http://news.bbc.co.uk/2/hi/science/nature/4648598.stm

Lederman, N. G. (1992). Students' and teachers' conceptions about the nature of science: A review of the research. *Journal of Research in Science Teaching*, *29*(4), 331–359. doi:10.1002/tea.3660290404

Lerner, L. S. (2000). *Good science, bad science: Teaching evolution in the states*. Washington, DC: Thomas B. Fordham Foundation.

Martin-Hansen, L. M. (2008). First-year college students' conflict with religion and science. *Science & Education*, *17*(4), 317–357. doi:10.1007/s11191-006-9039-5

Miller, J. D., Scott, E. C., & Okamoto, S. (2006). Public acceptance of evolution. *Science*, *313*, 765–766. doi:10.1126/science.1126746 PMID:16902112

National Academy of Science. (1998). *Teaching about evolution and the nature of science*. Washington, DC: National Academy Press.

National Academy of Science. (2008). *Science, evolution, and creationism*. Washington, DC: National Academy Press.

National Association of Biology Teachers. (2011). *NABT's statement on teaching evolution*. Retrieved from http://www.nabt.org/websites/institution/index.php?p=92

National Research Council. (1996). *National science education standards*. Washington, DC: National Academy Press.

National Research Council. (2007). *Taking science to school: Learning and teaching in grades K-8*. Washington, DC: National Academy Press.

National Research Council. (2012). *A framework for K-12 science education: Practices, crosscutting concepts, and core ideas*. Washington, DC: National Academy Press.

National Science Teachers Association. (2003). *NSTA position statement: The teaching of evolution*. Retrieved from http://www.nsta.org/pdfs/positionstatement_evolution.pdf

Osif, B. A. (1997). Evolution and religious beliefs: A survey of Pennsylvania high school teachers. *The American Biology Teacher*, *59*(9), 552–556. doi:10.2307/4450382

Pintrich, P. R., Marx, R. W., & Boyle, R. A. (1993). Beyond cold conceptual change: The role of motivational beliefs and classroom contextual factors in the process of conceptual change. *Review of Educational Research*, *63*(2), 167–199.

Plutzer, E., & Berkman, M. B. (2008). Evolution, creationism, and the teaching of human origins in schools. *Public Opinion Quarterly*, *72*(3), 540–553. doi:10.1093/poq/nfn034

Posner, G. J., Strike, K. A., Hewson, P. W., & Gertzog, W. A. (1982). Accommodation of a scientific conception: Toward a theory of conceptual change. *Science Education*, *66*(2), 211–227. doi:10.1002/sce.3730660207

Rutledge, M. L., & Warden, M. A. (2000). Evolutionary theory, the nature of science & high school biology teachers: Critical relationships. *The American Biology Teacher*, *62*(1), 23–31. doi:10.1662/0002-7685(2000)062[0023:ETTNOS]2.0.CO;2

Schwartz, R. S., Lederman, N. G., & Crawford, B. A. (2004). Developing views of nature of science in an authentic context: An explicit approach to bridging the gap between nature of science and scientific inquiry. *Journal of Research in Science Teaching*, *88*(4), 610–645.

Shankar, G., & Skoog, G. (1993). Emphasis given evolution and creationism by Texas high school biology teachers. *Science Education*, *77*(2), 221–233. doi:10.1002/sce.3730770209

Sinatra, G. M., Southerland, S. A., McConaughy, F., & Demastes, J. W. (2003). Intentions and beliefs in students' understanding and acceptance of biological evolution. *Journal of Research in Science Teaching*, *40*(5), 510–528. doi:10.1002/tea.10087

Solomon, K. (1987). Social influences on the construction of pupils' understanding of science. *Studies in Science Education*, *14*, 63–82. doi:10.1080/03057268708559939

Staver, J. R. (1999). When public understanding of science thwarts standards-based science education. *Electronic Journal of Science Education, 3*(4). Retrieved from http://ejse.southwestern.edu/article/viewArticle/7613/5380

Strike, K., & Posner, G. J. (1992). A revisionist theory of conceptual change. In R. A. Duschl & R. J. Hamilton (Eds.), *Philosophy of Science, Cognitive Psychology, and Educational Theory and Practice* (pp. 147–176). New York, NY: State University of New York.

Verhey, S. D. (2005). The effect of engaging prior learning on student attitudes toward creationism and evolution. *Bioscience*, *55*(11), 991–1003. doi:10.1641/0006-3568(2005)055[0996:TEOEPL]2.0.CO;2

Werth, A. (2012). Avoiding the pitfall of progress and associated perils of evolutionary education. *Evolution: Education & Outreach*, *5*(2), 249–265. doi:10.1007/s12052-012-0417-y

This work was previously published in "Approaches and Strategies in Next Generation Science Learning" edited by Myint Swe Khine and Issa M. Saleh, pages 52-63, copyright 2013 by Information Science Reference (an imprint of IGI Global).

Chapter 51
The Role of Authentic Science Research and Education Outreach in Increasing Community Resilience:
Case Studies Using Informal Education to Address Ocean Acidification and Healthy Soils

Cynthia Hall
West Chester University, USA

Joniqua Howard
University of Puerto Rico, Puerto Rico

Regina Easley
University of South Florida, USA

Trina Halfhide
University of South Florida, USA

ABSTRACT

Active, multi-dimensional learning is needed to establish higher-level scientific inquiry. Researchers who are engaged in scientific discovery are a valuable resource to communicate the link between science, society, and sustainability. Nontraditional settings like faith-based organizations and hobbies can play an important role in fostering greater scientific understanding. This chapter highlights the role that community structure (social, racial, and economic demographics) plays in developing successful project components by considering various theoretical frameworks to communicate sustainability principles to underserved communities. The researchers in these case studies presented the topics of ocean acidification and healthy soil to inner-city communities in Tampa, FL and Philadelphia, PA by utilizing authentic science research activities. Learners maximized the opportunities to construct new hypotheses and improve decision-making related to environmental stewardship behaviors and food security issues. A secondary but transformative outcome was increased interest in STEM fields among youth in cities with traditionally low performing schools.

DOI: 10.4018/978-1-4666-7363-2.ch051

ORGANIZATIONAL BACKGROUND

Education is critical for achieving environmental and ethical awareness, values and attitudes, skills and behaviour consistent with sustainable development and for effective public participation in decision-making. Both formal and non-formal education are indispensable to... sustainable development, United Nations Agenda 21: Chapter 36

Education plays a vital role in developing sustainable healthy communities with well-informed citizenry. The years 2005-2014 have been termed by the United Nations as the Decade of Education for Sustainable Development with an overarching mission "to integrate the principles, values, and practices of sustainable development into all aspects of education and learning" (UNCED, 1992). The overall goal of the initiative is to involve individuals and stakeholders (i.e. youth, educators, media, business owners) in collectively improving the global quality of life by establishing economic, societal, environmental, and political cohesion (UNCED, 1992). It is essential to recognize the challenges of implementing such an ambitious plan. For educators and citizens in the United States, a major hurdle is reconciling the current state of the U.S. education system into a system that effectively brings an understanding of sustainable development (both in definition and implications) and a capacity to equip emerging leaders with tools to solve the challenge of reducing global poverty, restoring and maintaining healthy natural resources, and ensuring access to political and economic stability.

A major challenge within the United States' public school system stemming from federal policy (United States Department of Education) is the increased focus on standards-based reforms and choice-testing as a means of performance evaluation. Within the current context of performance-based testing, sustainability concepts are rarely addressed (Rauch, 2002). Educators and school officials have limited time and resources to expound on topics which are not directly related to the content of the state-issued standardized tests and therefore introducing sustainable development concepts into the curriculum often becomes unfeasible (Au, 2007). Moreover, the emphasis on testing requires students to recite facts rather than develop the critical thinking skills necessary for basic living and scientific literacy (Au, 2007). In the current setting where standardized testing is a reality, creative approaches and partnerships are needed to provide opportunities for students to fully develop their capacity to solve broad local, regional and/or global problems (Kim, 2008). It is essential to address the current concerns about the U.S. education system from both the student and the educator perspective.

For students, the current system of education can perpetuate student disengagement from the learning process particularly among low income and minority students (Kodrzycki, 2002). The problem is exacerbated by inadequately trained teachers which lack specialization and/or real-world experience in the subject areas that they teach (Loucks-Horsley & Matsumoto, 1999; Roehrig & Luft, 2006) and rigidity in the curriculum that does not reflect the natural dynamic process of learning. In this environment, it can be challenging for educators to fully accommodate students with varying interests and learning styles. Additionally, overly-structured education models limit both student and educator creativity and imagination (Kim, 2011) thereby hindering the critical thinking process (Lloyd, 1999). There is often a perceived relationship between school outcomes and the availability of funding for underserved urban schools, in particular (Hanushek, 1997; Hedges, Laine, & Greenwald, 1994). Even though per pupil expenditure, class size, and access to technology are all factors which can influence student performance (Valadez & Durán, 2007), it is equally important to assess the role pedagogy, motivation (of both the educator and the student) and empowerment play in cultivating an academic culture which fosters student achievement.

Several studies have examined a direct link between reading engagement and minority student achievement (Cummins, 2007; Sealey-Ruiz, 2011) and interest in scientific fields both in the United States and abroad (McConney, Oliver, Woods-McConney, & Schibeci, 2010). For example, based on the observations of McConney et. al (2010), even though Australian indigenous students have a generally higher interest in scientific inquiry, their lower literacy and reading comprehension rates preclude them from pursuing degrees at the secondary school level in STEM (science, technology, engineering, and mathematics) fields. Cummins (2007) addresses issues related to literacy by suggesting that active involvement is needed to teach literacy since it is the basic foundation for education. He further notes students who are disengaged from reading rarely develop the vocabulary needed to master the scientific content. Thus if change in our society is to occur, we must evaluate and implement effective teaching and outreach practices to include culturally and socially (Taines, 2010) relevant inquiry-based learning, authentic science, and community based participatory programs which introduce students to scientific investigations and thus terminology in context (Strickland-Dixon, 2011).

Informal, or outside of the classroom, education initiatives in inner-cities have become increasingly important as public and charter schools have faced the challenges of larger class sizes, overworked teachers and staff, increased rates of violence in and around the school, and decreased budgets (Dee & Fu, 2004; Geske, Davis, & Hingle, 1997; Milam, Furr-Holden, & Leaf, 2010). Informal educational settings provide students and community members with access to educational and cultural activities that may not be readily available through the public school (Belle, 1982; Martin, 2004). After-school programs, religious institutions, and community centers are common forums which urban youth can access a variety of programs to extend learning outside of the classroom.

Globally, it is recognized that as development increases, changes in natural resources and the human environment are challenging ways in which we manage national security, economic, and human health issues (Lubchenco, 1998). It has become quintessential to pursue sustainable practices that will not further compromise current resources to ensure that future generations meet their own needs. These goals can be established through a horizontal relationship using the five pillars of sustainability described by McConville and Mihelcic (2007): economical, societal, cultural, environmental, and political cohesion. Education of sustainability concepts is the cornerstone to joining all five sectors to secure a better future. Informal education can be used to help develop sustainable, healthy communities by:

1. Building essential relationships to address public policy concerns;
2. Broadening the public's scientific literacy;
3. Promoting healthy living and food security;
4. Protecting natural resources and the environment; and
5. Strengthening cultural practices.

This chapter will outline ways in which authentic science can be used to introduce sustainability concepts through informal education. We will first focus on the theoretic construct of the education system and will then highlight case studies which served to either help improve community aesthetics, increase community pride, and/or raise awareness of topics related to sustainability. Through these examples the investigators were able to optimize each situation for the benefit of the learner by relating the new information with a past experience. The informal setting benefitted the learners by directly addressing a sustainability issue in a comfortable and culturally relevant atmosphere thereby fostering organic exchanges between parent and child, between friends, and even among strangers.

What the future holds in store for individual human beings, the nation, and the world depends largely on the wisdom with which humans use science and technology. And that, in turn, depends on the character, distribution, and effectiveness of the education that people receive. -Science for All Americans, (AAAS, 1990).

Introducing a concept such as sustainability can be daunting, even for the academically-privileged. Sustainability denotes a systematic change to an activity or habit that has been deemed conventionally wasteful. Such changes may be perceived as adopting an alternative energy source for homes or designing a city with improved waste and resource management. For long-term planning, the semantics of terming processes as sustainable, earth-friendly, or the more accessible term "green" (Yanarella, Levine, & Lancaster, 2009), are less important than the adoption of behaviors that minimize waste of resource, money, and/or time. It should be understood that such practices not only help preserve the environment but are economically beneficial for individuals, businesses, and governments. Education is essential to communicate the ideals of sustainability such that it becomes a readily accessible and realistic lifestyle for everyone.

Sustainability research spans a broad range of scientific fields including engineering, chemistry, economics, and the earth sciences. For example, discourse on fossil fuel consumption and carbon dioxide emissions relate to the combustion of organic substances within the context of chemistry but also have strong implications in the fields of economics and public policy. Similarly, the development of alternative energy sources such as solar power involve an understanding of energy transfer processes which necessitate a background in physics, engineering, and materials science. One of the biggest challenges to students at both the high school and college level is recognizing practical applications to abstract concepts covered in general science courses and understanding the concepts at a broad system level (Ben-Zvi-Assaraf & Orion, 2010). Scientists who are actively involved in research on any given sustainability issue can effectively help communicate the context and relevance of their research to students and community members. Their efforts will be magnified with the incorporation of effective teaching practices and a targeted goal of increasing scientific and sustainability literacy.

Introducing Sustainability Concepts to Communities with Inquiry Based Learning and Authentic Science

An extensive body of pedagogical literature dedicated to understanding student learning styles and methods of evaluation to assess learning exists, however there is no single approach that can meet the needs of all students globally. Science learning can be facilitated not only by educators, but caregivers (Crowley, Callanan, Tenenbaum, & Allen, 2001), peers, mentors (Haney, Lumpe, & Czerniak, 2003), and community members (Roth & Lee, 2004) all play a critical role in developing scientific acumen within students. Individuals thrive in environments that acknowledge their experiences, cultural/spiritual beliefs, and community (Howard, 2010). According to Simpson (2002) indigenous leaders advocate education to answer the questions of injustices that they often encounter within their communities but the education curriculum does not appropriately include the language, content, perspectives, and philosophies of the indigenous culture. For an effective informal education plan, it is necessary to incorporate all sectors of society to build a holistic approach.

Authentic science research is a multi-dimensional learning approach which engages students through interactive activities to develop higher-level scientific understanding and inquiry. Research oriented programs can be implemented at all levels of education and within the public sector (McCall & Minang, 2005) through formal curriculum, community based participatory programs (García & Brown, 2009; Venville, Sheffield, Rennie, & Wallace, 2008) and research assistantships, for example (Zion et al., 2004). Students are transformed into "researchers" by experiencing real-world representations of scientific enterprise (Elmesky & Tobin, 2005; Taines, 2011) and in some instances increased success with standardized testing has been noted (Geier et al., 2008). Authentic science driven activities involve task-directed responsibilities which are often developed and/or implemented by well-skilled scientific professionals and with a defining feature of open-ended inquiry (Jime´nez-Aleixandre, 2002). Researchers have found that youth can be effective/positive initiators of change particularly when participatory and active learning approaches are used (Cammarota & Fine, 2008).

Authentic science activities should inherently increase the participant's scientific literacy particularly if the educator is aware of the process through which scientific understanding evolves. Scientific literacy is a widely defined concept whereby the goal of the educator is to help cultivate a body of conscious learners that are able to fuse together scientific concepts with daily life activities and technologies to make well-informed decisions (AAAS, 1990; Nelson, 1999). With any discussion aimed at increasing scientific literacy, it is essential to emphasize the purpose, methodologies, and the limitations of the scientific process. Table 1 highlights fundamental concepts which can be included in any scientific outreach activity to promote sustainability (IOC/UNESCO, IMO, FAO, & UNDP, 2011), scientific (Colucci-Gray, Camino, Barbiero, & Gray, 2006), and ocean literacy (Steel, Smith, Opsommer, Curiel, & Warner-Steel, 2005; Walker & Keener-Chavis, 2005).

Table 1. Goals of science, ocean, and sustainability literacy in relation to concepts in ocean acidification

Broad Concepts	Science Literacy	Ocean Literacy	Sustainability Concepts	Ocean Acidification Concepts
Science	Understand the nature, goals, and limits of science.*	The ocean is a unique, defining feature shaping habitats and providing life-sustaining water.#	Understand rates of natural recharge and spatial scales of global processes.	Oceanic inorganic carbon cycle in seawater.
	Understand the process of collecting and interpreting numerical data.*	Land features (sand and sedimentary rocks) originate from the ocean. The ocean influences the Earth's geologic features.		Mineral formation (calcite, aragonite, and magnesium calcites).
Technology	Understand the use and limitations of technology in the real-world scientific research process.*	Only 5% of the ocean has been explored; its vitally important resources can be understood through the use of new technologies.	Understand the limits of technology to solve human induced problems.	pH and technologies used to measure pH in the ocean and their limitations.
Governance	Knowledge needed for intelligent participation in science-based social issues.**	An understanding of ocean governance.	Concept of the global commons. Public vs. Private ownership of common resources.	State and federal agencies help monitor and manage ocean resources.
Habitat		Understand the ocean is the largest of source of oxygen in the atmosphere.	Understand the extent of Earth's history and that humans are an integral part of the ecosystem. The Earth is a materially closed system.	Marine food webs and role of primary producers as food sources and oxygen producers.
		Diversity and distribution of life in the ocean is influenced environmental factors.		

Table 1. Continued

Broad Concepts	Science Literacy	Ocean Literacy	Sustainability Concepts	Ocean Acidification Concepts
Economics	Understand the risk and benefits of science.**	Understand the economic and societal value of ocean resources.	Fossil fuel consumption and carbon emissions affect ocean ecosystems and economics.	Business/scientific partnerships can minimize the economic impact of ocean acidification.
Human Need and Influence	Scientists can be an information source on issues relating to science and technology.*	Understand our dependence on the ocean for life (water source, food source, recreation, and influencing weather).	Recognition of human capacity to make large, long-lasting impacts on the global ecosystem.	Systems view of ocean-atmosphere interactions and the influence of human activity on geologic, biological, physical, and chemical phenomena in ocean ecosystems.
Systems Thinking	An appreciation of the inter-relationships between science, technology, and society.*	Understand the ocean's influence on weather and climate (i.e., ocean circulation, El Niño, and heat transfer) and as a dominant driver in the Earth's water and carbon cycles.#	Understanding global feedback systems and interconnectedness.	
Responsibility/ Action	Humans can use scientific knowledge to solve problems.**		Understanding of humans responsible for maintaining a healthy environment.	

Ref. *Thomas and Durant (1987); **Norris and Phillips (2003); #Steel et al. (2005); Sustainability Concepts (Zen & Palmer, 1999)

Addressing the Positionality and Role of the Scientist

Another significant outcome of informal education initiatives is the development of authentic, trusting relationships between scientists and the public (Grorud-Colvert, Lester, Airamé, Neeley, & Gaines, 2010; Minkler, 2005). Although the public generally views scientists as trustworthy and knowledgeable of environmental and community health related problems, there is often a disconnection between scientists and the general public (Rhoads, Wilson, Urban, & Herricks, 1999; Wynne, 2006). Scientists publish their data in scientific journals, present recent findings at scientific meetings, and tend to network, both socially and professionally, with other scientists and academics. The presented case studies exemplify the ability of scientists to use authentic science to transcend existing barriers in comfortable and informal settings. Both of these examples have proven to impact youth by increasing their comfort level of specific scientific concepts and possibly influencing their future career choices.

While scientists, in these case studies are bringing a certain level of knowledge to their audience, it is essential to acknowledge that an authentic and transformative education experience should not be a one-way exchange of information (Santos, 2009). The community can serve as a great resource to the science professional (Cope, McLafferty, & Rhoads, 2011). For example, researchers have noted that Inuit populations have an unsurpassable understanding of sea ice conditions which should not be overlooked by scientists who are just beginning to embark on research efforts in the Arctic Ocean (Laidler & Elee, 2008; Laidler, Elee, Ikummaq, Joamie, & Aporta, 2010).

The participating scientist should never overlook or underestimate the intrinsic value of the community (Barton, 2001). Insight into the processes of scientific inquiry, relationship building, and information transfer (Matthews, 2011) can be gained from community members. Education models which value participant experience and viewpoints not only increase motivation within the learner but increase community empowerment (Fetterman, 1994). Through the process of equal exchange, community transformation is possible.

CASE STUDIES

The Role of Educational Outreach on Food Security and Healthy Living

Environmental health outreach programs are a necessity in urban areas where inner-city residents daily experience differential and inadequate access to quality food and health-care and increased exposure to toxic pollutants (McMichael, 2000). Higher levels of both absolute and relative poverty particularly among minority groups, often limit the ability of these individuals to move away from such stressful conditions (Todd, 1996). Rates of obesity have been linked to a lack of access to healthy food outlets as well as proximity to fast-food restaurants (Pearce, Blakely, Witten, & Bartie, 2007). Neighborhoods that are low-income and primarily African-American are often targets of marketing campaigns for unhealthy living choices (Grier & Kumanyika, 2010; Lewis et al., 2005). Increasing knowledge and awareness of issues surrounding food security through the use of authentic science will improve the overall health of individuals in low- income, inner-city neighborhoods.

Food security can be emphasized in many different ways to urban communities. The World Health Organization defined food security based on three pillars: food availability, food access, and food use. A multiplicity of factors influence global food security (Binns, Maconachie, & Tanko, 2003; Carvalho, 2006; Cordell, Drangert, & White, 2009; Welch & Graham, 2000). Within the context of inner-cities in the U.S., food use is a primary concern which encompasses knowledge of basic nutrition and access to sanitary water and food sources. Community-based participatory research has been shown to effectively educate community members on sources of mercury from fish (Halfhide, 2009; Howard, 2010) and outcomes of such studies yielded insights into the role that consumption patterns (food use) play in exposure to toxic metals among urban populations (Holloman & Newman, 2010, 2012). The following case studies highlight the factors which influence food security with an emphasis on local implications for their audiences in Tampa, FL and Philadelphia, PA. The case studies presented provide context to increase community resilience through informal science education. Both researchers considered their topic's interconnectivity to formal education and its relevance or translation to the community in their approach.

Case Study 1: Ocean Acidification and Sustainability to a Faith-Based Audience

The topic of ocean acidification (concomitant decreases in ocean pH due to increasing levels of atmospheric carbon dioxide) encompasses a wide range of scientific knowledge including: chemistry (Millero, 2007; Morse, Arvidson, & Lüttge, 2007), biology (Doney, Fabry, Feely, & Kleypas, 2009; Kleypas et al., 2006), and marine ecology (Abbasi & Abbasi, 2011; Doney et al., 2012). In discussing ocean acidification, there is an opportunity to relate concepts which will not only increase scientific and ocean literacy but also to promote ideas related to sustainability (Table 1). UNESCO has highlighted that ocean acidification should indeed be a concern in efforts to achieve global sustainability because of the risk that it plays in undermining the health of vital food sources (IOC/UNESCO et al., 2011). Through effective dialogue and demonstrations, audiences in this case study are exposed to current research efforts to understand long-term effects of changing pH on the physiology and productivity of organisms such as marine phytoplankton, corals, and economically important species such as clams, oysters, and lobsters which form calcium carbonate shells.

Opportunities with Faith-Based Audiences

The researcher (an oceanographer) in this case study was an active participant in a nondenominational congregation in inner-city Tampa, Florida, for a span of approximately six years. The congregation consisted of over 600 active members (predominately African American and Latino) with a consistent attendance of nearly 40 teenagers and ~60 elementary and middle school students each week. During that time, strong relationships developed between members of the congregation and several students completing doctoral degrees in various STEM fields. The doctoral students were actively involved in mentoring activities, weekly tutorial sessions, and other special events which lead to opportunities to exchange ideas on their scientific research with various members of the church community. As a unit, the church community engendered a healing space to foster individual growth by emphasizing the significance of a strong work ethic and by frequently recognizing youth for their academic and leadership achievements.

In one instance, two high school students (one male and one female) were chosen through a formal application process advertised through the church to participate as volunteers in summer research at the University of South Florida (USF). Under the direction of three graduate students, their summer months (which normally consisted of a few weeks of camp and several weeks of television and video games) were transformed into a meaningful scientific adventure. During that time, they collected sediment samples from the field for general sedimentology analysis (grain size and porosity). In addition to assisting in laboratory and field work, they also participated in field trips to local environmental sites such as a wastewater treatment facility and a phosphate mine along with undergraduate level REU (Research Experience for Undergraduates) students. They learned how Arc-GIS can be applied to generate field maps and to use Excel to analyze and plot the data from their studies. The culmination of the summer ended with poster presentations both to an audience at USF alongside the REU students and to the members of the congregation. The success of this program hinted that the church members were highly interested in introducing opportunities for careers in science to the youth within the community.

During another instance, members of the church's education department suggested that the annual back to school celebration should involve activities which introduce the youth to careers in science. The concept of ocean acidification was first introduced during that time to an audience of elementary and middle school students using authentic science. First a short discussion was lead on acids and bases by showing the students examples of household items which were acidic, basic, or neutral. Since the students live in Florida, the discussion easily lead to fun summer activities such as going to the beach. The researcher discussed how increased levels of carbon dioxide from cars and factories and other land-based gases influence the organisms that live in the ocean. An interactive activity involving the use of chalk to represent corals was used to examine the process by which calcium carbonate shells weaken under acidic conditions (in this case vinegar was used) versus more neutral conditions (water). Students were asked to draw conclusions on the health of the coral based on factors such as the strength of the acid (pure vinegar versus diluted vinegar) and the size and thickness of the coral fragment.

While the long-term impact of the activity was not documented, one student in particular became interested in the research and asked for help with a middle school science fair project several months later. The student and the researcher designed a study to examine variations in pH throughout the

Tampa Bay area in November 2010. The initial meetings consisted of a general introduction to the concept of pH and methods of analysis. In order to maintain simplicity and minimize the cost of the study, the student prepared a batch of red cabbage dye for pH determination and also purchased aquarium test strips to measure pH, nitrate, nitrite, chlorine, and alkalinity. The graduate student provided access to a digital thermometer and handheld GPS unit. Samples were collected over one weekend at 13 different sites which included the Hillsborough and Alafia Rivers, the Tampa Bay estuary, and the Gulf of Mexico (GOM) (Table 2).

The student was introduced to a wide-range of scientific ideas, terminology, and practices during this process. The concept of pH was covered both within the context of ocean acidification and biochemistry/medical sciences. The student, who aspires to pursue a career in the medical field, learned through self-initiated independent research that the concept of pH is used in the medical community. The student also became familiar with differences between riverine, estuarine, and oceanic water bodies. An additional focus was placed on using the geographic information program Google Earth to construct a site map using latitude and longitude values. The student was introduced to methods of recording scientific data through the use of a field notebook and to data analysis techniques through the use of Microsoft Excel to develop both summary tables and plots of the data. An additional emphasis was placed on the organization of the components (rationale, research questions, study area, methods, and results) of the study into a poster presentation.

Approaching Adult Learners

The researcher also explored methods of communicating ocean acidification to an adult audience (within the faith community but outside the previously mentioned congregation) through a single community talk. The researcher addressed the audience from a positivist perspective due to time and location constraints with the goal of achieving an overall understanding of ocean acidification and oceanographic research. The researcher explained the effect of geological and physical processes on pH and the ensuing impact on marine organisms. The discussion of methodology included processes and equipment involved in sample collection on oceanographic vessels: Niskin bottles and rosettes with CTD (Conductivity, Temperature, and Depth) and other sensors attached. In addition the researcher highlighted others ways of collecting oceanographic data including satellites, moorings, and floats. Sources of error and spatial coverage were mentioned. One example involves pH measurements. Since there are several ways to measure and/or calculate pH, the adult talk included an overview of direct methods for examining pH using glass electrodes and spectrophotometry and the indirect method of calculating pH from other carbon dioxide (CO_2) system parameters. While the talk was not an interactive session, the audience members who were not familiar with oceanographic research expressed particular interest in the data collection process. The talk maintained a goal of increasing scientific literacy.

Study Outcomes and Insights

One of the most exciting aspects of these interactions was the genuine interest shown by the student who wanted to design a science fair project based on the mini-workshop conducted at the church. The motivation clearly demonstrates that when science is introduced in a non-school environment, students can become excited enough to create their own investigations when the resources are available (Roth, 2002). Subsequently, the researcher organized a field trip for seven middle school students to attend a marine technology showcase at the Mote Marine Lab in Sarasota, FL. The

Table 2. Samples collected by a middle school student in the Tampa Bay Region to examine chemical variations in water quality

Site	Classification	Distance from GOM	Description	Nitrate	Nitrate	Total Hardness	Total Chlorine	Total Alkalinity	pH	Temperature
1	River	42.27	Hillsborough River; clear water; 1ft deep	20	0	150	6	300	7.8	24.7
2	River	37.56	Hillsborough River; Rotary Riverfront Park; 33ft above sea level	0	3	0	0	300	7.8	24.28
3	River	28.9	Downtown Tampa; low tide; murky water	0	10	300	4	80	7.2	26.26
4	Estuary	25.41	Bay to Bay Blvd.; Near downtown Tampa	40	3	80	0	40	8.4	27.59
5	Estuary	19.21	Near Howard Franklin Bridge; Mangrove area; sea foam in water	40	40	150	4	120	7.8	29.04
6	Estuary	11.71	USF ST Pete Bayboro harbor	0	5	0	0	0	7.8	26.21
7	Estuary	6.69	Isla del sol; off highway 93	0	5	150	0.5	120	7.5	25.54
8	Estuary	2.7	off Pinellas Byway near Mullet Key; boat dock	0	0	40	0	40	7.2	25.68
9	Estuary/Gulf of Mexico	0	Fort de Soto	0	0	40	0	20	7.8	25.74
10	River	27.18	Alafia River; Williams Park	0	0	300	0	120	8.4	NA
11	River	47.43	Hillsborough River State Park	0	0	75	0	180	6.8	18.33
12	River	30.19	Alafia River; Riverview Civic Center	0	0	300	1	80	7.2	18.51
13	River	31.76	Alafia River; Boyette Springs Park	0	0	150	0	0	6.8	17.43

students participating in this event were excited to operate state-of-the-art AUV's (autonomous underwater vehicles) equipped with cameras and to participate in activities on marine pollution, sea turtle migration, and electrochemical sensors. During this time, most of them expressed interest in marine science, engineering, and astronomy.

While the adult talk was not designed to be interactive, the investigator learned the role that conceptual framework plays in communicating to audiences at all levels of education. Future outreach activities will strive for a more interactive approach when presenting ocean acidification to adult audiences (I. Thomas, 2009). In terms of youth outreach, these experiences have demonstrated that when given a choice to participate, the most confident students will step forward and express their interest in activities such as summer research opportunities. The students who were mentored as summer research assistants and in designing the science fair project more likely benefitted more from the experiences. Planned group activities like the back to school science bash and the marine technology field trip were effective in introducing scientific concepts to a broader range of students.

Case Study 2: Healthy Soil Makes Healthy Communities

Broad Concept of Healthy Soil

The second case study addresses healthy living in the urban environment in relation to soil, the foundation of terrestrial ecosystems. Soil is the place where all four spheres on Earth (geosphere, biosphere, atmosphere, and hydrosphere) interact to form the basis of the global food chain. The soil is the home and food source of all land vegetation which lies at the base of the terrestrial trophic structure. Healthy soil is vital to the well-being of every organism that resides on land (and even aquatic species), regardless of their diet.

Feeding the Soil with Compost

Terrestrial plants rely on photosynthesis for growth and increasing biomass. Aside from carbon dioxide, water, and sunlight, nutrients are required for optimal growth to enhance nutritional components of the plant and influence the time required to produce a harvest. Nitrogen, phosphorus, and potassium (NPK) heavily influence plant growth and are major components of plant fertilizers. Commercial fertilizers are composed of synthetic, inorganic NPK compounds and metals that can be added to soil to enhance the growth rate, appearance, and productivity of plants (Soumaré, Tack, & Verloo, 2003). However, these NPK nutrients are only partially taken up by the plants and a large portion is dissolved into groundwater and eventually transported into the watershed (Baligar & Bennett, 1986; Turtola & Yli-Halla, 1999). NPK nutrients are nutritional requirements for all living organisms, not just terrestrial plants.

In the aquatic system, NPK nutrients are taken up by marine algae causing increased growth of these organisms. Eutrophication, characterized by high levels of NPK, can produce algal blooms which stress aquatic ecosystems by blocking the penetration of sunlight through the water column. Under these conditions, submerged vegetation is unable to photosynthesize. In addition, after the nutrients are consumed and depleted by the algal population, the algal population begins to die off. Bacterial degradation of the organic matter (from the dying algae), depletes dissolved oxygen and creates conditions that are uninhabitable for species which require oxygen (heterotrophs). The sequence of events is the primary driver in the formation of "dead zones" in aquatic environments (Ryther & Dunstan, 1971).

One way to combat depleted soil without the overuse of inorganic plant fertilizers is to use compost. Compost is formed from the decomposition of organic material, whereby the nutrients of biotic matter are slowly released back into the soil. The addition of compost to soil is extremely beneficial for increased plant growth (Soumaré et al., 2003) but it has benefits beyond the garden.

The organic material that is composted includes the scraps of food that would normally be discarded, such as apple cores, banana peels, broccoli stems, and other leftover food. Coffee grounds and filters can also be included in the compost pile along with tea bags, paper, and leaf litter, all items which take up substantial space in residential trash bins and municipal waste facilities. It is estimated that creating a compost pile or bin can substantially reduce municipal waste created by individuals (Bartelings & Sterner, 1999). The organic material naturally decomposes into soil, which can be donated to a community garden or school, if the individual has no use for it.

Soil Contamination

For any living organism, consuming a nutritious diet must be balanced by avoiding toxins to maintain optimal health. Soil toxins can originate from constituents previously added to the soil and can persist over time. Toxic heavy metals and organic compounds are of particular concern to soil contamination. For example, cadmium can be introduced into fertilizer during the manufacturing process as a byproduct of phosphate mining. In urban areas, heavy metals and organic compounds originate from multiple sources: industrial activity, heavy automobile traffic, components in commercially available fertilizers, and high population density (Li, Poon, & Liu, 2001).

The Community Garden as a Classroom

Addressing soil health can be a challenge in inner cities, where exposed soil is quite limited. Most of the soil that is exposed is usually contained within city parks and other public spaces. Urban dwellers with a yard at their residence often opt to cover the dirt with concrete for ease of maintenance. There are few opportunities for students to learn about the importance of soil health within the school system. Earth science is being increasingly removed from curriculums and state standards due to decreased funding for public schools. Field trips to natural areas may not be realized for many children during formative years where attitudes and habits are developed. Children and even adults in urban environments are often not aware or interested in soil and its role in our local ecosystem. They seldom can communicate how soil impacts health through agriculture and the formation of dust particles in the atmosphere.

A Philadelphia-based martial arts group hosts an annual youth conference where participants are exposed to the philosophies and movements of the martial art, Capoeira Angola (an African-Brazilian martial art). The theme of the 2011 youth conference was Freedom Fighters, which included presentations and discussions about the historical contexts of figures such as Harriet Tubman and Marcus Garvey. The Healthy Soil Makes Healthy Communities workshop fit into the theme of Freedom Fighters by relating to community members the concept that personal and racial freedom are intrinsically connected with growing one's own food and ultimately, food security for the entire community. The culminating activity of the conference was the groundbreaking for the annual spring garden. The community-based garden is maintained by the members of the martial arts group and is used to feed the families of these group members. Another important component to the youth conference was free admission to the two-day intensive workshop, making the art itself and the information imparted during the weekend accessible and available to the local community of West Philadelphia. Regional visitors from other cities including New York City, Trenton, NJ, Baltimore, MD, and Washington, DC are also offered free admission.

In 2011, the case writer was invited to give a lecture about soil and how it relates to the garden. This was an opportunity to inform the conference attendees, which included youth ranging in age from 2-18 as well as the parents and teachers of the children, about several topics related to soil health and sustainability. Various issues and barriers were seemingly apparent with regards to composing a lecture/workshop to a diverse group within the structure of a martial arts conference. The attention of the attendees was likely to wane when transitioning from the martial art to the basics of soil science. The case writer brainstormed strategies to engage the group on specific topics which would prove to be relevant and interesting to such a diverse group.

An authentic science research experience is ideal for engaging students of all types and ages. The challenge was developing and executing an open-inquiry learning experience within the 30-mintute time limitation of the proposed lecture. One potential way to engage a non-scientific audience is with cool, state-of-the-art technology. A handheld X-Ray Fluorescence (XRF) Analyzer (accessed through the university) was chosen as the tool for simple analysis of heavy metals, most of which are considered toxic to soil and human health. The device had a fast analysis time of less than 60 seconds and did not involve destructive techniques.

A lecture based on XRF analysis of garden soil was developed for the authentic science research experience to involve the entire group. The lecture introduced various components of soil and the investigator followed up with the audience by posing a question on the soil components. After receiving various answers, the lecturer picked up a handful of soil to examine it. The investigator emphasized the point that visual inspection by the naked eye is often a poor determinant of soil health and the eyes are often unable to detect soil components. Additional emphasis was placed on understanding soil components prior to participant exposure, particularly children. The lecture continued by providing an overview of the ideal composition of healthy soil and the role of each of these components plays in supporting plant growth.

The next focus for the discussion was on the compost bin maintained by the martial arts academy over the course of several years. The audience was engaged in the activity, even before the experimental section commenced. Instant connections were made between the content of the talk and the actual environment where the discussion was taking place. Standing in the soil and seeing some of the younger children playing in the yard during the lecture was an actualization of the potential impact of soil on children's health (after 20 minutes or so, most of the children were covered in dirt!). Talking about compost in theory does not equate to experiencing a compost bin that contains several stages of decomposition including freshly deposited egg shells, banana peels, and fresh fertile soil created from deposits of organic material over the previous 9 months. The setting made the presentation realistic and accessible to the audience members.

The lecture proceeds to the discussion of unhealthy soil. The audience was then asked about the types of things that they would not want to be present and/or abundant in healthy soil. Before reviewing their answers, the XRF was introduced. A brief summary was given on the capabilities and purpose of the instrument. Some of the audience members had indicated that toxic chemicals would not be an ideal component of healthy soil, so the presentation of the XRF was seen as a direct response to the concerns of the group. While the instrument was calibrating, the lecturer transitioned into the experiment with an audience fully aware of the problem to be addressed. The group was able to construct a reasonable hypothesis to test with the available technology.

There is a public perception that Philadelphia is an industrialized city that has been subjected to pollution from a variety of sources. This perception was made apparent by the hypothesis proposed by the group. Their hypothesis stated that the soil would contain high concentrations of lead (Pb) and smaller but measurable amounts of mercury (Hg) and arsenic (As). The lecturer had no influence on the hypothesis; the parents in the audience directed the formulation of the research objectives.

The soil in the garden area was analyzed using the XRF. Triplicate analyses were performed on four different sections of the garden area. The results showed elevated levels of lead in one area with respect to the EPA standard of 400 ppm in residential soil where children are exposed. Mercury and arsenic concentrations were below the detection limits of the XRF (Table 3). The audience concluded that the soil is not so "dirty" after all. At the conclusion of the day, ground was broken for the annual spring garden. The participants worked hard and quickly developed a new appreciation for soil and its role in their own health and in the health of the ecosystem.

As shown in Table 3, the group hypothesized that lead (Pb), arsenic (As), and mercury (Hg) would be present in the soil at the martial arts academy in Philadelphia. The handheld X-Ray Fluorescence (XRF) used in this study provides concentrations for heavy elements (Titanium and above) but only the elements of interest are listed. Pb concentrations were significantly high in this area. Location 2 is a > 100 year old structure which presumably contains lead-based paint on its exterior. Locations 3 and 4 contained significant levels of Pb but were below the EPA standard (400 ppm).

Table 3. Heavy metal concentrations in Philadelphia community garden soils analyzed using a handheld XRF

Sample Location	Average Pb (ppm)	Average As (ppm)	Average Hg (ppm)
1	5 ± 4	5 ± 3	-11 ± 4
2	2959 ± 54	-430 ± 35	-30 ± 9
3	203 ± 9	2 ± 9	2 ± 5
4	220 ± 11	-12 ± 10	11 ± 6

Study Outcomes and Insights

The 2011 community garden at the martial arts school in Philadelphia was highly successful, yielding abundant quantities of kale, tomatoes, peppers, and corn, among other herbs and vegetables. The owners of the martial arts school requested that the entire property be tested with the XRF. The results indicated that adjacent to the 100 year old structure, there were levels of lead exceeding 1000 ppm (EPA standard is 400 ppm where children are exposed). Following the soil analyses, the owners restricted access of the contaminated area. Also, one of the participating families decided to have the soil tested at their residence as well as begin a garden. The Healthy Soil Makes Healthy Communities workshop was effective in improving scientific literacy in the audience members and by influencing its participants to make well-informed decisions.

SOLUTIONS, RECOMMENDATIONS, AND FUTURE DIRECTIONS

These case studies highlight successful strategies for building bridges between scientists and the broader community through informal settings. With these interactions, solid relationships were established, ideas were exchanged, and lives were changed. These case studies highlight that even in areas which have reputable low graduation rates, particularly for African American males (for 2007/8: 35% in Hillsborough County, FL; 21% in Pinellas County, FL; and 28% in Philadelphia, PA (Jackson, 2010)), both parents and students have a substantial amount of interest in scientific inquiry and insight (Hagay et al., 2012). We suggest that further consideration needs to be made in the manner (Ferguson, 2008; Mueller & Bentley, 2009) and frequency through which science is taught to these audiences (Tate, 2001). Based on our experiences, sustained and long-term relationships are most effective for reaching underserved and/or underrepresented communities (Ben-Zvi-Assaraf

& Orion, 2010) where relationship building is culturally and socially important.

While a great deal of emphasis has been placed on urban and inner-city areas, there is also a need to address the concerns of those residing in rural areas. In the United States, approximately 21.0% of the population resides in rural areas (with population densities less than 1,000 people per square mile) and in some of these areas, for example, within the states of Virginia and North Carolina, they contain high percentages of African Americans (> 30% of the population). Dopica and Gacia-Vazquez (2011) report tremendous success in teaching environmental science and sustainability concepts within rural communities in Spain by utilizing long-standing community traditions to emphasize resource management. We suggest that the American South may be a great setting to help expand efforts to engage communities with authentic science thereby reinforcing sustainable practices (Morris & Monroe, 2009).

We anticipate that with greater implementation of these ideas, education efforts will be expanded to other audiences such as small business owners (Gadenne, Kennedy, & McKeiver, 2009) which are often overlooked. In the current political context where the importance of environmental education is often undermined by various groups (Sharpe & Breunig, 2009), all citizens should be well-informed of the role that sustainable practices play in improving quality of life. Along the same line, an understanding of these principles, could potentially lead, not necessarily to increased efforts at the public policy level (and therefore increased regulations) but to genuine lifestyle and value changes among citizenry (Gadenne, Sharma, Kerr, & Smith, 2011). As the Decade of Education for Sustainable Development closes, we hope that our contribution to the legacy (Combes, 2005; Mulà & Tilbury, 2009) inspires creative and non-traditional approaches to increase long-term participation in STEM fields by underrepresented populations.

REFERENCES

AAAS. (1990). *Project 2061: Science for all Americans*. New York, NY: Oxford University Press.

Abbasi, T., & Abbasi, S. A. (2011). Ocean acidification: The newest threat to the global environment. *Critical Reviews in Environmental Science and Technology*, *41*(18), 1601–1663. doi:10.1080/10643389.2010.481579

Au, W. (2007). High-stakes testing and curricular control: A qualitative metasynthesis. *Educational Researcher*, *36*(5), 258–267. doi:10.3102/0013189X07306523

Baligar, V. C., & Bennett, O. L. (1986). NPK-fertilizer efficiency – A situation analysis for the tropics. *Nutrient Cycling in Agroecosystems*, *10*(2), 147–164. doi:10.1007/bf01074369

Bartelings, H., & Sterner, T. (1999). Household waste management in a Swedish municipality: Determinants of waste disposal, recycling and composting. *Environmental and Resource Economics*, *13*(4), 473–491. doi:10.1023/A:1008214417099

Barton, A. C. (2001). Capitalism, critical pedagogy, and urban science education: An interview with Peter McLaren. *Journal of Research in Science Teaching*, *38*(8), 847–859. doi:10.1002/tea.1035

Belle, T. J. (1982). Formal, nonformal and informal education: A holistic perspective on lifelong learning. *International Review of Education*, *28*(2), 159–175. doi:10.1007/BF00598444

Ben-Zvi-Assaraf, O., & Orion, N. (2010). Four case studies, six years later: Developing system thinking skills in junior high school and sustaining them over time. *Journal of Research in Science Teaching*, *47*(10), 1253–1280. doi:10.1002/tea.20383

Binns, J. A., Maconachie, R. A., & Tanko, A. I. (2003). Water, land and health in urban and peri-urban food production: The case of Kano, Nigeria. *Land Degradation & Development*, *14*(5), 431–444. doi:10.1002/ldr.571

Cammarota, J., & Fine, M. (2008). Youth participatory action research: A pedagogy for transformational resistance. In J. C. M. Fine (Ed.), *Youth participatory action in motion* (pp. 1–12). New York, NY: Routledge.

Carvalho, F. P. (2006). Agriculture, pesticides, food security and food safety. *Environmental Science & Policy*, *9*(7–8), 685–692. doi:10.1016/j.envsci.2006.08.002

Colucci-Gray, L., Camino, E., Barbiero, G., & Gray, D. (2006). From scientific literacy to sustainability literacy: An ecological framework for education. *Science Education*, *90*(2), 227–252. doi:10.1002/sce.20109

Combes, B. P. Y. (2005). The United Nations decade of education for sustainable development (2005–2014): Learning to live together sustainably. *Applied Environmental Education & Communication*, *4*(3), 215–219. doi:10.1080/15330150591004571

Cope, M. A., McLafferty, S., & Rhoads, B. L. (2011). Farmer attitudes toward production of perennial energy grasses in East Central Illinois: Implications for community-based decision making. *Annals of the Association of American Geographers. Association of American Geographers*, *101*(4), 852–862. doi:10.1080/00045608.2011.575320

Cordell, D., Drangert, J.-O., & White, S. (2009). The story of phosphorus: Global food security and food for thought. *Global Environmental Change*, *19*(2), 292–305. doi:10.1016/j.gloenvcha.2008.10.009

Crowley, K., Callanan, M. A., Tenenbaum, H. R., & Allen, E. (2001). Parents explain more often to boys than to girls during shared scientific thinking. *Psychological Science*, *12*(3), 258–261. doi:10.1111/1467-9280.00347 PMID:11437311

Cummins, J. (2007). Pedagogies for the poor? Realigning reading instruction for low-income students with scientifically based reading research. *Educational Researcher*, *36*(9), 564–572. doi:10.3102/0013189X07313156

Dee, T. S., & Fu, H. (2004). Do charter schools skim students or drain resources? *Economics of Education Review*, *23*(3), 259–271. doi:10.1016/j.econedurev.2003.10.001

Doney, S. C., Fabry, V. J., Feely, R. A., & Kleypas, J. A. (2009). Ocean acidification: The other CO_2 problem. *Annual Review of Marine Science*, *1*(1), 169–192. doi:10.1146/annurev.marine.010908.163834 PMID:21141034

Doney, S. C., Ruckelshaus, M., Emmett Duffy, J., Barry, J. P., Chan, F., & English, C. A. et al. (2012). Climate change impacts on marine ecosystems. *Annual Review of Marine Science*, *4*(1), 11–37. doi:10.1146/annurev-marine-041911-111611 PMID:22457967

Dopico, E., & Garcia-Vazquez, E. (2011). Leaving the classroom: A didactic framework for education in environmental sciences. *Cultural Studies of Science Education*, *6*(2), 311–326. doi:10.1007/s11422-010-9271-9

Elmesky, R., & Tobin, K. (2005). Expanding our understandings of urban science education by expanding the roles of students as researchers. *Journal of Research in Science Teaching*, *42*(7), 807–828. doi:10.1002/tea.20079

Ferguson, R. (2008). If multicultural science education standards' existed, what would they look like? *Journal of Science Teacher Education*, *19*(6), 547–564. doi:10.1007/s10972-008-9108-5

Fetterman, D. M. (1994). Empowerment evaluation. *Evaluation Practice*, *15*(1), 1–15. doi:10.1016/0886-1633(94)90055-8

Gadenne, D., Kennedy, J., & McKeiver, C. (2009). An empirical study of environmental awareness and practices in SMEs. *Journal of Business Ethics*, *84*(1), 45–63. doi:10.1007/s10551-008-9672-9

Gadenne, D., Sharma, B., Kerr, D., & Smith, T. (2011). The influence of consumers' environmental beliefs and attitudes on energy saving behaviours. *Energy Policy*, *39*(12), 7684–7694. doi:10.1016/j.enpol.2011.09.002

García, C. E. R., & Brown, S. (2009). Assessing water use and quality through youth participatory research in a rural Andean watershed. *Journal of Environmental Management*, *90*(10), 3040–3047. doi:10.1016/j.jenvman.2009.04.014 PMID:19500900

Geier, R., Blumenfeld, P. C., Marx, R. W., Krajcik, J. S., Fishman, B., Soloway, E., & Clay-Chambers, J. (2008). Standardized test outcomes for students engaged in inquiry-based science curricula in the context of urban reform. *Journal of Research in Science Teaching*, *45*(8), 922–939. doi:10.1002/tea.20248

Geske, T. G., Davis, D. R., & Hingle, P. L. (1997). Charter schools: A viable public school choice option? *Economics of Education Review*, *16*(1), 15–23. doi:10.1016/S0272-7757(96)00039-8

Grier, S. A., & Kumanyika, S. (2010). Targeted marketing and public health. *Annual Review of Public Health*, *31*(1), 349–369. doi:10.1146/annurev.publhealth.012809.103607 PMID:20070196

Grorud-Colvert, K., Lester, S. E., Airamé, S., Neeley, E., & Gaines, S. D. (2010). Communicating marine reserve science to diverse audiences. *Proceedings of the National Academy of Sciences of the United States of America*, *107*(43), 18306–18311. doi:10.1073/pnas.0914292107 PMID:20427745

Hagay, G., Baram-Tsabari, A., Ametller, J., Cakmakci, G., Lopes, B., Moreira, A., & Pedrosa-de-Jesus, H. (2012). The generalizability of students' interests in biology across gender, country and religion. *Research in Science Education*, 1–25. doi:10.1007/s11165-012-9289-y

Halfhide, T. (2009). *Mercury perception, community awareness and sustainability implications for the Tampa Bay region, Florida.* M.S., University of South Florida. Retrieved June 27, 2012, from http://scholarcommons.usf.edu/etd/3689/

Haney, J. J., Lumpe, A. T., & Czerniak, C. M. (2003). Constructivist beliefs about the science classroom learning environment: Perspectives from teachers, administrators, parents, community members, and students. *School Science and Mathematics*, *103*(8), 366–377. doi:10.1111/j.1949-8594.2003.tb18122.x

Hanushek, E. A. (1997). Assessing the effects of school resources on student performance: An update. *Educational Evaluation and Policy Analysis*, *19*(2), 141–164. doi:10.3102/01623737019002141

Hedges, L. V., Laine, R. D., & Greenwald, R. (1994). An exchange: Part I: Does money matter? A meta-analysis of studies of the effects of differential school inputs on student outcomes. *Educational Researcher*, *23*(3), 5–14. doi:10.3102/0013189x023003005

Holloman, E. L., & Newman, M. C. (2010). A community-based assessment of seafood consumption along the lower James River, Virginia, USA: Potential sources of dietary mercury exposure. *Environmental Research, 110*(3), 213–219. doi:10.1016/j.envres.2010.01.002 PMID:20129605

Holloman, E. L., & Newman, M. C. (2012). Expanding perceptions of subsistence fish consumption: Evidence of high commercial fish consumption and dietary mercury exposure in an urban coastal community. *The Science of the Total Environment, 416*(0), 111–120. doi:10.1016/j.scitotenv.2011.10.003 PMID:22225823

Howard, J. A. (2010). *Mercury in the environment: Field studies from Tampa, Bolivia, and Guyana.* Ph.D., University of South Florida. Retrieved June 27, 2012, from http://scholarcommons.usf.edu/etd/3465

IOC/UNESCO. IMO, FAO, & UNDP. (2011). A blueprint for ocean and coastal sustainability. Paris, France: IOC/UNESCO.

Jackson, J. H. (2010). *Yes we can: The Schott 50 state report on public education and black males: Schott Foundation for Public Education.* Retrieved June 27, 2012, from http://www.soros.org/sites/default/files/schott_report-20100817.pdf

Jimenez-Aleixandre, M.-P. (2002). Knowledge producers or knowledge consumers? Argumentation and decision making about environmental management. *International Journal of Science Education, 24*(11), 1171–1190. doi:10.1080/09500690210134857

Kim, K. H. (2008). Underachievement and creativity: Are gifted underachievers highly creative? *Creativity Research Journal, 20*(2), 234–242. doi:10.1080/10400410802060232

Kim, K. H. (2011). The creativity crisis: The decrease in creative thinking scores on the Torrance Tests of creative thinking. *Creativity Research Journal, 23*(4), 285–295. doi:10.1080/10400419.2011.627805

Kleypas, J. A., Feely, R. A., Fabry, V. J., Langdon, C., Sabine, C. L., & Robbins, L. L. (2006). *Impacts of ocean acidification on coral reefs and other marine calcifiers. A guide for future research.* Report of a workshop sponsored by NSF, NOAA & USGS. Technical report. Retrieved June 27, 2012, from http://www.ucar.edu/communications/Final_acidification.pdf

Kodrzycki, Y. K. (2002). *Education in the 21st century: Meeting the challenges of a changing world.* Paper presented at the New England Economic Review. Retrieved June 27, 2012, from http://www.bos.frb.org/economic/conf/conf47/conf47.pdf

Laidler, G. J., & Elee, P. (2008). Human geographies of sea ice: Freeze/thaw processes around Cape Dorset, Nunavut, Canada. *The Polar Record, 44*, 51–76.

Laidler, G. J., Elee, P., Ikummaq, T., Joamie, E., & Aporta, C. (2010). Mapping Inuit sea ice knowledge, use, and change in Nunavut, Canada (Cape Dorset, Igloolik, Pangnirtung). In I. Krupnik, C. Aporta, S. Gearheard, G. J. J. Laidler, & L. K. Kielsen Holm (Eds.), *SIKU: Knowing our ice* (pp. 45–80). Netherlands: Springer. doi:10.1007/978-90-481-8587-0_3

Lewis, L. B., Sloane, D. C., Nascimento, L. M., Diamant, A. L., Guinyard, J. J., Yancey, A. K., & Flynn, G. (2005). African Americans' access to healthy food options in South Los Angeles restaurants. *American Journal of Public Health, 95*(4), 668–673. doi:10.2105/AJPH.2004.050260 PMID:15798128

Li, X., Poon, C.-S., & Liu, P. S. (2001). Heavy metal contamination of urban soils and street dusts in Hong Kong. *Applied Geochemistry*, *16*(11–12), 1361–1368. doi:10.1016/S0883-2927(01)00045-2

Lloyd, G. (1999). Two teachers' conceptions of a reform-oriented curriculum: Implications for mathematics teacher development. *Journal of Mathematics Teacher Education*, *2*(3), 227–252. doi:10.1023/A:1009965804662

Loucks-Horsley, S., & Matsumoto, C. (1999). Research on professional development for teachers of mathematics and science: The state of the scene. *School Science and Mathematics*, *99*(5), 258–271. doi:10.1111/j.1949-8594.1999.tb17484.x

Lubchenco, J. (1998). Entering the century of the environment: A new social contract for science. *Science*, *279*(5350), 491–497. doi:10.1126/science.279.5350.491

Martin, L. M. W. (2004). An emerging research framework for studying informal learning and schools. *Science Education*, *88*(S1), S71–S82. doi:10.1002/sce.20020

Matthews, J. (2011). Hybrid pedagogies for sustainability education. *Review of Education, Pedagogy & Cultural Studies*, *33*(3), 260–277. doi:10.1080/10714413.2011.585288

McCall, M. K., & Minang, P. A. (2005). Assessing participatory GIS for community-based natural resource management: Claiming community forests in Cameroon. *The Geographical Journal*, *171*(4), 340–356. doi:10.1111/j.1475-4959.2005.00173.x

McConney, A., Oliver, M., Woods-McConney, A., & Schibeci, R. (2010). Bridging the gap? A comparative, retrospective analysis of science literacy and interest in science for indigenous and non-indigenous Australian students. *International Journal of Science Education*, *33*(14), 2017–2035. doi:10.1080/09500693.2010.529477

McConville, J. R., & Mihelcic, J. R. (2007). Adapting life-cycle thinking tools to evaluate project sustainability in international water and sanitation development work. *Environmental Engineering Science*, *24*(7), 937–948. doi:10.1089/ees.2006.0225

McMichael, A. J. (2000). Health and the urban environment in a world of increasing globilization. *Bulletin of the World Health Organization*, *78*(9), 1117–1126. PMID:11019460

Milam, A., Furr-Holden, C., & Leaf, P. (2010). Perceived school and neighborhood safety, neighborhood violence and academic achievement in urban school children. *The Urban Review*, *42*(5), 458–467. doi:10.1007/s11256-010-0165-7 PMID:21197388

Millero, F. J. (2007). The marine inorganic carbon cycle. *Chemical Reviews*, *107*(2), 308–341. doi:10.1021/cr0503557 PMID:17300138

Minkler, M. (2005). Community-based research partnerships: Challenges and opportunities. *Journal of Urban Health*, *82*(2Suppl 2), ii3–ii12. doi:10.1093/jurban/jti034 PMID:15888635

Morris, J. E., & Monroe, C. R. (2009). Why study the U.S. South? The nexus of race and place in investigating black student achievement. *Educational Researcher*, *38*(1), 21–36. doi:10.3102/0013189X08328876

Morse, J. W., Arvidson, R. S., & Lüttge, A. (2007). Calcium carbonate formation and dissolution. *Chemical Reviews*, *107*(2), 342–381. doi:10.1021/cr050358j PMID:17261071

Mueller, M. P., & Bentley, M. L. (2009). Environmental and science education in developing nations: A Ghanaian approach to renewing and revitalizing the local community and ecosystems. *The Journal of Environmental Education*, *40*(4), 53–64. doi:10.3200/JOEE.40.4.53-64

Mulà, I., & Tilbury, D. (2009). A United Nations decade of education for sustainable development (2005–14). *Journal of Education for Sustainable Development*, *3*(1), 87–97. doi:10.1177/097340820900300116

Nelson, G. (1999). Science literacy for all in the 21st century. *Educational Leadership*, *57*(2), 14–17.

Norris, S. P., & Phillips, L. M. (2003). How literacy in its fundamental sense is central to scientific literacy. *Science Education*, *87*(2), 224–240. doi:10.1002/sce.10066

Pearce, J., Blakely, T., Witten, K., & Bartie, P. (2007). Neighborhood deprivation and access to fast-food retailing: A national study. *American Journal of Preventive Medicine*, *32*(5), 375–382. doi:10.1016/j.amepre.2007.01.009 PMID:17478262

Rauch, F. (2002). The potential of education for sustainable development for reform in schools. *Environmental Education Research*, *8*(1), 43–51. doi:10.1080/13504620120109646

Rhoads, B. L., Wilson, D., Urban, M., & Herricks, E. E. (1999). Interaction between scientists and nonscientists in community-based watershed management: emergence of the concept of stream naturalization. *Environmental Management*, *24*(3), 297–308. doi:10.1007/s002679900234 PMID:10486041

Roehrig, G. H., & Luft, J. A. (2006). Does one size fit all? The induction experience of beginning science teachers from different teacher-preparation programs. *Journal of Research in Science Teaching*, *43*(9), 963–985. doi:10.1002/tea.20103

Roth, W. M. (2002). Taking science education beyond schooling. *Canadian Journal of Science. Mathematics and Technology Education*, *2*(1), 37–48. doi:10.1080/14926150209556497

Roth, W. M., & Lee, S. (2004). Science education as/for participation in the community. *Science Education*, *88*(2), 263–291. doi:10.1002/sce.10113

Ryther, J. H., & Dunstan, W. M. (1971). Nitrogen, phosphorus, and eutrophication in the coastal marine environment. *Science*, *171*(3975), 1008–1013. doi:10.1126/science.171.3975.1008 PMID:4993386

Santos, W. L. P. D. (2009). Scientific literacy: A Freirean perspective as a radical view of humanistic science education. *Science Education*, *93*(2), 361–382. doi:10.1002/sce.20301

Sealey-Ruiz, Y. (2011). Dismantling the school-to-prison pipeline through racial literacy development in teacher education. *Journal of Curriculum and Pedagogy*, *8*(2), 116–120. doi:10.1080/15505170.2011.624892

Sharpe, E., & Breunig, M. (2009). Sustaining environmental pedagogy in times of educational conservatism: A case study of integrated curriculum programs. *Environmental Education Research*, *15*(3), 299–313. doi:10.1080/13504620902807543

Simpson, L. (2002). Indigenous environmental education for cultural survival. *Canadian Journal of Environmental Education*, *7*(1), 13–25.

Soumaré, M., Tack, F. M. G., & Verloo, M. G. (2003). Effects of a municipal solid waste compost and mineral fertilization on plant growth in two tropical agricultural soils of Mali. *Bioresource Technology*, *86*(1), 15–20. doi:10.1016/S0960-8524(02)00133-5 PMID:12421002

Steel, B. S., Smith, C., Opsommer, L., Curiel, S., & Warner-Steel, R. (2005). Public ocean literacy in the United States. *Ocean and Coastal Management*, *48*(2), 97–114. doi:10.1016/j.ocecoaman.2005.01.002

Strickland-Dixon, K. (2011). Curriculum construction: Conflicts and constraints that promote the underachievement of African American students in urban schools. *Journal of Curriculum and Pedagogy*, *8*(2), 112–116. doi:10.1080/15505170.2011.624032

Taines, C. (2010). Educational or social reform? Students inform the debate over improving urban schools. *Education and Urban Society*, *44*(3), 247–273. doi:10.1177/0013124510392566

Taines, C. (2011). "What the real world in schools is like": Urban youth in dialogue about educational inequality. *The Urban Review*, *43*(3), 414–435. doi:10.1007/s11256-010-0169-3

Tate, W. (2001). Science education as a civil right: Urban schools and opportunity-to-learn considerations. *Journal of Research in Science Teaching*, *38*(9), 1015–1028. doi:10.1002/tea.1045

Thomas, G., & Durant, J. (1987). Why should we promote the public understanding of science? In M. Shortland (Ed.), *Scientific Literary Papers: A Journal of Research in Science, Education, and Research* (pp. 1–14). Oxford, UK: Department for External Studies, University of Oxford.

Thomas, I. (2009). Critical thinking, transformative learning, sustainable education, and problem-based learning in universities. *Journal of Transformative Education*, *7*(3), 245–264. doi:10.1177/1541344610385753

Todd, A. (1996). Health inequalities in urban areas: a guide to the literature. *Environment and Urbanization*, *8*(2), 141–152. doi:10.1177/095624789600800220

Turtola, E., & Yli-Halla, M. (1999). Fate of phosphorus applied in slurry and mineral fertilizer: accumulation in soil and release into surface runoff water. *Nutrient Cycling in Agroecosystems*, *55*(2), 165–174. doi:10.1023/A:1009862227026

UNCED. (1992). *United Nations Conference on Environment and Development*. Retrieved June 27, 2012, from http://www.un.org/esa/dsd/agenda21/

Valadez, J. R., & Durán, R. P. (2007). Redefining the digital divide: Beyond access to computers and the internet. *High School Journal*, *90*(3), 31–44. doi:10.1353/hsj.2007.0013

Venville, G., Sheffield, R., Rennie, L. J., & Wallace, J. (2008). The writing on the wall: Classroom context, curriculum implementation, and student learning in integrated, community-based science projects. *Journal of Research in Science Teaching*, *45*(8), 857–880. doi:10.1002/tea.20245

Walker, S. H., & Keener-Chavis, P. (2005). The path to ocean sciences literacy: Essential steps along the way. *Marine Technology Society Journal*, *39*(4), 20–32. doi:10.4031/002533205787465977

Welch, R. M., & Graham, R. D. (2000). A new paradigm for world agriculture: Productive, sustainable, nutritious, healthful food systems. *Food and Nutrition Bulletin*, *21*(4), 361–366.

Wynne, B. (2006). Public engagement as a means of restoring public trust in science – Hitting the notes, but missing the music? *Public Health Genomics*, *9*(3), 211–220. PMID:16741352

Yanarella, E. J., Levine, R. S., & Lancaster, R. W. (2009). Research and solutions: "Green" vs. sustainability: From semantics to enlightenment. *Sustainability: The Journal of Record*, *2*(5), 296–302. doi:10.1089/SUS.2009.9838

Zen, E.-a., & Palmer, A. R. (1999, 10/30/99). *Guidelines for sustainability literacy: The intricacies of living in an interactive world.* Retrieved June 27, 2012, from http://bcn.boulder.co.us/basin/local/sustainguide.htm

Zion, M., Slezak, M., Shapira, D., Link, E., Bashan, N., & Brumer, M. et al. (2004). Dynamic, open inquiry in biology learning. *Science Education*, *88*(5), 728–753. doi:10.1002/sce.10145

This work was previously published in "Cases on the Diffusion and Adoption of Sustainable Development Practices" edited by Helen E. Muga and Ken D. Thomas, pages 376-402, copyright 2013 by Information Science Reference (an imprint of IGI Global).

Chapter 52
Teaching Political Science Students to Find and Evaluate Information in the Social Media Flow

Megan Fitzgibbons
McGill University, Canada

ABSTRACT

The advent of social media necessitates new pedagogical approaches in the field of political science, specifically in relation to undergraduate students' critical thinking and information evaluation skills. Instead of seeking out traditional static pools of knowledge, researchers and researchers-in-training now interact with information in an amorphous stream of production and consumption. Socially created information is now firmly integrated in the basic subject matter of political science, as manifested in primary sources in the field, scholars' communication practices, and the emergence of collective and distributed expertise. Existing models of information evaluation competencies do not address these realities of participatory authorship and decentralized distribution of information. Thus, in order to educate "information-literate" students in political science, educators must foster an understanding of how information is produced and how to critically evaluate individual information sources in the context of academic tasks.

INTRODUCTION

On May 7, 2011, the *Guardian* newspaper in the United Kingdom printed an interview with a "heroine of the Syrian revolt": Amina Abdullah, purportedly a Syrian blogger. Praised as an outspoken, courageous activist who contributed to the protest movement through her blog, "A Gay Girl in Damascus," Amina developed a strong following around the world (Marsh, 2011) and demonstrated the power of Web 2.0 tools in furthering activism. Many were alarmed, then, when the *Guardian* reported one month later that Amina had been taken hostage, ostensibly by Syrian security forces (Hassan, 2011). Clearly, it would seem, dictatorial governments recognize and fear the power of social media. The saga took a different twist a few days later, however, when it was revealed that Amina and her blog were a hoax propagated by a middle-aged American

DOI: 10.4018/978-1-4666-7363-2.ch052

PhD student living in Britain (Addley, 2011). The widely reported story became emblematic of social software's power not only to share information but also to deceive. Skills in evaluating information now more than ever underlie the ability to make sense of the world.

The discipline of political science is uniquely implicated in the rise of social media. The data that constitute fodder for study in the field are now highly diffuse. Primary materials pour out from sources ranging from politicians' tweets and blogs to documents "leaked" by Internet activists. Political action–from e-government to political protest–frequently takes place via networked conduits. Secondary sources, too, are no longer simply traditional research articles and professional commentary, as pundits of all stripes and academics alike spread their messages through websites and interactive media. Indeed, anyone with access to the relevant technology can be a content creator and thus a political actor. Therefore, instead of being able to rely on familiar pools where information collects, researchers and researchers-in-training now function in a constantly moving flow of information production and consumption (Boyd, 2010). In the complexity of this environment, it follows that skills in evaluating information are fundamental to the practice of political science.

The word "expert" is generally applied to someone who is a master of a particular technical skill or domain of knowledge. Expertise is often tied to recognized credentials as well as the means to publicly share knowledge, either through published work, via vetted media channels, or in classrooms and lectures. With regard to published information, then, expertise is linked to oversight and editorial review. The notion of credibility is closely tied to expertise, as experts by definition are understood to be trustworthy. In the past, information had to come from particular channels in order to have the label of expertise applied. In the Web 2.0 environment, however, anyone with a piece of information to share can contribute to a larger "hive mind" of knowledge that can be accessed online.

This chapter explores connections between Web 2.0 tools and political science, as manifested in primary sources in the field, scholars' communication practices, and the emergence of collective and distributed expertise. Techniques and conceptual models for teaching information evaluation skills are then explored in light of the ubiquity of the information produced and disseminated through social, networked conduits. Analysis focuses on ways in which educators in political science can facilitate students' skills for evaluating information, which, in the flow of social media, is now an iterative, constant process that has changed the meaning of expertise. Educators must therefore more broadly educate students both in understanding how information is produced and how to critically evaluate individual information sources in the context of their academic tasks. Educators are no longer the only experts on whom students can, or indeed should, rely. The discussion primarily refers to the education of undergraduate university students, but the principles can apply to many student-teacher scenarios.

BACKGROUND

Goals of Political Science Education

Political science has been broadly defined as "the study of human behavior relating primarily to the operations of government, the state, and in principle other outcomes deemed important by actors" (Polsby, 2001). The ideas of power and conflict–among and within states–are also foundational themes. The field is complex, spanning a history that has seen the rise of normative, behavioralist, and post-behavioralist approaches (Gunnell, 2011) as well as splintering into a variety of sub-fields. Current research is increasingly interdisciplinary, with sociology, psychology, and communication studies contributing new frameworks for understanding how people and societies relate to each other and to governmental systems.

At the university level, students learn methodological and theoretical approaches to making sense of political phenomena, such as state formation, war, civil society, and electoral politics. As in other social sciences, courses are designed to introduce foundational thought in the field and encourage students to apply theories to explain events of the past and present while simultaneously developing an understanding of the field's unique methodologies of scholarship. They are also generally encouraged to foster a spirit of inquiry and question the phenomena that they observe in the world. In addition, the study of political science is sometimes seen as training for careers in politics (local, national, or international), diplomacy, public policy, international development, public administration, and law. Beyond the focus of "pure" academia, then, political science students are often exposed to practical dimensions of governments and develop transferable skills for employment in related fields.

Finally, the development of a civic sensibility is frequently seen to be an underlying goal of political science education. In an analysis of the links between democracy and political education in the United Kingdom, it has been argued that educators should increasingly value student-centered, constructivist, experiential learning approaches and seek out pedagogical methods that emphasize participation in the object of study (Sloam, 2008). Although academic political scientists are sometimes accused of detachment from the real-world workings of nations, the political science or governmental studies classroom is often singled out as a natural place to foster political engagement and rekindle the interest of generations that are perceived to be apathetic.

The study of the workings of governmental apparatus and of human behavior is fundamentally tied to the processes by which information is produced and communicated. Regardless of the ultimate goal or the sub-field of political science–methods, theory, comparative politics, domestic politics, international relations, or civic engagement–information is implicated in research and study. Educators, then, need to ask: what skills do students need in order to evaluate information in political science?

The Challenge of Critical Evaluation

The Internet is said to be a great democratizer in information dissemination, acting as an infinite conduit to information of all types. More than ever, information seekers access information via Web search engines and social networking sites, but the actual function and performance of these tools are not well understood by the general public. Indeed, reference librarians can testify that, anecdotally, it is common for people to believe that a given piece of information itself is actually *in*, rather than *retrieved by,* the search engine. Web search engines' results pages tend to obscure documents' types and origins from casual searchers. In the online environment, everything is seen through the vehicle of a Web browser, where visual cues as to documents' reliability and purpose are less marked.

There has been a lot of dialogue in recent years around the notion that today's young people are "wired" differently, with the label "Google" or "Net" generation often applied to describe their reliance on the Internet for information and communication. While the idea that "Google is making us stupid" (Carr, 2008) smacks of hyperbole, it is true that there have been significant shifts–over time–in how people access information for academic and personal purposes. One noteworthy aspect of this change is that information seekers, especially those who have had Internet access for their entire lives, often lack an understanding of the context of the information that they retrieve, as many types of material are undifferentiated when accessed through Web search engines (Rowlands, et al., 2008). The traditional experts on how to identify information types–professors and librarians–are no longer the sole intermediaries between students and the universe of information (Davis, 2003; Robinson & Schlegl, 2005). Web search tools now play a significant role in determining the information that students find.

Eli Pariser, self-described as an "Internet organizer and disorganizer," has become influential in raising awareness of the reality that many Internet services, including the powerhouses of Google, Facebook, Microsoft, Yahoo, and major news websites, manipulate Internet users' search experiences using invisible algorithms (Pariser, 2011). Rather than retrieving an objective picture of the vast range of the Web with a given search request, Internet users instead operate within invisible "filter bubbles," or personalized pockets of information that Web services shape based on users' past behavior. The manipulation is largely based on users' social network connections as well as geography and previous searches. For example, a person who comments frequently on left-wing discussion forums and has a large number of Facebook contacts who frequently share articles on left-leaning magazines' websites would likely retrieve documents that favorably portray a left-wing political candidate whose name was searched in a Web search engine. This has obvious implications for students when searching for politically charged information via their preferred Web search tools if they lack the skill or inclination to evaluate their results thoroughly. Giant Web service providers, Pariser argues, have become the new gatekeepers to information.

Researchers employed in institutions of higher education, the traditional experts in their fields, generally rely on tacit knowledge for critical thinking tasks and therefore often take the issue for granted (Randall & Kellian, 2009). Students lack experts' specialized mental models of how information is produced and organized, which means that they must employ more structured strategies for evaluating information. Countless authors in the fields of education and information studies have explored the difficulties that students currently face in critically evaluating the information, especially that which is found online. Models of information-seeking behavior that consider the entirety of students' research process are instructive for educators. In particular, new explorations of how students make judgments about authority can elucidate strategies for facilitating skill development in the social media world.

One particularly significant contribution, given its broad scope and multifaceted approach, is Project Information Literacy at the University of Washington. The ongoing study consists of a large body of nuanced data on students' information-seeking behavior, but one general conclusion thus far is that "research seems to be far more difficult to conduct in the digital age than it did in previous times" (Head & Eisenberg, 2009). This might be self-evident to anyone who has experienced the recent shifts in information paradigms, but it is a truth that should prompt educators to explicitly pursue the development of their students' abilities in finding and using information. These skills cannot be taken for granted.

Head and Eisenberg (2010) have found that students generally select fairly conservative paths to finding information by using course readings first, recognizing this as a useful first step for identifying appropriate materials. They also found that students often take a collaborative approach to evaluating information, with the majority seeking the opinions of friends and friends. Fewer than half reported seeking assistance from professors or librarians. Social software, of course, certainly enhances the ease with which acquaintances' opinions can be gathered.

In the area of critical evaluation, it has been found that students frequently show an awareness of the need to evaluate information found on the Internet and actively apply basic criteria to judge the currency and authorship of a given document (Biddix, Chung, & Park, 2011; Head & Eisenberg, 2010; Meola, 2004; Metzger, Flanagin, & Zwarun, 2003). The problem, though, is that their skill sets are not necessarily strong enough to make sense of information that is created and distributed via social and networked channels. Randall and Kellian (2009), for example, found that although students recognize the need for authoritative sources, they have difficulty detecting bias. Head and Eisenberg (2010) likewise found that students had more difficultly in determining the credibility or "believability" of a resource found online than with the task of general evaluation (which includes judging the relevance and appropriateness of the source). Moreover, time pressures are a reality for students, leading them to privilege efficiency and accessibility over credibility in their information seeking (Biddix, et al., 2011).

Using a comparison of undergraduate political science students who completed a research training workshop with a control group, Dolowitz (2007) argues that students do in fact require training to understand the information landscape and to appropriately find, use, evaluate, and integrate print and digital resources in their writing assignments. Participants were presented with sets of resources that fell into a range of categories related to authority and accessibility: traditionally authoritative sources like subscription article indexes, newly free government data, social media like blogs and message forums, and tools for communicating with scholars (e.g., email). Students were not found to be particularly adept at understanding which tools were relevant for which purpose, as evidenced by their unsophisticated use of online research tools in particular, as well as their inability to differentiate academic information from other categories.

In an online environment where context is uncertain, credibility is often obscured. Therefore, students are required to go beyond the simple dichotomies of true/false and biased/unbiased information. They have to develop critical thinking skills that allow them to take into account the purpose and origin of texts. If students are to check for authority–information that originates from an "expert"–what are they to do when the nature of expertise is undergoing change? Since the creation and dissemination of information via Web 2.0 channels is now integral to research in political science, students who are being educated in the discipline must understand, in a subject-specific way that differs from that of other disciplines, how information is produced, and be able to evaluate it critically. The point here is not that political science students are different from other undergraduates, but that their *discipline* and its resources require different evaluation techniques.

EVALUATING POLITICAL SCIENCE INFORMATION IN A WORLD OF CHANGING EXPERTISE

Social media are now integral to the study of political science, both as an object of study and as a conduit for information dissemination. The field of electoral politics is most directly connected to Web 2.0 communication issues, but the technologies are also relevant in international relations, comparative politics, and even political theory (e.g., in discussions of the democratization of information creation).

Three principle categories of information have been redefined by the rise of social media:

- Primary sources
- Professional scholarship
- Contributions based on collective and distributed expertise

It is not insignificant that the Internet has granted (and continues to grant) access to vastly greater quantities of information than ever before, increasing the body of information to be searched and studied. The evolutions represented in each of these categories must also be understood in the context of the general information explosion, which itself has the effect of obscuring information authority and expertise. As discussed, the information search experience is also intertwined with social media, as evidenced by Pariser's (2011) "filter bubbles."

Primary Sources: The "Data" of Political Science

Political scientists have always studied original source documents, such as leaders' speeches and writings, election counts, public opinion polls, official government statements, and so on. Now, however, the source, credibility, origin, and provenance of such documents are often dubious. In many countries, politicians' personal communications have a strong presence in the realm of Web 2.0, as can be evidenced by a Web search for a given politician's name. However, personal communications can be spoofed, are not necessarily preserved or organized for future study, and are disseminated on a wide range of disconnected platforms, making it difficult to collect the full body of data. We need new methodologies for analyzing these ephemeral documents and for examining the public's reactions to them. The format in which politicians disseminate information has also changed, with, for example, 140-character messages on Twitter and images on the photo-sharing website Flickr. It should be noted that social media give politicians and public figures vastly more control over the portrayal of their message. Web 2.0 tools can facilitate direct communication streams with the public that no longer need to be filtered or spun through mass media. For example, Japanese politicians have affirmed the use of alternative social channels such as the video-sharing site Nico Nico Douga for press conferences in order to bypass mass media interpretations and commentary (McCargo & Hyon-Suk, 2010; Nagata, 2011). National governments and international organizations also engage in communication via Web 2.0 tools. For example, the public and researchers alike can get information updates via the United Nation's Facebook page. The relationship between these updates and other official communication channels is not necessarily explicit. Moreover, Facebook visitors can write their own posts on the UN's page and read other users' comments–adding a layer of commentary that is accessed directly alongside the primary source but usually does not identify the commentator. Numerical data are currently being produced and made available at astronomic rates, but the curator of the data is not always clear. The *Guardian* newspaper's World Government Data website is one of the many impressive tools to emerge in recent years that bring together datasets from disparate sources to be viewed and analyzed through one interface. While the source of the data is always noted, the question of reliability can be obscured if the *Guardian* is perceived to be the content provider. Interestingly, in addition to accessing data, users can comment on blog posts from the site's contributors, post links from the database directly into other social media sites, and even rate the quality of a given data set. Which begs the question: should users then trust the dataset with the highest user ratings? The Internet has also made possible the dissemination of previously inaccessible documents, some voluntarily released and some involuntarily. Recent years have seen a strong rise in transparency movements, which are dedicated to holding leaders and governments accountable for their action through the free sharing of information. While there are many players in the movement, the case of WikiLeaks has been the most dramatic in terms of implications for the nature of expertise in the study of politics. Founded by the now-notorious Julian Assange in 2006, WikiLeaks is a nonprofit media organization that collects highly sensitive and secret documents (in all formats including video) from anonymous sources and releases them–unfiltered–online. From there, the information is further distributed through a vast range of channels, such as mainstream news agencies, individuals' online social networks, and sites that mirror WikiLeaks' own content. The organization's name evokes a collaborative approach to collecting and disseminating information, but in fact, "expertise" in the provision of information is determined by WikiLeaks' leaders. Leaked documents have yielded groundbreaking data for analysis by researchers, including classified documents related to the US-led war in Iraq, private diplomatic cables with exchanges among leaders of several nations, and secretive international treaty negotiations (e.g., ACTA). The very existence of organizations like WikiLeaks has become an object of study, raising questions about the nature of government transparency and the ethics of Internet activism. It has been argued that with regard to the transparency movement, it both helps and hurts that we are living in a time of radical uncertainty about the "official" version of the truth. All kinds of "authoritative" claims made by leading public figures in recent years have turned out to be little more than thin air (Sifry, 2011, p. 16). The irony is, however, that the sources of the WikiLeaks information "dumps" themselves are kept anonymous, causing critics to doubt their authenticity. The WikiLeaks case represents a twofold truth for anyone seeking to evaluate authority in political science: official documents and statements cannot necessarily be trusted, and at the same time, the source and provenance of any primary document has to be questioned. In sum, networked dissemination of information has dramatically changed how political scientists and students access the primary data of their field, whether they are communications crafted by leaders and governments or information that is presented and curated collaboratively. More significantly, it has altered the criteria through which authenticity and expertise can be judged, as the source of the information itself now has a social dimension.

Professional Scholarship In the realm of secondary sources, academic experts are communicating outside of the usual information channels of the past. International relations scholars Carpenter and Drezner (2010) provide a useful survey of the use of Web 2.0 technologies in their field specifically, seeking to fill the "decided lack of scholarly discussion" (p. 256) on the profound impact of new media in teaching and research in the field. In the traditional world of academic publishing, scholars can engage in indirect discussion via letters to the editor, critical reviews of others' work, and citation. Now, however, discussions and works in progress are visible to the world. Informal conversations among scholars take place via blogs, microblogging services (e.g., Twitter), social networking sites (e.g., Academia.edu, Facebook, and LinkedIn), electronic mailing lists, and online discussion forums. Many widely read and trusted professional publications like *Foreign Policy*, the *Economist*, and *Foreign Affairs* include blog-style columns from regular contributors on which readers can freely comment. As in other fields, it has been argued that communication via social media furthers the research enterprise by enabling collaboration and exposing ideas to criticism before formal publication (Carpenter & Drezner, 2010). Moreover, transparency and open communication can circumvent some of the problematic practices of traditional peer review, where ghostwriting, plagiarism, nepotism, and falsified data can be uncovered, despite the near-sacredness of the institution. Academic communication now circumvents the traditional academic publishing model of peer review with the rise of social publishing sites and repositories like the Social Science Research Network (SSRN). Such services allow users to upload working papers, conference presentations, and versions of published articles for which they hold copyright, which readers can download, tag, and comment on. In addition, research groups make collaborative wikis available to the public, revealing their work and inviting commentary before publication. The popular search engine Google Scholar indexes this gray literature in addition to traditionally published materials. From the user's perspective, then, the search engine melds together works in many formats and stages of completion from writers who vary in their expertise: from graduate students' conference posters to master's theses to think-tank working papers to the most seminal journal articles ever published in the field. Information seeking has a social dimension as well. The notion of "scholarly impact" is also being redefined to include an individual's name recognition and level of participation in public dialogue (Smith, 2011). Many academics participate in social networks, and in addition to discussing issues, they use these channels to share information with others. Social networks are also a source of answers. Instead of seeking information independently, users frequently solicit book recommendations or advice on particular topics, trusting their network of contacts to provide a quick answer that is "good enough" for the purpose. Beyond personal contacts, there are also a number of Web 2.0 tools for sharing scholarly information. CiteULike and Mendeley, for example, are citation management programs that allow users to create profiles, share stored citations, and track the popularity of papers among other users. Many other services have features for generating recommendations for their users, such as the online retailer Amazon, which is well known for leveraging user data to promote further sales. Several tools take advantage of the idea in an academic setting–for example, the bX recommender service that many libraries use to recommend related articles to users.

Whether information is retrieved through personal or algorithmic social networks, the title of "expert" no longer belongs to a single individual. Collective intelligence has a real impact on which information sources are found and used. In the scholarly milieu, these socially based methods of information retrieval further the "filter bubble" phenomenon in which the information gatekeepers are now the Web services themselves (Pariser, 2011). For novices as well as experts, it can be difficult to separate the materials that adhere to the standards of the academic genre from those that constitute more informal communication or work in progress. At the same time, serious scholars cannot ignore Web 2.0 dissemination of information if they want to stay abreast of trends and have an impact in their field.

Contributions Based on Collective and Distributed Expertise

Social media are frequently linked to meaningful political participation (e.g., Rheingold, 2008). Indeed, popular social software like YouTube, Twitter, and Facebook have been in the forefront of democratic action around the globe, including US labor strikes, the contested Iranian elections of 2009, and the Arab Spring revolutionary movements in 2011. Much scholarship has been dedicated to studying the impact of Web 2.0 technology in elections and social movements, with the general conclusion being that the tools are immensely powerful in communicating messages and mobilizing citizens (e.g., Attia, Aziz, Friedman, & Elhusseiny, 2011; Jaeger, Paquette, & Simmons, 2010; Schuff, Mandviwalla, Williams, & Wattal, 2010). Those who study politics must now often take into account social communication technologies as one of the factors that influence political behavior.

Web 2.0 tools have also increased lateral communication among voting publics, leading to the rise of "opinion leaders" in online environments. Scholars of network analysis,[1] which has a long tradition in communication studies and political science, pursue the question of how people determine expertise and trust in a social setting. The public often determines expertise according to perceived social position and knowledge (Ahn, Huckfeldt, Mayer, & Ryan, 2008; Huckfeldt, 2001; McClurg, 2006). In social networks, then, single voices can rise in prominence, based not on their credentials, but on their ability to use social software effectively in relation to the community's values. In some cases, the simple fact that information is accessible online is more important than the perceived expertise of the creator (Kaid & Postelnicu, 2006). Moreover, in today's media-saturated world, citizens tend to rely on mass media pundits, especially ones that fit their pre-formed views, in order to avoid information overload. A pundit's success, again, is determined by a public following, which can be made or broken by a social media presence. News reporting is becoming more and more participatory, with major networks across the globe inviting audiences to submit their own reports, photographs, comments, and videos to supplement or even replace the agency's content. It is common practice for news organizations to have Twitter accounts and Facebook pages both to push out information and also to invite commentary. Overall, this movement has changed the nature of expertise in news reporting, as the line between reporters and viewers is increasingly blurred. At the same time, as in the idea of "filter bubbles," user-driven content can lead people to be exposed only to what they want to see, rather than to a larger world of information that is encountered serendipitously or chosen by a gatekeeper. Moreover, the anonymity of participation in political action in a Web 2.0 environment allows hoaxes to be propagated quite easily, as in the case of the blogger "Amina."

Social models of knowledge creation are increasingly influential in the world of information in general, as typified by the Web 2.0 giant Wikipedia. Reference works like dictionaries and handbooks are by their very definition produced by experts for the benefit of non-experts. However, Wikipedia is entirely produced by volunteer contributors, whose expertise is self-defined. The idea, then, is that everyone can contribute in order to amass a broad and varied compendium of knowledge, which, according to the oft-quoted description, is "an idea that could never work in theory, only in practice" (Cohen, 2011). The value of every Wikipedia article must be taken on its own terms, evaluated in the context of the task at hand. For those studying political science, Wikipedia represents a source of factual information on broad and minute topics alike–especially those related to current events–with the caveat that entries should be double-checked for accuracy. In addition, as with other Web 2.0 sources, Wikipedia provides evidence of popular conceptions of events and documents the issues that tend to mobilize participation in knowledge creation.

Societal relations, by definition, have an impact on political action, but in the world of Web 2.0, "society" spans geographic boundaries and creates conditions of anonymity that cannot exist in face-to-face communication. New experts who would not otherwise have a public voice can emerge. Content creators' views are accessible by all, but at the same time, but they are often mixed with traditionally published or otherwise vetted sources. The separation of categories of materials was more evident in the past, as "alternative" sources had to be specifically sought out. This amalgamation of material types necessitates careful examination of authorship and purpose before the information can be understood in its proper context.

In sum, Web 2.0 technologies and the rise of networked politics have changed:

1. The nature of primary sources in political science,
2. The manner in which academic information is communicated, and
3. The ways in which secondary information and commentary are created and shared.

The new, multifaceted networked channels through which political information is created and disseminated have broadened the field while increasing its complexity. It follows, then, that students likewise encounter difficulties in discerning expertise when faced with the task of evaluating information in the context of political science.

Information Literacy

Although the idea has been explored in many fields, the concept of "information literacy" has been championed by librarians over the past few decades. Information literacy can be defined as the ability to find, evaluate, and use information effectively (American Library Association, 1989). In the context of this paper, it should be emphasized that this definition includes an understanding of how information is produced and accessed. Traditionally, educators with a library background have tended to emphasize information-seeking aspects of information literacy–that is, the technical skills required for crafting a search strategy, navigating article indexes, and deciphering citations. However, the documents that buttress the theory of information literacy emphasize the full spectrum of information seeking and use.

Many organizations and individual institutions have developed standards for defining and measuring information literacy skills. The one most often cited is the Association of College and Research Libraries' (ACRL) Information Literacy Competency Standards for Higher Education, which consists of a system of overarching standards, performance indicators, and outcomes that act as evidence of information literacy skills. According to the five overarching standards, information literate students:

- Determine the extent of information needed
- Access the needed information effectively and efficiently
- Evaluate information and its sources critically
- Incorporate selected information into one's knowledge base
- Use information effectively to accomplish a specific purpose
- Understand the economic, legal, and social issues surrounding the use of information, and access and use information ethically and legally (Association of College and Research Libraries, 2000, "Information Literacy Defined")

Many institutions and individuals have adopted these standards to scaffold their information literacy education efforts. In addition, though, current theory in information literacy emphasizes the importance of the *disciplinary context* in students' successful development of the relevant competencies.

To this end, a document like the Political Science Research Competency Guidelines developed by the Law and Political Science Section of ACRL is quite useful in its elaboration of specific information literacy competencies via concrete and examples of specific competencies that students should develop in the context of the discipline (Association of College and Research Libraries, Law and Political Science Section, 2008). The ACRL document links the overarching standards with specific tasks that political science students might be required to complete, such as accessing demographic statistics from a government database, searching for articles in subject-specific indexes, or critiquing the methodology of public opinion surveys.

To take a broader example, the Quality Assurance Agency for Higher Education in the United Kingdom has created a set of "subject benchmarks" for the study of politics and international relations at the university level (Quality Assurance Agency for Higher Education, 2000). The document details learning outcomes that holders of undergraduate diplomas are expected to have reached. In addition to subject knowledge, graduates should develop skills in retrieving and critically evaluating political information and employ appropriate technologies for finding information and presenting their own ideas.

Information-literacy training based on competency standards has been widely applied in many institutions and research studies. The literature on political science education, too, testifies to several recent instances of this approach (Barberio, 2004; Dolowitz, 2007; Driver, Jette, & Lira, 2008; Marfleet & Dille, 2005; Olsen & Statham, 2005; Robinson & Schlegl, 2005; C. Stevens & Campbell, 2007; C. R. Stevens & Campbell, 2006, 2009; Thornton, 2006, 2008, 2010; Williams & Evans, 2008; Williams, Goodson, & Howard, 2006). However, a single-direction model of information creator/consumer is firmly ingrained in the codified approaches mentioned here. It is true that the ability to evaluate information underlies many of the competencies, for example in the indicators that refer to an understanding of the difference between "popular" and "scholarly" sources. In addition, the ability to select appropriate information for the task at hand is included. Students are meant to learn to recognize "authority," and the Law and Political Science Section of ACRL in particular advises that an information-literate student "consults with professors" and "consults with local experts on campus, in government, or in non-profit agencies" when faced with the task of validating information (2008, p. 10). In other words, traditional sources of authority are recommended. Traditional categories of information sources are emphasized, but the notions of dispersed, participatory authorship and revolutionary dissemination of primary information related to politics are not addressed. Further, social media are not mentioned in any of the standards related to accessing information. The exhortation to evaluate is there, but the task has new implications in a world in which expertise has been redefined by social media. In the current environment of socially shaped information flow, the standard categories of information do not necessarily apply. Moreover, the unique circumstances of political information must be taken into account when developing specific educational approaches.

Recommendations for Information Evaluation in Political Science

In order to further students' development of relevant information-literacy skills, competency standards must be brought to bear in a context-based, process-oriented manner that involves authentic situations and participatory learning. The old model of gatekeepers organizing information resources according to reliability and purpose no longer apply in the world of Web-based information retrieval, so students must develop strategies for dealing with information that is created and accessed through social channels. The competency standards' frameworks are generally broad enough to encompass new models of information dissemination, but the specific methods of teaching and assessing students' abilities need to be revisited in light of the importance of social information creation and retrieval.

Indeed, the approach of content-as-object in the information literacy standards does not take into account the complexities of information retrieval in the current age, when the actual channels of information themselves are implicated in how students conceptualize their learning (as argued by Whitworth, 2009). Moreover, the reliance on categorization of materials according to published format can be limiting, when, as has been argued, much of the subject of study involves the actual flow of information in addition to the final format. It is therefore recommended that educators abandon resource-centered training and focus instead on helping students to think critically in context (Head & Eisenberg, 2010).

Requiring students to cite a certain number of material types in their assignments is a common approach for furthering students' understanding of publishing and information dissemination. Robinson and Schlegl (2005) explore the effectiveness of this method, ultimately finding that providing grade incentives for citing certain material types is more effective than simply teaching students about the appropriateness of different types of information. Although they advise that professors *not* ban the use of the Internet, they note that students who are not graded on the inclusion of scholarly resources tend to rely more heavily on Web-based materials, to the possible neglect of more contextually suitable materials. However, they do not address the idea of information created collaboratively or through social media.

Explicit instruction on information types continues to be relevant, as beginning undergraduate students in particular do not necessarily understand the basic model of peer review and how a journal differs from a magazine. Despite the fact that they daily experience the sensation of information flowing around them, they should learn that certain types of information tend to collect in particular information resources. Dolowitz (2007) presents a training module designed to teach students to differentiate between different types of information sources, emphasizing heavily skills in categorizing information according to its purpose. Although the focus of the module was on identifying information types, it was significant that students were also taught to seek out individual experts and communicate via email, discussion groups, blogs, instant messaging, and Web conferencing. The implication here is that personal–as opposed to published–information has a real value for the research process. Pedagogy that eliminates discussion limits students' abilities to separate academic information from non-academic information and also their understanding of the larger process of scholarly communication. Exchanges through informal channels have long been an aspect of scholarly communication (e.g., letters to journals and conferences or colloquia), but until recently, this communication was not accessible to undergraduates. As researchers-in-training, it makes sense that students should become aware of informal professional discourse.

"Checklist" approaches to information evaluation are also quite common, as can be seen by examining nearly any instruction librarian's arsenal of information-literacy teaching tools.[2] These checklists generally require students to use external clues to evaluate the currency, purpose, authority, and validity of a source, with the aim of separating the proverbial wheat from the chaff. However, the checklist formula has come under criticism for its oversimplification of the evaluation task, guiding students toward binary judgments without prompting them to consider the larger context of the source (Dahl, 2009; Meola, 2004; Metzger, 2007). A more nuanced approach can be seen in evaluation tools that take a matrix approach.[3] A matrix-style worksheet requires students to rate an information source on a scale of appropriateness for their task according to the piece's authorship, approach, genre, and currency. Sources are evaluated according to a continuum of factors that often require students to seek out additional contextual information about documents and their creators. Students have to consider, for example, if a given document is reliable vis-à-vis its factual basis, or whether it follows standards required for the academic genre, or why it might be effective as evidence to support their own argument.

Indeed, strategies that emphasize context over absolute judgments are more relevant when working with information found via social media. In this environment, it is crucial for students to understand how particular content came to be created and why this matters in terms of larger patterns of power relations and communication. Networked channels can lead to information that is politically charged by its very existence, and thus students must evaluate both the content and how it came to be created in the first place. Critical judgment is a process of iterative steps related to the task at hand. In political science, for example, the appropriateness of an article or website depends on whether primary evidence or secondary commentary is sought. The relative judgment might subsequently change as a result of further investigation, for example, if a student learns more about a particular political figure's bias or communication track record. Along these lines, exercises in comparing documents can be quite effective for helping students recognize differences in purpose and credibility (Meola, 2004).

The traditional annotated bibliography can also be a very useful instruction tool for context-focused evaluation if students are required to explain who wrote a document and why, and why it is therefore relevant for their task. Preparing students to understand the different types of evidence and argumentative strategies in their discipline also promotes critical evaluation skills (Fitzgerald & Baird, 2011). Activities that incorporate evaluation of methods can also promote a focus on the entire writing process rather than mechanical information retrieval skills.

Head and Eisenberg (2010) report that few students in their study employed Web 2.0 tools for managing their own coursework. As social networks are now the channels through which political information is created and disseminated, students should be given more opportunities to explore the tools critically and understand their implications for the study of politics beyond students' own personal use. Another important strategy is to create authentic environments in which students both find information and also produce information themselves. For example, course assignments can be designed around Wikipedia: students can create their own entries, correct errors in existing articles, critically evaluate articles' accuracy, and engage in discussion with other contributors (as proposed in Carpenter & Drezner, 2010). The evaluation lesson here is twofold. First, students experience firsthand how easily information can be created and distributed by non-experts (such as themselves), underlining the need for critical evaluation of any type of information. Second, it demonstrates how information creation can be collaborative and potentially empowering in a social or political sense. Furthermore, some would argue, students can learn the values of citizenship by participating in politics as content creators (Sloam, 2008).

User feedback and "peer review" are defining features of Web 2.0 services. The "experts" behind user-supplied information and product ratings can range from mechanical usage statistics, to information collected by site creators, to judgments made by user communities. From Amazon's rating system, to Facebook's famous "Like" button, to retweets on Twitter, user-driven assessments are valued to help Internet users identify trustworthy or valuable information. Students who likewise seek out this type of judgment aid for academic information might be drawn to tools like Google Scholar's "cited by" function and journal publishers' "most viewed" article lists. Such ratings can be quite useful for prompting new directions in research, but it is important to recognize that there are limitations to the wisdom of the crowds. Students must understand that any given document should be evaluated on its own merit, regardless of how it was discovered in social channels.

One theme underlying all these recommendations is the value of inquiry-guided and participatory learning. Beginning with the influential 1998 report of the Boyer Commission on Educating Undergraduates in the Research University, there has been an ongoing movement in Europe and North America to better integrate research into undergraduate education. Beyond simply ingesting information and memorizing facts, the goal of this approach is for students to develop general and subject-specific skills for engaging in knowledge discovery, critical evaluation, and communication of ideas. For students to develop the skills needed to become researchers, they must mimic the research process of professionals. For political scientists, social media are an intrinsic part of this practice. Rather than being told that certain types of information are good or bad, students can go much farther in developing higher-order thinking skills when they experience the vagaries of the information environment firsthand. The reality is that today's students have always accessed information via the Internet and its complicated static and social flows of information (and presumably always will). Controlled classroom learning that focuses on the traditional roles of professors and librarians as information experts will not be transferrable to the world beyond the classroom. Thus, students are better served when educators assume the role of facilitators and guides that prompt them to question what they find and illustrate the implications of social communication channels for the enterprise of academia and political engagement in the world at large.

FUTURE RESEARCH DIRECTIONS

While many studies have examined students' and other groups' behavior in finding and evaluating information, more research is needed to investigate the effect, if any, of "filter bubbles" and Web 2.0 phenomena–like user ratings on young people's information behavior. It would likewise be interesting to investigate the same question among professional researchers, with comparisons made across disciplines.

Inquiry-based approaches to education have been widely discussed for many years, but more work is needed to assess their effectiveness in particular contexts, specifically in terms of information evaluation. Because the entrée to information-literacy skills development is so subject dependent, further work could focus on whether regional, national, and institutional differences affect how students evaluate information. Similarly, more systematic research on disciplinary differences in information use practices, especially with regard to Web 2.0 and the question of evaluation is needed.

In order to develop specific learning objects and lessons plans that foster skills in understanding expertise and evaluating information, new models for assessing credibility that take into account users' tasks and motivation beyond absolute judgments are needed (as recommended by Metzger, 2007).

Finally, more research is needed to investigate how experts in the field of political science themselves judge expertise. The academic profession is undergoing generational shifts, and it is important to better understand how social software will continue to shape academic communication in coming years. The next step, then, would be to compare this to how students evaluate information, and the gaps found therein could inform further pedagogical recommendations.

CONCLUSION

Students are now expected to differentiate between different types of expertise when they are seeking information, and sometimes even to participate in content creation themselves. Given the changing nature of expertise, evaluating information in the context of political science is a daunting task. Traditional gatekeepers to information are no longer relevant as the sole determinants of the information that students access. Dahl (2009) effectively summarizes the challenge of information-literacy training in a world of changing communication modes:

Applying narrow definitions of words such as "peer-reviewed," "scholarly" and "academic," and "expert" to the research task reinforces boundaries that are being pushed and, in some cases, eroded by the very people who engage in the task of scholarly investigation and communication (p. 159).

Expertise is defined by disciplinary context, ergo, information literacy has to be developed in the context of the subject, the task at hand, and the specific information resources required. This is not to say that the competencies are not transferrable, but the development of information literacy skills is most effectively approached via subject-based entry points. Social software and networked communication's integration with the study of politics–both in terms of the subject matter and methodology–has influenced the meaning of expertise and authority, demanding that subject-specific evaluation skills be taught in political science education.

In order to be successful, students must learn about the primary sources of political science–the data to be studied–and how this is more than ever the product of communication via social networks, which can empower citizens while simultaneously creating a false illusion of comprehensibility. They must also understand that academic communication happens via informal Web 2.0 channels in which traditional publishing models are only part of the picture. Finally, new experts now contribute commentary and data relevant to the study of politics, adding a further layer of complexity to the task of information evaluation. The solution to fostering students' development of these skills goes beyond the basic frameworks of information literacy competencies and binary checklists for evaluation. Students can explore the importance of information evaluation by experiencing first-hand the uncertainty of information overload. If they are given inquiry-based prompts instead of discrete tasks, they will be motivated to seek out wider ranges of sources that require more nuanced evaluation skills.

Political science education, like all disciplines, benefits from collaboration among different types of educators. As Meola (2004) points out, librarians cannot claim to be the sole experts in compiling a finite body of vetted information resources in today's information environment. However, they can advise learners and researchers on how information is produced and distributed, as well as the technical mechanics of information retrieval. They understand the theory and practice of information literacy and how holistic approaches should be taken to develop the fundamental skills of lifelong learning. University faculty are experts in their relative fields and are responsible for helping students learn the subject content of a course. As has been described, though, critical evaluation of information is not divorced from the subject matter itself.

With the guidance of librarians and professors, students can learn to use techniques like comparison and context-based judgments to evaluate how an information source was produced and why it may relevant for their purposes. As Whitworth (2009) argues, "the idea that all one needs to do to learn about something is to retrieve information that has already been created by others is actually no more than the old behaviourist approach to education reconstructed for the Internet age" (p. 103). For all their demonstration of information literacy competencies, students must still read, create knowledge, and communicate it to others in order to learn, and the task of distinguishing "expertise" must happen within this context.

REFERENCES

Addley, E. (2011, June 13). Gay girl in Damascus hoaxer acted out of vanity. *The Guardian*. Retrieved from http://www.guardian.co.uk/world/2011/jun/13/gay-girl-damascus-tom-macmaster

Ahn, T. K., Huckfeldt, R., Mayer, A. K., & Ryan, J. B. (2008, August). *Availability and the centrality of experts in the communication of political information*. Paper presented at the Annual Meeting of the American Political Science Association, Boston, MA.

American Library Association. (1989). *Presidential Committee on Information Literacy: Final report*. Retrieved June 16, 2011, from http://www.ala.org/ala/mgrps/divs/acrl/publications/whitepapers/presidential.cfm

Association of College and Research Libraries. (2000). *Information literacy competency standards for higher education*. Retrieved June 29, 2011, from http://www.ala.org/ala/mgrps/divs/acrl/standards/informationliteracycompetency.cfm

Association of College and Research Libraries. Law and Political Science Section. (2008). *Political science research competency guidelines*. Retrieved June 16, 2011, from http://www.acrl.org/ala/mgrps/divs/acrl/standards/PoliSciGuide.pdf

Attia, A. M., Aziz, N., Friedman, B., & Elhusseiny, M. F. (2011). Commentary: The impact of social networking tools on political change in Egypt's "revolution 2.0.." *Electronic Commerce Research and Applications*, *10*(4), 369–374. doi:10.1016/j.elerap.2011.05.003

Barberio, R. P. (2004). The one-armed bandit syndrome: Overuse of the Internet in student research projects. *PS: Political Science & Politics, 37*(2), 307-311. doi: 10.1017.S1049096504004275

Biddix, J. P., Chung, C. J., & Park, H. W. (2011). Convenience or credibility? A study of college student online research behaviors. *The Internet and Higher Education*, *14*(3), 175–182. doi:10.1016/j.iheduc.2011.01.003

Boyd, D. (2010). Streams of content, limited attention: The flow of information through social media. *EDUCAUSE Review*, *45*(5), 26–36. Retrieved from http://www.educause.edu/EDUCAUSE+Review/EDUCAUSEReviewMagazineVolume45/StreamsofContentLimitedAttenti/213923

Boyer Commission on Educating Undergraduates in the Research University. S. S. Kenny, chair. (1998). Reinventing undergraduate education: A blueprint for America's research universities. Stony Brook, NY: State University of New York–Stony Brook.

Carpenter, C., & Drezner, D. W. (2010). International relations 2.0: The implications of new media for an old profession. *International Studies Perspectives*, *11*(3), 255–272. doi:10.1111/j.1528-3585.2010.00407.x

Carr, N. (2008). Is Google making us stupid?: Why you can't read the way you used to. *Atlantic Monthly*, *56*, Retrieved from http://www.theatlantic.com/magazine/archive/2008/07/is-google-making-us-stupid/6868/

Cohen, N. (2011, May 23). *Wikipedia*. Retrieved June 29, 2011, from http://topics.nytimes.com/top/news/business/companies/wikipedia/index.html

Dahl, C. (2009). Undergraduate research in the public domain: The evaluation of non-academic sources online. *RSR. Reference Services Review*, *37*(2), 155–163. doi:10.1108/00907320910957198

Davis, P. M. (2003). Effect of the web on undergraduate citation behavior: Guiding student scholarship in a networked age. *portal. Libraries and the Academy*, *3*(1), 41–51. doi:10.1353/pla.2003.0005

Dolowitz, D. P. (2007). The Big E: How electronic information can be fitted into the academic process. *Journal of Political Science Education*, *3*(2), 177–190. doi:10.1080/15512160701338338

Driver, D., Jette, K., & Lira, L. (2008). Student learning identities: Developing a learning taxonomy for the political science classroom. *Journal of Political Science Education*, *4*(1), 61–85. doi:10.1080/15512160701816135

Fitzgerald, J., & Baird, V. A. (2011). Taking a step back: Teaching critical thinking by distinguishing appropriate types of evidence. *PS: Political Science & Politics*, *44*(3), 619–624. doi:10.1017/S1049096511000710

Gunnell, J. G. (2011). Political science, History of. In G. T. Kurian (Ed.), *The Encyclopedia of Political Science*. Retrieved from http://library.cqpress.com/

Hassan, N. (2011, June 7). Syrian blogger Amina Abdallah kidnapped by armed men. *The Guardian*. Retrieved from http://www.guardian.co.uk/world/2011/jun/07/syrian-blogger-amina-abdallah-kidnapped

Head, A. J., & Eisenberg, M. B. (2009). *Finding context: What today's college students say about conducting research in the digital age*. Retrieved from http://projectinfolit.org/pdfs/PIL_Progress-Report_2_2009.pdf

Head, A. J., & Eisenberg, M. B. (2010). *Truth be told: How college students evaluate and use information in the digital age*. Retrieved from http://projectinfolit.org/pdfs/PIL_Fall2010_Survey_FullReport1.pdf

Huckfeldt, R. (2001). The social communication of political expertise. *American Journal of Political Science*, *45*(2), 425–438. doi:10.2307/2669350

Jaeger, P. T., Paquette, S., & Simmons, S. N. (2010). Information policy in national political campaigns: A comparison of the 2008 campaigns for President of the United States and Prime Minister of Canada. *Journal of Information Technology & Politics*, *7*(1), 67–82. doi:10.1080/19331680903316700

Kaid, L. L., & Postelnicu, M. (2006). Credibility of political messages on the Internet: A comparison of blog sources. In M. Tremayne (Ed.), *Blogging, citizenship, and the future of media* (pp. 149–164). New York, NY: Routledge.

Marfleet, B. G., & Dille, B. J. (2005). Information literacy and the undergraduate research methods curriculum. *Journal of Political Science Education*, *1*(2), 175–190. doi:10.1080/15512160590961793

Marsh, K. (2011, May 7). A gay girl in Damascus becomes a heroine of the Syrian revolt. *The Guardian*. Retrieved from http://www.guardian.co.uk/world/2011/may/06/gay-girl-damascus-syria-blog

McCargo, D., & Hyon-Suk, L. (2010). Japan's political tsunami: What's media got to do with it? *The International Journal of Press/Politics*, *15*(2), 236–245. doi:10.1177/1940161210361588

McClurg, S. D. (2006). The electoral relevance of political talk: Examining disagreement and expertise effects in social networks on political participation. *American Journal of Political Science*, *50*(3), 737–754. doi:10.1111/j.1540-5907.2006.00213.x

Meola, M. (2004). Chucking the checklist: A contextual approach to teaching undergraduates website evaluation. *portal. Libraries and the Academy*, *4*(3), 331–344. doi:10.1353/pla.2004.0055

Metzger, M. J. (2007). Making sense of credibility on the web: Models for evaluating online information and recommendations for future research. *Journal of the American Society for Information Science and Technology*, *58*(13), 2078–2091. doi:10.1002/asi.20672

Metzger, M. J., Flanagin, A. J., & Zwarun, L. (2003). College student web use, perceptions of information credibility, and verification behavior. *Computers & Education*, *41*(3), 271–290. doi:10.1016/S0360-1315(03)00049-6

Nagata, K. (2011, June 10). New media keep old media honest. *The Japan Times Online*. Retrieved from http://search.japantimes.co.jp/cgi-bin/nb20110610a1.html

Olsen, J., & Statham, A. (2005). Critical thinking in political science: Evidence from the introductory comparative politics course. *Journal of Political Science Education*, *1*(3), 323–344. doi:10.1080/15512160500261186

Pariser, E. (2011). *The filter bubble: What the Internet is hiding from you*. New York, NY: Penguin Press.

Polsby, N. W. (2001). Political science: Overview. In J. S. Neil & B. B. Paul (Eds.), *International encyclopedia of the social & behavioral sciences* (pp. 11698–11701). Amsterdam, The Netherlands: Elsevier. doi:10.1016/B0-08-043076-7/01283-3

Quality Assurance Agency for Higher Education. (2000). *Subject benchmark statements: Politics and international relations*. Retrieved June 16, 2011, from http://www.qaa.ac.uk/Publications/InformationAndGuidance/Pages/Subject-benchmark-statement-Polictics-and-international-relations.aspx

Randall, M., & Kellian, C. (2009). How do you know that?: An investigation of student research practices in the digital age. *portal. Libraries and the Academy*, *9*(1), 115–132. doi:10.1353/pla.0.0033

Rheingold, H. (2008). Using participatory media and public voice to encourage civic engagement. In W. L. Bennett (Ed.), *Civic life online: Learning how digital media can engage youth* [PDF version]. (pp. 97–118).

Robinson, A. M., & Schlegl, K. (2005). Student use of the Internet for research projects: A problem? Our problem? What can we do about it? *PS: Political Science and Politics*, *38*(2), 311–315.

Rowlands, I., Nicholas, D., Williams, P., Huntington, P., Fieldhouse, M., & Gunter, B. et al. (2008). The Google generation: The information behaviour of the researcher of the future. *Aslib Proceedings*, *60*(4), 290–310. doi:10.1108/00012530810887953

Schuff, D., Mandviwalla, M., Williams, C., & Wattal, S. (2010). Web 2.0 and politics: The 2008 U.S. presidential election and an e-politics research agenda. *Management Information Systems Quarterly*, *34*(4), 669–688.

Sifry, M. L. (2011). *Wikileaks and the age of transparency*. Berkeley, CA: Counterpoint.

Sloam, J. (2008). Teaching democracy: The role of political science education. *British Journal of Politics and International Relations*, *10*(3), 509–524. doi:10.1111/j.1467-856X.2008.00332.x

Smith, D. (2011, June 28). How do you solve a problem like clay shirky? Or, Silicon Valley discovers impact factor. Retrieved from http://scholarlykitchen.sspnet.org/2011/06/28/how-do-you-solve-a-problem-like-clay-shirky-or-silicon-valley-discovers-impact-factor/

Stevens, C., & Campbell, P. (2007). The politics of information literacy: Integrating information literacy into the political science curriculum. In T. Jacobson & T. Mackey (Eds.), *Information literacy collaborations that work* (pp. 123–146). New York, NY: Neal-Schuman.

Stevens, C. R., & Campbell, P. J. (2006). Collaborating to connect global citizenship, information literacy, and lifelong learning in the global studies classroom. *RSR. Reference Services Review*, *34*(4), 536–556. doi:10.1108/00907320610716431

Stevens, C. R., & Campbell, P. J. (2009). Collaborating with librarians to develop lower division political science students' information literacy competencies. *Journal of Political Science Education*, *4*(2), 225–252. doi:10.1080/15512160801998114

Thornton, S. (2006). Information literacy and the teaching of politics. *LATISS: Learning and Teaching in the Social Sciences*, *3*(1), 29–45. doi:10.1386/ltss.3.1.29_1

Thornton, S. (2008). Pedagogy, politics and information literacy. *Politics*, *28*(1), 50–56. doi:10.1111/j.1467-9256.2007.00310.x

Thornton, S. (2010). From "scuba diving" to "jet skiing"? Information behavior, political science, and the Google generation. *Journal of Political Science Education*, *6*(4), 353–368. doi:10.1080/15512169.2010.518111

Whitworth, A. (2009). *Information obesity*. Oxford, UK: Chandos. doi:10.1533/9781780630045

Williams, M. H., & Evans, J. J. (2008). Factors in information literacy education. *Journal of Political Science Education*, *4*(1), 116–130. doi:10.1080/15512160701816234

Williams, M. H., Goodson, K. A., & Howard, W. G. (2006). Weighing the research paper option: The difference that information literacy skills can make. *PS: Political Science & Politics*, *39*(3), 513–519. doi:10.1017/S1049096506060793

ADDITIONAL READING

Andersen, J. (2006). The public sphere and discursive activities: Information literacy as sociopolitical skills. *The Journal of Documentation*, *62*(2), 213–228. doi:10.1108/00220410610653307

Badke, W. (2010). How stupid is Google making us? *Online*, *34*(6), 51–53.

Bos, A. L., & Schneider, M. C. (2009). Stepping around the brick wall: Overcoming student obstacles in methods courses. *PS: Political Science & Politics*, *42*(2), 375–383. doi:10.1017/S1049096509090519

Boys, J. D., & Keating, M. F. (2009). The policy brief: Building practical and academic skills in international relations and political science. *Politics*, *29*(3), 201–208. doi:10.1111/j.1467-9256.2009.01356.x

Davis, P. M. (2002). The effect of the web on undergraduate citation behavior: A 2000 update. *College & Research Libraries*, *63*(1), 53–53.

Deardorff, M. D. (2005). Assessment through the grassroots: Assessing the department via student peer evaluation. *Journal of Political Science Education*, *1*(1), 109–127. doi:10.1080/15512160590907649

Dillon, J., & McKeel, W. (2005). Political nihilism, alternative media and the 2004 presidential election. *SIMILE: Studies in Media & Information Literacy Education*, *5*(2), 1–12. doi:10.3138/sim.5.2.003

Ernst, H. R., & Ernst, T. L. (2005). The promise and pitfalls of differentiated instruction for undergraduate political science courses: Student and instructor impressions of an unconventional teaching strategy. *Journal of Political Science Education*, *1*(1), 39–59. doi:10.1080/15512160590907513

Forte, A. (2009). *Learning in public: Information literacy and participatory media.* Unpublished doctoral dissertation, Georgia Institute of Technology. Retrieved from http://hdl.handle.net/1853/29767

Foster, N. (2007). *Studying students: The undergraduate research project at the University of Rochester*. Chicago, IL: Association of College and Research Libraries.

Greenhow, C., & Robelia, B. (2009). Informal learning and identity formation in online social networks. *Learning, Media and Technology*, *34*(2), 119–140. doi:10.1080/17439880902923580

Greer, C. F. (2005). How local television stations used the web to report the second anniversary of the 9/11 terrorist attacks. *SIMILE: Studies in Media & Information Literacy Education*, *5*(3), 1–10. doi:10.3138/sim.5.3.002

Head, A. J., & Eisenberg, M. B. (2011). How college students use the web to conduct everyday life research. *First Monday*, *16*(4). Retrieved from http://firstmonday.org/htbin/cgiwrap/bin/ojs/index.php/fm/article/view/3484/2857

Heaney, M. T., & McClurg, S. D. (2009). Social networks and American politics. *American Politics Research*, *37*(5), 727–741. doi:10.1177/1532673X09337771

Huckfeldt, R. (2001). The social communication of political expertise. *American Journal of Political Science*, *45*(2), 425–438. doi:10.2307/2669350

Ishiyama, J. (2002, August). *Participation in undergraduate research and the development of political science students.* Paper presented at the Annual Meeting of the American Political Science Association, Boston, MA.

Ishiyama, J., Breuning, M., & Lopez, L. (2006). A century of continuity and (little) change in the undergraduate political science curriculum. *The American Political Science Review*, *100*(4), 659–665. doi:10.1017/S0003055406062551

Jenkins, H., Clinton, K., Purushotma, R., Robison, A. J., & Weigel, M. (2006). *Confronting the challenges of participatory culture: Media education for the 21st century*. Retrieved from http://digitallearning.macfound.org/site/c.enJLKQNlFiG/b.2029291/k.97E5/Occasional_Papers.htm

Livingstone, S. (2008). Internet literacy: Young people's negotiation of new online opportunities. In T. McPherson (Ed.), *Digital youth, innovation, and the unexpected* (pp. 101–122). Cambridge, MA: MIT Press.

Marks, M. P. (2008). Fostering scholarly discussion and critical thinking in the political science classroom. *Journal of Political Science Education*, *4*(2), 205–224. doi:10.1080/15512160801998080

Metzger, M. J., & Flanagin, A. J. (2011). *Credibility and digital media @ UCSB*. Retrieved June 16, 2011, from http://www.credibility.ucsb.edu/index.php

Omelicheva, M. Y. (2005). Self and peer evaluation in undergraduate education: Structuring conditions that maximize its promises and minimize the perils. *Journal of Political Science Education*, *1*(2), 191–205. doi:10.1080/15512160590961784

Richey, S. (2009). Hierarchy in political discussion. *Political Communication*, *26*(2), 137–152. doi:10.1080/10584600902851419

Rieh, S. Y., & Hilligoss, B. (2008). College students' credibility judgments in the information-seeking process. In T. McPherson (Ed.), *Digital youth, innovation, and the unexpected* (pp. 49–72). Cambridge, MA: MIT Press.

Stevens, C. R., & Campbell, P. J. (2006). Collaborating to connect global citizenship, information literacy, and lifelong learning in the global studies classroom. *RSR. Reference Services Review*, *34*(4), 536–556. doi:10.1108/00907320610716431

Teslik, L. H., & Schrage, E. (2009). *New media tools and public diplomacy*. Retrieved June 18, 2011, from http://www.cfr.org/public-diplomacy/new-media-tools-public-diplomacy/p19300

Thornton, S. (2009). Lessons from America: Teaching politics with the Google generation. *Enhancing Learning in the Social Sciences (ELiSS)*, *1*(2), 1-20. Retrieved from http://www.eliss.org.uk/PreviousIssues/Volume1Issue2/ViewArticle/tabid/72/itemid/68/pubtabid/242/repmodid/411/Default.aspx

Tremayne, M. (2006). *Blogging, citizenship, and the future of media*. New York, NY: Routledge.

Trudeau, R. H. (2005). Get them to read, get them to talk: Using discussion forums to enhance student learning. *Journal of Political Science Education*, *1*(3), 289–322. doi:10.1080/15512160500261178

Warschauer, M., & Grimes, D. (2007). Audience, authorship, and artifact: The emergent semiotics of web 2.0. *Annual Review of Applied Linguistics*, *27*, 1–23. doi:10.1017/S0267190508070013

Williams, V. C. (2006). Assuming identities, enhancing understanding: Applying active learning principles to research projects. *Journal of Political Science Education*, *2*(2), 171–186. doi:10.1080/15512160600669015

Wilton, S., & Polkinghorne, S. (2010). Research is a verb: Exploring a new information literacy-embedded undergraduate research methods course. *Canadian Journal of Information and Library Science*, *34*(4), 457–473.

KEY TERMS AND DEFINITIONS

Blog: A website used to quickly publish short texts, news, commentary, etc. Usually displayed in reverse chronological order. Often similar in content to a diary or newspaper column.

Filter Bubbles: A term coined by Eli Pariser to describe the limited pockets of information to which Internet users are exposed as a result of Web services' invisible algorithms that display information based on users' behavior.

Information Literacy: The skills required to find, evaluate, and use information effectively.

Microblog: Web services that allow users to post very short messages (usually one or two sentences), images, or videos that other users can view in real time. Twitter is the best-known microblogging service.

Political Science: A field of inquiry that focuses on systems of governments, power relations within and among nations, and human behavior in relation to these systems.

Primary Source: A text or recording (in any medium) that is studied as the data or evidence of an event or phenomenon. The designation of "primary" is relative to the purpose of study.

Social Media: Web-based applications that allow users to create and share content, often in a collaborative manner. Examples of social media include social networks, photo- and video-sharing websites, wikis, message forums, social bookmarking websites, blogs, and microblogging services. Also known as Web 2.0 technologies.

Social Networks: Websites that allow users to create profiles and form contacts with other users to communicate and share information in various media. Widely used social networks include Facebook, LinkedIn, Academia.edu, and MySpace.

Secondary Source: Commentary on or analysis of a text or event, usually involving interpretation of an original source, artifact, or observation.

ENDNOTES

1 For recent contributions, see the September 2009 special issue of *American Politics Research* (volume 3, number 5).

2 See, for example, the materials compiled in the DMOZ open project directory: http://www.dmoz.org/search?q=web+evaluation

3 See for example De Montfort University's Information Sources Matrix: http://www.library.dmu.ac.uk/Images/Selfstudy/ISEM-Leaflet.pdf

This work was previously published in "Social Software and the Evolution of User Expertise" edited by Tatjana Takševa, pages 180-200, copyright 2013 by Information Science Reference (an imprint of IGI Global).

Chapter 53
Subject Specialization and Science Teachers' Perception of Information and Communication Technology for Instruction in Kwara State

Michael Ayodele Fakomogbon
University of Ilorin, Nigeria

Mosiforeba Victoria Adegbija
University of Ilorin, Nigeria

Rachael Funmi Adebayo
Landmark University, Nigeria

Ahmed Tajudeen Shittu
Al-Hikmah University, Nigeria

Oloyede Solomon Oyelekan
University of Ilorin, Nigeria

ABSTRACT

This study examined Kwara State secondary school science teachers' perception of ICT for instruction based on their area of specialization. Participants were 630 science teachers of Biology, Chemistry, Physics and Mathematics from both public and private senior secondary schools in 12 Local Government Areas in Kwara State. The data collected through questionnaires tagged Science Teachers Information and Communication Technology Perception Questionnaire (STICOTEPQ) were analysed using frequency count, mean, chi-square and Analysis of Variance (ANOVA) statistics. Findings revealed that (i) the science teachers' perception of ICT was generally positively low and (ii) there was significant difference in the secondary school science teachers' perception of ICT for instruction based on their area of specialization{$F(3,252) = 50.323$, $p<0.05$}with the Biology teachers having the highest level of perception while the Mathematics teachers had the lowest perception. The implication of this study was that should the secondary school science teachers remain at this level of perception of ICT for instruction, the probability that ICT will be used for classroom instruction in Kwara State is very low. It was therefore recommended that ICT should be fully integrated into science teacher education at the Colleges of Education and the Universities and that Kwara State Government should formulate a programme of ICT training and retraining of teachers currently in service as a way of improving their perception of the use of ICT for instruction.

DOI: 10.4018/978-1-4666-7363-2.ch053

INTRODUCTION

Science education plays important role in the industrial and technological development of a nation (Ajewole 2003). Realizing the importance of scientific literacy in personal decision making, which eventually will have impact on the society, involvement in civic and cultural affairs and in economic productivity (Dickey & Singer, 2004), the Nigeria National Policy on Education made science a compulsory subject both at the lower and upper basic education (UBE) levels. At the senior secondary level, science is broken down into three components – Physics, Chemistry and Biology out of which each student must choose to offer at least one (Federal Republic of Nigeria-FRN, 2004). Mathematics is not officially classified as a science subject because it is central to all subjects especially science subjects. This is the reason why it is a compulsory subject for all the students at the Junior and Senior Secondary levels of education in Nigeria.

Researchers in Science education have confirmed that this area of study has been faced with many challenges such as student truancy and rote learning which lead to poor performances in the Secondary School Leaving Certificate Examination (SSCE). This has largely been attributed to poor teaching method by science teachers, abstractness of concept and large student population (Bamigbala, 2000, Seymour, 2001, Ajewole, 2003, Oyelekan, 2009). Most of the traditional teaching method employed by the science teachers are rather teacher-centered than learner-centered. This makes the abstractness of the concept stronger and the environment dull and boring since the learners are normally not given room to be involved in the teaching and learning process. In order to realize the vision of science education, emphasis should be placed on inquiry, scientific literacy and focus on understanding rather than surface knowledge (Dickey & Singer, 2004). As a result, there is the need for a teaching method in which the learners will be involved thereby arousing their interest, motivating them to learn and helping them in problem solving. This is the essence of electronic education (e-education). Information and Communication Technology, that is, all range of hardware and software which can be used to collect, store, process, transmit and present the scientific concept to be taught and learned in order improve learning. ICT involves use of computers, internet and electronic mail. ICT in education is referred to as e-learning and it has a great potential to complement and supplement traditional teaching and learning process.

The introduction of the new technology to education with emphasis on ICT makes it possible to review the teaching and learning processes especially in sciences. Several researchers have indicated potentials of ICT in education regarding improvement of learning when ICT is employed (Dawson, Forster & Reid, 2006, Bakar & Mohammed, 2000, Ofsted, 2009 and Yusuf & Balogun, 2011). Science teachers should therefore be encouraged to integrate ICT into classroom activities. For instance, Heinich, Molenda, Russel and Smaldino (2002) reported that ICT acts as an instructor-independent media which acts as a catalyst for change in the whole instructional environment; Ofsted (2009) and Yusuf & Balogun (2011) opined that ICT improves student achievement and removes boredom and superficial questioning; while Arnold, Padilla and Chart (2009) opined that ICT helps the teachers and learners develop Knowledge. Yuen, Lee, Law and Chan (2008) opined that ICT helps to foster the development of students' 21st Century skills in becoming life-long learners and ability to work and learn from experts and peers in a global community. Kwache (2007) opined that the management of education and the traditional process of teaching and learning have been seriously challenged by the new waves of technology.

There are many factors affecting the secondary school science teachers' integration of ICT into teaching and learning. These include their background knowledge of ICT during training.

Ololube (2006) reported that the use of ICT had not been part of the teacher training curriculum in Nigeria. Other factors include commitment to professional learning, background in formal computer training and teachers' beliefs about ICT (Mumtaz, 2000 and Yusuf & Balogun, 2011). According to Goodywm, Adams and Clarke (1997) the teachers' belief is central to ICT integration. They in their work classified teachers into 3 groups based on their feelings: the fearful, the unresolved and optimistic. Adebayo (2008) reported that science teachers did not integrate ICT into classroom instruction because of their negative perception of ICT in instruction.

The perception of a user of ICT will result in the user's attitude to ICT. Hence attitude may not be adequately measured without the understanding of the perception of the user. Adediwura and Tayo (2007) defined perception as "the reaction elicited when an impression is perceived from without after making connections with other materials in the consciousness (memory)" (p.165). Gulbahar and Guven (2008) reported that there was a strong correlation between teachers' attitude towards ICT in education and their perception of the advantage of the use of ICT. In other words the attitude of the teachers is hinged on their perception of the new technology (Kwache, 2007).

Different people from different fields have given different definitions to the word perception. Encarta (2010) defined perception as any neurological process of acquiring and mentally interpreting information from the senses. In this study, science teachers' perception of ICT is defined as the science teachers understanding and interpretation of what they have heard, learnt and seen about ICT as an instructional tool. Teachers' perception is so important such that it can have a serious effect on the teaching and learning process (Yuen, Lee, Law & Chan, 2008). One of the major variables that can affect the perception of ICT by the science teachers is the teachers' quality. According to Ololube (2006), the quality of teachers is a key prediction of students' learning. This quality includes area of specialization, that is, whether or not the science teachers are trained in the area of the particular subject they are teaching. Specialization helps a teacher to be well equipped to teach the subject effectively and meaningfully (Okonkwo, 2000 and Ukah, 2007).

The main purpose of this study was to investigate the secondary school science teachers' perception of ICT for instruction based on specialization. Specifically, the study examined

1. The Kwara State secondary school science teachers' perception of ICT for instruction
2. The Influence of subject specialization on the Kwara state secondary school science teachers' perception of ICT for instruction

Research Questions

In this study, the following research questions were answered:

1. What is the Kwara State secondary school teachers' perception of ICT for instruction?
2. What is the influence of area of specialization on secondary school science teachers' perception of ICT for instruction?

Research Hypothesis

The second research question gave rise to the following research hypothesis:

There is no significant difference in the Kwara State secondary school science Teachers' perception of ICT for instruction based on area of specialization.

Research Methodology

This study was a descriptive research of survey type.

The target population was all the senior secondary school science teachers in Kwara State. Six hundred and thirty science teachers were

Table 1. Distribution of respondents by area of specialization

Area of Specialization	Frequency of Teachers	Percentage
Biology	239	37. 94
Physics	137	21.75
Chemistry	172	27.30
Mathematics	82	13.02
Total	630	100

purposively sampled from the randomly selected 157 secondary schools from 13 Local government areas (LGAs) out of the 16 LGAs in Kwara State.

The instrument used is a questionnaire titled Science Teachers' Information and Communication Technology Perception Questionnaire (STICOTEPQ) adapted from Gulbahar and Guven (2008). This consisted of two sections A and B. Section A was on participants demographic data consisting of 7 questions; while section B was on their perception of ICT for instruction consisting of 24 questions. The section B, although adapted from Gulbahar and Guven (2008), was different in response mode. While that of the Gulbahar and Guven (2008) was of 3-Likert scale: Good, Average and None; the adapted one was of 4-Likert scale: Strongly agreed (SA), Agreed (A), Disagree (D) and Strongly disagreed (SD). The rating scales were scored 4, 3, 2 and 1 respectively. Section B was further divided into D1 and D2. D1 contained statements on teacher's perception of the impact of the use of ICT for instruction while D2 contained statements on the teachers' perception of the impact of using ICT on students' achievement. A typical question in D1 reads 'ICT use for instructional purposes may not be as useful as traditional and electronic resources (books, overhead projectors etc.)'. A typical question in D2 reads 'Using ICT for instruction will assist in developing the independent learning skills of our students'.

Three science educators from the Department of Science Education, University of Ilorin, carried out the face and the content validity of section B. The reliability test was carried out using test-retest method. The questionnaire has administered and re-administered to 30 science teachers at three weeks interval. Using Pearson Moment Correlation Coefficient. A reliability value of 0.85 was obtained for the instrument.

RESULTS

The results obtained from the study are presented as follows in Table 1.

Table 1 reveals the frequency and the percentage of the science teachers involved in the study based on their area of specialization. The Biology teachers had the highest frequency and percentage (239, 37.94%); while mathematics had the lowest frequency and percentage (82, 13.02%).

Research Question 1: What is the Kwara State Secondary School Science Teachers' perception of ICT for instruction?

The result of the analysis related to this question is presented on Table 2.

The result reveals that the science teachers' perception of ICT for instruction was positive but low with overall mean of 2.37.

From the result presented on Table 2, the overall rating of the perception of the science teachers on the use of ICT for instruction stands at 2.37 out of a total value of 5. This score translates into a low level of perception. Hence, one could conclude from the result that Kwara State secondary school teachers' perception of ICT for instruction is low.

Research Question 2: What is the Kwara State Secondary School Science Teachers' perception of ICT for instruction based on area of specialization?

Table 2. Mean of the Kwara state secondary school teachers' perception of ICT for instruction

Item	Mean($\bar{X}$)	Remarks
1	2.16	Positive perception
2	2.15	" "
3	2.14	" "
4	2.19	" "
5	2.29	" "
6	2.13	" "
7	2.12	" "
8	2.45	" "
9	2.58	" "
10	2.54	" "
11	2.78	" "
12	2.75	" "
13	2.73	" "
14	2.12	" "
15	2.67	" "
16	2.06	" "
OVERALL		
MEAN	2.37	

KEY
High Perception $\bar{X} \geq 3.5$
Moderate Perception $3.0 \leq \bar{X} < 3.5$
Low Perception $2.0 \leq \bar{X} < 3.0$
Wrong Perception - $\bar{X} < 2.0$

This research question gave rise to the following research hypothesis:

Research Hypothesis: There is no significant difference in the Science teachers' perception of ICT for instruction based on area of specialization.

The hypothesis was rejected. In other words, there was a significant difference in the secondary school science teachers' perception of ICT for instruction based on area of specialization.

The hypothesis was tested using Chi-square analysis which involved item by item analysis at 0.05 significance level. As revealed on the table, the hypothesis was rejected for 10 items and accepted for 6 items. This reveals that while there was a significant difference in the science teachers' perception on the potentials of ICT such as – "using ICT will enhance better class room practices and procedures" and "the use of ICT makes management of class time effective" there was no significant difference in the Science teachers' perception on negative statements such as "ICT tools does not necessarily improve my teaching" "I do not have the time to become familiar with ICT resources" This meant that for those positive statements, the perception of the science teachers was influenced by their area of specialization while for the negative statements perception was not influenced by their area of specialization. Over all, the hypothesis was rejected with a chi-square result as revealed on Table 3.

Table 3. The overall chi-square analysis on the secondary school science teachers' perception of ICT for instruction based on area of specialization

	Biology	Physics	Chemistry	Mathematics	Total
SD	870	583	562	242	2257
D	1195	779	925	401	3300
A	1138	523	884	447	2992
SA	592	307	380	222	1501

χ^2Cal=88.3589, P- Value =0.000, χ^2 0.95, 9=16.91898

Table 4 presents the result obtained in respect of the only hypothesis put to test in this study.

The result presented on Table 4 indicates that there was significant difference in the science teachers' perception of ICT for instruction based on their area of specialization $F(3,252)=50.323$, $P<0.05$. Hence the hypothesis is rejected. To know where the significant difference lies, the Duncan Multiple Range Test was carried out. The result obtained is presented on Table 5.

Table 5 shows the direction of the significant difference based on area of specialization. The result reveals that the Biology teachers had the highest level of positive perception followed by Chemistry teachers Physics teachers while the mathematics teachers had the lowest level of positive perception of ICT for instruction. This is also represented in the Bar Chart in Figure 1.

DISCUSSION

Several studies have established the important role played by ICT as an instructional tool in secondary schools, especially in sciences. However, studies still show that some teachers' do not employ ICT for instructional delivery. The results obtained from the data analysis in this study revealed that the Secondary School Science teachers' perception was generally positive compared to teachers in other countries (Faggiano & Fassano 2004, Yuen 2008, Albrini 2006). The positive perception is an important indication that the science teachers' in Kwara State had the willingness to use ICT as an instructional tool; since perception is an important step in ICT integration into teaching and learning environment.

The result further revealed that the Biology teachers had the highest level of positive perception followed by Chemistry, Physics and Mathematics teachers respectively. This is an indication that the Biology teachers are the most prepared for the use of ICT as an instructional tool. For science teachers to introduce ICT as an instructional tool into the classroom environment, they must be enthusiastic about the use of ICT as an instructional tool. The level of the mathematics teachers' perception is in support of Faggiano and Fassano's (2004) finding. They in the course of their study had to promote the reflection of the mathematics teachers and

Table 4. ANOVA result on science teachers' perception of ICT for instruction based on area of specialization

	Sums of Squares	df	Mean Squares	F	Sig.
Between Groups	50712.641	3	16904.214	50.323	.05
Within Groups	84650.344	252	335.914		
Total	135363.0	255			

Table 5. The Duncan Multiple Range Test (DMRT) result on science teachers' perception of ICT for instruction based on area of specialization

Area of Specialization	N	1	2	3	4
Mathematics	82	20.500			
Physics	137		34.2500		
Chemistry	172			42.9844	
Biology	239				50.323
Sig.		1.000	1.000	1.000	1.000

Figure 1. Bar chart of the significant difference in science teachers' perception of ICT based on area of specialization

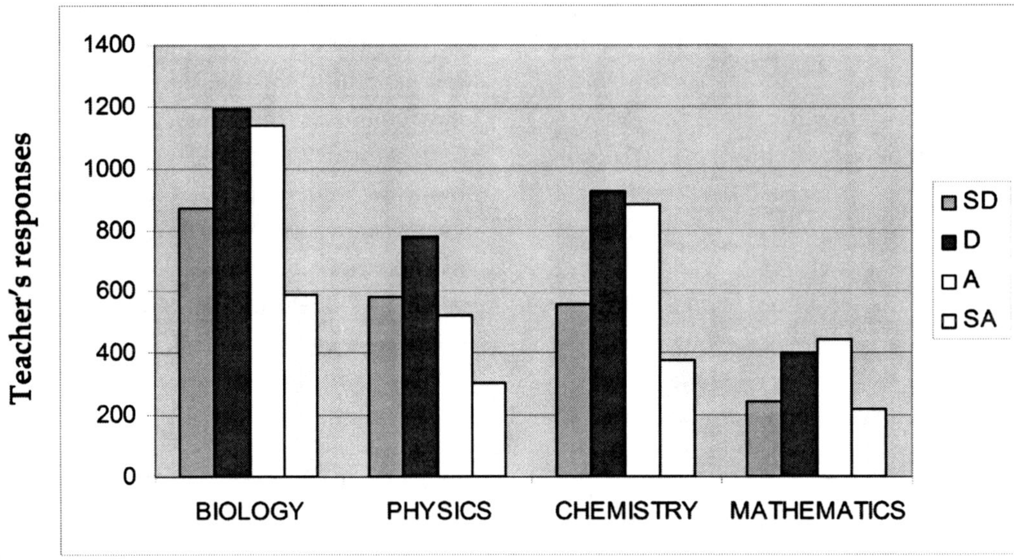

potentials of the appropriately used technology tools in order to improve on the perception of the mathematics teachers.

CONCLUSION

In this study, it was discovered that the secondary school science teacher's perception of ICT for instruction was positive but at a low level. This finding revealed that the perception of the Secondary School Science teachers must have to still be worked on for effective integration of ICT into teaching and learning process in Kwara state. In addition, science student- teachers will need hands-on experience since they would produce highly placed citizen in future. Hence there is the need to expose them to more courses in ICT. The limitation of this study is that the study is restricted to Kwara State of Nigeria only. In addition, other secondary school subjects apart from sciences were not examined in this study. This could be good areas of study with regard to teachers' ICT use and perception.

RECOMMENDATIONS

Based on the findings of this study, the following recommendations are made:

1. In order to make teachers aware of the inherent potentials of using ICTs for instruction, ICT should be fully integrated into science teacher education at Colleges of Education and the Universities. This will make prospective teachers see the use of ICT for instruction as a norm in the teaching-learning process and their perception of its use for instruction will be positively high.
2. Kwara State Government should formulate a programme of ICT training and retraining of teachers currently in service as a way of improving their perception of the use of ICT for instruction, hence, they will be able to utilize ICT to improve their professional practice.

REFERENCES

Adebayo, F. A. (2008). Usage and challenges of information communication technology (ICT) in teaching and learning in Nigerian universities. *Asian Journal of Information Technology*, *7*(7), 290–295.

Adediwura, A. A., & Tayo, B. (2007). Perception of teachers' knowledge, attitude and teaching skills as predictor of academic performance in Nigerian secondary schools. [from http://www.academicjournals.org]. *Educational Research Review*, *2*(7), 165–171. Retrieved January 3, 2009

Ajewole, G. A. (2003). ICT and biology education: implication for secondary schools. *Proceedings of the 44th Annual Conference of the Science Teachers Association of Nigeria* (pp. 128–130).

Albirini, A. (2006). Cultural perceptions: The missing element in the implementation of ICT in developing countries. *International Journal of Education and Development Using ICT*, *2*(1), 49-65. Retrieved February 10, 2013, from http://ijedict.dec.uwi.edu/viewarticle.php?id=146&layout=html

Arnold, R. S., Padilla, M. J., & Tunhikorn, C. B. (2009). The development of pre-service science teachers' professional knowledge in utilizing ICT to support professional lives. *Eurasia Journal of Mathematics. Science and Technology Education*, *5*(2), 91–101.

Bakar, A. B. R., & Mohammed, S. (2008). Teaching using information and communication technology. Do trainee teachers have the confidence? *International Journal of Education and Development Using ICT*, *4*(1). Retrieved November 3, 2012 from http://ijedict.dec.uwi.edu

Bamigbala, P. A. (2000). Factors influencing students' perception of difficult concepts in senior secondary school physics in Osun state. An M. Ed. research proposal submitted to the Department of Curriculum Studies and Educational Technology, University of Ilorin, Nigeria.

Dawson, V., Forster, P., & Reid, D. (2006). Information Communication Technology in a science education unit for preservice science teachers; students' perceptions of their ICT skills, knowledge and pedagogy. *International Journal of Science and Mathematics Education*, *4*(2), 345–363. doi:10.1007/s10763-005-9003-x

Dickey, E., & Singer, J. (2004). Technology in mathematics and science instruction. In M. D. Roblyer (Ed.), *Integrating educational technology to teaching/curriculum* (3rd ed., pp. 253–270). Pearson Merrill-Prentice Hall.

Encarta. (2010). *Perception.*

Faggiano, E., &Fasano, M. (2004). *Teachers' perceptions and usage of ICT: An issue for educators to deal with.* Retrieved on November 8, 2008, from Tsg.icmell.org/document/get/229

Federal Republic of Nigeria. (2004). National policy on education (4th ed.). Lagos: NERDC press.

Goodywn, A., Adams, A., & Clarke, S. (1997). The great god of the future: The views of current and future English Teachers on the place of IT in literacy English. *Education*, *31*(2), 54–62.

Gulbahar, Y., & Guven, I. (2008). A survey on ICT usage and the perceptions of social studies teachers in Turkey. *Journal of Educational Technology & Society*, *11*(3), 37–51.

Heinich, R., Molenda, M., Russel, J. B., & Smaldinow, S. E. (2002). *Instructional media and technologies for learning* (7th ed.). Merril Prentice–Hall.

Kwache, P. Z. (2007). The imperatives of ICT for teachers in Nigeria higher education. *MERLOT Journal of Online Learning and Teaching*, *3*(4), 395–399.

Mumtaz, S. (2000). Factors affecting teachers' use of information and communications technology: A review of the literature. *Journal of Information Technology for Teacher Education*, *9*(3), 319–342. doi:10.1080/14759390000200096

Ofsted. (2009). *The importance of ICT: Information and communication technology in primary and secondary schools.* Retrieved on March 5, 2011 from www.ofsted.gov.uk

Okonkwo, S. (2000). Relationship between some school and teacher variables and students achievement in Mathematics. *Journal of Science Teachers Association of Nigeria*, *35*(1&2), 43–49.

Ololube, P. N. (2006). Appraising the relationship between ICT usage and integration and the standard of teacher Education programs in a developing economy. *International Journal of Education and Development using ICT*. *2*(3), 1-18. Retrieved September 9, 2009, from http://ijedict.dec.uwi.edu

Oyelekan, O. S. (2009). *The development and validation of a computer instructional package on electrochemistry for Nigerian secondary schools.* Unpublished Ph.D. thesis submitted to the Department of Science Education, University of Ilorin, Nigeria.

Seymour, E. (2001). Tracking the process of change in US undergraduate education in science, mathematics, engineering and technology. *Journal of Science Education*, *86*, 79–105. doi:10.1002/sce.1044

Ukah, I. M. (2007). Teachers perception of teacher and pupil factors affecting childrens' acquisition of creative thinking skills through handicraft studies; implication for technology education. *Journal of the Science Teachers Association of Nigeria*, *42*(1&2), 78–84.

Yuen, A. H. K., Lee, M. W., Law, N., & Chan, A. (2008). Factors predicting impact of ICT – use on students: An exploration of teachers' perceptions. In *the Proceedings of ICR.*

Yusuf, M. O., & Balogun, M. R. (2011). Student-teachers' competence and attitude towards information and communication technology: A case study in a Nigerian University. *Journal of Contemporary Educational Technology*, *2*(1), 18–36.

This work was previously published in the International Journal of Information and Communication Technology Education (IJICTE), 10(1); edited by Lawrence A. Tomei, pages 25-34, copyright 2014 by IGI Publishing (an imprint of IGI Global).

Chapter 54
Enhancing Diversity in STEM Interdisciplinary Learning

Reginald A. Blake
New York City College of Technology, City University of New York, USA

Janet Liou-Mark
New York City College of Technology, City University of New York, USA

ABSTRACT

The Science, Technology, Engineering, and Mathematics (STEM) disciplines have traditionally been woefully unsuccessful in attracting, retaining, and graduating acceptable numbers of Underrepresented Minorities (URMs). A new paradigm of STEM practices is needed to address this vexing problem. This chapter highlights a novel interdisciplinary approach to STEM education. Instead of being siloed and mired in their respective STEM disciplines, students integrate real world, inquiry-based learning that is underpinned by a strong foundation in mathematics and a myriad of other pillars of STEM activities. These activities include Peer-Assisted Learning Workshops, Mentoring Programs, Undergraduate Research Experiences, STEM Exposure Trips, Conference Participation, and Peer Leadership. This strategy enhances STEM education among URMs by purposefully connecting and integrating knowledge and skills from across the STEM disciplines to solve real-world problems, by synthesizing and transferring knowledge across disciplinary boundaries, and by building critical thinking skills in a manner that is relevant to their experiences and yet transformative.

ORGANIZATION BACKGROUND

New York City College of Technology (City Tech) is a minority-serving institution and designated college of technology of the City University of New York, currently offering STEM baccalaureate and associate degree programs in Applied Mathematics, Chemical Technology, Computer Information Systems, Computer Science, Telecommunication and various Engineering Technologies. Although it is a federally designated Hispanic Serving Institution (HSI), of the 15,000 plus student population, 31.7% of students identified themselves as Hispanic, 34.0% as Black (non-Hispanic), 18.5% as Asian/Pacific Islander, 11.0% as White, 0.5% as Native American, and 4.3% as other. Sixty-seven percent (67%) are the first in their families to attend college. Fifty-two percent (52%) reported a

DOI: 10.4018/978-1-4666-7363-2.ch054

household income of less than $30,000 and 74% of incoming first-year students received need-based financial aid. Sixty-eight percent (68%) are the first in their families to attend college (New York City College of Technology, 2011).

U.S. STEM PLIGHT AND MINORITIES

The United States remains in grave danger of losing its global competitive edge in Science, Technology, Engineering, and Mathematics (STEM). Recent concerns were propelled into the national spotlight with the *American Competitiveness Initiative* of the Bush Administration in 2006 and the *Rising Above the Gathering Storm* report of the National Research Council (2008). Yet the concerns are nothing new. It was in 1980 that Congress first mandated the National Science Foundation (through the Science and Engineering Equal Opportunities Act) to diligently seek to develop the STEM talents of the country's citizenry irrespective of gender, ethnicity, race and economic background. The 32-year battle that begun last century against poor representation among students and in the science and technology workforce continues on. Many–if not most–of our 21st century challenges will coalesce around STEM. Solutions to these challenges will require a new scientific workforce armed with a skill-set that engenders technological sophistication and interdisciplinary thinking. Therefore, it is critical to train and engage a diverse workforce in STEM and to provide foundational STEM education for the nation's citizenry in all of its inherent diversity. This is absolutely critical, and it must be seen as an urgent matter of national security.

To maintain its global leadership and competitiveness in STEM, the United States must invest in research, encourage innovation, and grow a strong and talented science and technology workforce (Expanding Underrepresented Minority Participation, 2011). A strong, diverse STEM workforce is essential in keeping America innovative and competitive. According to the book, *Expanding Underrepresented Minority Participation* (2011), STEM will lead the projected growth of the U.S. labor market at rates that outpace and out-accelerate all other sectors in the coming years, thus making minority participation in STEM education at all levels (K – 20) a national priority. These daunting projections are being made in conjunction with other projections that show that minorities are the fastest growing segment of the U.S. population and that minorities are grossly underrepresented in STEM. In the past, U.S. economic growth and its global leadership have been inextricably linked to access and attainment in STEM education. Therefore, as we embark on a new century, it behooves the federal government, industry, and K – 20 institutions to re-think and to find innovative strategies to significantly boost minority participation in STEM.

All demographic studies indicate that the country's current minority population will this century become its majority. Yet statistical data (Huntoon & Lane, 2007; and www.nsf.gov/statistics/degrees/) confirm recalcitrant under-representation of minorities in STEM disciplines. In 2005, African Americans, Hispanic Americans, Native Americans (American Indians and Alaskan Natives), Native Pacific Islanders (Polynesians or Micronesians), and persons with disabilities constituted nearly 30% of the general population, but earned only 17% of the total number of bachelor's degrees granted in STEM disciplines. According to Shirley Ann Jackson (2002), the U.S. would not experience a talent gap and a downward spiral in attracting, retaining, and graduating in STEM if women and minorities became active STEM participants in proportion to their percentage of the general population. Chubin and Babco (2003) emphasized that even though underrepresented minority students make up 30 percent of the school-age population, White non-Hispanic and Asians dominate the U.S. engineering workforce, African Americans, Latinos, and American Indian have minimal participation. Although women make up about half of the school-age population, they represent only about 10 percent of the STEM workforce (Chubin & Babco, 2003). Our nation's current STEM "lay-of-the-land" is thus unsustainable and untenable. If our nation is to remain globally competitive, intentional and intrusive steps are immediately needed to ensure adequate participation of Underrepresented Minorities (URM) in STEM. Moreover, Tsui (2007) makes the point that the general recommendations/strategies of STEM may not work for URM. Therefore, new and innovative paradigms of STEM education may be needed. Tsui further suggests that the disproportionately low participation of underrepresented minorities in STEM is directly linked to barriers that are of a cultural (social expectations for different groups), structural (historical laws and regulations that barred the entry of minorities into education and employment), and institutional nature (discriminatory policies and practices). A preponderance and flurry of recent studies (Association of American Colleges and Universities, 2007; Arrison, 2009; Atkinson, 2010; Augustine, 2007; Bond, Schaefer, Rejeski, & Nichols, 2008; California Council on Science and Technology, 2009; California Council on Science and Technology, 2007; College Board, 2008; Fairweather, 2008; Mestre, 2008) all highlight the need for the U.S. to seriously address its STEM crisis, for the handwriting is on the wall, and time is running out. However, what is needed to redress this crisis is not just reformation. What is needed is a radical and innovative revamping and reconstruction of STEM. This necessary STEM revolution must squarely and intentionally focus on the underrepresented minority component of the nation. Below, Figures 1 and 2 highlight the STEM problem. The figures show how out of step URM STEM occupation data are with the nation's demographics, for although African Americans and Hispanics comprised 27.7% of the population in 2008, they only account for about 7% of the science and engineering occupations in 2006 (Science and Engineering occupation data change very little from year-to-year).

Figure 1. Resident population of the United States by sex and race/ethnicity
Source: (2008) Women, Minorities, and Persons with Disabilities in Science and Engineering (www.nsf.gov/statistics/wmpd)

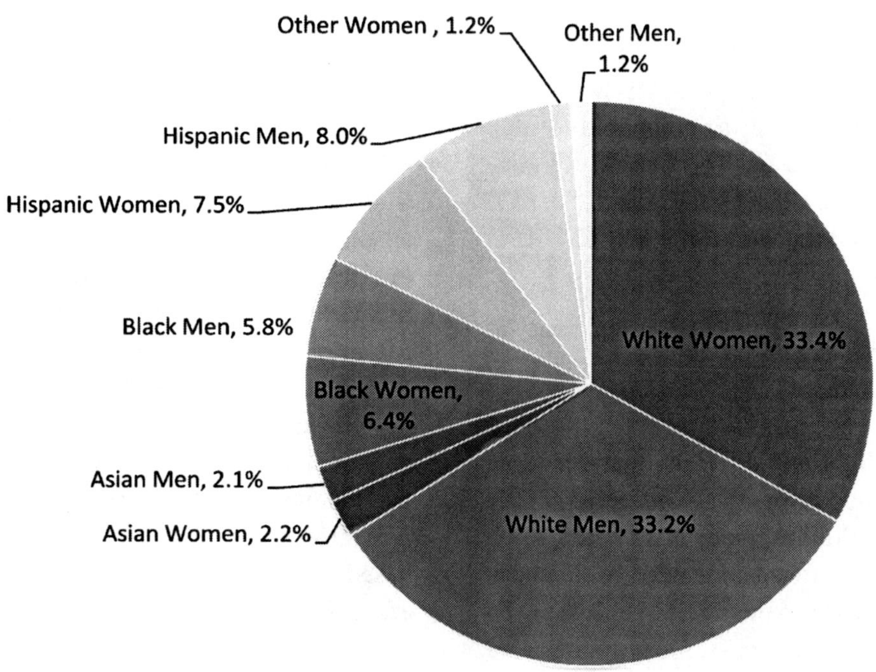

Figure 2. Scientists and engineers in science and engineering occupations
Source: (2006) Women, Minorities, and Persons with Disabilities in Science and Engineering (www.nsf.gov/statistics/wmpd)

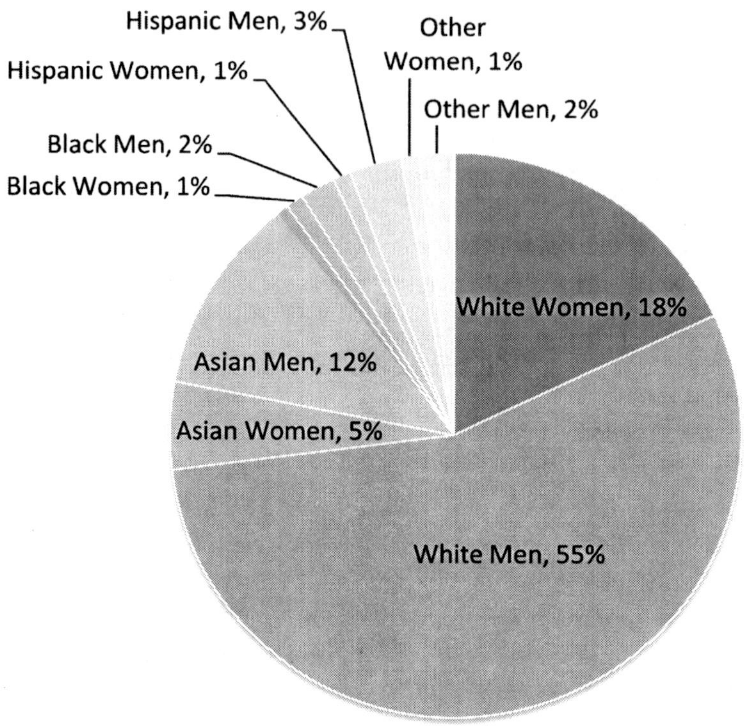

Moreover, to further highlight the STEM diversity problem, in a National Center for Education Statistics study, Huang (2000) reported that 46% of White and Asian American college students finished their STEM degree programs within five years of initial enrollment as compared to only 26.8% of their URM peers. The Bureau of Labor and Statistics projects that by the year 2050, 85% of the entrants into the workforce will be URMs and women.

The Black Male Initiative (BMI) Program

The national plight and dilemma that decouple STEM success and diversity are reflected at City Tech. At the institution, there is an acute, critically evident, and transparent need to diversify and radically transform the STEM landscape. To this end, City Tech has created a new coordinated initiative, the Black Male Initiative (BMI), to attract, retain, and graduate African American and Hispanic male students in the STEM disciplines. STEM was selected as the area of focus not only because of the national crisis, but also because the STEM disciplines are among the college's flagship and strongest programs, and these areas of study are areas in which African-Americans and Hispanics are notably underrepresented. Increasing the numbers of URM students who succeed in STEM disciplines is important not only to the economic vitality of the New York metropolitan area, but to the nation as a whole. Inasmuch as the purpose of this project is to strengthen the engagement and success of African American and Hispanic male students in STEM fields, the project also serves as a prototype for future cohort-based initiatives that address the needs of other educationally underserved populations that have not been advantaged equally in higher education.

Interdisciplinary Approach to STEM Success

To ameliorate and enhance STEM success among BMI students at City Tech, a bold venture that utilizes interdisciplinary learning as its primary transformative tool was launched. Interdisciplinary learning modalities were introduced to combat the STEM plight because they:

- Are associated with high impact practices and higher rates of student-faculty interaction;
- Enhance critical thinking ability;
- Improve writing skills;
- Lend themselves to a greater appreciation for diversity and persistence;
- Recognize disciplinary strengths, processes, limitations, and perspectives;
- Are designed to connect and integrate knowledge and skills from across the disciplines to solve real-world problems;
- Synthesize and transfer knowledge across disciplinary boundaries, even beyond the STEM disciplines, in the context of novel situations;
- Produce agile, flexible, reflective thinkers who are comfortable with complexity and uncertainty, and can apply their knowledge to respond appropriately and positively;
- Are inclusive in emphasizing that other factors–cultural, political, ethical, historical, and economic–must also be considered when addressing the complex problems of this century;
- Promote the universal nature and deep structure of science, as well as the relationship of STEM disciplines to other disciplines;
- Prepare students to become lifelong learners;
- Create integrative thinkers who can solve problems in ethically and socially responsible ways; and
- Afford students the ability to think critically, communicate effectively, and work collaboratively with others within diverse disciplines, cultures and communities.

The interdisciplinary approach to STEM learning in the BMI program finds its vitality, vibrancy, and validity in the following programmatic activities depicted in Figure 3: Peer Assisted Learning Workshops, Peer Leadership, Mentoring Program, Undergraduate STEM Research Experiences, STEM Exposure Trips, and STEM Conferences.

A brief description of each of the symbiotic, inter-dependent components and activities of Figure 3 follows.

Peer Assisted Learning (PAL) Workshops

Interdisciplinary themes reflected in Peer Assisted Learning (PAL) workshops underpin that mathematics is indeed the language of science, and it is the unifying thread that is common to all STEM disciplines. Too often City Tech BMI

Figure 3. Components and activities of STEM interdisciplinary learning

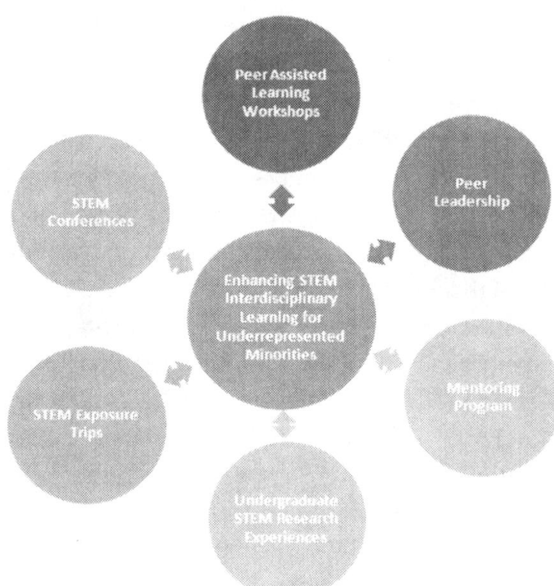

students have difficulty using mathematical concepts across contexts to explore and examine the world they live in. To them, the principles of mathematics exist in a vacuum, and they fail to see the relevancy of this universal language.

Unequivocally, City Tech data show that mathematics is the main gatekeeper that prohibits participation and inhibits success in STEM among African American and Hispanic students. Most of these students either drop out of school or are forced to change majors because of their inability to comprehend and communicate this all-important language of mathematics. According to Davis-Lowe (2006) and Strayhorn (2008), academic supported communities are vital to the success of underrepresented students who venture along STEM pathways. Moreover, research indicates that peer assisted learning improves student success, retention, graduation, and transfer rates among all students, with particular gains for underrepresented students (Arendale, 2004; Davis & Finelli, 2007; Peterfreund, Rath, Xenos, & Bayliss, 2008). Therefore, to assist the students in remedying this vexing problem, a Peer-Assisted Learning (PAL) program was initiated at City Tech. Leveraging the program's main goals of increasing the persistence and retention rates of underrepresented students in the learning of mathematics, the program readily embraced the importance of enhancing interdisciplinary learning.

The collaborative atmosphere of the PAL workshops provides diverse learners from wide-ranging disciplines an environment where mathematical concepts are reinforced. The workshop participants comprise of undergraduates majoring in various STEM disciplines including Applied Mathematics, Chemistry, and the vast array of Engineering Technologies. Adopted from the Peer-Led Team Learning (PLTL) student-centered instructional model originally developed for Chemistry (Gosser, et al., 2001), the PAL workshop model was redesigned to enhance the learning of mathematics. A typical PAL workshop consists of eight to ten students working with an undergraduate peer leader for one hour per week for thirteen weeks. As the team recognizes the strength of each member and the prior knowledge that each participant possesses, they work collectively as a unit on modules that are designed to foster thinking skills, enhance problem-solving abilities, and increase computational proficiency of the course material (Arrington, 2008; Quitadamo, Brahler, & Crouch, 2009). These modules highlight key mathematical concepts and provide real-world examples conducive to interdisciplinary learning.

At City Tech, the organizational composition of PAL workshops is designed for students either enrolled in a course where workshops are mandatory or offered as an independent academic support (Liou-Mark, Dreyfuss, & Younge, 2010). Workshops which are integrated into a course are available in the first two credit bearing mathematics courses–MAT 1175: Fundamental of Mathematics and MAT 1275: College Algebra and Trigonometry. Because of the consistently dismal pass rates of 52.4% and 59.2% for MAT 1175 and MAT 1275, respectively (New York City College of Technology, 2011), this additional hour of workshop fosters an academic support system to increase student success rates primarily for first- or second-year students. Many of these students enter college academically underprepared and lacking the necessary skills required for success. By providing them with a culture of collaboration and by prompting them to integrate and transfer their knowledge, these students are better positioned to succeed in their academic life, especially in the STEM disciplines.

For students enrolled in upper divisional mathematics courses (Precalculus and the Calculus 1, 2, 3 sequence), stand-alone PAL workshops draw students from a cross section of STEM majors to discuss and debate ideas centered on key mathematical concepts. As students integrate this level of engagement in their learning, the knowledge they attained in their respective disciplines provides a stimulating platform for interdisciplinary learning.

Table 1. Grade distribution of MAT 1175 PAL workshop attendees, PAL workshop embedded in course and comparison group

Grades	Cohort 1 (PAL Workshop Attendees) n=22	Cohort 2 (PAL Workshop Integrated in Course) n=23	Comparison Group (n=20)
ABC	77.3% (17)	91.3% (21)	45.0% (9)
ABCD	86.4% (19)	95.7% (22)	65.0% (13)
F	9.1% (2)	0.0% (0)	20.0% (4)
Withdraws	4.5% (1)	4.3% (1)	15.0% (3)

Several studies have been piloted on the effectiveness of PAL workshops on increasing the persistence and pass rates of students enrolled in lower divisional mathematics courses. The results from the Coordinate Undergraduate Education's AIM for Success pilot project (2009-2010) for MAT 1175, showed students achieved a higher pass rate after attending optional PAL workshops and an even higher pass rate with mandatory, embedded workshops, when compared to a comparison group (see Table 1). Although the sample size is small, the results alluded to workshop participants having a higher persistent rate than the comparison group.

A pilot study conducted during the spring 2011 semester showed PAL workshops for a MAT 1275 course had a positive effect on the uniform departmental final exam grades and course grade. The average departmental final exam grade was statistically significantly higher for the PAL workshop attendees compared with the workshop non-attendees. Moreover, the mean course grade was higher for the PAL workshop attendees compared with the non-attendees. However, the difference was not statistically significant (see Table 2).

Because of the low pass rates, it is necessary for students to repeat the course. In order to assess the effectiveness of PAL workshops on students who are repeaters of MAT 1275, final grades from PAL workshop attendees were compared to the Comparison Group of non-workshop attendees in spring 2011 (see Table 3). The mean course grade for the Repeaters who were workshop attendees was statistically significantly higher than the Comparison Group. Thus, indicating that the treatment of having PAL workshops may have helped the repeaters improve their course grade and persist in the course.

Students from different STEM disciplines who coalesce in meaningful discussions with their peers increase their capacity to understand multiple viewpoints. They learn to appreciate how each discipline uses different problem solv-

Table 2. Grade distribution of MAT 1275 of PAL workshop attendees and non-attendees

Grades	MAT 1275 PAL Workshop Attendees (n=24)	MAT 1275 Non-Attendees (n=13)
ABC	79.2% (19)	61.5% (8)
ABCD	87.5% (21)	69.2% (9)
F	12.5% (3)	23.1% (3)
Withdraws	0.0% (0)	7.7% (1)

Table 3. Grade distribution of MAT 1275 repeaters of PAL workshop attendees and the repeaters comparison group

Grades	MAT 1275 Repeaters PAL Workshop Attendees (n=15)	MAT 1275 Repeaters Comparison Group (n=30)
ABC	46.7% (7)	30.0% (9)
ABCD	66.7% (10)	46.7% (14)
F	26.7% (4)	23.3% (7)
Withdraws	6.7% (1)	30.0% (9)

ing approaches and how to integrate the ideas. If the PAL workshops can promote a passion for learning, then a natural outcome would be significant improvements in students' performance, persistence, and retention.

Peer Leadership

Undergraduates who facilitate workshops in mathematics or science empower themselves to gain expertise in both disciplinary and interdisciplinary areas as they develop their leadership skills (Micari, Gould, & Lainez, 2010). In recognition of the students' disciplinary strengths, the Peer Leadership experience challenges undergraduates, especially the BMI students, to expand their knowledge through the facilitation of STEM workshops. Through guiding workshops in subject areas outside of their disciplines, peer leaders learn to establish connections of the content at hand to other subject areas, create scenarios relevant to the workshop topics, and develop a multi-pronged approach to finding solutions.

The PAL Leadership program, moreover, addresses the retention of students, namely underrepresented minorities, in the STEM disciplines. Peer leaders become role models who "humanize" and can demystify STEM challenges. This is a role crucial to establishing a productive learning community (Washburn & Miller, 2004). Peer leaders are not expected to be content experts, but are undergraduates who have successfully completed the course and have been trained in group dynamics. Recruitment is based on their STEM majors, GPA, and interpersonal and communication skills. The peer leaders play an indispensable role in providing guidance for the group by ensuring that the team actively engages itself with the appropriate mathematics or science concepts and with each other, thereby building commitment and confidence among the students and constructing meaningful deliberations and discussions. They understand the challenges, nuances, and misconceptions of the course material. They, however, are trained not to give away answers, but rather offer well-timed assistance when the group members find themselves in difficult situations as they solve the problems collaboratively. The critical view of being an integral part of their peer's success in mathematics and science motivates them to persist and expand their knowledge in these areas.

The training of peer leaders is conducted through a one-credit course. The goals of the course include the successful implementation of pedagogical techniques in workshops, the application of learning theories to the workshop experience, and the establishment of a community of practice among peer leaders from diverse backgrounds and majors. Undergraduates in STEM are seldom trained in teaching and learning pedagogy, which this course uniquely provides. To further promote the understanding of learning, the peer leaders are provided with a first-hand experience in conducting a mini-research project on a theory of learning. Moreover, this project is then transferred into a poster presentation; thus giving students a first-hand experience on how to design and format a research poster presentation.

The Peer Leadership project has provided leadership skills and opportunities for our students for the past three years and continues to be a growing vibrant community of student leaders. The outcomes for students involved in this program have been promising. As of the fall 2011 semester, a total of 53 students have participated or are participating as a mathematics or science peer leader in the program. Of the 53 students, 35.8% (19) have graduated (16 students with a bachelor's degree and 3 with an associate degree), seven transferred to another institution to pursue a STEM degree, and 58.5% (31) are still persisting in their STEM majors or Liberal Arts and Sciences degree. Of the 16 students who graduated with a bachelor's degree, six are in graduate school, six are applying to graduate school, three are in medical or pharmacy school, and one is working in a STEM-related job. The three students who graduated with an associate degree are continuing with their bachelor's degree in a STEM discipline. From the group, 69.8% (37 out of 53) of the peer leaders have or are conducting research in a STEM discipline of which 59.5% (22) of the students are from underrepresented minority groups in STEM. Approximately 73.6% (39 out of 54) of all the peer leaders have made oral or poster presentations at national and regional STEM conferences. As a result of this research experience, four of the underrepresented minority peer leaders have co-authored articles with their mentors in peer-reviewed journals.

Weekly leadership seminars designed to provide additional academic support is currently being piloted at the institution. It is designed to enhance the leadership development of a network of peer leaders. The focus of the weekly meetings address several competencies required at all levels of leadership. The core competencies are as follows:

Communicating

- Expressing oneself effectively both orally and in written form
- Introducing oneself formally
- Negotiating by using a win-win approach in order to maintain relationships
- Listening to others

Teamwork

- Allocating responsibilities to appropriate individuals
- Organizing resources
- Steering team members towards achieving goals

Interpersonal Skills

- Promoting integrity
- Building respect and trust

Self-Direction

- Establishing goals, timelines, and deliverables

Networking

- Building relationships with peers from other disciplines
- Creating diverse teams to implement projects

Retaining students in the STEM disciplines, particularly underrepresented minorities, through this Peer Leadership program is currently being refined. Topics such as integrity, diversity, and professional leadership skills are also covered. Thus far, the promising result of URM students developing a culture to pursue advanced degrees in STEM is encouraging. It may be imperative to consider leadership and interdisciplinary learning as factors to consider as essential mechanism to increase and continue the STEM education pipeline.

Mentoring Program

Mentoring does not take place by happenstance. Successful mentoring must be planned, intentional, and strategic. Mentoring is a process of establishing, cultivating, nurturing, and developing a mentee to realize his or her full potential as a scholar. The Mertz Model of Mentoring (Mertz, 2004) believes that the spectrum of mentoring can be represented in a six-leveled pyramid.

- Base-Levels 1 and 2: These first two levels involve role modeling and general support that includes guidance and advice. These mentoring levels may involve many mentees and is often time consuming.
- Mid-Levels 3 and 4: As one ascends the pyramid to these two levels, the mentoring process becomes more focused. It now embraces the role of academic advisor and includes activities such as sharing content knowledge and progress assessment.
- Top-Levels 5 and 6: At these upper levels of the mentoring pyramid, the mentor

is fully vested in the success of the mentee. The number of mentees significantly decreases, and a shared bond of responsibility and trust develops. The mentor now views the mentee's success (or failure) as a direct reflection of him/herself. The mentor's reputation is now at stake, so time, effort, resources, contacts, and knowledge are readily shared with the mentee to ensure mentee success.

In our mentoring program component, all levels of Mertz's pyramid are being used. For us, mentoring is an intentional commitment of sustained engagement and advocacy for students, and each student in our targeted population finds himself or herself at some one of the levels of the pyramid. The mentoring program has three subcomponents: Research Mentoring, Faculty Mentoring, and Peer Mentoring. Each sub-component is briefly described in the following.

Research Mentoring: Since our BMI program is a STEM-designated program, many of our students are given the opportunity to conduct STEM research with a faculty member who is conducting research. Oftentimes, the research area is outside of the students' majors. For example, we may have an architectural student conducting research with a geo-physicist or an applied math student working with a professor in mechanical engineering. At this stage of mentoring, the mentoring process is at levels 5 and 6 of the pyramid.

Faculty Mentoring: This second component of the mentoring scheme straddles levels 2, 3, and 4 of the pyramid. Here mentoring is not research-centered, and the faculty mentor need not be a STEM faculty. Here the faculty mentor provides counseling and academic advising and monitors the mentee's academic progress. Not only are the BMI students provided prescriptive advising within their majors, but they are also given developmental advising from a core group of faculty from various STEM and non-STEM disciplines. Being mentored by a disciplinary and non-disciplinary mentor promotes a holistic approach to their learning.

Peer Mentoring: At this level of mentoring (level 1 on the pyramid), advanced/senior students mentor new/fledgling students. At this stage of mentoring, the experienced students act as a role model for the less-advanced mentee. Moreover, the peer mentor equips the mentee with the tools, the "know-how," the skills, and the resources necessary to navigate the college environment, courses, and professors. They help the mentees to spot and to avoid academic pitfalls, and they are often the first line of defence against mentee deviation from the academic rigor that is necessary for academic success. It is usually the case that because of similar ages and experiences, the mentee relates better to–and confides more in–the peer mentor than in the faculty or the research mentor. Therefore, the peer mentor plays an extremely critical role in the success of the mentee. Since their mentoring roles are so pivotal, special training is usually given to these mentors. Once again, this type of mentoring takes place within the milieu of interdisciplinary cross-pollination as the peer mentor and the mentee are often from different STEM majors.

Undergraduate Research Experiences

The value of research experiences for undergraduate students has long been accepted and verified. Researchers have delved into this issue from many angles and have almost in unison proclaimed the many benefits of undergraduate research experiences. A plethora of solid good evidence and support for this view exists. Here are a few: Dillner et al. (2011) reported that professional societies and educational organizations not only recognize the importance of undergraduate research experiences, but they have also used their influence to create initiatives and incentives for academic institutions to infuse research experiences in course curricula and to promote capstone experiences for senior students. In a long-term study of ten-week research programs and with more than 1,135 students participating in the study via an online Likert survey, Lopatto (2007) found that undergraduate research enhances the educational experience of science undergraduates, attracts and retains talented students to careers in science, and acts as a pathway for minority students into science careers. Lopatto (2007) also found that undergraduate research supports active learning and science career decisions. Moreover, students reported gains in independence, intrinsic motivation to learn, and active participation in courses taken after the summer undergraduate research experience. At Rice University's Spend a Summer with a Scientist program, Alexander et al. (2000) reported that of the sixty-two minority students who conducted Summer research projects with faculty role models, 67% of them continued on to graduate school, and 33% of them completed their undergraduate degrees and joined the STEM workforce. The authors of the study believe that the comprehensive, meaningful research experience coupled with working under the tutelage of role models and other undergraduates in a "learning community" were the critical features of the program's success. In a recent study by Bauer and Bennett (2003) 986 alumni respondents from a research institution were placed in one of three categories (category 1: those alumni who participated in a university organized research program; category 2: those alumni who found their own research mentors and had a research experience separate from that which was organized by the university; and category 3: those alumni who had no undergraduate research experience at all.). Category 1 had 418 respondents, Category 2 had 213 respondents, and Category 3 had 355 respondents. The survey showed that alumni in Categories 1 and 2 were more likely to go on to graduate school (80 percent, and 71 percent versus 59 percent, respectively. The researchers also reported that the alumnae of categories 1 and 2 reported:

- Greater satisfaction with their undergraduate experience;
- Increases in intellectual curiosity;
- Expansive research skills; and
- Better communication skills.

The literature is replete with studies (e.g., Campbell & Skoog, 2004; Chandra, Stoecklin, & Harmon, 1998; Chaplin, Manske, & Cruise, 1998; Foertsch, Alexander, & Penberthy, 2000; Hurtado, et al., 2009; Nagda, et al., 1998; Nnadozie, Ishimaya, & Chon, 2001; Russell, Hancock, & McCullough, 2007; Seymour, et al., 2004; Shellito, et al., 2001; Summers & Hrabowski, 2006) that highlight the tremendous benefits of undergraduate research. In general, this type of research experience provides (among many other intangibles):

- Critical preparation for graduate school acceptance;
- Paid summer research internships, academic year, and summer research (paid or unpaid–part of a course);
- Opportunities to learn and to enhance research skills;
- A safe, short-term mechanism of gauging whether the investment in graduate school is worthwhile and whether there is real personal interest in pursuing a graduate degree;
- A means of determining and honing skills in the area of research that the student is passionate about;
- Opportunities to increase in confidence;
- Motivation;
- Opportunities to make both internal and external presentations; and
- The chance to interact with and learn from other students, particularly from those who may come from a different discipline.

At City Tech, we provide a range of opportunities for URMs to engage in interdisciplinary STEM research. The following is a brief description of each of these research programs.

National Science Foundation Research Experiences for Undergraduates (REU): The City Tech REU program is entitled Research Experience for Undergraduates in Satellite and Ground-Based Remote Sensing at NOAA-CREST_2. This project continues to provide undergraduate students an opportunity to conduct full-time, collaborative satellite and ground-based remote sensing research with faculty mentors from within the National Oceanic and Atmospheric Administration Cooperative Remote Sensing Science and Technology Center (NOAA-CREST) consortium. STEM students are recruited from the City University of New York's twenty-three campus wide institutions. For fifteen weeks (nine weeks in the Summer, three weeks in the Fall semester, and three weeks in the Spring semester) these students become part of a community of research scholars who are actively engaged in state-of-the-art remote sensing applications. They participate in research group meetings, seminars, oral and poster presentations at local, regional, and national conferences, and some of them will sail across the Atlantic Ocean conducting in situ remote sensing research. Through the project's unique and innovative endeavors many URMs who would otherwise not pursue STEM education altogether or would not pursue STEM education through to the graduate school level now choose to follow the STEM corridor to either graduate education or to careers in STEM.

Louis Stokes Alliance for Minority Participation (LSAMP) Program: The New York City Louis Stokes Alliance for Minority Participation (NYCLSAMP) is an alliance of 17 CUNY Colleges and the CUNY Graduate Center. The Alliance goal is to substantially increase the number of underrepresented minority students pursuing and graduating with baccalaureate degrees in STEM. From its inception in November 1992, the Alliance has been at the forefront of a concerted effort to increase annual minority STEM enrollment and graduation in the City University of New York. The Alliance provides academic scholarship support to CUNY students majoring in the STEM disciplines, making research an integral part of STEM education in New York City.

LSAMP Program activities include:

- Collaborative Learning Approach to STEM Education
- Restructured Gatekeeper Courses in Chemistry, Physics and Mathematics
- Curriculum Coordination and Articulation across the City University of New York
- Faculty Research Initiation and Articulation Program
- Research Assistantships for LSAMP Scholars
- Undergraduate and Graduate Research Fellowships
- Peer and Faculty Mentoring
- Science and Engineering Learning Centers at CUNY campuses
- The Urban University Series Conference
- NASA Summer and Academic Year Collaborations
- Bridge to the Doctorate
- Brookhaven National Labs Research and Teaching Opportunities for Students and Faculty
- Brookhaven National Labs Summer Participation for Community College Students
- International Training and Education

Earthwatch: Earthwatch is a unique initiative that supports long-term scientific field research geared towards tackling the world's most pressing environmental and cultural issues and focuses its support on applied research where their citizen science model can most effectively make a significant difference in the following central, global ecological, and cultural challenges:

- Ensuring the sustainability of coastal, forested, agricultural and freshwater ecosystems through optimization of multiple ecosystem services;

- Managing protected areas and species at the landscape and seascape levels to enhance biodiversity and provide local and regional ecosystem services;
- Restoring, revitalizing, and conserving our global cultural heritage through a better understanding of socio-cultural links to natural resource use, environmental management, and bio-cultural diversity;
- Addressing the challenge of climate change and its threat to global sustainability.

City Tech's BMI students have travelled to Mexico and to Canada to participate in Earthwatch Summer research projects.

New York City Research Initiative (NYCRI): The NYCRI program is sponsored by the National Aeronautics and Space Administration (NASA) Education Office. NYCRI research teams are assigned to one of 12 colleges in the New York City Metropolitan Area or at the NASA Goddard Institute for Space Studies (GISS). The NYCRI Summer Research project provides teams of high school and undergraduate students and faculty to work alongside graduate students and the principal investigators (lead scientists) of NASA funded research projects at universities or at GISS under the mentorship of a GISS scientist. Summer enrichment experiences include content and research seminars, team oral research reports, visits to various research laboratories and informal education institutions, participation in local and national research summits and a final research summit with participants from other government agencies, such as the National Science Foundation, National Oceanic and Atmospheric Administration, the United States Department of Education, and the United States Department of Defense.

Honors and Emerging Scholars Program: City Tech's Honors Scholars program encourages students who have demonstrated high academic achievement to undertake honors-level work in any appropriate course. Interdisciplinary learning is supported by offering students an option to receive honor's credit in courses within as well as outside their majors. Scholars conduct research in their particular area of interest and present their findings at the institution's bi-annual poster presentation event.

The Emerging Scholars Program provides funding for student researchers who assist faculty members with his/her research or other scholarly endeavors. The purpose of the program is to help students develop a close relationship with a faculty member via research and to promote a practical understanding of material learned in courses. This is an interdisciplinary program in which students are encouraged to explore other STEM areas under the guidance of a faculty mentor.

To support Honors and Emerging Scholars in the different stages of their research, a series of four workshops are offered every semester. Students from a cross section of STEM and non-STEM disciplines engage themselves in meaningful discussions regarding either their disciplinary or interdisciplinary research projects. The first workshop, *Writing Abstracts for Research Projects*, assists students in developing a strong abstract for their project. Students work in groups to discuss the components of a good abstract. In the second workshop, *Advancing Library Research Techniques*, Scholars explore library resources and services for honors course work and research projects. They learn advanced search techniques and strategies to find the best information for their projects and assignments. The third workshop, *Developing and Delivering Effective Research Presentations,* provides students with the best practices for effectively communicating their research project during a poster session. The session discusses criteria for choosing visual aids, elements of effective delivery, and resources available to the students as they construct their posters. The last workshop, *Designing a Research Poster Presentation*, is designed to give students the necessary skills to create posters for display at research conferences. During the workshop, students learn how to edit content for a poster and how to structure that content to make visually pleasing, well-presented detailed posters.

Class Projects: The Integrated Research Strategies of LSAMP Phase IV across CUNY engaged LSAMP Research Faculty Mentors by forming a core of faculty participants who engage students in Integrated Research Strategies at the class level. These faculty mentors develop CUNY wide discipline-specific approaches to integrate research strategies into first year STEM coursework that lead to research being an integral component of the CUNY STEM experience. The integration of research into the introductory STEM curriculum develops a synergy between introductory science courses and those experiences that lead students to select research and research related educational and career pathways. Many of the students in the targeted population are involved in this novel paradigm of infusing research into STEM courses.

STEM Exposure Trips

Extending the learning environment outside the classroom exposes students to current technology, state-of-the-art equipment, leading scientists and those in the forefront of STEM. Lectures and workshops from these science-related facilities provide students a first-hand experience in the interdisciplinary nature of science. At City Tech, several field trips have been sponsored during the academic year to broaden the students' perspectives on STEM. The following facilities are where students have visited.

Brookhaven National Laboratories hosts major research centers that students are able to tour and talk to leading scientists in the fields. The Center for Functional Nanomaterials fabricates and studies nanoscale materials. Nanomaterials offer different chemical and physical properties than bulk materials, and have the potential to form the basis of new technologies. The RIKEN BNL Research Center focuses on the physics program of the Relativistic Heavy Ion Collider, hard Quantum Chromodynamic, and relativistic heavy ion physics. The Computational Science Center provides computational science capabilities through the use of computers for researchers in biology, chemistry, physics, applied mathematics, medicine, and nanoscience. The Center for Radiation Chemistry Research exploits pulse radiolysis techniques to study chemical reactions and other phenomena by subjecting samples to pulses of high-energy electrons.

National Oceanic and Atmospheric Administration Cooperative Remote Sensing Science and Technology Center (NOAA CREST) is a multidisciplinary center led by the City College of the City University of New York. The Center's research and training focuses on all aspects of remote sensing including: sensor development, satellite remote sensing, ground-based field measurements, data processing, and analysis, modeling, and forecasting. CREST educates and trains students in the field of environmental remote sensing by conducting research consistent with NOAA's Science Goals and Missions. One aspect of the recruitment and training focuses on mentoring undergraduates into scientific and engineering disciplines in which they can specialize in remote sensing. By visiting the Center, students not only learn about the latest geoscience developments, but also summer internships with NOAA and with industrial partners, and school-year research assistantships

National Aeronautics and Space Administration Goddard Institute for Space Studies (NASA GISS) is a laboratory in the Earth Science Division of NASA's Goddard Space Flight Center and a unit of the Columbia University Earth Institute. Students learn about research conducted by leading scientists at the Center on global change, which is an interdisciplinary initiative addressing natural and man-made changes in our environment. A key objective of GISS research is prediction of atmospheric and climate changes in the 21st century. Programs for undergraduates interested in this research are also promoted.

Trips to museums such as the *New York Hall of Science, Liberty Science Center*, and the *Museum of Natural History* help students visualize and expand their understanding of STEM concepts. The guided tours encourage students to learn new areas outside their disciplines, and the interactive exhibits demonstrate STEM principles in an engaging manner.

STEM Conferences

Participation in national and local STEM conferences is a vital component to the enculturation of undergraduate researchers in the professional scientific community, especially for underrepresented minority scientists, mathematicians, and technical engineers (Mabrouk, 2009). At professional organizational meetings, students embrace a vision of themselves in a greater academic society, consequently providing them with a motivation to continue in their areas of study. These conferences provide students with an opportunity to present their research results, expand their knowledge in the field, connect with disciplines to others disciplines, obtain updated industry information, and network with peers, faculty, and administrators at other institutions as well as industry leaders.

The opportunity for students to present their research to peers and faculty at conferences affirm and validate their significance in their fields. In addition to gaining a sense of the wider research environment, students not only learn from their own research projects, but also what their fellow peers have discovered through theirs. These presentations provide them with a chance to see the vast array of research taking place in other fields and disciplines as well as to be exposed to a wide variety of research methodologies and scholarly activities currently being conducted in other fields different from their own. Most often, conferences with a student focus will offer supportive workshops on employment and summer research opportunities, scholarships for students who are underrepresented in STEM, and how to apply to graduate school. Some will even have career development sessions on how to develop presentation and interview skills.

Attending conferences have had a profound impact on undergraduates in persisting with their STEM disciplines and in opening the pipeline for advanced studies. The students have participated in a wide range of conferences starting with institutional-sponsored conferences and extending to ones at the national level. For many of our undergraduates, their first conference experience is at the institution's biannual Honors and Emerging Scholars Poster Presentation or at the annual Faculty Research Conference. These two events highlight the work of their peers with their faculty mentors and the research currently being explored by our faculty. By attending these conferences, students broaden their perspectives on how interdisciplinary in nature many research projects really are.

Through federal and institutional funding, support has been provided for students to participate in local, regional, and national conferences. Some of the conferences are highlighted.

The New York City Louis Stokes Alliance for Minority Participation (NYC LSAMP)'s Urban University Series Conference: The Urban University Series Conference sponsored by NYC LSAMP coordinates an annual conference for minority students. In 2011, NYC LSAMP co-sponsored an Einsteins in the City Conference in commemoration of the 100-year anniversary of the Einstein Papers. Students performing interdisciplinary research were invited to present their latest results to a diverse, multi-cultural, and multi-disciplinary audience of young researchers. Papers in all the major disciplines, including science, engineering, social science, architecture, humanities, the arts, and education were presented.

New York State Mathematics Association of Two-Year Colleges (NYSMATYC): NYSMATYC is an organization designed specifically to serve as a means to discuss issues relevant to two-year college mathematics faculty. It is the first two-year college mathematics organization in the United States. This professional association is devoted exclusively to the unique problems of teaching freshmen and sophomore college level mathematics. NYSMATYC sponsors regional and annual meetings that undergraduates are invited to attend and present mathematical projects relevant to first- and second-year students.

National Organization for the Professional Advancement of Black Chemists and Chemical Engineers (NOBCChE): The primary purpose of NOBCChE is to initiate and support local, regional, national, and global programs that assist faculty and students of color in fully realizing their potential in academic, professional, and entrepreneurial pursuits in chemistry, chemical engineering, and related fields. The Organization promotes careers in science and technology and encourages college students to pursue graduate degrees in the STEM disciplines. At their annual conference, students are encouraged to present their research at the poster sessions.

Mathematical Association of America (MAA): MAA is the largest mathematics professional society in the nation. One of the focuses of the organization is undergraduate education. The mission of the organization is "to advance the mathematical sciences, especially at the collegiate level." The MAA supports learning in the mathematical sciences by encouraging: 1) effective curriculum, teaching, and assessment at all levels; 2) research, scholarship, and its exposition at all appropriate levels and venues, including research by undergraduates; and 3) providing resources and activities that foster scholarship, professional growth, and cooperation among teachers, other professionals, and students.

American Meteorological Society (AMS): AMS promotes the development and dissemination of information and education on the atmospheric and related oceanic and hydrologic sciences and the advancement of their professional applications. The society actively promotes the development and dissemination of information and education on the atmospheric and related sciences through its programs and services. Students are encouraged to engage in discussions, present their research, and network with other undergraduates, faculty, and professionals.

Emerging Researchers National (ERN) Conference in STEM: The Emerging Researchers National (ERN) Conference in STEM is hosted by the American Association for the Advancement of Science, Education and Human Resources Programs and the National Science Foundation (NSF) Division of Human Resource Development. The conference is aimed at college and university undergraduate and graduate students who participate in programs funded by NSF, including underrepresented minorities and persons with disabilities. The objectives of the conference are to help undergraduate and graduate students to enhance their science communication skills and to better understand how to prepare for science careers in a global workforce. The general format for the conference include student poster and oral presentations and activities that focus on graduate school, career preparation, summer research opportunities, fellowships, and international research and education opportunities.

SOLUTIONS AND RECOMMENDATIONS

A truly holistic view on this nation's STEM crisis will clearly acknowledge that there are a plethora of reasons to explain the current state of affairs. Clearly, particularly for URMs, the STEM crisis cannot be divorced from the host of other problems extant in today's society. However, with that being said, optimism yet abounds, for unique opportunities exist that hold the potential and the promise of stemming the STEM crisis. The following, therefore, are some issues that should be addressed if the nation's STEM talent pool among URMs is to be fully developed to produce first-rate STEM students, a world-envied STEM workforce, and a STEM literate public.

Cultivate a STEM Talent Pool

Creating access and success in STEM among URMs is an achievable and noble endeavour. It is even a matter of national security. If seriously and intentionally implemented, the following steps could lead to full URM participation in the STEM advancement of this nation:

- Teach STEM early and often – begin to introduce basic STEM concepts in kindergarten and follow through with STEM initiatives and exposure throughout the K – 16 stages; Create pre-college STEM programs for elementary and secondary school URMs to assist in the development of the skills necessary to qualify for college admission while simultaneously prepare pre-freshman URMs in STEM enrichment, orientation and transition programs. This model has been devised and is being administered at the New Jersey Institute of Technology (MSCU, 2002). Another fine example is that of the University of Arizona that has a comprehensive Early Academic Outreach program called The Mathematics, Science, and Engineering Achievement (MESA) program. Its primary goal is to work with public school URMs in grades 6 – 12 to increase the number of these students in the STEM fields by providing learning skills workshops, tutoring programs, and summer enrichment programs (MSCU, 2002).
- Seek and encourage parental and community support for STEM learning. URMs are closely tied to their families and their communities. Both can be indispensable partners in advancing the STEM cause among students.
- Provide top-level administrative vision and support for STEM advancement among URMs at the institution. STEM must be treated as an institutional priority with champions from the upper echelons of administration.
- Encourage faculty (URMs and non-URMs) to become part of this STEM revolution.
- Make an all-out effort to recruit and retain more URMs in STEM,
- Provide students with paid research internships and scholarships, especially in the first three years (Murashkin, et al., 2004),
- Place students in STEM learning communities to avoid the discouragement of student isolation –this approach recognizes the need for diverse student interaction, and it should further provide the programming, the support, and the structure to facilitate it.
- Provide URMs with STEM role models by hiring and retaining URM STEM faculty.
- Create incentives to encourage all STEM faculty to promote and to provide undergraduate research opportunities for URM students.
- Provide special, intentionally targeted academic (bridge programs, peer workshops, mentoring, tutoring, intrusive advising, etc.) and cultural (BMI-type programs, diversity awareness, or multicultural sensitivity programs) support services. Since there is a high correlation between first-term performance and 5-year graduation rates, Bridge programs can be an indispensable component to URMs STEM success (Matthews, 2006).
- Stress and create opportunities for undergraduate STEM research. Perhaps the most potent means of attracting and retaining STEM students, of advancing graduate school ambitions, and of replenishing the national STEM workforce is to actively

and intentionally engage URMs in undergraduate research (Hurtado, et al., 2009).

- Provide students with the opportunity to participate in STEM Exposure trips.
- Provide students with the opportunity to participate in STEM Conferences.
- Place special emphasis on changing mathematics from being a "gatekeeper" discipline to being a "gateway" discipline. After all, mathematics is the language of STEM. High attrition rates are usually associated with mathematics and other "gatekeeper" courses, and this problematic issue affects URMs more so than other students. Studies (Schneider, 2000; Vetter, 1994) indicate that substandard high school preparation and its ensuing weak scholastic foundation may be the cause.

Embrace Interdisciplinary STEM Learning

Interdisciplinary learning can be a potent antidote to remedy the STEM paralysis if the following steps are taken:

- Introduce, enforce, and support the institutionalization interdisciplinary STEM programs at all levels of learning and at the very earliest stages of learning;
- Create and involve industry partnerships in interdisciplinary STEM course oversight and implementation;
- Require that a learning specialist participate on the implementation team to provide guidance on effective interdisciplinary STEM pedagogical techniques;
- Seek and obtain institutional buy-in for interdisciplinary STEM learning from the top administrators at the institution;
- Institutions need a vision of undergraduate interdisciplinary STEM learning;
- Attract (perhaps through campus workshops) faculty champions (particularly chairpersons) to become participants on an active interdisciplinary STEM learning task force;
- Have clearly defined interdisciplinary STEM learning goals;
- Dispel the notion that interdisciplinary STEM initiatives are less demanding and less rigorous than conventional uni-modal disciplinary efforts;
- Make clear the advantages of the interdisciplinary STEM learning curricula;
- Develop special programs that are specifically geared to ensuring access and success of URMs in STEM through the implementation of interdisciplinary learning. For example, institute science-themed seminars, workshops, conferences, and competitions that integrate STEM fields like engineering, biology, chemistry, mathematics, and physics and are designed to enhance students' quantitative and qualitative literacy and their critical thinking and problem solving skills;
- Create Capstone interdisciplinary STEM courses;
- Create interdisciplinary STEM learning communities;
- Create extra-curricular interdisciplinary STEM activities.

As we venture further in this young millennium, and as knowledge becomes more systematic, more integrative, and more applied, there is an imperative that creates a demand for students to acquire STEM skills, despite their majors. As we grapple with global, national, and local issues of environmental sustainability, global food and health crises, renewable energy, medicine, ethics, and economic viability, the critical need for STEM interdisciplinary knowledge becomes more and more apparent. The traditional STEM approaches and pathways are now inadequate. A paradigm shift in STEM learning and application is inevitable. Educators and industry must, therefore, forge a vibrant nexus that will serve to propel this nation back to its global leadership role and dominance in a fertile world where interdisciplinary STEM initiatives thrive. However, to accomplish this goal, all hands must be on deck. This nation can no longer afford to neglect its URM STEM talent pool. If it continues to do so, it will do it at its own self-induced peril.

ACKNOWLEDGMENT

This work was partially supported by NSF REU Grant #1062934, NSF STEP Grant #0622493, and City Tech's Black Male Initiative. The authors would like to sincerely thank Dean Pamela Brown, Ms. Andrea DePetris, Dr. A. E. Dreyfuss, Dean Sonja Jackson, Ms. Sonia Johnson, Dr. Reneta D. Lansiquot, Ms. Connie Lu, Mr. Kurt Sealy, Ms. Laura Yuen-Lau, and all the peer leaders and mentors for partnering with them in this critical and rewarding project.

REFERENCES

Alexander, B., Foertsch, J., Daffinrud, S., & Tapia, R. (2000). The "spend a summer with a scientist" (SaS) program at Rice University: A study of program outcomes and essential elements 1991-1997. *Council on Undergraduate Research Quarterly*, *20*(3), 127–133.

American Competitiveness Initiative. (2006). *Website*. Retrieved from http://georgewbush-whitehouse.archives.gov/stateoftheunion/2006/aci

Arendale, D. R. (2004). Pathways of persistence: A review of postsecondary peer cooperative learning programs. In I. M. Duranczyk, J. L. Higbee, & D. B. Lundless (Eds.), *Best Practices for Access and Retention in Higher Education*. Retrieved from http://www.cehd.umn.edu/CRDEUL/pdf/monograph/5-a.pdf

Arrington, C. A., Hill, J. B., Radfar, R., Whisnant, D. M., & Bass, C. G. (2008). Peer mentoring in the general chemistry and organic chemistry laboratories: The Pinacol rearrangement: An exercise in NMR and IR spectroscopy for general chemistry and organic chemistry laboratories. *Journal of Chemical Education*, *85*(2), 288–290. doi:10.1021/ed085p288

Arrison, T. (Ed.). (2009). *Rising above the gathering storm two years later: Accelerating progress toward a brighter economic future: Summary of a convocation*. Washington, DC: National Academies Press. Retrieved from http://www.nap.edu/catalog.php?record_id=12537

Association of American Colleges and Universities. (2007). *College learning for the new global century: A report from the national leadership council for liberal education and America's promise*. Washington, DC: AACU. Retrieved from http://www.aacu.org/leap/documents/GlobalCentury_final.pdf

Atkinson, R. (2010). *Eight ideas for improving the America COMPETES act*. Washington, DC: Information Technology and Innovation Foundation. Retrieved from http://www.itif.org/files/2010america-competes.pdf

Augustine, N. R. (2007). *Is America falling off the flat earth?* Washington, DC: National Academies Press. Retrieved from http://www.nap.edu/catalog.php?record_id=12021

Bauer, K. W., & Bennett, J. S. (2003). Alumni perceptions used to assess undergraduate research experience. *The Journal of Higher Education*, *74*, 210–230. doi:10.1353/jhe.2003.0011

Bond, J., Schaefer, M., Rejeski, D., & Nichols, R. (2008). *OSTP 2.0: Critical upgrade*. Washington, DC: Woodrow Wilson Center for International Scholars. Retrieved from http://wilsoncenter.org/news/docs/OSTP%20Paper1.pdf

California Council on Science and Technology. (2007). *Shaping the future: California's response to rising above the gathering storm.* Sacramento, CA: CCST. Retrieved from http://www.ccst.us/publications/2006/GSTFrecs.pdf

California Council on Science and Technology. (2009). *Creating a well-prepared science, technology, engineering, and mathematics (STEM) workforce: How do we get from here to there?* Sacramento, CA: CCST. Retrieved from http://www.ccst.us/publications/2009/2009CalTAC.pdf

Campbell, A., & Skoog, G. (2004). Preparing undergraduate women for science careers. *Journal of College Science Teaching*, *33*(5), 24–26.

Chandra, U., Stoecklin, S., & Harmon, M. (1998). A successful model for introducing research in an undergraduate program. *Journal of College Science Teaching*, *28*(2), 116–118.

Chaplin, S., Manske, J., & Cruise, J. (1998). Introducing freshmen to investigative research: A course for biology majors at Minnesota's University of St. Thomas. *Journal of College Science Teaching*, *27*(5), 347–350.

Chubin, D. E., & Babco, E. L. (2003). *Walking the talk in retention-to-graduation: Institutional production of minority engineers-A NACME analysis.* Commission on Professionals in Science and Technology. Retrieved from www.cpst.org/WalktheTalk.pdf

College Board. (2008). *Coming to our senses: Education and the American future, report of the commission on access, admissions and success in higher education.* New York, NY: College Board. Retrieved from http://professionals.collegeboard.com/profdownload/coming-to-our-senses-college-board2008.pdf

Davis, C. G., & Finelli, C. J. (2007). Diversity and retention in engineering. *New Directions for Teaching and Learning*, *111*, 63–71. doi:10.1002/tl.287

Davis-Lowe, E. (2006). Fostering STEM diversity. *Diversity Committee, Oregon Pre-Engineering and Applied Sciences Initiative.* Retrieved from http://opas.ous.edu/Committees/Resources/Staff_papers/Fostering_STEM_Diversity.pdf

Dillner, D. K., Ferrante, R. F., Fitzgerald, J. P., & Schroeder, M. J. (2011). Integrated laboratories: Laying the foundation for undergraduate research experiences. *Journal of Chemical Education*, *88*(12), 1623–1629. doi:10.1021/ed101180f

Fairweather, J. (2008). *Linking evidence and promising practices in science, technology, engineering, and mathematics (STEM) education: A status report.* Paper presented at NRC workshop on Evidence on Selected Promising Practices in Undergraduate Science, Technology, Engineering, and Mathematics (STEM) Education. Washington, DC. Retrieved from http://nrc51/xpedio/groups/dbasse/documents/webpage/0465551.pdf

Foertsch, J., Alexander, B., & Penberthy, D. (2000). Summer research opportunity programs (SROPs) for minority undergraduates: A longitudinal study of program outcomes 1986-1996. *Council of Undergraduate Research Quarterly*, *20*(3), 114–119.

Gosser, D., Cracolice, M., Kampmeier, J., Roth, V., Strozak, V. S., & Varma-Nelson, P. (2001). *Peer-led team learning: A guidebook.* Upper Saddle River, NJ: Prentice Hall.

Huang, G., Taddese, N., & Walter, E. (2000). *Entry and persistence of women and minorities in college science and engineering education (No. NCES 2000601).* Washington, DC: National Center for Education Statistics.

Huntoon, J. E., & Lane, M. (2007). Diversity in the geosciences and successful strategies for increasing diversity. *Journal of Geoscience Education*, *55*, 447–457.

Hurtado, S., Cabrera, N. L., Lin, M. H., Arellano, L., & Espinosa, L. L. (2009). Diversifying science: Underrepresented student experiences in structured research programs. *Research in Higher Education*, *50*(2), 189–214. doi:10.1007/s11162-008-9114-7

Jackson, S. A. (2002). *The quiet crisis: Falling short in producing American scientific and technical talent*. Building Engineering and Science Talent. Retrieved from www.bestworkforce.org/PDFdocs/Quiet_Crisis.pdf

Liou-Mark, J., Dreyfuss, A. E., & Younge, L. (2010). The implementation of peer assisted learning workshops in precalculus: An approach to increasing student success. *Journal of Mathematics and Computer Education*, *44*(3), 249–259.

Loppato, D. (2007). Undergraduate research experiences support science career decisions and active learning. *Journal of Cell Biology Education. CBE Life Sciences Education*, *6*(4), 297–306. doi:10.1187/cbe.07-06-0039 PMID:18056301

Mabrouk, P. A. (2009). Survey study investigating the significance of conference participation to undergraduate research students. *Journal of Chemical Education*, *86*(11), 1335–1340. doi:10.1021/ed086p1335

Matthews, F. L. (2006). The envy of its peers? *Diverse Issues in Higher Education*, *23*(15), 26–31.

Mertz, N. T. (2004). What's a mentor, anyway? *Educational Administration Quarterly*, *40*(4), 541–561. doi:10.1177/0013161X04267110

Mestre, J. (2008). *Learning goals in undergraduate STEM education and evidence for achieving them.* Paper presented at NRC Workshop on Evidence on Selected Promising Practices in Undergraduate Science, Technology, Engineering, and Mathematics (STEM) Education. Washington, DC. Retrieved from www.nationalacademies.org/bose/PP_Mestre_STEM%20White%20Paper.pdf

Micari, M., Gould, A. K., & Lainez, L. (2010). Becoming a leader along the way: Embedding leadership training into a large-scale peer-learning program in the STEM disciplines. *Journal of College Student Development*, *51*(2), 218–230. doi:10.1353/csd.0.0125

Minnesota State Colleges & Universities. (2002). *A summary of "best practices" for recruitment and retention of students of color.* St. Paul, MN: MNSCU. Retrieved from http://www.studentaffairs.mnscu.edu/BestPractices.pdf

Muraskin, L., Lee, J., Wilner, A., & Swail, W. S. (2004). *Raising the graduation rates of low-income college students.* Washington, DC: Pell Institute for the Study of Opportunity in Higher Education. Retrieved from http://www.pellinstitute.org/gradrates/Pell_Web.pdf

Nagda, B., Gregerman, S., Jonides, J., Von Hippel, W., & Lerner, J. S. (1998). Undergraduate student-faculty research partnerships affect student retention. *Review of Higher Education*, *22*, 55–72. doi:10.1353/rhe.1998.0016

National Academies. (2008). *Rising above the gathering storm: Energizing and employing America for a brighter economic future*. Washington, DC: The National Academies Press.

National Science Foundation. (2006). *Women, minorities, and persons with disabilities in science and engineering*. Retrieved from www.nsf.gov/statistics/wmpd

National Science Foundation. (2008). *Women, minorities, and persons with disabilities in science and engineering*. Retrieved from www.nsf.gov/statistics/wmpd

New York City College of Technology. (2011a). *Facts 2010-11*. Retrieved from http://www.citytech.cuny.edu/files/aboutus/facts.pdf

New York City College of Technology. (2011b). *Website*. Retrieved from http://air.citytech.cuny.edu/air/page713243539.aspx

Nnadozie, E., Ishimaya, J., & Chon, N. (2001). Undergraduate research internships and graduate school success. *Journal of College Student Development*, *42*(2), 145–156.

Peterfreund, A. R., Rath, K. A., Xenos, S. P., & Bayliss, F. (2008). *The impact of supplemental instruction on STEM courses: Results from San Francisco State University*. Retrieved from http://www.sagefoxgroup.com/articles_conferences/Impact_of_SI_STEM_2008.pdf, Accessed 1/3/11

Quitadamo, I. J., Brahler, C. J., & Crouch, G. J. (2009). Peer-led team learning: A prospective method for increasing critical thinking in undergraduate science courses. *Science Educator*, *18*(1), 29–39.

Russell, S. H., Hancock, M. P., & McCullough, J. (2007). Benefits of undergraduate research experiences. *Science*, *316*, 548–549. doi:10.1126/science.1140384 PMID:17463273

Schneider, B. (2000). Explaining the unrealized aspirations of racial and ethnic minorities. In G. Campbell, R. Denes, & C. Morrison (Eds.), *Access Denied: Race, Ethnicity, and the Scientific Enterprise* (pp. 174–187). Oxford, UK: Oxford University Press.

Seymour, E., Hunter, A., Laursen, S., & De Antoni, T. (2004). Establishing the benefits of research experiences for undergraduates in the sciences: First findings from a three-year study. *Science Education*, *88*(4), 493–534. doi:10.1002/sce.10131

Shellito, C., Shea, K., Weissmann, G., Mueller-Solger, A., & Davis, W. (2001). Successful mentoring of undergraduate researchers: Tips for creating positive student research experiences. *Journal of College Science Teaching*, *30*, 460–465.

SPISU. (2011). *Expanding underrepresented minority participation*. Washington, DC: The National Academies Press. Retrieved from http://www.spisu.iastate.edu/litFiles/31/2/National%20Academies%20Report%20-%20Expanding%20URM%20Participation%20-%20Sept%202010.pdf

Strayhorn, T. L. (2008). The role of supportive relationships in facilitating African American males' success in college. *NASPA Journal*, *45*(1), 26–48.

Summers, M., & Hrabowski, F. (2006). Preparing minority scientists and engineers. *Science*, *11*(5769), 1870–1871. doi:10.1126/science.1125257 PMID:16574853

Tsui, L. (2007). Effective strategies to increase diversity in STEM fields: A review of research literature. *The Journal of Negro Education*, *76*(4), 555–581.

Vetter, B. M. (1994). The next generation of scientists and engineers: Who's in the pipeline? In W. Pearson Jr & A. Fechter (Eds.), *Who Will Do Science? Educating the Next Generation*. Baltimore, MD: Johns Hopkins University Press.

Washburn, M. H., & Miller, S. G. (2004). Retaining undergraduate women in science, engineering, and technology: A survey of a student organization. *Journal of College Student Retention*, *6*(2), 155–168. doi:10.2190/NDXH-YM83-TKWY-4E6C

KEY TERMS AND DEFINITIONS

Black Male Initiative (BMI): A City University of New York initiative intended to increase, encourage, and support the inclusion and educational success of underrepresented groups in higher education, in particular black males. All programs and activities of the BMI are open to all academically eligible students, faculty, and staff, without regard to race, gender, national origin, or other characteristic.

Faculty Mentoring: An advisory relationship that focuses on guiding a student's academic progress and fosters the student's career development.

Peer-Assisted Learning: A pedagogical approach that utilizes peer-led group activities to enhance student understanding of STEM concepts.

Peer Leadership: The process by which an undergraduate student facilitates and promotes collaboration among peers to accomplish learning goals and to enhance STEM problem solving skills.

Peer Mentoring: A process in which a more experienced student encourages, advises, and assists a less experienced student to develop his or her potential within a STEM content area.

Research Mentoring: The practice of engagement when an undergraduate student and faculty collaborate to conduct research and share new knowledge within a STEM discipline.

STEM Majors: Students who are enrolled in Applied Mathematics, Chemical Technology, Computer Information Systems, Computer Science, Telecommunication, or any of the Engineering Technologies within the associate and baccalaureate degree programs at City Tech.

Underrepresented Minorities (URM): STEM students who have the ethnic and gender designation of African Americans, Hispanic Americans, Native Americans, Southeast Asians, Pacific Islanders, and women.

This work was previously published in "Cases on Interdisciplinary Research Trends in Science, Technology, Engineering, and Mathematics" edited by Reneta D. Lansiquot, pages 237-267, copyright 2013 by Information Science Reference (an imprint of IGI Global).

Chapter 55

Improving Novice Programmers' Skills through Playability and Pattern Discovery:
A Descriptive Study of a Game Building Workshop

Thiago Schumacher Barcelos
Instituto Federal de Educação, Ciência e Tecnologia de São Paulo, Brazil & Universidade Cruzeiro do Sul, Brazil

Roberto Muñoz Soto
Universidad de Valparaíso – Escuela de Ingeniería Civil Informática, Chile

Ismar Frango Silveira
Universidade Cruzeiro do Sul, Brazil & Universidade Presbiteriana Mackenzie, Brazil

ABSTRACT

Game design and development has already been discussed as a viable, motivating alternative to introduce Computer Science concepts to young students. In this sense, it would be useful to obtain a deeper understanding of which skills could be developed in these activities and how such skills could be useful in future careers. This chapter presents the design and evaluation of a Game Building Workshop aimed at introducing the fundamentals of structured programming to students. The games produced by students during 12 weeks were evaluated and the results confronted with students' questions and comments made along the workshop meetings and a final interview. The results indicate that students explored novel programming concepts in order to add features that were not initially planned for the proposed games. These additional features solve playability issues that are highly influential to the experience of the students as game players. Students also reused previously applied solutions to solve similar problems that appeared in subsequent activities. This is an indication that students developed or exercised analogy and abstraction skills during the workshop activities.

DOI: 10.4018/978-1-4666-7363-2.ch055

INTRODUCTION

Along the last decade, students' enrollment rates in undergraduate courses related to Computer Science (CS) and Information Technology (IT) have been decreasing in various countries. Muratet et al. (2009) identified a 25% decrease in the number of students enrolled in Computer Science careers in a French university between 2005 and 2009. Crenshaw et al. (2008) reported that the interest of North American students for CS courses declined by 50% during the '00 decade. Hernandez et al. (2010) compared the student enrollment rate in CS courses of a top Brazilian university to the same rate in a university in the United States and found that the same tendency for a diminishing interest appears in both countries.

It has been argued (Hernandez et al., 2010) that student enrollment in computing courses is historically related to the spread of technological advances, such as the personal computer and the Internet. Based on this argument, it would be surprising to witness a lack of students' interest for the field of computing nowadays, given the high degree of technological achievements in the past few years, besides of their pervasiveness. After all, children and teenagers deal with a growing variety of interactive computational devices, such as tablets, cell phones and portable videogames. However, the increasing exposure to such gadgets does not seem to stimulate students to pursue technology-related careers. One possible reason that has been pointed out is that students are not properly exposed to Computer Science concepts during the basic educational levels, since CS is not considered as a Basic Science in any curricula (Barcelos & Silveira, 2012; Carter, 2006). High dropout rates in introductory courses may also be related to the lack of motivating didactic strategies (Rizvi, Humphries, Major, Jones, & Lauzun, 2011).

On the other hand, the presence of some skills in specific knowledge areas may contribute to the academic success of CS and IT students. Skills related to Mathematics are often mentioned to have an influence in student achievement in CS/IT. Such studies are not new; Campbell and McCabe (1984) analyzed a sample of 256 freshmen of an north American university to conclude that SAT scores in Math, high school rank and average grade high school grades in Math are predictors for the group of students who continued in the course after two semesters. More recently, Wilson e Shrock (2001) found a similar correlation between the number of Math disciplines taken by students during high school and their performance in an introductory CS discipline. Beaubouef (2002) presents an extensive discussion about the importance of Math topics that are related to many subjects studied in CS. It may be convenient, though, to identify which are the high-level skills related to Math that should be mastered by students in order to improve their academic performance in CS/IT. As a consequence, college education would be able to deliver better professionals to the labor market.

Abstraction is, for instance, a relevant skill that is present in the field of Mathematics and that is equally important to Computing. A discussion about this topic is presented by Kramer (2007), who argues that building and understanding abstract models is crucial to object-oriented design and programming; also, selecting the most important aspects of a given real situation and leaving out unnecessary details is important to the activities of a requirements engineer. On the other hand, the process of problem solving in Math involves skills related to abstraction, such as analogy, generalization and specialization (Polya, 2004).

This chapter presents the design of a Game Building Workshop aimed at teaching the fundamentals of programming to students enrolled in Computing-related courses. The workshop was designed based on previous evidence (Bayliss & Strout, 2006; Leutenegger, 2006) that the domain of game design and construction might have a relevant impact on students' motivation. Besides that, our objective was also to understand how students might develop higher-order skills, such as abstraction, by dealing with tasks related to game design, which are presented at progressive levels of complexity. The workshop structure and its activities are presented, also discussing its pedagogical foundations. In order to analyze its impacts, an experimental evaluation of the workshop activities was conducted in the first semester of 2013. Based on the observation of lab activities and analysis of the games produced by students, the role of students' motivation is discussed, as well as how traces of pattern recognition skills could be identified on students' works.

BACKGROUND

Game Design as a Teaching Strategy

Building games has been advocated as a teaching strategy to develop skills in fields such as computational thinking (Settle, 2011), logic reasoning (Souza & Dias, 2012) and programming fundamentals (Hernandez et al., 2010). Games are often seen as a motivating resource for students of the new generations, defined by Prensky (2004) as *digital natives*. These students, born after the 1990 decade, usually grew up dealing with several interactive devices such as cell phones, videogames and computers. Hence, games are one kind of digital media that students are very familiar with, supposedly bringing a unlimited potential for its use in educational activities. Peppler and Kafai (2009) argue that games are so integrated into modern culture that students should develop a *fluency* in games, that is, a deeper understanding of its mechanics and how they are built. In addition to that, Salen (2007, p. 302) states that digital games are "entry points for many young people into digital literacy, social communities, and tech-savvy identities."

It is important to point out that, in a general sense, digital games are no more than interactive systems with special requirements related to the way the interaction with its user takes place. The interaction with the player must provide increasing challenges and unpredictable outcomes, but also allow the player to feel "in control" in order to explore the world of the game (Barcelos, Carvalho, Schimiguel, & Silveira, 2011). In their classic work about user interface development, Hix & Hartson (1993) argue that a regular user perceives the system interface as the whole system; there lies the importance of good user interface design. Clearly, this also applies to the design of digital games. Salen (2007) has already mentioned that digital game design is a process of "reflection-in-action," as discussed by Schön & Bennett (1996): the designer must take into consideration the possibilities and constraints that emerge during the design process. According to these authors, this process may only take place when the designer actively deals with the materials. In the case of software design, the "materials" are not tangible; in this case, prototyping must be used to identify and reflect about the design possibilities.

Several authoring and programming environments were developed in the last years to facilitate the process of creating games. These environments include Scratch (MIT Media Lab, 2012), Alice (Carnegie Mellon University, 2013), GameMaker (YoYo Games, Ltd., 2014) and Kodu Game Lab (Microsoft Research, 2014), among others. These environments allow children and teenagers to create their own games in a less complex way than what was possible with traditional programming languages. Some research works describe the learning outcomes obtained by young students who are introduced to these game creation environments. Maloney et al. (2008) analyze 536 projects made in Scratch by children and teenagers in an after-school computer clubhouse to conclude that the participants successfully incorporated in their project concepts of user interaction, loops and message synchronization, among others. Participants were not explicitly instructed to use Scratch and frequently opted to use this environment, despite the fact there were other activities available. The authors argue that the simplified programming mechanics and the immediate feedback may have stimulated the participants to use Scratch. Stolee and Fristoe (2011) analyze games developed with Kodu Game Lab that were made available at an online community. Concepts such as variables, control flow, Boolean logic and rudimentary instantiation of objects could be identified in the games. Hernandez et al. (2010) describe a didactic experience with freshman in an introductory discipline of an Information Technology undergraduate course. Students built games using GameMaker environment and applied the concepts of variables, event handling and loops. The approval rate of students and the number of delivered exercises both increased after this experience, which may be an indicative that students were more motivated to learn introductory programming concepts by building games.

Experiences like the one described by Hernandez et al. (2010) indicate that undergraduate students may also benefit from didactic strategies that incorporate game building. Malan & Leitner (2007) describe a programming fundamentals course that was modified to present the basic programming structures to students using Scratch; afterwards, Java syntax and structures were introduced. Most students perceived that this approach helped them to deal with the complexity of the Java syntax, especially those that had no previous programming experience. According to the authors of such study, Scratch environment allowed students to focus almost exclusively on the logical aspects of the developed projects. The code in Scratch is made out of blocks that only "snap" together in syntactically correct ways; hence, students do not have to deal with syntax issues when they start coding. Other experiences where digital game creation was incorporated into introductory programming courses at the undergraduate level are described by Rizvi et al. (2011), Leutenegger (2006) and Bayliss & Strout (2006). The evaluation of such initiatives is usually based on the improvement of overall grades or the self-perception of students about their motivation with the course. However, few studies have dealt with the question of specific skills and contents that might have been learned by students. These studies are mostly focused on Computer Science outreach programs (Franklin et al., 2013; Adams & Webster, 2012; Denner, Werner, & Ortiz, 2012) and include some kind of assessment of programming skills acquisition. However, little attention has been given to how higher-level cognitive skills related to these programming activities would be useful to students that will continue their studies in CS or IT.

Programming-Related Skills

Several authors have researched about skills differences between novice and expert programmers. These studies often have a cognitivist basis and have been developed since the late 1970's. For example, according to McKeithen et al. (1981), the differences between novice and expert programmers do not lie only in their capacity to process large quantities of significant information. This has been demonstrated in other domains than programming, such as chess (Waters, Gobet, & Leyden, 2002), web navigation (Chen, Fan, & Macredie, 2006), music (Colley, Banton, Down, & Pither, 1992) or physics (Chi, Feltovich, & Glaser, 1981; Larkin, McDermott, Simon, & Simon, 1980). Apart from information processing skills, there is evidence that the differences may reside in the way novices and experts gather and organize information (Lister, Simon, Thompson, Whalley, & Prasad, 2006). Although it is expected that experts possess more knowledge than novices, Bateson et al. (1987) indicate that experts also have a more sophisticate and flexible knowledge organization. A representative example from an experiment in the domain of chess is described by Chase & Simon (1973): expert chess players have better information recall when pieces are associated to a strategy; however, their recalling is much worse when pieces are randomly distributed on the board, because the known abstract models of attack and defense are no longer present. This example is revisited by Bateson et al. (1987) and Lister et al. (2006), who argue that expert programmers tend to organize and recall their knowledge using higher-order strategies.

A similar argument is made by Weiser & Shertz (1983), who state that expert programmers are no different from experienced physics as far as their problem-solving skills are considered. The authors argue that both professionals represent a given problem in terms of general principles that do not appear in the general statement of the problem. Also, they identified that programming skills were related to the time necessary to write a new program in order to make it functional, to the number of times a program is submitted before completion, the number of known languages, and the familiarity to some programming concepts. On the other hand, skills were not related to the time needed to debug, modify or understand a new program.

The previous findings are related to those reported by Shneiderman & Mayer (1979), who argue about differences between novice and expert programmers and about the differences between the syntactic and semantic memories in the domain of computer programming. Semantic memory is relatively independent from the syntactic rules of some programming language, and semantic knowledge can be constructed by constantly engaging in problem-solving activities (Bateson et al., 1987). Besides that, semantic knowledge is stored as general and meaningful sets of information (Shneiderman & Mayer, 1979). Bateson et al. (1987) indicate that expert programmers use the syntactic memory and a high-level knowledge plan to guide their programming activities. Previously, Shneiderman (1976) had already identified that expert programmers can perform better than novices on recalling structured programs, although the former ones may introduce syntactic mistakes that do not compromise the output of the program (in other words, its semantic aspect).

Soloway et al. (1982) suggest that expert programmers use more than syntactic and semantic knowledge when writing code to solve a problem; they would also rely on a high level plan based on previous knowledge and reflection. According to these authors, a plan is a procedure or set of activities in which the key elements of the process have been summarized and represented in an explicit form. When faced with a new problem, an expert retrieves from her knowledge base the possible plans that were useful in similar situations and incorporates those plans into the solution. Possible plans are classified in three categories (Soloway et al., 1982):

- **Strategic Plans:** That specify a global strategy to define an algorithm.
- **Tactical Plans:** That specify a local strategy to solve a problem.
- **Execution Plans:** That rely on characteristics of a specific programming language to implement tactical and strategic plans.

Based on the previous findings about the way experts solve problems, it would be useful for educators to have a systematic way to "communicate" these skills to novices, perhaps through a set of patterns. For instance, software patterns have been proven to be a powerful tool to represent knowledge about best practices used by experts in the area of software design and development (Muller, Haberman, & Averbuch, 2004). One of main advantages of patterns is the possibility to reuse and adapt them to solve new problems, as long as the pattern matches the relevant characteristics of previous problems for which it has been proven to be useful. According to East et al. (1996), students must develop and apply two skills in order to properly use patterns. The first skill is being able to choose the most appropriate model or schema among similar ones. The second one is being able to adapt and modify patterns to suit the solution to the problem that is being currently solved.

Both of these skills are related to aspects of abstraction. Polya (2004) defined that aspects related to abstraction (analogy, generalization and specialization) are crucial for a Math student to succeed in problem-solving tasks. A similar definition in the context of programming skills acquisition is presented by Haberman & Muller (2008), who defined abstraction by the following features:

- Generalization of specific examples;
- Identification, extraction, and isolation of essential components; and
- Ignoring or holding back irrelevant details.

Abstraction skills related to pattern identification and usage must be obtained through various experiences, similar to the development of semantic memory (Bateson et al., 1987); also, adequate teacher support may be necessary to help students identify which might be the relevant aspects of some problem (Haberman & Muller, 2008; Clancy & Linn, 1999).

Hence, game design activities have already been used to motivate and engage students; previous results show that students may successfully deal with fundamental programming concepts through some game engine and afterwards make a "smoother" transition to conventional programming languages. On the other hand, based on previous research (Bateson et al., 1987; Soloway et al., 1982; Shneiderman & Mayer, 1979) it is expected that a novice programmer will increasingly rely on higher-order semantic plans when he/she becomes more experienced. These plans may be defined by using analogy and abstraction (Polya, 2004; East et al., 1996); in this chapter, this particular hypothesis is tested in the context of digital game building activities. Considering that most students that take part in such activities may be classified as digital natives, as stated by Prensky (2004), it will also analyzed how their previous experience with digital games, if any, may influence their programming skills development.

GAME BUILDING WORKSHOP

Structure

The Game Building Workshop was designed to be incorporated into a introductory course on fundamentals of programming to freshmen in a professional education career in Informatics offered by the São Paulo Federal Institute of Education, Science and Technology, Brazil. This type of career is offered in Brazil to students that are enrolled at the same time in regular high school or that have already completed this educational level. It is a 3-semester long course aimed at preparing students to the job market in a short time. Learning the fundamentals of programming and the syntax of a first programming language is usually a significant hurdle to students; hence, the issues with high failure and dropout rates mentioned before are also present in this course. The Game Building Workshop was initially designed as a strategy to deal with these issues. At the same time, a secondary goal for the workshop was that its activities were partially compliant to curriculum guidelines related to the development of computational thinking skills. The CSTA (Computer Science Teachers Association) Standards for Computer Science Education at the K-12 level (CSTA, 2011) define that students should be exposed to the principles of construction of algorithms; also, they should be able to use some fundamental programming structures (looping behavior, conditional statements, logic, expressions, variables and functions) between the last years of middle school and the first years of high school. This way, the workshop activities should be flexible enough to be also offered in the future to students that are enrolled only in regular high school education.

The workshop activities are distributed along 12 weeks; each weekly meeting lasts about 2.5 hours, when students explore concepts related to game design (sprite animation, collision detection, keyboard and mouse control, for instance) and also to programming fundamentals (e. g., variables, conditional and loop structures, messages). Each activity is related to the construction of interaction mechanisms that are present in a digital game. Scratch was the tool used to develop the games due to the similarity of its programming structures to those used by traditional, structured programming languages.

In the first meetings of the workshop, students are introduced to the Scratch environment, to the concept of variables and to basic conditional and looping structures. Starting in the fourth meeting, students begin to implement full-featured games. The implemented games are: a digital version of the Rock-Paper-Scissors hand game, a war simulation game (named *Point-and-Shoot* from now on) and clones of the famous games Breakout and Pacman. In Rock-Paper-Scissors game, students define keystrokes for each one of the three possible actions of both players and define how the outcomes of the game will be displayed. In Point-and-Shoot game, students build an algorithm that defines a linear trajectory for a bullet, triggered by a keypress, and the possible conditions for the bullet to explode (hitting an enemy target or a wall). Breakout rules also involve defining a linear trajectory, but now of a ball, with the additional rule that it has to bounce when it hits a brick or a player-controlled paddle. In Pacman students must implement controls for main character's movements, constrained by the paths of a maze, and also define the movements of two ghost enemies. One of the ghosts should chase the Pacman and the other one should follow a fixed path inside the maze. This last game is the Workshop's final project and its results are part of students' evaluation (Figure 1).

The workshop follows a Problem-Based-Learning (PBL) perspective. According to Merril (2002), PBL is a student-centered strategy; they should work in a collaborative way to find a solution to a problem proposed by the teacher. The teacher's role is to help students to mobilize their own previous knowledge and then build new knowledge in order to solve the problem. This new knowledge is gradually built, based on empirical experimentation. In each Workshop meeting, teacher is meant to present a previously

Figure 1. Examples of games produced during the Game Building Workshop

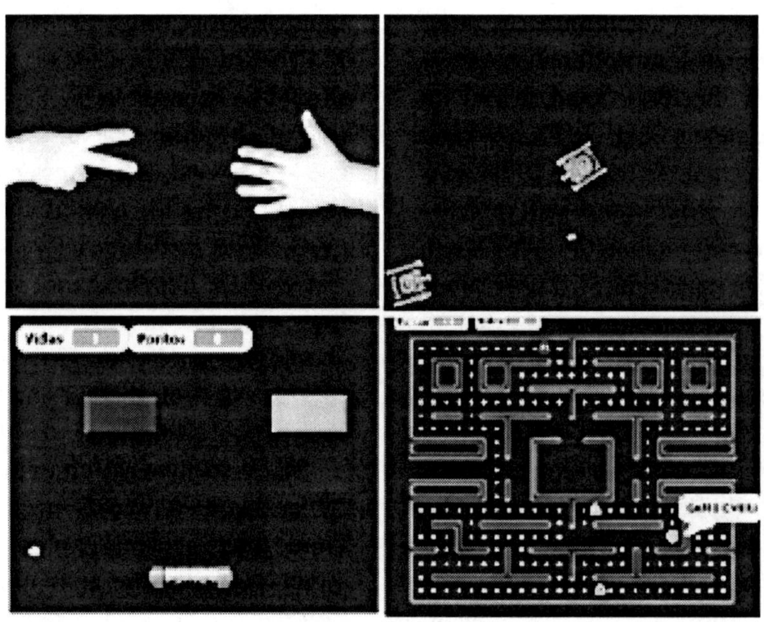

implemented version of the game to illustrate the features that should be implemented by the students. Then, students begin to work and the teacher may intervene when asked by them.

In order to define Workshop activities and their organization according to the schedule of the meetings, four guidelines were defined, presented and justified in the following.

The motivation for all workshop activities is the implementation of some game's features: This guideline is grounded on the game fluency development advocated by Peppler and Kafai (2009), who argue that digital games are a cultural artifact for the new generations; hence, when students are engaged in game building activities, these activities may provide opportunities for critical thinking and meaningful learning.

The sequence of activities must propel students to building full-featured games: This guideline is based on the Use – Modify – Create framework introduced by Lee *et al.* (2011). In many of the activities, students are initially asked to interact with small programs. Once students are familiarized with the programming structures used, they are invited to introduce some modifications to the original features of the programs. Eventually, students will be confident enough to create their own games from scratch using skills they have acquired on previous activities. One should notice that this strategy is not entirely "linear": a student may be a "creator" at some stage and then be a "user" in a following activity in order to understand a novel concept. Nevertheless, the structure of Workshop activities is designed to allow students to be "creators" more and more often in subsequent activities.

Activities must progressively require that students explore new concepts and, at the same time, reuse concepts that were previously explored by them: As students must learn new concepts in order to accomplish new tasks and also use concepts that were previously learned, it is expected that the activities present a challenging level that is always proportional to the students' skills. This guideline is based on the flow theory, discussed by Nakamura & Csikszentmihalyi (2009). According to the authors, flow is a state of deep concentration and engagement in a given activity, when higher productivity levels could be reached; when one's skill level matches the level of challenge required by a given activity, he/she is said to be in the flow state. This is applicable when students are working autonomously; however, teacher support is still needed when a student is not able to accomplish a task otherwise. These situations are defined by Vygotsky (1978) as representing the Zone of Proximal Development (ZPD) of an individual. While some activities demand that students reuse simple concepts on their own, other activities will require the help of the teacher, since students would be probably working at their ZPD.

Game mechanics should be simple and bring references to the world of "real" games that are meaningful to students: This guideline is also grounded on the game fluency (Peppler & Kafai, 2009) that students should develop. Besides this, building game elements that are simple are crucial to manage students' expectations, as it would not be viable to ask students to build complex games (including 3D elements or complex algorithms, for instance). Also, when students are challenged to build a known game element, this might help them to engage in a "reflection-in-action" process (Schön & Bennett, 1996) as their previous experience with a similar game may be a reference for the quality level desired for the given element.

In each meeting of the Workshop, students have to deal with a new programming concept. At the same time, concepts introduced in previous meetings should be necessary again in subsequent activities to allow students to identify some key situations and apply the concepts accordingly. This organizing strategy is compliant to the third guideline presented before.

The activity scheduling for the Game Building Workshop is presented in Table 1 each activity is presented together with the new concepts that are introduced with it.

Table 1. Activity schedule for the Game Building Workshop

Week	Activity	Novel Concepts Introduced
1	Control two sprites with the keyboard and identify their collision	Algorithm, statement, simple conditional structure (IF), logical conditions
2	Fishing game	Variables and operations Iteration structures – "infinite loop"
3	Guess the number	Iteration structures with logical conditions Relational conditions and expressions
4	Rock-Paper-Scissors	Synchronization through message broadcasting
5-6	Point-and-Shoot	OR Boolean operator IF...ELSE conditional structure
7-8	Breakout	--
9-11	Pacman	AND Boolean operator
12	Final projects presentation	--

It must be noted that even the fundamental concepts of programming are introduced in the first four weeks through the implementation of the mechanics of small games. Variables and iteration structures are implemented in a fishing game (2nd week), in which the player must perform mouse clicks on sprites that represent fishes in order to get points. The conditional execution of statements is introduced through a guessing game (3rd week): the player has a limited number of attempts to guess a number that was randomly chosen by the game. In this occasion, the computer generation of random numbers is also discussed with the students due to its importance to the game mechanics that they will build in the following meetings.

Due to the adoption of the PBL perspective, a computer laboratory is the preferred environment to the Workshop activities. Students are encouraged to work in pairs using the computer in order to stimulate discussions and exchange of ideas. As expected, the domain of digital games naturally brings a motivational appeal to students, as well as it is also expected that they would share and compare their results with their peers.

Experimental Design

The Game Building Workshop experiment was first offered during the first semester of 2013 to a group of 40 freshmen of São Paulo Federal Institute. The types of problems to be solved by students, as described in the previous section, would be quite different from those usually found in a programming fundamentals course. Besides that, one could expect that the domain of digital games is familiar to students, considering the age range that typically enroll in the course (usually students from 15 to 20 years old). Hence, building digital games could be an intrinsically motivating activity to these students. If this hypothesis is correct, students' motivation would be observed on their attitudes towards the activities or on the artifacts produced by them. Based on these perceptions, the experimental design was defined according to a qualitative perspective; thus, three research questions guided its definition:

1. Which strategies are used to implement the features proposed by the teacher?

The identification of possible common strategies used by students is relevant to understand how they may improve their expertise in programming by solving problems that gradually demand some previous knowledge.

2. Which quality aspects of digital games are valued by students?

Considering that students probably had some previous experiences with digital games, this question tries to identify which game features are most appreciated by students. Although the Workshop activities were initially not designed to incorporate these characteristics, the main intention was to verify if students' motivation – or the games implemented by them – would be affected in any way by their preferences as gamers.

3. Do students implement additional features that were not required by the activities?

This question was added as a complement to Question 2, after a preliminary analysis of the implemented games showing that a high number of students added extra features to their games. An in-depth analysis of these extra features and the strategies used to implement them could clarify the role of the game mechanics on students' learning process.

Initially, students answered a survey to allow the identification of their previous experience with digital games. They also identified through a 5-point Likert scale how 19 quality aspects for digital games influenced their user experience. The quality aspects are derived from the playability heuristics described by Barcelos et al. (2011) and were presented in a non-technical language. The original list of heuristics is presented in Table 2.

All games implemented by students during the workshop were collected for analysis. In order to identify how gameplay issues possibly affected

Table 2. Playability heuristics extracted from Barcelos et al. (2011)

H1	Game controls must be clear, customizable and physically comfortable; their response time must be immediate.
H2	The player must be able to customize video and audio settings.
H3	The player must easily find information about the game status and score.
H4	The game must allow the player to develop skills that should be necessary later on in the gameplay.
H5	The game must provide a clear tutorial.
H6	All visual features must be understandable to the player.
H7	The player must be able to save the current state of the game to continue playing later on.
H8	Visual layout and menus must be intuitive and organized to allow the player to focus only on gameplay.
H9	The game story must be rich and engaging.
H10	The graphics and sound track must spark the player's interest.
H11	The digital characters and the environment must look realistic and consistent.
H12	The main goal of the game must be immediately clear to the player.
H13	The game must propose secondary additional goals along with its main goal.
H14	The game must present many challenges and allow different strategies.
H15	Design of the game pace must consider the player's fatigue and attention levels.
H16	The challenge must be adjustable to the player's skill.
H17	The player must be rewarded for his/her achievements in a clear and immediate way.
H18	Artificial intelligence (AI) must provide unexpected challenges and surprises to the player.
H19	The game must provide hints, but not too many.

the students' implementation strategies, games were separated in two categories:

1. Games that incorporated only the minimum features required in the activity; and
2. Games that incorporated extra features.

This separation was possible because game design activities were part of the regular program of the course and, as such, students always had to complete a minimum set of features in order to get credits for each activity. Then, a heuristic evaluation of all extra features was performed. Heuristic evaluation was originally defined by Nielsen and Molich (1990) as a method to identify usability issues in an interface by means of a systematic inspection, based on a set of quality criteria – the usability heuristics. However, in this context, the method was adapted to allow the identification of playability criteria associated to the extra features incorporated to games by students. Two of the authors of this paper independently analyzed the games, associating at least one of the 19 playability heuristics to each extra feature. Finally, the results of the analysis were checked against each other to identify discrepancies; the combined results were then compared to the answers given by students in the survey.

All activities of the workshop took place in a computer laboratory. The observation of the class activity was conducted as an educational ethnography. Goetz & LeCompte (1984) indicate that, in this research modality, researcher should bear in mind some important aspects of the group that will be observed, such as: individual profiles, the place where the group meets, their interactions and the motivation for them. In order to facilitate the latter analysis of the observed events, audio recordings of all workshop meetings were captured with previous consent of the participants. The objective was to identify students' doubts and comments about each activity and their interactions with the teacher and with each other. The analysis was conducted using the audio recordings' transcripts. During the Workshop's last meeting, students presented their Pacman final projects to the teacher and were interviewed, based on the following questions:

1. Which strategies were used to implement the game characters;
2. Which features were harder to implement;
3. Which features students would like to add to the game, if they continued to develop them.

The interviews were also recorded and transcribed for later analysis. It must be pointed out that specific providences should be taken in order to assure the validity of a qualitative research. Creswell & Miller (2000) present nine different validation procedures that could be used depending on the audience of the study, the availability of individuals to perform additional verifications, and the costs of using them. The option of this work was to use the triangulation of methods, defined as the comparison between data obtained by different sources. In the case of this study, students' profile, obtained by the initial survey, the analysis of the games produced by students, the observation of class activities and the final interview are the data sources included in the triangulation.

Results

Students' Profile

Thirty students answered the initial survey. Most of them (23 students) are 15, 16 or 17-year old and are regularly enrolled in High School. The remaining ones are older students who are able to enroll in professional education courses. Seventy-seven percent of students declared to be regular gamers (47% claim to spend from 1 to 5 hours a week playing games and another 27% claim to spend from 5 to 15 hours a week with this activity; 3% of them spend less than an hour a week playing). The group has a relative experience playing games: 66% of them have played games for three of more years.

The quality criteria most often cited by students as highly influential to their user experience (i.e. rated 4 or 5 in the Likert scale) are shown in Figure 2. They are related to saving the game state (H7 – 92.9%), the presence of clear visual features (H6 – 89.3%), the quality of the game story (H9 – 85.7%), variety of challenges and strategies (H14), challenges provided by AI (H18 – 78.6%), comfort and response time of controls (H1 – 75%) and graphics and sound quality (H10 – 75%).

Additional Features in Games and Association to Playability Criteria

During the workshop, students developed 29 Rock-Paper-Scissors games, 22 Point-and-shoot games, 29 Breakout games and 21 PacMan games. The number of games had some variation from activity to activity due to the absence of students in some meetings; besides, some students opted to work in pairs in some activities and alone in others. opted to look for extra features only in these games, which were the Workshop's last four activities; the mechanics of the first games were very simplified, since they targeted to introducing students to programming fundamentals. It was found out that many students spontaneously added more features to the games than the ones that were explicitly proposed in the activities. We could observe that students made frequent comments during the activities about their intention to improve the mechanics and visual aspects of the games to suit their preferences. The first references to this intention could be observed in the second activity (Fishing game), when a student indicates that the displayed fish should change its position in a quicker way (the student's speech is translated to English, as the original transcript is in Brazilian Portuguese):

Figure 2. Distribution of answers to the question: "How relevant are the following criteria for you to enjoy a game?"

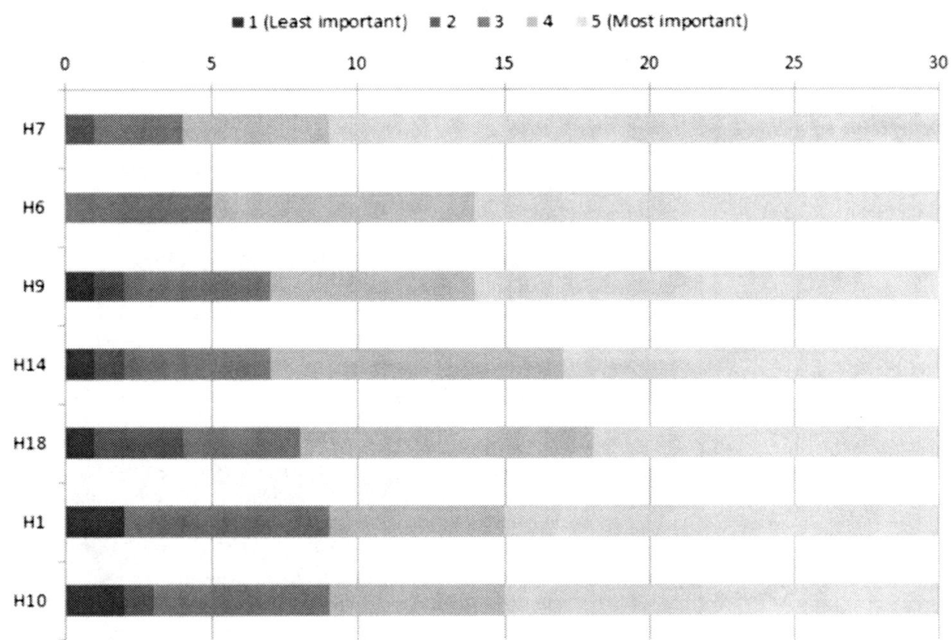

The fish should become quicker when the player gets less points, 20 [points] is too much... I think I should try to lower this [the points limit] from 20 to 10

In the original description of the activity, students were asked to implement a fish that changes its position in the screen. The player should perform mouse clicks on the fish to get points, and when the player gets 20 points, the fish would change its position in a quicker way. That student considered that it took too long for the game to get "harder" and identified that lowering the limit that defined a quicker repositioning of the fish would be the solution.

Another representative example was identified when students were building the Rock-Paper-Scissors. A limitation of the Scratch environment was identified by one student:

There will be a playability problem here. Look at this: the first player can make his move, and the second one can just wait for a little while to make his move... So, it's possible to cheat! Isn't there a way to solve this?

This student identified that, initially, the actions of both players in the game were not synchronized in any way; in order to add this feature, students were required to use the message broadcasting and synchronization mechanisms provided by Scratch. However, students would only be introduced to these features in the following meetings. This example is representative of a tendency that could be identified in other episodes: some game features were very relevant to students, even though they required the use of more advances programming techniques.

Consonant with the PBL methodology, the students were encouraged to experiment with new programming structures and with the Scratch environment to implement the desired extra features. In addition, they tended to add more extra features to games during subsequent meetings. Figure 3 shows the proportion of games with minimum features and extra features in each activity.

Two of the extra features that were more often found in each game are presented in Table 3. Additionally, the playability heuristics that were associated to each extra feature during the heuristic evaluation are also presented. It is noteworthy that most extra features are associated to the presenta-

Figure 3. Proportion of games with minimum and extra features

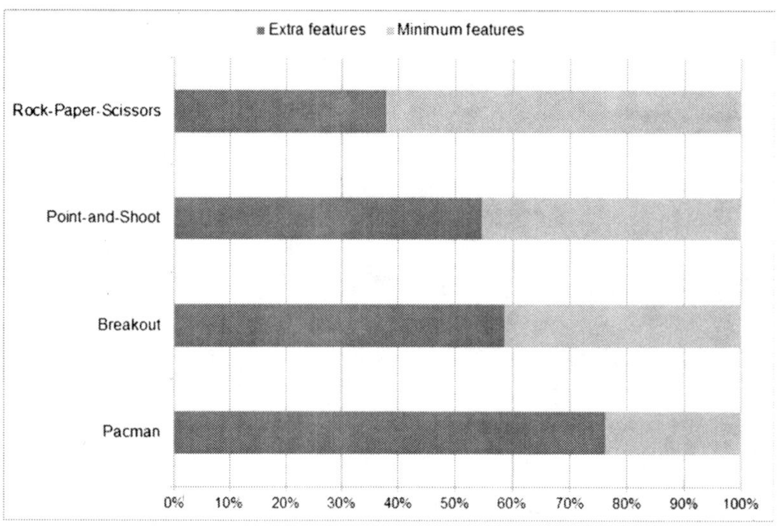

Table 3. Most implemented extra features, playability heuristics associated and programming concept(s) needed

Game	Feature	Heuristics	Programming Concept(s) Needed
Rock-Paper-Scissors	Win/draw message in screen background	H3; H6	Busy loop or Message broadcast
	Score	H3; H6	Variables
Point-and-Shoot	Allow new shot	H1	Message broadcast
	Game Over Screen	H3; H6	Message broadcast
Breakout	Fade out of paddle when losing ball	H6; H10	Loop with defined number of iterations
	More bricks to be destroyed	H14	None (sprite duplication is sufficient)
Pacman	Display number of lives	H3; H6	Variables or Busy loop or Message broadcast
	Sound effects	H10	None (only correct usage of "play sound" command)

tion of visual features (H6), often associated to the game status and score (H3), but some features are also associated to the response time of controls (H1) and presence of graphical and sound effects that would spark the player's interest (H10). In the last column, the programming concepts that should be used to implement each extra feature are presented.

The most implemented extra feature in Rock-Paper-Scissors game was a message shown in screen background that indicated which player won the match or if there was a draw – 9 from 29 games (31%) presented this feature. Although the activity actually required that students implemented the message, the most straightforward method would be to use a Scratch's "say" command, linked to one of the sprites. Each element that is displayed on the screen (named in Scratch as the "stage") is a sprite, and every command must be attached to a sprite or to the stage. If the code is associated to the stage, for the purpose of this feature it must continuously verify the state of other sprites; this is an additional challenge to students, as it will be discussed later. The most implemented extra feature in the Point-and-Shoot game (10 from 22 games, near the half part) was allowing the player to immediately shoot again while the last shot bullet was still heading towards the opposite side of the screen. This feature overcomes a limitation of Scratch that will also be discussed in the following section. In the Breakout game, the most frequent extra feature (9 from 29 games) was an animation that shows the paddle fading out when it misses the ball. In Figure 4 an example of the extra features in Rock-Paper-Scissors and Breakout are presented.

In PacMan, the most implemented extra feature (8 from 21 games, more than a third part) was an

Figure 4. Example of extra features added to Rock-Paper-Scissors and Breakout

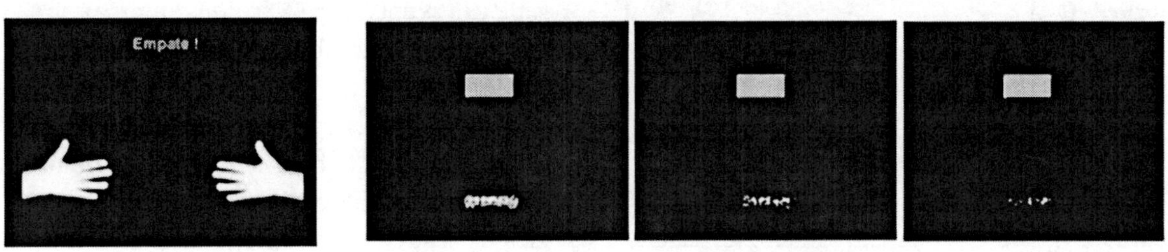

indication of the number of remaining lives and sound effects (7 games) for events such: the beginning of the game, the moment when PacMan "eats" a yellow dot, and the collision of Pacman with a ghost. It is important to point out that students were not expected to create their own sound effects and graphics during the workshop; hence, the same set of images and sounds was made available to them, even though its use was not required in every activity. Students had Internet access during the activities and were allowed to search for additional items as they desired.

The transcription of the final interviews with students reveal their intentions to improve the mechanics and visual aspects of the PacMan games they developed if they had more time to work on them. A content analysis was performed on the students' speech as defined by Bardin (1977) in order to identify their intentions. A few representative examples are pointed out:

As the game level rises, I would make it more difficult, I would add more ghosts

Oh, I would add that 'little power' so that it [the PacMan] could defeat its opponents

I would make it [the game] harder by making them [the ghosts] move faster

The categories defined from the content analysis are presented in Table 4, in decreasing order of the number of occurrences in students' speech. During the process of determining these categories, each one of them was associated to the playability heuristics that most closely expressed the meaning of the extra features suggested by students.

Table 4. Students' intention to add features to Pacman and associated heuristics

Number of Occurrences	Categories	Heuristics
8	More difficulty levels	H14/H16
7	Give more powers to Pacman	H14/H16
7	Give more powers to enemies	H14/H16
5	Add rewards	H17

The first category, "More difficulty levels," was associated to the students' ideas related to making PacMan or the ghosts faster as the player successfully eats all the white dots on the maze. The second category, "Give more powers to PacMan," mostly grouped the students' references to a feature that is present in the original game: the "power pills" that, when eaten by PacMan, causes the ghosts to move in a slower pace and allows the PacMan to eat the ghosts to get extra points. In the final interview, one student even mentioned his failed attempt to actually implement this feature:

I managed to make it [the power pill], when the PacMan touched it, the ghosts started to glow... But then I had to change a lot of things, I erased it and had to start all over again [...] But it worked like this: when he ate the pill, he started to glow... But then he would not change back.

Students also wished that the enemies has more powers, such as "faster movements," or simply that the game had more enemies (as the students were asked to implement only two ghosts, due to time limitations). Finally, the fourth category, "Add rewards," grouped references to additional rewards that should be added to the game in the students' opinion. In this case, another feature of the real game was frequently mentioned: the fruit that appear near to the center of the maze at specific time intervals, allowing the player to get extra points. Apart from being mentioned in the students' intentions, this feature was actually added in two of the games built by the students.

Twenty-two excerpts of the students' speeches were associated to heuristic H14 (the game must present many challenges and allow different strategies). These excerpts were related to adding more levels to the game or adding more powers to the main character or its enemies. The same occurrences could also be associated to heuristic H16 (challenge must be adjustable to the player's skill); should more levels and more skillful enemies be added to the game, it would demand additional skills from the player. Besides that, five excerpts were associated to heuristic H17 (The player must be rewarded for his/her achievements in a clear and immediate way) when students mention that the game should provide bonus features, such as the little bonus fruit that are available in the original Pacman game when the player reaches a given score.

Problem-Solving Strategies and Patterns

In order to analyze the possible recognition and usage of patterns by students to solve problems, it was necessary to identify the strategy used to implement each of the subtasks demanded in each game. In the scope of this analysis, a subtask is the most atomic goal that should be accomplished by students in order to add a new functionality to the game. For example, in Point-and-Shoot, making the bullet fire in a linear trajectory when a key is pressed is considered as a subtask; detecting the collision of the bullet with an enemy target and implementing an animation that shows the target "exploding" is another subtask. As mentioned before, students were required to complete a given number of subtasks in each meeting to get credit for the activity.

The workshop activities required the completion of a total of 38 subtasks; 12 of them were related to the preliminary activities of the workshop, 5 to Rock-Paper-Scissors, 8 to Point-and-Shoot, 5 to Breakout and 8 to Pacman. It is reasonable to expect that the students' skills would be more consolidated in the last weeks of the workshop; so, the identification of patterns is focused on the subtasks of the last two games, Breakout and Pacman. The analysis of these games allowed to identify that the usage of sprites to calculate distances and positioning of game elements appears to be a strategy widely recognized and used by students.

A subtask related to the positioning of game elements is present in Breakout: whenever the ball cannot be reached by the paddle anymore, the player must lose one life and the ball must be repositioned over the paddle for a new turn. The strategies used by students to implement this subtask are presented in Table 5. Most students opted to create a new, "artificial" sprite that covers the bottom of the screen, and the collision of the ball with this sprite is used to identify that it cannot be reached by the paddle (Figure 5). Checking the y-coordinate of the ball would be a more straightforward strategy, as it did not require the creation of an additional sprite, but surprisingly it was adopted by a smaller number of students.

In PacMan a similar strategy was used by most students to detect when the main character of the game or an enemy ghost reached the end of the side tunnels that are present in the middle of the maze.

Table 5. Strategies to implement the subtask "detect when the paddle misses the ball" in Breakout

Strategy	Number of Games	% of Games
Collision with "artificial" sprite	17	58.6%
Verification of y-coordinate	5	17.2%
Subtask not implemented	7	24.1%
Total	29	100.0%

Figure 5. Example of sprite used to detect the position of the ball

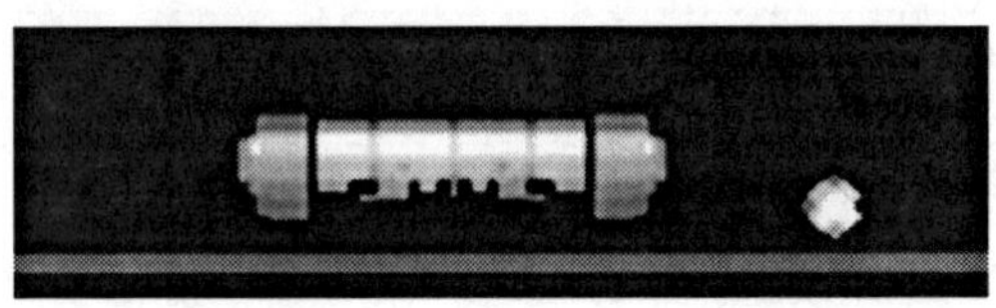

One subtask requested that students implemented the classic rule stating that the PacMan shall appear at the opposite side of the maze when it reaches the end of each tunnel (Figure 6). Again, most students opted to create an additional sprite and detect the collision of the PacMan with this sprite instead of checking the x and y coordinates of the PacMan sprite. The statistics for this subtask are displayed in Table 6.

Another subtask in PacMan is related to the implementation of the ghost that should follow a fixed path inside the maze. In this case, a prevailing strategy could not be identified; however, the three most used strategies all rely on programming features previously experienced by students. The first strategy used an atomic command of Scratch, named *glide*. It requires a pair of (x, y) coordinates and an interval in seconds as parameters and, when executed, moves the sprite to the specified position, in a linear trajectory whose duration is specified by the given time interval. Using this command is very convenient to solve this subtask; however, in previous similar situations, students had trouble using it, especially when it was necessary to interrupt the movement of the sprite due to its collision to other sprite (refer to the Discussion section for further details). This might be a possible explanation for the existence of the other two strategies. They are both similar because they are based on a loop with a stop condition: reaching a position given by a pair of coordinates, or reaching an "artificial" target sprite, identical to the one used in the previously described strategies. The loop step moves the sprite a few steps ahead at a time. Some coding options are shown for each of the three strategies in Figure 7. Table 7 shows the statistics for this subtask, where it is possible to notice that the three strategies were equally present in the implemented games.

Figure 6. Example of sprite used to detect the position of PacMan

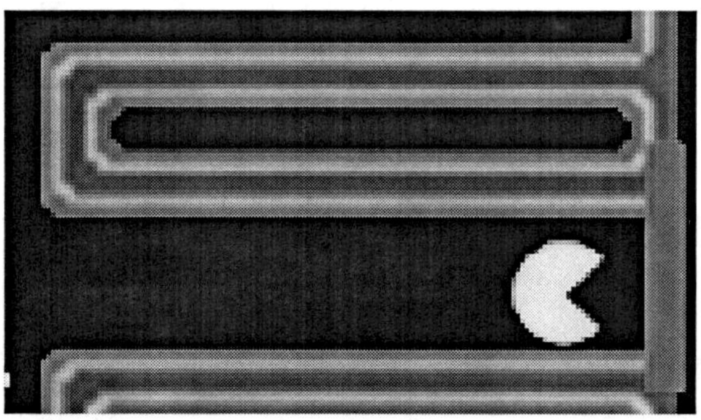

Table 6. Strategies to implement the strategy "detect when PacMan reaches the end of the tunnel"

Strategy	Number of Games	% of Games
Collision with "artificial" sprite	14	66.7%
Verification of x and y coordinates	3	14.3%
Range verification of x-coordinate	2	9.5%
Subtask not implemented	2	9.5%
Total	21	100.0%

Discussion

The motivation of students to implement additional features in games seemed to be closely linked to playability aspects. Many students implemented

Table 7. Strategies to implement the subtask "implement the ghost that follows a fixed path inside the maze"

Strategy	Number of Games	% of Games
Use of *glide* command	6	28.6%
Use of loop – stop condition with coordinate	6	28.6%
Use of loop – stop condition with sprite collision	6	28.6%
Other	1	4.8%
Subtask not implemented	2	9.5%
Total	21	100.0%

Figure 7. Strategies to implement the subtask "implement the ghost that follows a fixed path inside the maze"

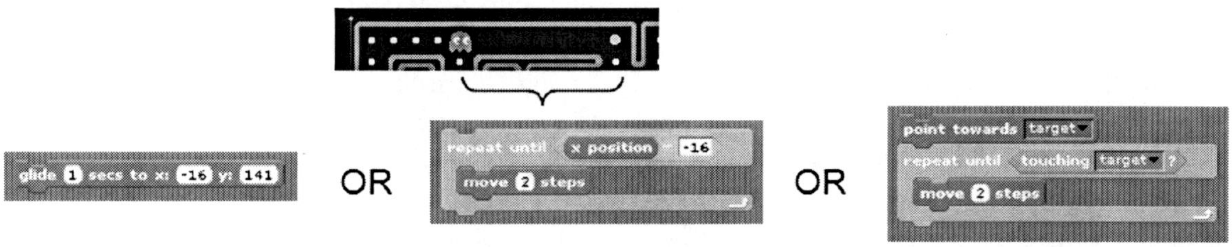

similar features, and a possible explanation to this fact is the social nature of the activity: students naturally interacted with each other in the computer laboratory and compared the functionalities of the games they implemented. In the audio transcripts of the workshop activities it was possible to identify several moments when students questioned some of the interaction mechanisms that restrained gameplay and tried to improve them on their own. These events were more frequent when the activities started to involve the implementation of game mechanics that were reasonably more complete, that is, since students started to implement Rock-Paper-Scissors.

It is possible to infer, based on the heuristic evaluation of extra features and the survey answered by students, that most features were associated to quality aspects that students also consider as absolutely relevant to their own experience as game players. The additional features were often associated to the display of visual features (H6), mostly related to the game status and score (H3), but features related to the response time of game controls (H1) and presence of graphic and sound effects (H10) were also identified. This is an indication that the "quality" of the proposed activities, in terms of its interactive features, has an impact in the students' motivation to engage in such activities.

In spite of that, when the features students would like to add to PacMan (if they continued to develop the game) were analyzed, it was noticed that students did not necessarily restrained their intentions to the quality aspects that were initially cited in the survey. By interpreting the students' speech in the final interviews, it is possible to associate heuristics H16 and H17 to their intention to add new difficulty levels, bonus items and additional powers to PacMan and its enemies. These heuristics were not frequently cited in the survey: they were, respectively, 16th and 17th in the decreasing order of the relevant heuristics in the students' opinion. On the other hand, heuristic H14, which is related to the presence of several challenges and different strategies, can also be associated to most features suggested by students, and this heuristic was the fourth most cited in the survey. These discrepancies indicate that, initially, students may not have realized the importance of making the game harder when the player improves her skills (expressed by H16) or the importance of providing clear and immediate rewards (expressed by H17). However, after being involved with several game design issues for several weeks, it seems that students begin to consider these particular issues as relevant. It is important to point out that, as a whole, heuristics H14 (that was already cited in the initial survey), H16 and H17 can be associated to level design, which is a crucial feature of game design (Juul, 2005). The practical experiences during the workshop seemed to make the students more aware of how growing levels of difficulty could be implemented in the game.

While students seemed to have expanded their own conceptions about quality in games, it is also true that most additional features found in the games built by students are similar to features that can be found in "real" games. It is the case of full-screen "game over" and "win" messages, sprite animations, score and lives control, previously mentioned in the heuristic evaluation of the games. In final interview, students also revealed their intention to add features of the real PacMan game in their projects. Based on this information, it could be argued that students were not engaged in a truly creative design process, but they could have copied known features instead. Nevertheless, this strategy is valid, considering that the main objective of the workshop was to help students develop programming skills, and not to develop an original design for a game. The critical reflection of students about the features that seemed to be "missing" in the games they were developing is consonant with the game fluency introduced by Peppler & Kafai (2009) and with the fourth guideline for the definition of the workshop activities ("The game mechanics should be simple and yet bring references to the world of *real* games that are meaningful to students"). Based on their previous gaming experience, students not only identify that "something is missing" in the game design, but also develop strategies to incorporate the desired functionalities. Sometimes these strategies even demanded that students dealt with programming concepts that were relatively new to them at the time they were needed.

Other aspect that emerged after several meetings was the reuse of some programming structures to solve problems similar to those that were already present in previous meetings. The problems were basically related to sprite animation and collision detection. The students produced some solutions that can be considered as "unusual" when they are seen from the point of view of conventional programming languages that do not provide an interactive authoring environment, such as Scratch. Nevertheless, it is possible to identify the influence of some particular functionalities of Scratch in the unconventional solutions provided. It could be argued that, motivated by desire to create better games, students sometimes had to deal with new concepts to implement additional features; at other times, they had to identify when solutions previously known by them were applicable, either empirically – engaging in a "reflection-in-action" process (Schön & Bennett, 1996) – or with the help of the teacher, working at their Zone of Proximal Development (Vygotsky, 1978). The occurrences of both situations will be discussed in the following subsections.

Exploring New Programming Structures to Add New Features

An analysis of the games' most frequent extra features reveals that students in fact had to deal with and understand novel abstract concepts in order to implement them. In Table 3, programming concepts that were necessary to implement these additional features have already been presented. This section will discuss how students dealt with those concepts. In order to implement the win/draw message displayed in the screen background of Rock-Paper-Scissors, students had to use either a busy wait loop to continuously verify the game state or a message broadcast mechanism. Scratch provides a *forever* loop that allows the implementation of busy wait loops in a fairly easy way. Figure 8 presents an example of such a loop to check the number of the *costume* (i.e., the graphical depiction) currently displayed by the sprite that represents each player in the game. It is interesting to point out that this strategy has not been used or presented to students before, although they have used the *forever* and *if* structures separately in previous activities.

In Point-and-Shoot game, students had to overcome a limitation of the Scratch environment to allow the player to immediately shoot again. In this case, the most straightforward implementation is to associate the loop that defines the trajectory

Figure 8. Busy wait loop implemented by a student in Rock-Paper-Scissors. "Jogador" is a Portuguese word for "player"

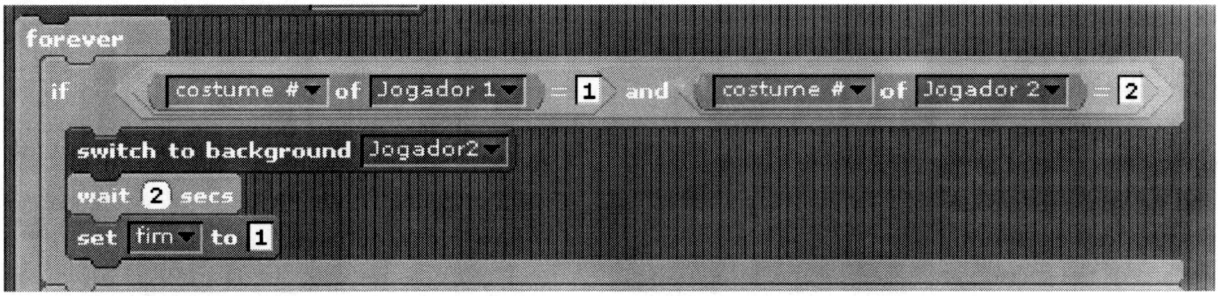

of the bullet to a key press event. However, this strategy prevents the player from shooting again while the bullet has not yet collided with the wall or the enemy. Some students identified that this might be a huge disadvantage to the player if a shot is missed. In addition to that, the bullet is represented by a single sprite to simplify the implementation; so, two or more bullets cannot be presented at the same time, introducing a large latency time between shots. Considering the simple strategy mentioned above, a new key press event cannot be detected while the current loop is still executing; technically speaking, each block of commands in Scratch is executed as a separate thread, whose execution cannot be interrupted by other threads. However, there is one exception to this behavior: a thread can be interrupted when its trigger event is a message. This exception was identified by many students that solved the issue by associating the loop that defines the trajectory of the bullet to a "message received" event. This way, the loop execution can be immediately interrupted when a new message is received, thus allowing a new shot. The message broadcast mechanism was also used to implement a "game over" screen: a message is sent to the screen background (which has the same behavior as a sprite) to change its depiction to the desired message. The concept of message synchronization had been presented to students one week earlier, and these extra features seem to indicate that students were confident enough to explore the new concept.

Identifying Patterns to Solve Similar Problems

In two situations, students had to deal with similar problems in order to implement game features: identifying the position of a given sprite on the screen – in Arkanoid, when the ball is positioned at the same height or lower than the paddle, and in PacMan, when the main character of the game reaches the end of the tunnel. Scratch provides special variables that store the value of the x and y coordinates of a sprite at all times; however, a significant parcel of students preferred to rely on the strategy of positioning a new sprite at the position to be detected, and check for the collision of this new sprite with the desired sprite.

Why would students rather reuse a more laborious solution? One possible explanation is that positioning a sprite on a desired position of the screen is predominantly a visual operation that does not demand a higher-order abstraction about the Cartesian coordinates of the sprite. In the scope of Mathematics Education, the visual level is the first and less sophisticated level of geometrical reasoning, according to the van Hiele model (Jones, 2002). As mentioned before, most students that attended the workshop are enrolled in the second or third year of Brazilian high school education. Usually, topics of Analytic Geometry are first covered in the third and last year of High School (Brasil, 2002); hence, it could either be assumed that many students had limited experience with Analytic Geometry concepts or that these students were not confident enough to explore the programming structures provided by Scratch to identify the positioning of sprites on Cartesian plane.

In either case, it is possible to identify that operations related to the creation of new sprites and collision detection were extensively used in the previous activities. Indirectly, this is a consequence of the second guideline for the workshop activities ("Activities must progressively require that students explore new concepts and, at the same time, reuse concepts that were previously explored by them"). Figure 9 presents a block diagram depicting the relationship between all subtasks that involve the concept of sprite collision. It is possible to notice that this concept was necessary in three subtasks in Point-and-Shoot, which may have influenced the students to use collisions to overcome their limitations with Geometry concepts in the following activities.

During the process of game building, it is often necessary to move sprites in a step-by-step fashion to allow collision detection with other moving sprites. This need led students to recognize possible uses for iterative structures with stop conditions. The first occasion when this structure was absolutely necessary was in Point-and-Shoot. A loop that defines the trajectory of the bullet and the enemy tank must be interrupted only when a collision with another sprite is detected, so it is not possible to initially define the number of loop iterations. At this point, the availability of Scratch's "glide" command was confusing for several students. This happened because, although it is possible to define a linear trajectory with a single *glide* command, its execution cannot be stopped. In this case, a feature of Scratch was a hurdle to solving the problem, in contrast to the last situation, when the concept of sprite collision allowed students to overcome their difficulties. Here, students are required to understand and apply a loop structure similar to the one shown in Figure 10.

A similar strategy could be used in Breakout to move the ball and detect its collision to the paddle and the blocks. At this moment, the ball must bounce off the obstacle. In this occasion, an example of loop usage to move the ball was presented to all students at the beginning of the activity, so no disparate solutions were identified. Then, in PacMan, students must define a fixed path inside the maze for one of the ghosts. Students did not follow a single strategy in this

Figure 9. Relationship between subtasks involving the concept of collision. Subtasks in blue explicitly involved collision; subtasks in green were solved by students using collision.

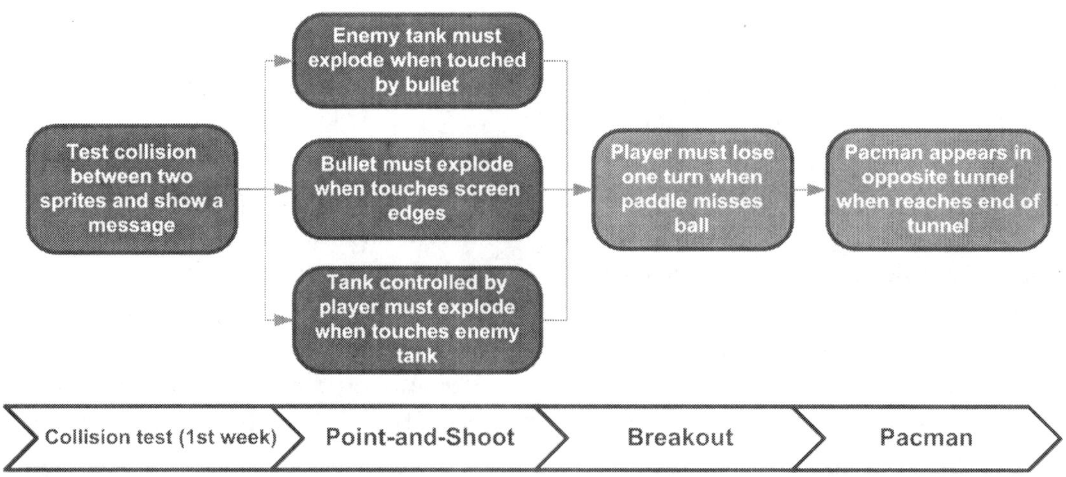

Figure 10. Example of a loop structure with a stop condition based on collision

case. As presented in Table 7, in twelve games students reused the strategy of using a loop with a stop condition. This may indicate that students identified a similarity to the problems previously solved in Point-and-Shoot and Breakout. This hypothesis is confirmed by one of the students in the final interview. When questioned about how he implemented the ghost, his answer was:

I did just like we have done before with the tank game [Point-and-shoot]... He [the ghost] moves some steps forward until he reaches the right position, and then he turns... down, left, right, wherever he must go

In six of the games where this strategy was applied, the coordinates of the position where the ghost should change his direction were used in the stop condition. However, in the other six games, sprites were used to define the positions between which the ghost should move; this is additional evidence to the difficulties of students to deal with Cartesian coordinates. In other six games, students preferred to use Scratch's "glide" command – in this case, the collision is detected by the PacMan sprite and not the ghost sprite, so the strategy turns out to be effective.

Figure 11 shows that moving sprites by using loops with collision detection was also a frequent strategy in previous activities, particularly in Point-and-Shoot. In 12 out of 21 PacMan games, students reused a previous strategy using loops to detect collisions. However, several students still chose a strategy that did not require implementing a loop. This is consonant to the findings of Ginat (2004), who identified that novice programmers tend to find it difficult to identify the boundaries for a loop; as mentioned before, this difficulty seems to have been circumvented by some students by using a "artificial" sprite to indicate the loop boundary (Figure 12).

Figure 11. Relationship between subtasks involving the concept of loop. In grey subtasks, simpler loops are used. Blue subtasks demand loops with stop conditions. Students reused previous strategies to solve the green subtask.

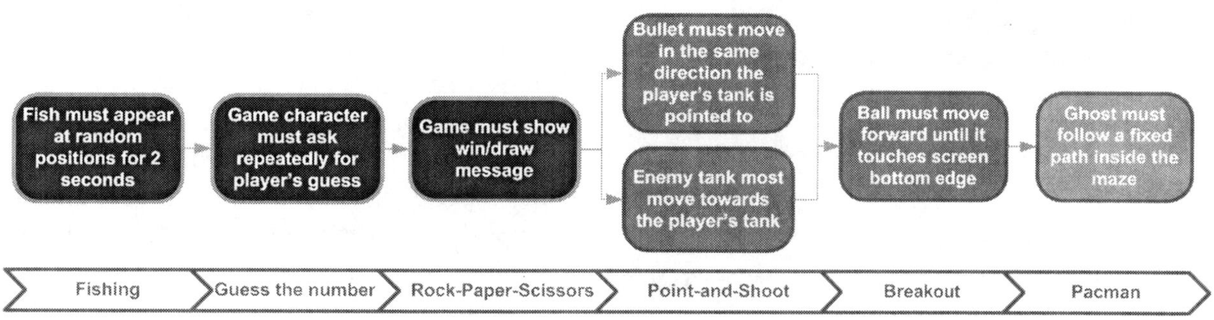

Figure 12. Strategy used by a student to implement the ghost that follows a fixed path. The red dots are sprites, used to define the stop condition of the loop used to move the ghost.

FUTURE RESEARCH DIRECTIONS

It is important to point out that the features identified in the students' games do not necessarily indicate that students understood all of the programming concepts that were applied to implement them. Hence, in future works it is necessary to evaluate how students' knowledge about key concepts has evolved during the workshop. This could be accomplished by assessing students with pre and posttests related to programming knowledge; however, it is often the case that students have no previous knowledge about programming when they begin their studies. Further works will analyze the performance of the students who attended the workshop in the following disciplines related to programming in comparison to those students who have not attended the workshop, that is, who took part in a programming fundamentals course with traditional activities.

Although it was not feasible to evaluate students' previous knowledge about programming in this case, all students should have developed higher-order problem solving skills in Mathematics classes. Hence, a test has been developed based on the Brazilian official Math examinations in order to evaluate problem solving skills of students who start the workshop. The goal of such test was to verify if the skills that students appear to have developed or exercised during the workshop, such as analogy and generalization, could be transferred to the Math domain. Such skills are also crucial to the study of Software Engineering (Kramer, 2007) in undergraduate CS/IT courses, so it would be relevant to apply the workshop to freshman of undergraduate courses to identify if the workshop activities may incentive students to further study topics that appeared in their first projects, even in an incipient way, such as encapsulation, successive refining and process synchronization.

Some Math topics were also used during the workshop activities. For instance, students had to deal with inequalities, arithmetical operations and positioning of sprites on the Cartesian plane. One could expect that students with a greater knowledge of these topics would be more resourceful in the workshop activities; on the other hand, the workshop activities may create opportunities for students to learn these Math topics in a meaningful way. Further works are necessary to assess the impact of the workshop activities on the students' knowledge of Math topics.

CONCLUSION

Digital games are a part of everyday life of many prospective CS or IT students. Therefore, it would be natural that these students expect to build games as a part of their activities in the course. Unfortunately, the initial activities in the course are often not motivating, and this may discourage students to continue studying in the area. Game design and construction have already been discussed as a viable and motivating alternative to introduce Computer Science concepts to students. However, it is important to understand which skills students may acquire through building games and how these skills can be useful as they continue their studies.

In order to investigate this issue, a Game Building Workshop was designed to be offered to students of a professional education course in Brazil. The definition of the Workshop activities was conducted by four guidelines, based on previous research results. Hence, the activities were heavily based on the construction of digital games that were similar to "real" games. Besides this, the activities required that students explored the same programming concepts several times. The selection and organization of activities seem to have influenced the students' motivation and attitude towards some of the proposed tasks. All the games produced by students were evaluated in a period of 12 weeks and the results confronted with the students' questions and comments during the workshop meetings and during a final interview. Thus, the present investigation produced three main contributions:

- Students tended to implement additional features in the games in order to overcome limitations identified by them. These limitations were related to playability aspects that were previously classified by students as very relevant to their experience as game players. This indicates that the motivation of students led them to explore new topics to implement these additional features.
- Students reused solutions for previously solved problems in similar contexts that appeared in subsequent activities. The sequence of the workshop activities was designed to make students recall and use previously seen concepts, although students were not explicitly instructed to search for patterns. Based on the previous contribution, it could be argued that the domain of game design and the features of the programming environment may have contributed to make students exercise higher-order skills that are important for a CS/IT professional.
- The impact of the workshop activities on the students' motivation to solve quite complex problems indicates that the used guidelines have a potential to direct the definition of game building workshops oriented towards different target audiences and aimed at developing different skills.

In regular introductory courses, abstraction skills are also exercised; however, they may be initially neglected because of the limited scope of the activities. Game design may stimulate students to deal with bigger problems right from the start and, most important, develop skills to cope with them. After all, a "fun reward" awaits for those who succeed.

REFERENCES

Adams, J. C., & Webster, A. R. (2012). What do students learn about programming from game, music video, and storytelling projects? In *Proceedings of the 43rd ACM technical symposium on Computer Science Education* (pp. 643–648). New York: ACM. doi:10.1145/2157136.2157319

Barcelos, T. S., Carvalho, T., Schimiguel, J., & Silveira, I. F. (2011). Análise comparativa de heurísticas para avaliação de jogos digitais. In *Proc. IHC+CLIHC 2011* (pp. 187–196). Pernambuco, Brazil: SBC.

Barcelos, T. S., & Silveira, I. F. (2012). Teaching computational thinking in initial series. In *Proceedings of CLEI 2012*. Medellín: Centro Latinoamericano de Estudios en Informática.

Bardin, L. (1977). *Content analysis*. Lisbon: Edições 70.

Bateson, A. G., Alexander, R. A., & Murphy, M. D. (1987). Cognitive processing differences between novice and expert computer programmers. *International Journal of Man-Machine Studies*, *26*(6), 649–660. doi:10.1016/S0020-7373(87)80058-5

Bayliss, J. D., & Strout, S. (2006). Games as a "flavor" of CS1. *SIGCSE Bulletin*, *38*(1), 500–504. doi:10.1145/1124706.1121498

Beaubouef, T. (2002). Why computer science students need math. *SIGCSE Bulletin*, *34*(4), 57–59. doi:10.1145/820127.820166

Brasil. Ministério da Educação. (2002). *PCN+ Ensino Médio: Orientações Curriculares Complementares aos Parâmetros Curriculares Nacionais*. Brasília: MEC/SEB.

Campbell, P. F., & McCabe, G. P. (1984). Predicting the success of freshmen in a computer science major. *Communications of the ACM*, *27*(11), 1108–1113. http://doi.acm.org/10.1145/1968.358288

Carnegie Mellon University. (2013). *Alice - An educational software that teaches students computer programming in a 3D environment*. Retrieved August 15, 2013, from http://www.alice.org

Carter, L. (2006). Why students with an apparent aptitude for computer science don't choose to major in computer science. *SIGCSE Bulletin*, *38*(1), 27–31. doi:10.1145/1124706.1121352

Chase, W. G., & Simon, H. A. (1973). Perception in chess. *Cognitive Psychology*, *4*(1), 55–81. doi:10.1016/0010-0285(73)90004-2

Chen, S. Y., Fan, J.-P., & Macredie, R. D. (2006). Navigation in hypermedia learning systems: Experts vs. novices. *Computers in Human Behavior*, *22*(2), 251–266. doi:10.1016/j.chb.2004.06.004

Chi, M. T. H., Feltovich, P. J., & Glaser, R. (1981). Categorization and representation of physics problems by experts and novices. *Cognitive Science*, *5*(2), 121–152. doi:10.1207/s15516709cog0502_2

Clancy, M. J., & Linn, M. C. (1999). Patterns and pedagogy. *SIGCSE Bulletin*, *31*(1), 37–42. doi:10.1145/384266.299673

Colley, A., Banton, L., Down, J., & Pither, A. (1992). An Expert-Novice Comparison in Musical Composition. *Psychology of Music*, *20*(2), 124–137. doi:10.1177/0305735692202003

Crenshaw, T. L., Chambers, E. W., Metcalf, H., & Thakkar, U. (2008). A case study of retention practices at the University of Illinois at Urbana-Champaign. In *Proc. SIGCSE 2008* (pp. 412–416). New York: ACM. doi:10.1145/1352135.1352276

Creswell, J. W., & Miller, D. L. (2000). Determining validity in qualitative inquiry. *Theory into Practice*, *39*(3), 124–130. doi:10.1207/s15430421tip3903_2

Denner, J., Werner, L., & Ortiz, E. (2012). Computer games created by middle school girls: Can they be used to measure understanding of computer science concepts? *Computers & Education*, *58*(1), 240–249. doi:10.1016/j.compedu.2011.08.006

East, J. P., Thomas, S. R., Wallingford, E., Beck, W., & Drake, J. (1996). Pattern-based programming instruction. In *Proceedings of the ASEE Annual Conference and Exposition*. Washington, DC: ASEE.

Franklin, D., Conrad, P., Boe, B., Nilsen, K., Hill, C., & Len, M., … Waite, R. (2013). Assessment of computer science learning in a scratch-based outreach program. In *Proceeding of the 44th ACM technical symposium on Computer science education* (pp. 371–376). New York, NY: ACM. doi:10.1145/2445196.2445304

Ginat, D. (2004). On Novice Loop Boundaries and Range Conceptions. *Computer Science Education*, *14*(3), 165–181. doi:10.1080/0899340042000302709

Goetz, J. P., & LeCompte, M. D. (1984). *Ethnography and qualitative design in educational research*. Orlando, FL: Academic Press.

Haberman, B., & Muller, O. (2008). *Teaching abstraction to novices: Pattern-based and ADT-based problem-solving processes*. IEEE. doi:10.1109/FIE.2008.4720415

Hernandez, C. C., Silva, L., Segura, R. A., Schimiguel, J., Ledon, M. F. P., Bezerra, L. N. M., & Silveira, I. F. (2010). Teaching Programming Principles through a Game Engine. *CLEI Electronic Journal*, 1–8.

Hix, D., & Hartson, H. (1993). Developing user interfaces: ensuring usability through product and process. New York: John Wiley & Sons, Inc.

Jones, K. (2002). Issues in the Teaching and Learning of Geometry. In *Aspects of Teaching Secondary Mathematics: perspectives on practice* (pp. 121–139). London: Routledge.

Juul, J. (2005). *Half-real: Video Games between Real Rules and Fictional Worlds*. MIT Press.

Kramer, J. (2007). Is abstraction the key to computing? *Communications of the ACM*, *50*(4), 36–42. doi:10.1145/1232743.1232745

Larkin, J., McDermott, J., Simon, D. P., & Simon, H. A. (1980). Expert and Novice Performance in Solving Physics Problems. *Science*, *208*(4450), 1335–1342. doi:10.1126/science.208.4450.1335 PMID:17775709

Lee, I., Martin, F., Denner, J., Coulter, B., Allan, W., & Erickson, J. et al. (2011). Computational thinking for youth in practice. *ACM Inroads*, *2*(1), 32–37. doi:10.1145/1929887.1929902

Leutenegger, S. T. (2006). A CS1 to CS2 bridge class using 2D game programming. *Journal of Computing Sciences in Colleges*, *21*(5), 76–83.

Lister, R., Simon, B., Thompson, E., Whalley, J. L., & Prasad, C. (2006). Not seeing the forest for the trees: novice programmers and the SOLO taxonomy. In *Proceedings of the 11th annual SIGCSE conference on Innovation and technology in computer science education* (pp. 118–122). New York, NY: ACM. doi:10.1145/1140124.1140157

Malan, D. J., & Leitner, H. H. (2007). Scratch for budding computer scientists. *SIGCSE Bulletin*, *39*(1), 223–227. doi:10.1145/1227504.1227388

Maloney, J. H., Peppler, K., Kafai, Y., Resnick, M., & Rusk, N. (2008). Programming by choice: urban youth learning programming with scratch. In *Proceedings of the 39th SIGCSE technical symposium on Computer science education* (pp. 367–371). New York, NY: ACM. doi:10.1145/1352135.1352260

McKeithen, K. B., Reitman, J. S., Rueter, H. H., & Hirtle, S. C. (1981). Knowledge organization and skill differences in computer programmers. *Cognitive Psychology*, *13*(3), 307–325. doi:10.1016/0010-0285(81)90012-8

MIT Media Lab & the Lifelong Kindergarten Group. (2012). Scratch. *Scratch*. Retrieved April 27, 2012, from http://scratch.mit.edu

Merril, D. (2002). A Pebble-in-the-Pond Model For Instructional Design. *Performance Improvement, 41*(7), 41–46. doi:10.1002/pfi.4140410709

Microsoft Research. (2014). *Kodu Game Lab Community*. Retrieved January 16, 2014, from http://www.kodugamelab.com/

Muller, O., Haberman, B., & Averbuch, H. (2004). (An almost) pedagogical pattern for pattern-based problem-solving instruction. *ACM SIGCSE Bulletin, 36*(3), 102. doi:10.1145/1026487.1008025

Muratet, M., Torguet, P., Jessel, J.-P., & Viallet, F. (2009). Towards a serious game to help students learn computer programming. *International Journal of Computer Games Technology,* 3:1–3:12. doi:10.1155/2009/470590

Nakamura, J., & Csikszentmihalyi, M. (2009). Flow theory and research. In *Oxford Handbook of Positive Psychology* (2nd ed., pp. 195–206). Oxford, UK: Oxford University Press.

Nielsen, J., & Molich, R. (1990). Heuristic evaluation of user interfaces. In *CHI '90: Proceedings of the SIGCHI conference on Human factors in computing systems* (pp. 249–256). New York: ACM. http://doi.acm.org/10.1145/97243.97281

Peppler, K., & Kafai, Y. (2009). Gaming Fluencies: Pathways into Participatory Culture in a Community Design Studio. *International Journal of Learning and Media, 1*(4), 45–58. doi:10.1162/ijlm_a_00032

Polya, G. (2004). *How to solve it: a new aspect of mathematical method.* Princeton, NJ: Princeton University Press.

Prensky, M. (2004). *Digital Game-Based Learning*. Washington, DC: McGraw-Hill Pub. Co.

Rizvi, M., Humphries, T., Major, D., Jones, M., & Lauzun, H. (2011). A CS0 course using Scratch. *Journal of Computing Sciences in Colleges, 26*(3), 19–27.

Salen, K. (2007). Gaming literacies: A game design study in action. *Journal of Educational Multimedia and Hypermedia, 16*(3), 301–322.

Schön, D., & Bennett, J. (1996). Reflective conversations with materials. In T. Winograd, J. Bennett, L. De Young, & B. Hartfield (Eds.), *Bringing design to software*. New York: ACM Press.

Settle, A. (2011). Computational thinking in a game design course. In *Proceedings of the 2011 conference on Information technology education* (pp. 61–66). New York, NY: ACM. doi:10.1145/2047594.2047612

Shneiderman, B. (1976). Exploratory experiments in programmer behavior. *International Journal of Computer & Information Sciences, 5*(2), 123–143. doi:10.1007/BF00975629

Shneiderman, B., & Mayer, R. (1979). Syntactic/semantic interactions in programmer behavior: A model and experimental results. *International Journal of Computer & Information Sciences, 8*(3), 219–238. doi:10.1007/BF00977789

Soloway, E., Bonar, J., & Ehrlich, K. (1982). What do novices know about programming? In *Directions in Human-Computer Interaction* (pp. 87–122). New York: Ablex.

Souza, P. R. de A., & Dias, L. R. (2012). Kodu Game Labs: Estimulando o Raciocínio Lógico através de Jogos. In *Anais do 23º Simpósio Brasileiro de Informática na Educação*. Rio de Janeiro: SBC. Retrieved from http://br-ie.org/pub/index.php/sbie/article/view/1733

Stolee, K. T., & Fristoe, T. (2011). Expressing computer science concepts through Kodu game lab. In *Proceedings of the 42nd ACM technical symposium on Computer science education* (pp. 99–104). New York: ACM. doi:10.1145/1953163.1953197

The CSTA Standards Task Force. (2011). *CSTA K-12 Computer Science Standards*. New York: ACM Computer Science Teachers Association. Retrieved from http://csta.acm.org/Curriculum/sub/CurrFiles/CSTA_K-12_CSS.pdf

Vygotsky, L. S. (1978). Zone of Proximal Development. In *Mind in society: The development of higher psychological processes* (pp. 52–91). Oxford, UK: Harvard University Press.

Waters, A. J., Gobet, F., & Leyden, G. (2002). Visuospatial abilities of chess players. *British Journal of Psychology*, *93*(4), 557–565. doi:10.1348/000712602761381402 PMID:12519534

Weiser, M., & Shertz, J. (1983). Programming problem representation in novice and expert programmers. *International Journal of Man-Machine Studies*, *19*(4), 391–398. doi:10.1016/S0020-7373(83)80061-3

Wilson, B. C., & Shrock, S. (2001). Contributing to success in an introductory computer science course: a study of twelve factors. In *Proceedings of the thirty-second SIGCSE technical symposium on Computer Science Education* (pp. 184–188). New York: ACM. doi:10.1145/364447.364581

YoYo Games, Ltd. (2014). *GameMaker: Studio*. Retrieved January 16, 2014, from https://www.yoyogames.com/studio

ADDITIONAL READING

Adams, J. C., & Webster, A. R. (2012). What do students learn about programming from game, music video, and storytelling projects? In *Proceedings of the 43rd ACM technical symposium on Computer Science Education* (pp. 643–648). New York: ACM. doi:10.1145/2157136.2157319

Avraamidou, A., Monaghan, J., & Walker, A. (2012). Abstraction Through Game Play. *Technology. Knowledge and Learning*, *17*(1-2), 1–21. doi:10.1007/s10758-012-9189-2

Barcelos, T. S., Costa, G. C., Muñoz, R., & Silveira, I. F. (2014). Informal HCI: Fixing Playability Issues As A Strategy To Improve The Skills Of Novice Programmers. *IEEE Latin American Transactions*, *12*(1), 29–35. doi:10.1109/TLA.2014.6716489

Barr, V., & Stephenson, C. (2011). Bringing computational thinking to K-12: What is Involved and what is the role of the computer science education community? *ACM Inroads*, *2*(1), 48–54. doi:10.1145/1929887.1929905

Basawapatna, A., Koh, K. H., Repenning, A., Webb, D. C., & Marshall, K. S. (2011). Recognizing computational thinking patterns. In *Proceedings of the 42nd ACM technical symposium on Computer science education* (pp. 245–250). Presented at the SIGCSE 2011, New York: ACM.

Basawapatna, A. R., Koh, K. H., & Repenning, A. (2010). Using scalable game design to teach computer science from middle school to graduate school. In *Proceedings of the fifteenth annual conference on Innovation and technology in computer science education* (pp. 224–228). New York, NY, USA: ACM. doi:10.1145/1822090.1822154

Basawapatna, A. R., Repenning, A., Koh, K. H., & Nickerson, H. (2013). The Zones of Proximal Flow: Guiding Students Through a Space of Computational Thinking Skills and Challenges. In *Proceedings of the Ninth Annual International ACM Conference on International Computing Education Research* (pp. 67–74). New York: ACM. doi:10.1145/2493394.2493404

Bayliss, J. D., & Strout, S. (2006). Games as a "flavor" of CS1. *SIGCSE Bulletin*, *38*(1), 500–504. doi:10.1145/1124706.1121498

Bourgonjon, J., Valcke, M., Soetaert, R., & Schellens, T. (2010). Students' perceptions about the use of video games in the classroom. *Computers & Education*, *54*(4), 1145–1156. doi:10.1016/j.compedu.2009.10.022

Brennan, K., & Resnick, M. (2012). New frameworks for studying and assessing the development of computational thinking. In *Proceedings of the 2012 annual meeting of the American Educational Research Association*. Vancouver: American Educational Research Association.

De Souza, C. S., Garcia, A. C. B., Slaviero, C., Pinto, H., & Repenning, A. (2011). Semiotic traces of computational thinking acquisition. In *Proceedings of the Third international conference on End-user development* (pp. 155–170). Heidelberg: Springer-Verlag.

Denner, J., Werner, L., & Ortiz, E. (2012). Computer games created by middle school girls: Can they be used to measure understanding of computer science concepts? *Computers & Education*, *58*(1), 240–249. doi:10.1016/j.compedu.2011.08.006

Desuvire, H., & Wiberg, C. (2009). Game Usability Heuristics (PLAY) for Evaluating and Designing Better Games: The Next Iteration. In A. A. Ozok & P. Zaphiris (Eds.), *Online Communities, LTCS 5621* (pp. 557–566). Springer-Verlag.

Eckerdal, A. (2009). *Novice Programming Students' Learning of Concepts and Practise* (PhD Thesis). Uppsala University, Sweden.

Febretti, A., & Garzotto, F. (2009). Usability, playability, and long-term engagement in computer games. In *Proc. CHI '09* (pp. 4063–4068). New York, NY, USA: ACM. doi:10.1145/1520340.1520618

Federoff, M. (2002). *Heuristics and usability guidelines for the creation and evaluation of fun in video games* (Master's Dissertation). Indiana University.

Franklin, D., Conrad, P., Boe, B., Nilsen, K., Hill, C., & Len, M., … Waite, R. (2013). Assessment of computer science learning in a Scratch-based outreach program. In *Proceeding of the 44th ACM technical symposium on Computer science education* (pp. 371–376). New York, NY, USA: ACM. doi:10.1145/2445196.2445304

Goldberg, M. (2007, August 9). A complete history of Breakout. *Classic Gaming*. Retrieved October 10, 2013, from http://classicgaming.gamespy.com/View.php?view=Articles.Detail&id=395

Gorriz, C. M., & Medina, C. (2000). Engaging girls with computers through software games. *Communications of the ACM*, *43*(1), 42–49. doi:10.1145/323830.323843

Harel, I. (1991). *Children Designers: Interdisciplinary constructions for learning and knowing mathematics in a computer-rich school*. New Jersey: Ablex Publishing.

Henderson, P. B. (2011). MATHCOUNTS: Mathematical reasoning in computing education. *ACM Inroads*, *1*(3), 22–23. doi:10.1145/1835428.1835438

Hu, C. (2011). Computational thinking: what it might mean and what we might do about it. In *Proceedings of the 16th annual joint conference on Innovation and technology in computer science education* (pp. 223–227). New York, NY, USA: ACM. doi:10.1145/1999747.1999811

Kafai, Y. (2006). Playing and making games for learning: Instructionist and Constructionist Perspectives for Game Studies. *Games and Culture*, *1*(1), 36–40. doi:10.1177/1555412005281767

Kafai, Y. B., Franke, M. L., Ching, C. C., & Shih, J. C. (1998). Game Design as an Interactive Learning Environment for Fostering Students' and Teachers' Mathematical Inquiry. *International Journal of Computers for Mathematical Learning*, *3*(2), 149–184. doi:10.1023/A:1009777905226

Kuruvada, P., Asamoah, D. A., Dalal, N., & Kak, S. (2010). Learning Computational Thinking from Rapid Digital Game Creation. In *Proceedings of 2nd Annual Conference on Theoretical and Applied Computer Science* (pp. 31–35). Stillwater, OK, USA.

Malone, T. W. (1980). What makes things fun to learn? Heuristics for designing instructional computer games. In *Proceedings of the 3rd ACM SIGSMALL symposium and the first SIGPC symposium on Small systems* (pp. 162–169). New York, NY, USA: ACM. doi:10.1145/800088.802839

McMaster, K., Rague, B., & Anderson, N. (2010). Integrating Mathematical Thinking, Abstract Thinking, and Computational Thinking. In *Frontiers in Education Conference (FIE) 2010* (pp. S3G–1–S3G–6). doi:10.1109/FIE.2010.5673139

Muller, O. (2005). Pattern oriented instruction and the enhancement of analogical reasoning. In *Proceedings of the first international workshop on Computing education research* (pp. 57–67). New York: ACM. doi:10.1145/1089786.1089792

Mustaro, P. N., Silva, L., & Silveira, I. F. (2011). Using games to teach design patterns and computer graphics. *Instructional Design: Concepts, Methodologies. Tools and Applications.*, *1*(1), 173.

Papert, S. (1980). *Mindstorms: children, computers and powerful ideas*. New York: Basic Books.

Papert, S. (1993). *The children's machine: rethinking school in the age of the computer*. New York: Basic Books.

Pinelle, D., Wong, N., & Stach, T. (2008). Heuristic evaluation for games: usability principles for video game design. In *CHI '08: Proceeding of the twenty-sixth annual SIGCHI conference on Human factors in computing systems* (pp. 1453–1462). New York, NY, USA: ACM. doi:10.1145/1357054.1357282

Pittman, J. (2011). The Pac-Man Dossier. Retrieved December 15, 2012, from http://home.comcast.net/~jpittman2/pacman/pacmandossier.html

Prensky, M. (2001). Digital natives, digital immigrants. *On the Horizon*, *9*(5), 1–6. doi:10.1108/10748120110424816

Resnick, M., Maloney, J., Monroy-Hernández, A., Rusk, N., Eastmond, E., & Brennan, K. et al. (2009). Scratch: Programming for all. *Communications of the ACM*, *52*(11), 60–67. doi:10.1145/1592761.1592779

Salen, K., & Zimmerman, E. (2003). *Rules of Play: Game Design Fundamentals*. Massachusetts: The MIT Press.

Schell, J. (2008). *The art of game design: a book of lenses*. Burlington, MA: Elsevier.

Selwyn, N. (2009). The digital native - myth and reality. *Aslib Proceedings*, *61*(4), 364–379. doi:10.1108/00012530910973776

Settle, A. (2011). Computational thinking in a game design course. In *Proceedings of the 2011 conference on Information technology education* (pp. 61–66). New York, NY, USA: ACM. doi:10.1145/2047594.2047612

Shneiderman, B. (2004). Designing for fun: How can we design user interfaces to be more fun? *Interaction*, *11*(5), 48–50. doi:10.1145/1015530.1015552

Soloway, E. (1986). Learning to program = learning to construct mechanisms and explanations. *Communications of the ACM*, *29*(9), 850–858. doi:10.1145/6592.6594

Wing, J. M. (2006). Computational thinking. *Communications of the ACM*, *49*(3), 33–35. doi:10.1145/1118178.1118215

Wu, M. L., & Richards, K. (2011). Facilitating computational thinking through game design. In *Proceedings of the 6th international conference on E-learning and games, edutainment technologies* (pp. 220–227). Berlin, Heidelberg: Springer-Verlag.

KEY TERMS AND DEFINITIONS

Analogy: According to Polya (2004), a problem-solving skill related to the identification of similarities between situations or objects. It is informally used even in everyday situations, but it can also be systematically used to solve problems. For instance, a strategy to solve a mathematical problem would be to find an analogous instance of the problem, with simpler constraints, for which a solution can be found.

Breakout: A popular arcade game initially developed by Atari in 1976. It consists of a ball that travels across the screen and bounces off the top and side walls of the screen. A line of bricks is presented on the upper side of the screen; when the ball hits a brick, it is destroyed. The ball must not touch the bottom of the screen. To prevent this from happening, the player controls a moving paddle to bounce the ball upwards.

Busy Wait Loop: It is the simplest form of process synchronization. A loop that continuously verify if some Boolean condition is fulfilled; when the condition is true, the code inside the loop is executed.

PacMan: A game developed by Nanco Ltd. and first released in 1980. The player controls PacMan through a maze, eating yellow dots. When all the yellow dots are eaten, Pac-Man is taken to the next stage. Four ghosts chase PacMan, each of them following a particular strategy. If an enemy touches Pac-Man, a life is lost and the PacMan itself withers and dies.

Pattern: In the context of software development, a pattern is a previously defined solution for a class of known problems. A software developer who is able to identify the main characteristics of the pattern in a given problem instance may reuse the solution specified in the pattern definition. Analogy skills are required to successfully use patterns.

Problem-Based Learning (PBL): A pedagogical perspective where learning takes place through the experience of problem-solving. Students should work in groups to identify what they already know, what they need to know, and how to access new information that may lead to resolution of the problem. The role of the teacher in a PBL environment is to facilitate learning by guiding and monitoring the learning process.

Sprite: From the latin *spiritus*. In computer graphics, a sprite is originally a 2D image that can be quickly drawn repeated times, leaving no traces of its previous appearances (hence the analogy with "spirit"). Sprite drawing was usually supported by several hardware and software optimizations in the first personal computers and videogame consoles. In Scratch, any object that can be individually manipulated on the screen is called a sprite.

This work was previously published in "Human Factors in Software Development and Design" edited by Saqib Saeed, Imran Sarwar Bajwa, and Zaigham Mahmood, pages 141-172, copyright 2015 by Information Science Reference (an imprint of IGI Global).

Chapter 56
Viewing the Implementation of the CCSS through the Lens of One Transformative District–University Partnership

P. Michael Lutz
California State University – Bakersfield, USA

ABSTRACT

The study described in this chapter is on a more-than-20-year collaboration between a university mathematics department and its local high school district. The joint effort has created multiple components (such as strengthening teachers' mathematics knowledge, developing teacher leadership, and increasing teachers' appreciation of the importance of engaging students in tasks with a high cognitive demand) that are facilitating the districts' current efforts to implement the California Common Core State Standards for Mathematics. It describes a partnership that has always been grounded in mutual respect.

INTRODUCTION

This chapter shares a case study of the history of collaboration between the California Mathematics Project (CMP) regional site in the Mathematics Department at California State University, Bakersfield (CSUB) and the Kern High School District (KHSD). It is written from a participant-observer perspective versus one of a "fly on the wall" and covers more than a quarter century. The relationship between the district and the university is ongoing, continually growing, and "grounded in mutual respect" (Greene & Gallagher, 2009, p. 1) with participants from both institutions believing that each side of the partnership brings unique and valuable expertise to the collaboration.

Of course when the collaboration began several decades ago, no one was thinking about the Common Core State Standards in Mathematics (CCSSM), but there is evidence that the district's current movement to implement the new standards is being facilitated with by-products of previous joint efforts in ways that can be referred to as transforming. The vocabulary has changed in some cases, but the work to build the foundation began decades ago.

DOI: 10.4018/978-1-4666-7363-2.ch056

The KHSD includes 18 comprehensive high schools, 6 alternative education campuses, 2 career technical education sites and serves 35,000 students. Kern County is largely agricultural and leads the state in oil and natural gas production. Approximately half the population is Hispanic, and CSUB is the only four-year institution within 100 miles.

A BRIEF HISTORY

The 1980s were a time of great change in mathematics education, especially in our state. Wilson (2003) provides an in-depth description of efforts to reform mathematics education in the state in the 1980s and 1990s and the subsequent curriculum "wars" that occurred. During those two decades and beyond, Margaret DeArmond, now retired, was a mathematics teacher in the district and vividly recalls the turmoil that occurred. In an interview on 17 October 2013, her words confirmed Wilson's observation that the 1985 *Mathematics Framework for California Schools, K-12* "advocated a less traditional, more progressive approach to mathematics education, moving away from rote memorization and the dominance of worksheets toward 'teaching for understanding." DeArmond recalled that the 1985 California Framework was "very different" and that one of the mathematics faculty at CSUB, Dr. Lee Webb, had been on the writing committee. She knew Dr. Webb as the organizer of a mathematics field day each year on the campus and as being very popular among the K – 8 teachers who had taken his mathematics classes as undergraduates.

The 1985 Framework was the first DeArmond had ever heard of the use of manipulatives and technology as tools for the teaching of mathematics. She was in her second decade as a local high school teacher and would soon become a mathematics teacher leader in the state, including being a strong advocate for the appropriate use of tools such as manipulatives and technology.

In 1989, the same year that the National Council of Teachers of Mathematics published its *Curriculum and Evaluation Standards for School Mathematics,* Dr. Joe Fiedler was hired by CSUB's Mathematics Department. He and DeArmond would soon become key colleagues in collaborative activities between KHSD and CSUB. In an interview, Fiedler recalls hearing of DeArmond first when he was presenting at a conference across the country in Washington, DC. When one of the participants learned where he lived, the participant said that he must know Margaret DeArmond, who had recently conducted an impressive workshop at his school. Not long after that Fiedler was in Baltimore, and a participant there shared a similar story. He was impressed that she was known so far from where she lived and worked.

DeArmond recalls the first time she met Fiedler. She had heard of him as being a mathematics professor and interested in the use of technology. One day when she was making a presentation in a local auditorium and stressing the importance of the use of technology, she told the audience that she had heard that Dr. Fiedler had made the statement "If a textbook has tables in the back, don't buy it." At that point, Fiedler, who was in the audience, stood up and shouted "I am here!"

DeArmond vividly recalls that she still did not know much about Fiedler's positions on current issues in California's mathematics education until the evening he came up to her after another presentation that she made locally and asserted, "I'm on your side!" The declaration planted the seed for a professional collaboration that continued to grow during the next two decades.

COGNITIVE DEMAND

As Hess states below, the Common Core State Standards for Mathematics call for raising the cognitive demand of the tasks that students are given, a focus that has existed in the KHSD-CSUB partnership for a decade and beyond.

At the heart of College and Career Readiness is the need to increase the level of rigor in our classrooms for all students. The Common Core State Standards are a step in the right direction. However, the standards alone will not bring rigor to our classrooms. The implementation of these standards requires practical tools to develop local curricula and assessments and to promote classroom discourse aligned to higher levels of cognitive demand.

Norman Webb's Depth-of-Knowledge (DOK) schema has become one of the key tools educators can employ to analyze the cognitive demand (complexity) intended by the standards, curricular activities, and assessment tasks. Webb (1997) developed a process and criteria for systematically analyzing the alignment between standards and test items in standardized assessments. Since then the process and criteria have demonstrated application to reviewing curricular alignment as well. The model categorizes assessment tasks by different levels of cognitive expectation, or depth of knowledge, required to successfully complete the task. Hess (2004-2012) further articulated the model with content specific descriptions for use by classroom teachers and organizations conducting alignment studies (Hess, 2013).

Beginning by at least 2000, the KHSD-CSUB partnership has emphasized the importance of engaging students in tasks that have a high cognitive demand. Prior to the emergence of the CCSS, the work in the partnership was based on the QUASAR research (Stein, Smith, Henningsen & Silver, 2000). The vocabulary has changed in Common Core, but a critical idea remains the same; student-learning increases when students are engaged in tasks with a high cognitive demand. Today, depending on their backgrounds/history, individuals in the partnership may make sense of cognitive demand via Webb's Depth of Understanding or QUASAR (Stein et al, 2000) or Bloom's Taxonomy (1956).

Following are brief descriptions of some of the partnership activities that include the cognitive demand focus.

Master of Arts in Teaching Mathematics

In 2001, Fiedler successfully created a Masters program intended for secondary mathematics teachers in the area. Between fall 2001 and fall 2009, 79 KHSD mathematics teachers enrolled in the program. Prior to that, local teachers who desired advance degrees were limited to getting degrees in education such as administration and curriculum. Fiedler's program includes eight courses. Seven of them are mathematics courses and one is an introduction to mathematics education research. This program was modeled on the highly successful Master of Arts and Summer Institute program of the 1970's and 1980's designed and implemented at Ohio State by Joanne and Jim Leitzel. The courses, designed to approach advanced mathematics from an elementary point of view, were first offered in September 2001 and make significant and consistent use of modern handheld technology, specifically the TI Voyage 200. This approach has allowed the courses to challenge fully credentialed teachers while supplying an alternate path into the profession and a mechanism for middle school teachers to upgrade their credential to become No-Child-Left-Behind compliant. The mathematics content courses are Discrete Mathematical Models, Statistics and Data Analysis, Numerical Approach to Calculus and Differential Equations, Geometric Linear Algebra, Number Theory and Cryptography, Dynamical Geometry, and History of Mathematics. Completion of a final project is required for the degree. The QUASAR research (Stein et al, 2000) with its focus on the importance of cognitive demand is emphasized in the Introduction to Mathematics Education Research course. Furthermore, the graduate students experience high cognitive demand tasks as learners in all of the mathematical content classes that also connect to the mathematics they teach.

BTSA: In 1997, California created the Beginning Teacher Support and Assessment (BTSA) program. It is a state-funded induction program intended to support the professional development of new teachers during the first two years of their teaching. Advanced subject matter content knowledge is to be especially included during the second year of the program.

By 2002, the district had made DeArmond the first-ever Resource Teacher for Mathematics in the district. She believed that the mathematics content part of the district's BTSA program could be improved and contacted Fiedler to discuss the issue. Together with the assistance of a couple of colleagues including this author, they convinced the district to support four all-day workshops that were held between January and May 2003. The district leaders were pleased enough with the program to expand the model to seven workshops the following year (Lutz, 2012). At the time of this writing, the program is in its twelfth year of existence.

The district provides the seven *Connected Mathematics Project* (CMP) units (http://connectedmath.msu.edu/) that are used for the curriculum materials for the workshops, the space for the meetings and the district's resource person for mathematics as one of the facilitators. CSUB's California Mathematics Project (CMP) regional site provides two or three faculty, including Fiedler and this author, and the technology for the participants. Each participant receives a Texas Instruments TI-73 graphing calculator and a CBR2 data collection device.

Each of the seven all-day workshops uses one of the CMP units. The units are selected each year based on what topics would be most appropriate for what the participants would be teaching during the year. In addition to learning more about tasks that have a high cognitive demand, the teachers experience activities in which the "grain size," as described by the writers of the CCSS, is larger than one lesson.

Mathematics Education: Connecting Research to Practice Conference

The conference on September 10-11, 2010 was held on CSUB's campus with financial support from Texas Instruments, the local Partnership for Teacher Quality Programs and CSUB's regional site of the California Mathematics Project. The welcome statement in the program stated:

This conference's theme is one that is reflected in most current and authoritative literature on mathematics education. That is, we in the United States need to do a better job of actively engaging ALL of our mathematics students in tasks that are appropriately challenging.

The opening keynote speaker was Dr. Ed Silver, who led the QUASAR project. There were more than 200 attendees at the conference.

Discourse, Mathematical Thinking, and Classroom Practice Workshop

As stated previously, Margaret DeArmond was the first mathematics resource person that the district had ever had. On August 9-11, 2010, she and the person who succeeded her, Leah Shields, conducted an institute in which they focused on cognitive demand. Participants examined the work of Stein et al (2000) as well as Stigler and Hiebert (1999). They read and discussed examples of task implementation, examined the textbooks and tasks that they used in their own teaching and analyzed the cognitive demand of each, and discussed how to raise the cognitive demand of the tasks that were not at that level.

Many of the teachers who participated in these activities are now working in the district to implement California's CCSS. The part of the process that requires increased cognitive demand seems to be understood by them.

THE STANDARDS FOR MATHEMATICAL PRACTICE: MP4: MODEL WITH MATHEMATICS AND MP5: USE APPROPRIATE TOOLS STRATEGICALLY

As was the case with cognitive demand, the partnership has emphasized mathematical modeling and the appropriate use of technology throughout its existence. Following is a brief summary of some of those events.

MPDIs: At the turn of the century, California provided strong support for the professional development of teachers. One form of that support for mathematics teachers was through Mathematics Professional Development Institutes (MPDIs). Many of these institutes were 40-hour summer workshops in which the teachers were provided a monetary stipend for attending. The content of all, or most, of the institutes held at CSUB were determined by conversations between DeArmond and Fiedler. DeArmond described the most immediate needs of the teachers while Fiedler made suggestions as to what the university could provide. The institutes often had a modeling focus through the use of technology such as Geometers' Sketchpad or Texas Instruments' handheld technology with data collection. Some, such as *Connecting Math and Science*, also had STEM goals. The planning was always a collaboration with the goal to provide whatever is most beneficial to the teachers and their students. From summer 2000 until summer 2004, the partnership held 20 MPDI workshops with a total enrollment of 423.

Two good examples of meeting teachers' needs are the *Discovering Algebra* (2002) and *Discovering Advanced Algebra* (2003) institutes. One of the district's high schools, Centennial, was considering adopting the Discovering Algebra textbook from Key Press. To assist them in their decision, Jerry Murdock, one of the authors of the textbook, was brought in for a 40-hour workshop. A general announcement was made to local teachers who might also have been interested in attending even though their schools were not considering the adoption. A few years later, having adopted the Discovering Algebra book, Centennial was considering adopting the new Discovering Advanced Algebra textbook. Jerry Murdock was called upon again. Both textbooks include modeling and the use of technology.

On another occasion, a few schools in the district were using the *Prentice Hall Algebra: Tools for a Changing World* (1998) textbook. Although many of the mathematics teachers in the schools had attended one or more of the partnerships activities in which grain size, modeling, use of technology, and cognitive demand were included, they believed that they could not use their existing textbook in such ways. Coincidentally, one of the authors, Allan Bellman, was known to the partnership leaders, who knew he believed and practiced the approach to teaching that the partnership advocated. He was brought in for a workshop to help the teachers be able to use the textbook in the more creative way. His workshop resolved the issues.

Master of Arts in Teaching Mathematics

One would expect that the Discrete Mathematical Models course, based solely on the title of the course, would have a modeling component, which it does. However, modeling is a strong component throughout the courses in the program as is the appropriate use of technology. Each course contains a "lab" session each week in which the students collaborate on challenging tasks. Most of the mathematics in the courses is high school mathematics at an advanced viewpoint. Accordingly, students deepen their understanding of the mathematics they are teaching and experience learning via a larger "grain size" than is often seen. In an online video, Phil Daro, one of the authors of the CCSSM states that it is wrong for teachers to focus on lesson-size pieces when planning (http://serpmedia.org/daro-talks/index.

html). He believes the single-lesson grain size is too small and that planning chapters makes more sense and is more effective.

More Workshops

A high percentage of the workshops include using appropriate tools strategically. In any workshop in this partnership that has such a component, the participants always receive the respective tools for their personal use. Between 2000 and 2007, 715 individual instructional tools were provided by CSUB's CMP regional site. That included Texas Instruments handheld technology, projection and data collection devices as well as Key Curriculum Press Sketchpad and Tinker Plots software and also manipulatives such as algebra tiles and fraction bars. In recent years, there have been workshops on Texas Instruments TI-Nspire technology that include the computer software and the handheld devices. The Nspire Navigator wireless classroom system is utilized during the workshops although the teachers' individual schools are left with the responsibility of providing that technology for the participants.

LEADERSHIP

The first step in developing teacher leaders is to retain them in the district. Although it is commonly reported that half of beginning teachers will no longer be in the profession after five years, an examination of the participants in the partnership's BTSA workshops during the first ten years of its existence reveals that two-thirds of the teachers were not only still teaching but were also still teaching in the district (Lutz, 2012, p. 83).

From 2007-2012, CSUB's California Mathematics Project regional site was one of ten sites to participate in the Supporting-Teachers-to-Increase-Retention (STIR) project (California Mathematics Project, 2012). At the end of the six years, 24 of the 27 participating teachers were still teaching mathematics in the state, which was the highest retention rate of all the ten sites.

Another step toward taking a leadership role is to participate in professional activities such as making conference presentations. Greene and Gallagher (2009) conducted an in-depth study of one of the district's schools and its collaboration with CSUB. When the collaboration began in 2000, the school had a history of high teacher turnover. However by the time of Greene and Gallagher's study in 2009, the high teacher turnover had ended and more than half of the eighteen mathematics teachers at the school had made one or more conference presentations.

TRANSITIONING TO THE COMMON CORE

On 2 August 2010, California's State Board of Education officially adopted an annotated version of the Common Core State Standards for Mathematics as the state's standards. In August 2011, Andy Hicks became the fourth mathematics resource teacher for the KHSD; a position he would hold for two years. The district's transition to the CCSS fell heavily on his shoulders. How could the district prepare for the new, radically different Smarter Balanced assessments in less than three years (http://www.smarterbalanced.org/smarter-balanced-assessments/)? He and Mark Balch, the Director of Instructional Services, believed it would take a well-thought-out plan with broad-based support. Hicks had taught in the district for more than a decade and had served as the chair of one of the district's school's mathematics department, so he knew there were good leaders among the more than 200 high school mathematics teachers in the system. The monthly meetings of the district's eighteen mathematics chairs would be an important vehicle for building consensus. He gave Fiedler and this author a standing invitation to attend those meetings as well as the many meetings and workshops that would follow. The long history of district-university collaboration was facilitating ongoing communication.

The district's mathematics department chairs received the support of the district administrators, who informed the mathematics chairs that they "are the district" when it comes to mathematics. That is, they had the authority and responsibility to come up with recommendations for the implementation of the new standards. Hicks and many others increased their attendance at conferences and workshops and searched the internet to learn more about the California Common Core State Standards (CCSSM).

As somewhat of an outside observer, this author believes that as time passed and the teachers learned more, cohesiveness among the mathematics teachers in the district increased. As would be expected with eighteen different schools, the philosophies in the district had a history of varying greatly. Yet in this process, they seem to be coming together.

The mathematics chairs heeded the advice of the Common Core Mathematics authors and decided the first step in the transition should be to focus on the eight Standards for Mathematical Practice. For many of their teachers, following the practices would mean a radical change in their teaching. There is a volume of research recommending that teachers only try to change a small part of their teaching each year. Stigler and Hiebert (1999) state that changing teaching means changing a culture, which is almost impossible to do. Hicks, Balch and the district mathematics chairs developed a three-year plan intended to move the district toward adoption without moving teachers too far out of their comfort zones. *Awareness* was the goal for 2011-2012, *Ramping-Up-the-Rigor* the goal for 2012-2013, and *Instruction* the goal for 2013-2014. Awareness would mean increasing everyone's general understandings regarding the CCSSM.

Ramping Up the Rigor involved encouraging teachers and assisting them in occasionally using tasks that had a higher cognitive demand than what they may have been using in recent years. To help teachers make the transition, the partnership jointly sponsored a weeklong workshop in August 2012 that included a couple of days with Heather Dallas from UCLA and a couple of days with Patrick Callahan, the mathematician for the California Mathematics Project.

Hicks, Balch, the mathematics department chairs, and Dr. Michael Zulfa, the Assistant Superintendent of Instruction realized that it would take several years before textbooks would be published that are properly aligned with the content and philosophy of the CCSSM. Many textbooks claim to have that trait, but a quick look at them is all that is needed to realize that the claim is false. This realization supported their decision to focus first on the Standards for Mathematical Practice and especially raising the rigor. Hicks saw the illustration in Figure 1 in a California Teachers Association publication (Wells, 2013, p. 37) and thought it illustrated the balanced classroom that the district was seeking.

One question that the district leaders are often asked is whether the district will be going with the Traditional U.S. alignment or the International alignment of courses. David Foster from the Silicon Valley Mathematics Initiative was brought in to provide more information regarding the two alignments. With that information and other research that the group had conducted, the department chairs reached consensus in April 2013 that the International model seemed to make the most sense and be in the best interests of the students, but the final decision should not be made until good curriculum materials become available. That is, the quality of the materials may determine which alignment would be best.

2013-2014

Instruction is the goal for 2013-2014. During the summer 2013, the district transitioned to its fifth mathematics resource teacher, Kyle Atkin, and brought 14 district teachers together for five-day

Figure 1. A balanced mathematics class

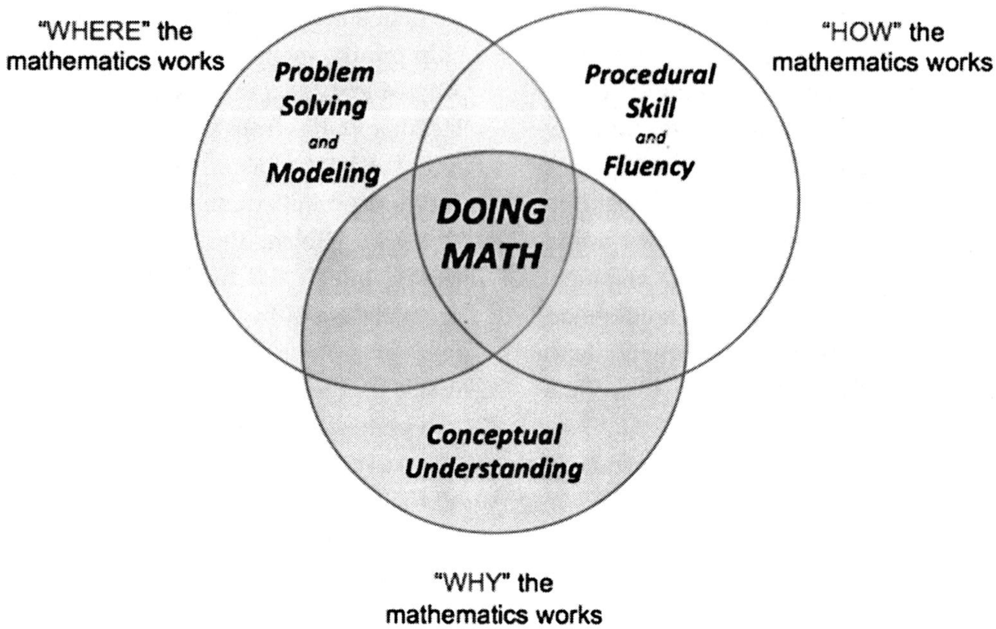

weeks for six weeks to create what they call *Anchor Tasks* (Heethuis & Hoyt, 2013). The teachers were chosen after going through an application and interview process and were paid their normal summer-school salaries. Their assignment was to create at least one anchor task for each of the four quarters of the 2013-2014 academic year in Algebra One and Geometry, a total of eight tasks. The department chairs had agreed that every school, every Algebra One class, and every Geometry class in the district would use each of the four anchor tasks for that course. Their Anchor Tasks need to meet seven criteria:

1. Develop Mathematical Intuition,
2. Explore the Task,
3. Model with Mathematics,
4. Apply the Mathematical Model,
5. Communicate Mathematical Reasoning,
6. Formalize the Mathematics, and
7. Extend the Mathematics.

The tasks were uploaded to a website for easy access by the teachers (Anchortasks.net). In the process of creating the anchor tasks, the group created many more than eight. Those, too, were uploaded, and teachers are encouraged to use them as they believe to be appropriate.

Very early in the first quarter of the year, the district used four instructional days to educate the respective teachers regarding the anchor task they would be using during the first quarter. Half of the teachers in each of the two groups were brought in each day with one of the teacher-writers of the particular task facilitating. At the time of this writing, the schools are in their second quarter, so all classes have completed their first task.

The preparation process for the second quarter was the same as the first with the additional activity of sharing details of the use of the first task. The overall response of the teachers was very positive at their second meeting. They described some of the details of their experience using the first anchor task and made suggestions of how they thought the respective task could be improved.

The district seems to have, once again, made a smooth transition to a new mathematics resource teacher. Atkin coordinated the summer anchor task writing activity, the first two BTSA workshops for 2013-2014, and the anchor-task professional development days. He had been one of the very active mathematics department chairs in the district prior to accepting the resource position.

THE FUTURE

Of course, it is much too early to know how well the implementation of the California Common Core State Standards for Mathematics will go, but it seems as if "so far, so good." However, the process is already creating benefits that should serve the district's students well for many years. The collaboration among the mathematics department chairs, the district's mathematics teachers, and the district administrators is admirable. The broader view of teaching that is being developed will serve all of the teachers and students well, with or without the CCSS.

The collaboration between the district and the university has contributed to the development of a cadre of young, forward-looking teacher leaders in mathematics who will most likely keep the district striving to increase the mathematical achievement of all students in the district. The teachers and the administrators are the ones moving the process forward with the university providing support as described by the teachers in the study by Greene and Gallagher (2009):

High school faculty repeatedly cited the approach and unique personalities of the mathematics professors as being critical to the success of the partnership. Nearly all the teachers indicated that the personal relationships and mutual respect they had built with university professors were key to the success of the partnership's work. The role of the university professors was described by an administrator as that of "an advisor and guide." She said it worked "because they didn't come in with a know-it-all attitude. It was 'let's work together to do...what's best for the high school.'" Another teacher described the interaction with a university professor as "the way to do it. He had a gentle hand. He didn't say 'this is what you should be doing.' We respected his opinion but he wouldn't chime in unless he was asked. But he did sometimes offer his two cents. He offered really good insights." University professors knew the high school teachers because they spent time with them in their classrooms, in their department meetings, and on numerous Saturday afternoons and summer days working to tailor the curriculum and instructional strategies to meet the needs of high school students. (p 4)

However, there is a cause for concern. The university faculty members who have done the most work in the partnership are nearing retirement, probably within the next five years. When new tenure-track professors are hired, will interest in and qualifications for participating in partnerships such as this be conditions for hiring at the university? The history of the university's role in the partnership may be best expressed as "For the most part, they did not interfere." It is hoped that there is an understanding that partnerships such as this benefit the university as well as the K-12 schools. For example, knowing the culture of the K-12 schools by spending time in the different environments that exist within each school provides valuable insight for those faculty who are preparing preservice teachers. And if teaching is a priority at the university, then K-12 campuses are excellent resources for learning more about teaching. It is easy to find informed discussions about teaching and learning among those teachers, especially regarding the California CCSSM.

REFERENCES

Bellman, A., Bragg, S. C., Chapin, S. H., Gardella, T. J., Hall, B. C., Handlin, W. G. Sr, & Manfre, E. (1998). *Algebra: Tools for a changing world.* Needham, MA: Prentice Hall.

Bloom, B. S. (1956). *Taxonomy of educational objectives, the classification of educational goals, handbook I*. NY: Cognitive Domain. David McKay.

California Department of Education. (1992). *Mathematics framework for California public schools kindergarten through grade twelve*. Sacramento, CA: Author.

California Mathematics Project. (2012). *Monograph: Mathematics teacher retention*. Los Angeles, CA: Author.

California State Department of Education. (1985). *Mathematics framework for California public schools kindergarten through grade twelve*. Sacramento, CA: Author.

Greene, L., & Gallagher, H. A. (2009). *Evaluation of the California Subject Matter Projects (CSMP): California Mathematics Project case study*. Menlo Park, CA: SRI International.

Heethuis, M., & Hoyt, M. (2013). *Anchor tasks for teaching the Common Core State Standards in secondary mathematics classrooms*. (Unpublished master's thesis). California State University, Bakersfield, CA.

Hess, K. (2013). *A guide for using Webb's depth of knowledge with Common Core State Standards*. Retrieved from http://www.palmbeachschools.org/academics/documents/WebsDepthofKnowledgeFlipChart_nobleeds_05142013_000.pdf

Hess, K. K. (2004-2012). *Center for Assessment, National Center for the Improvement of Educational Assessment, Inc.*. Retrieved from www.nciea.org

Lutz, M. (2012). *District-university collaboration is alive and well and living in California. Monograph of Psychology of Mathematics Education – North Amercia*. PME-NA.

Murdock, J., Kamischke, E., & Kamischke, E. (2003). *Discovering advanced algebra: An investigative approach*. Burlingame, CA: Key Press.

Stein, M. K., Smith, M. S., Henningsen, M. A., & Silver, E. A. (2000). *Implementing standards-based mathematics instruction: A casebook for professional development*. New York: Teachers College Press.

Stigler, J. W., & Hiebert, J. (1999). *The teaching gap*. NY: The Free Press.

Webb, N. (1997). *Criteria for alignment of expectations and assessments on mathematics and science education. Research monograph number 6*. Washington, DC: CCSSO.

Wells, F. (2013). Common core: Commonsense math standards. *California Educator*, *17*, 36–37.

Wilson, S. M. (2003). *California dreaming: Reforming mathematics education*. New Haven, CT: Yale University Press.

ADDITIONAL READING

Greene, L., & Gallagher, H. A. (2009). *Evaluation of the California Subject Matter Projects (CSMP): California Mathematics Project case study*. Menlo Park, CA: SRI International.

Heethuis, M., & Hoyt, M. (2013). *Anchor tasks for teaching the Common Core State Standards in secondary mathematics classrooms*. (Unpublished master's thesis). California State University, Bakersfield, CA.

Stein, M. K., Smith, M. S., Henningsen, M. A., & Silver, E. A. (2000). *Implementing standards-based mathematics instruction: A casebook for professional development*. New York: Teachers College Press.

Stigler, J. W., & Hiebert, J. (1999). *The teaching gap*. New York: The Free Press.

KEY TERMS AND DEFINITIONS

Cognitive Demand: A category used to describe the level of difficulty of tasks and activities.

Curriculum Development: The process of designing and creating a set of tasks and activities.

Mathematics Teaching: The process of designing, implementing, analyzing and refining mathematics instruction.

Professional Development: Programs and activities designed to support teachers' learning.

Teacher Change: The process of teachers adopting new instructional practices.

This work was previously published in "Professional Development Schools and Transformative Partnerships" edited by Drew Polly, Tina Heafner, Marvin Chapman, and Melba Spooner, pages 297-307, copyright 2015 by Information Science Reference (an imprint of IGI Global).

Index

D

E

F

G

M

N

O

P

Q

R

S

T

U

V

W

X

CPSIA information can be obtained at www.ICGtesting.com
Printed in the USA
BVOW06*1022140115
383197BV00005B/39/P